WITHDRAWN
FROM
STOCK

D0267520

Crystal Structures

CRYSTAL STRUCTURES

Second Edition

Ralph W. G. Wyckoff, *University of Arizona, Tucson, Arizona*

VOLUME 6

The Structure of Benzene Derivatives

Part 1.

Molecules Containing one Benzene Ring

INTERSCIENCE PUBLISHERS

a division of John Wiley & Sons

New York • London • Sydney • Toronto

The paper used in this book has pH of 6.5 or higher. It has been used because the best information now available indicates that this will contribute to its longevity.

10 9 8 7 6 5 4 3 2 1
Library of Congress Catalog Card Number 63-22897
SBN 471 96869 2
Printed in the United States of America

Preface

In recent years structures containing cyclic organic molecules have been determined at a steadily mounting rate. Instead of the single volume given over to them in the loose-leaf edition, it will now require more than three for even the brief descriptions attempted here. The first of these volumes, covering Part 1 of Chapter XV, deals with compounds having a single benzene ring per molecule.

An effort has been made to arrange these compounds in a way that will associate together those that are chemically related. The classification chosen, based on the number of substitutions in the ring, departs from that used in the loose-leaf edition. It is straightforward except for chelates, where the molecule can be taken either as the entire complex or as one of the chelated groups. For the present purposes it has seemed better to take the second alternative which permits grouping together, for instance, all the derivatives of salicylic acid. If this, perhaps arbitrary, step is borne in mind, it should not be too difficult to locate a compound in the text even without the use of the master table or the index.

The complexity of many of the organic compounds now being analyzed has made it necessary to omit packing drawings for most of the more recently determined structures. From many standpoints this seems regrettable but there are arguments favoring the omission. Such drawings lose value as structures become more complex, they are very costly to prepare and they occupy space which can more usefully be given to molecular drawings that state the principal bond dimensions.

As more and more complicated compounds are dealt with the critical evaluation that can be given in the kind of condensed description being attempted here becomes more restricted and the description itself becomes stereotyped. This has happened in the makeup of the present volume. It inevitably gives rise to a monotonous presentation but since a compendium of this sort would only be read in fragments this dullness in style is perhaps unimportant.

Most organic structures have their molecules distributed according to one or another of a very few space groups, notably C_{2h}^5 and C_{2h}^6. This means that, except for the very differently shaped cells needed, the molecular distributions fall into a limited number of groups. In the loose-leaf edition this fact was taken into account in grouping the compounds for description. It did not, however, lead to an easy comparison of chemically related structures. For this reason such crystallographic considerations have been abandoned in arranging the present chapter.

Contents of Volume 6, Part 1

Molecules Containing One Benzene Ring

Contents

Chapter XV

THE STRUCTURE OF BENZENE DERIVATIVES

A. COMPOUNDS BASED ON A SINGLE BENZENE RING

I. Benzene and its Addition Products

XV,aI,1. The crystal structure of solidified *benzene*, C_6H_6, has been determined through both x-ray and neutron diffraction. The symmetry is orthorhombic at all measured temperatures, with a tetramolecular unit having the edge lengths:

$$\text{At } -3°\text{C.: } a_0 = 7.460 \text{ A.}; \quad b_0 = 9.666 \text{ A.}; \quad c_0 = 7.034 \text{ A.}$$
$$\text{At } -195°\text{C.: } a_0 = 7.292 \text{ A.}; \quad b_0 = 9.471 \text{ A.}; \quad c_0 = 6.742 \text{ A.}$$

The space group is V_h^{15} (*Pbca*) all atoms being in the general positions:

$$(8c) \quad \pm(xyz;\ x+1/2, 1/2-y, \bar{z};\ \bar{x}, y+1/2, 1/2-z;\ 1/2-x, \bar{y}, z+1/2)$$

The parameters for the carbon atoms as finally determined by x-rays are:

Atom	x	y	z
C(1)	−0.0514	0.1386	−0.0057
C(2)	−0.1302	0.0481	0.1241
C(3)	−0.0793	−0.0901	0.1310

The neutron measurements, which also lead to hydrogen positions, were made at −55°C. and −135°C. They yield the following parameters, those in parentheses applying to the lower temperature:

Atom	x	y	z
C(1)	−0.0569 (−0.0607)	0.1387 (0.1393)	−0.0054 (−0.0069)
C(2)	−0.1335 (−0.1377)	0.0460 (0.0447)	0.1264 (0.1260)
C(3)	−0.0774 (−0.0770)	−0.0925 (−0.0958)	0.1295 (0.1325)
H(1)	−0.0976 (−0.1046)	0.2447 (0.2502)	−0.0177 (−0.0123)
H(2)	−0.2409 (−0.2458)	0.0794 (0.0781)	0.2218 (0.2241)
H(3)	−0.1371 (−0.1371)	−0.1631 (−0.1681)	0.2312 (0.2360)

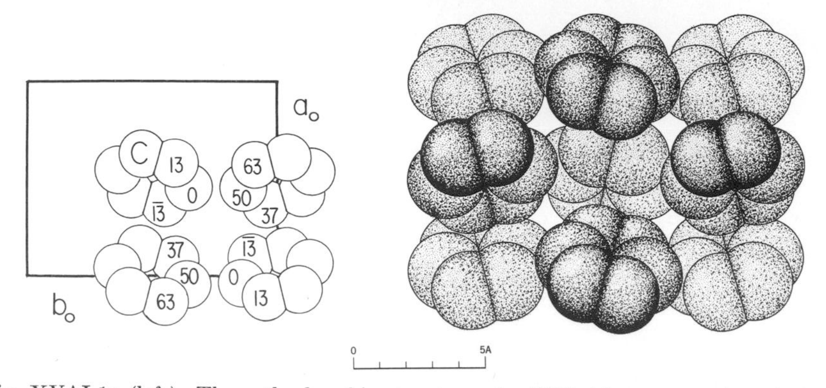

Fig. XVAI,1a (left). The orthorhombic structure of solidified benzene projected along its c_0 axis. Only the carbon atoms are shown.

Fig. XVAI,1b(right). A packing drawing of the orthorhombic benzene arrangement viewed along its c_0 axis.

The structure is shown in Figure XVAI,1. In its molecules, the C–C bonds are equal within the limit of experimental error; corrected for thermal motion, C–C = 1.398 A. by neutrons (1.392 A. by x-rays). The mean C–H = 1.077 A. at −55°C. and 1.090 A. at −135°C.

XV,aI,2. Crystals of the addition compound *benzene–chlorine*, $C_6H_6 \cdot Cl_2$, are monoclinic with a bimolecular unit of the dimensions:

$$a_0 = 7.41 \text{ A.}; \quad b_0 = 8.65 \text{ A.}; \quad c_0 = 5.65 \text{ A.}; \quad \beta = 99°30' \ (-90°\text{C.})$$

The space group C_{2h}^3 $(C2/m)$ puts atoms in the positions:

$$\text{C(1)}: (4h) \quad \pm(0\,u\,{}^1/_2;\ {}^1/_2,u+{}^1/_2,{}^1/_2) \qquad \text{with } u = -0.339$$

$$\text{Cl}: (4i) \quad \pm(u0v;\ u+{}^1/_2,{}^1/_2,v) \qquad \text{with } u = v = 0.1165$$

$$\text{C(2)}: (8j) \quad \pm(xyz;\ x\bar{y}z;\ x+{}^1/_2,y+{}^1/_2,z;\ x+{}^1/_2,{}^1/_2-y,z)$$

$$\text{with } x = 0.575,\ y = 0.080,\ z = 0.342$$

The resulting structure is shown in Figure XVAI,2. In its Cl_2 molecule, Cl–Cl = 1.99 A. This bond is normal to the plane of the benzene ring, the distance between chlorine and the center of the ring being 3.28 A.

The corresponding bromine compound, $C_6H_6 \cdot Br_2$, is isostructural with a cell of the dimensions:

$$a_0 = 7.75 \text{ A.}; \quad b_0 = 8.83 \text{ A.}; \quad c_0 = 5.94 \text{ A.}; \quad \beta = 99°18'$$

Parameters do not seem to have been published.

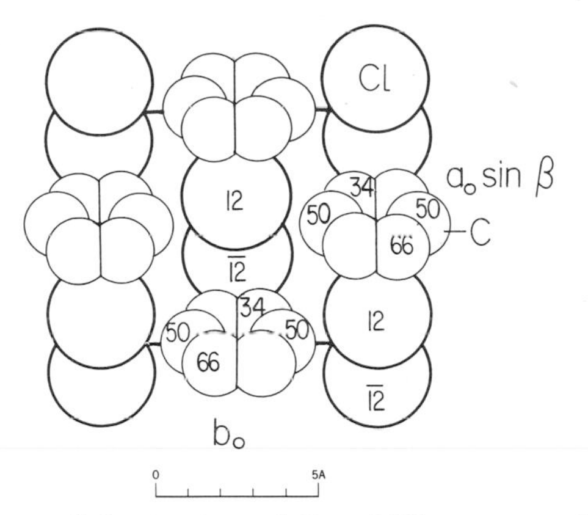

Fig. XVAI,2. The monoclinic structure of the addition compound benzene–chlorine projected along its c_0 axis.

XV,aI,3. There has been much debate concerning the correct structure for *chromium dibenzene*, $Cr(C_6H_6)_2$. Its cubic crystals have a tetramolecular unit of the edge length:

$$a_0 = 9.67 \text{ A.}$$

The space group is T_h^6 ($Pa3$) with the chromium atoms in:

$$(4a) \quad 000;\ 0\ {}^1/_2\ {}^1/_2;\ {}^1/_2\ 0\ {}^1/_2;\ {}^1/_2\ {}^1/_2\ 0$$

The other atoms have been placed in the general positions:

$$(24d) \quad \pm(xyz;\ x+{}^1/_2,{}^1/_2-y,\bar{z};\ z+{}^1/_2,{}^1/_2-x,\bar{y};\ y+{}^1/_2,{}^1/_2-z,\bar{x});\ \text{tr}$$

The following sets of parameters were chosen for them:

Atom	x	y	z
C(1)	0.1462(9)	0.1654(8)	−0.0200(9)
C(2)	0.0493(9)	0.0240(8)	0.2136(8)
H(1)	0.169	0.230	−0.114
H(2)	0.014	−0.038	0.302

These parameters lead to benzene rings in which the C–C distances alternate between 1.436 and 1.366 A., and this has raised the question of whether or not the existence of a nonregular benzene hexagon is thus proved. In this connection it has been suggested that the apparent lack of sixfold symmetry is better attributed to a measure of disorder in the structure.

XV,aI,4. Crystals of *chromium tricarbonyl–benzene,* $Cr(CO)_3 \cdot C_6H_6$, are monoclinic. Their bimolecular unit has the dimensions:

$$a_0 = 6.17(2) \text{ A.}; \quad b_0 = 11.07(4) \text{ A.}; \quad c_0 = 6.57(2) \text{ A.}$$
$$\beta = 101.5(1)°$$

The space group is C_{2h}^2 ($P2_1/m$) with atoms in the positions:

$$(2e) \quad \pm(u\ {}^1/_4\ v)$$
$$(4f) \quad \pm(xyz;\ x,{}^1/_2-y,z)$$

The parameters are those of Table XVAI,1.

TABLE XVAI,1

Parameters of the Atoms in $Cr(CO)_3 \cdot C_6H_6$

Atom	Position	x	y	z
Cr	(2e)	0.3319(2)	$^1/_4$	0.0225(2)
C(1)	(4f)	0.1804(8)	0.3119(4)	−0.2973(6)
C(2)	(4f)	0.3761(10)	0.3769(4)	−0.2273(6)
C(3)	(4f)	0.5753(8)	0.3142(4)	−0.1598(7)
C(4)	(2e)	0.5538(12)	$^1/_4$	0.2557(10)
O(4)	(2e)	0.6899(9)	$^1/_4$	0.4002(8)
C(5)	(4f)	0.1827(8)	0.3642(4)	0.1453(7)
O(5)	(4f)	0.0894(7)	0.4341(3)	0.2248(5)
H(1)	(4f)	0.028 —	0.361 —	−0.350 —
H(2)	(4f)	0.376 —	0.474 —	−0.227 —
H(3)	(4f)	0.728 —	0.363 —	−0.107 —

The structure is shown in Figure XVAI,3. It places the chromium atom over the center of a regular and planar benzene ring. In the ring, the mean C–C = 1.40 A.; the Cr–C(ring) distances lie between 2.216 and 2.230 A. In the $Cr(CO)_3$ group, Cr–C = 1.84 A. and C–O = 1.135 and 1.150 A.

XV,aI,5. Crystals of the complex *aluminum bromide–benzene,* $Al_2Br_6 \cdot C_6H_6$, are triclinic with a unimolecular cell of the dimensions:

$$a_0 = 6.85(2) \text{ A.}; \quad b_0 = 6.91(2) \text{ A.}; \quad c_0 = 9.00(2) \text{ A.}$$
$$\alpha = 104.6(5)°; \quad \beta = 103.1(5)°; \quad \gamma = 90.0(5)°$$

The space group C_i^1 ($P\bar{1}$) places atoms in the positions (2i) $\pm(xyz)$. Parameters are given in Table XVAI,2.

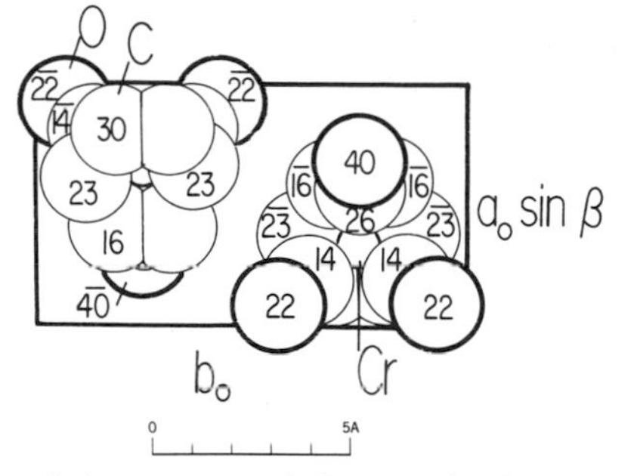

Fig. XVAI,3. The monoclinic structure of the complex benzene–chromium tricarbonyl projected along its c_0 axis.

TABLE XVAI,2
Parameters of the Atoms in $Al_2Br_6 \cdot C_6H_6$

Atom	x	y	z
Br(1)	0.1263	0.8839	0.1415
Br(2)	0.3946	0.2980	0.0628
Br(3)	0.9809	0.4223	0.2793
Al	0.1498	0.2084	0.1028
C(1)	0.644	0.136	0.447
C(2)	0.552	0.815	0.467
C(3)	0.654	0.912	0.418

The resulting structure is shown in Figure XVAI,4. In it, molecules of Al_2Br_6 (see **V,b8**) alternate with those of benzene. The average bond dimensions in the bromide molecule are shown in Figure XVAI,5.

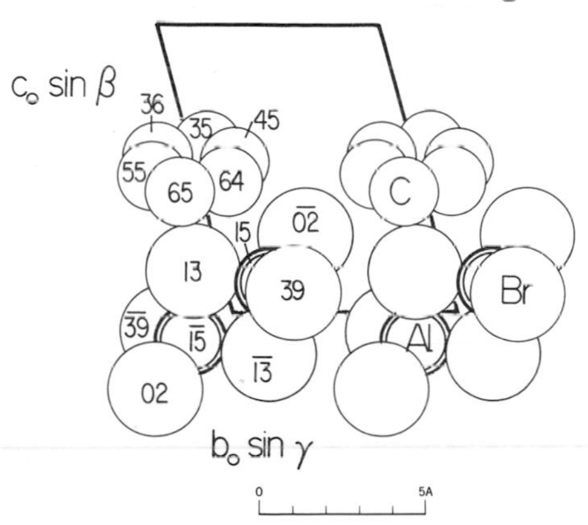

Fig. XVAI,4. The triclinic structure of the complex benzene–aluminum bromide projected along its a_0 axis.

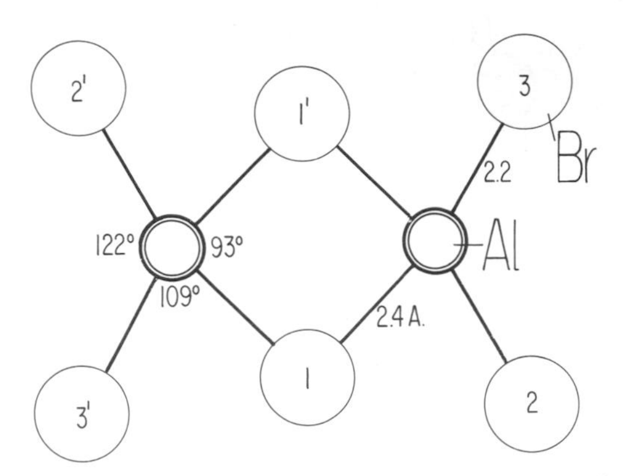

Fig. XVAI,5. Bond dimensions in the Al_2Br_6 portion of the complex $C_6H_6 \cdot Al_2Br_6$.

XV,aI,6. The complex *silver perchlorate–benzene*, $AgClO_4 \cdot C_6H_6$, forms orthorhombic crystals whose tetramolecular unit has the edge lengths:

$$a_0 = 8.35 \text{ A.}; \qquad b_0 = 8.02 \text{ A.}; \quad c_0 = 11.68 \text{ A.}$$

The space group was chosen as V_h^{17} (*Cmcm*) with atoms in the following partially disordered arrangement.

Ag: in half the positions of
(8*g*) $\pm(u\ v\ {}^1/_4;\ \bar{u}\ v\ {}^1/_4;\ u+{}^1/_2,v+{}^1/_2,{}^1/_4;\ {}^1/_2-u,v+{}^1/_2,{}^1/_4)$
with $u = 0.0313(3)$, $v = 0.0437(1)$

Cl: (4*c*) $\pm(0\ u\ {}^1/_4;\ {}^1/_2,u+{}^1/_2,{}^1/_4)$
with $u = 0.4468(3)$

O(1): in half the positions of
(16*h*) $\pm(xyz;\ \bar{x}yz;\ x,y,{}^1/_2-z;\ x,\bar{y},z+{}^1/_2;\ x+{}^1/_2,y+{}^1/_2,z;\ {}^1/_2-x,y+{}^1/_2,z;\ x+{}^1/_2,y+{}^1/_2,{}^1/_2-z;\ x+{}^1/_2,{}^1/_2-y,z+{}^1/_2)$
with $x = 0.0258(27)$, $y = 0.3422(18)$, $z = 0.1511(17)$

O(2): in half the positions of (16*h*)
with $x = 0.1468(13)$, $y = 0.5488(8)$, $z = 0.2647(9)$

C(1): (8*e*) $\pm(u00;\ u\ 0\ {}^1/_2;\ u+{}^1/_2,{}^1/_2,0;\ u+{}^1/_2,{}^1/_2,{}^1/_2)$
with $u = 0.1720(17)$

C(2): (16*h*) with $x = 0.0811(10)$, $y = 0.8874(7)$, $z = 0.0687(5)$

This is a structure which gives the carbon atoms of the benzene ring definite positions even though there is disorder in the distribution of the inorganic component. There has been some subsequent work suggesting disorder in the orientation of the benzene molecule also, and proposing further study.

XV,aI,7. Crystals of the complex *cuprous aluminum chloride–benzene*, $CuAlCl_4 \cdot C_6H_6$, are monoclinic with a tetramolecular unit of the dimensions:

$$a_0 = 8.59(1) \text{ A.}; \quad b_0 = 21.59(3) \text{ A.}; \quad c_0 = 6.07(1) \text{ A.}$$
$$\beta = 93°0(15)'$$

The space group C_{2h}^5 ($P2_1/n$) puts atoms in the positions:

$$(4e) \quad \pm(xyz;\ x+{}^1/_2,{}^1/_2-y,z+{}^1/_2)$$

The parameters are listed in Table XVAI,3.

TABLE XVAI,3

Parameters of the Atoms in $CuAlCl_4 \cdot C_6H_6$

Atom	x	y	z
Cu	0.1790(4)	0.1703(1)	0.0043(4)
Cl(1)	0.2319(7)	0.2838(2)	0.1031(7)
Cl(2)	0.3810(9)	0.4318(3)	0.0692(12)
Cl(3)	0.4200(6)	0.3270(3)	0.6351(6)
Cl(4)	0.1381(6)	0.1921(2)	0.6237(6)
Al	0.4145(6)	0.3398(2)	0.9839(7)
C(1)	0.473(4)	0.0896(16)	0.826(5)
C(2)	0.483(3)	0.1255(14)	0.989(5)
C(3)	0.379(5)	0.1229(19)	0.160(7)
C(4)	0.279(4)	0.0796(17)	0.154(5)
C(5)	0.261(4)	0.0403(19)	0.970(6)
C(6)	0.361(4)	0.0442(13)	0.819(4)

The structure is shown in Figure XVAI,6. The aluminum atoms are tetrahedrally surrounded by four chlorine atoms, with Al–Cl = 2.07–2.15 A. Around the Cu^+ atoms are three chlorines (Cu–Cl = 2.365, 2.398 and 2.555 A.) and two still nearer carbon atoms of a benzene molecule (Cu–C(3) = 2.15 A. and Cu–C(4) = 2.30 A.). If the sp^3 bond between these carbon atoms is thought of as providing a fourth coordination for Cu^+, the distance between the latter and the midpoint of the bond is 2.13 A.

XV,aI,8. Crystals of the monoclinic *silver tetrachloroaluminate–benzene*, $AgAlCl_4 \cdot C_6H_6$, have a tetramolecular unit of the dimensions:

$$a_0 = 9.09(3) \text{ A.}; \quad b_0 = 10.22(3) \text{ A.}; \quad c_0 = 12.73(3) \text{ A.}$$
$$\beta = 95°5(15)'$$

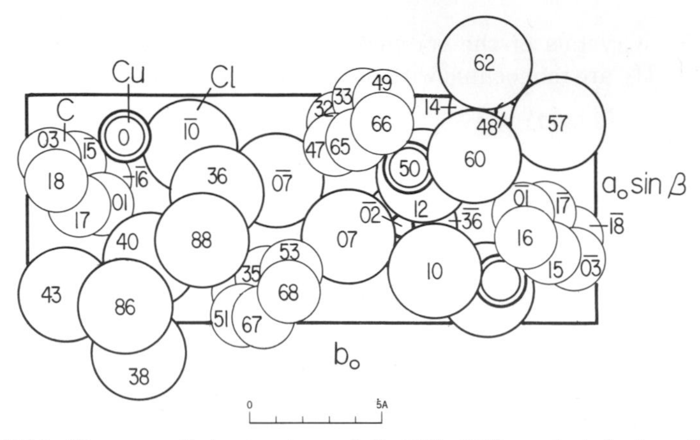

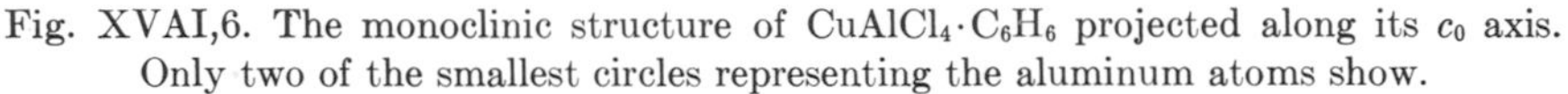
Fig. XVAI,6. The monoclinic structure of $CuAlCl_4 \cdot C_6H_6$ projected along its c_0 axis. Only two of the smallest circles representing the aluminum atoms show.

The space group is C_{2h}^5 ($P2_1/c$) with atoms in the positions:

$$(4e) \quad \pm(xyz;\ x,{}^1/_2-y,z+{}^1/_2)$$

The determined parameters are listed in Table XVAI,4.

TABLE XVAI,4

Parameters of the Atoms in $AgAlCl_4 \cdot C_6H_6$

Atom	x	y	z
Ag	0.1382(3)	0.2297(3)	0.0323(8)
Cl(1)	0.8257(14)	0.5069(14)	0.6290(30)
Cl(2)	0.2300(14)	0.4385(14)	0.6753(29)
Cl(3)	0.0472(16)	0.6076(16)	0.8651(31)
Cl(4)	0.0996(18)	0.7547(18)	0.6215(30)
Al	0.0574(16)	0.5767(16)	0.6998(36)
C(1)	0.542(6)	0.250(6)	0.112(11)
C(2)	0.490(5)	0.355(5)	0.155(10)
C(3)	0.353(6)	0.429(6)	0.108(12)
C(4)	0.348(6)	0.375(6)	0.001(11)
C(5)	0.446(5)	0.288(5)	0.947(11)
C(6)	0.514(6)	0.291(6)	0.481(10)

In this structure (Fig. XVAI,7) the silver atoms have an unusual fivefold coordination with four chlorine atoms at distances of 2.59, 2.77, 2.80, and 3.04 A. and a benzene carbon atom 2.57 A. away. The aluminum atoms are tetrahedrally surrounded by chlorine (Al–Cl = 2.13 A.).

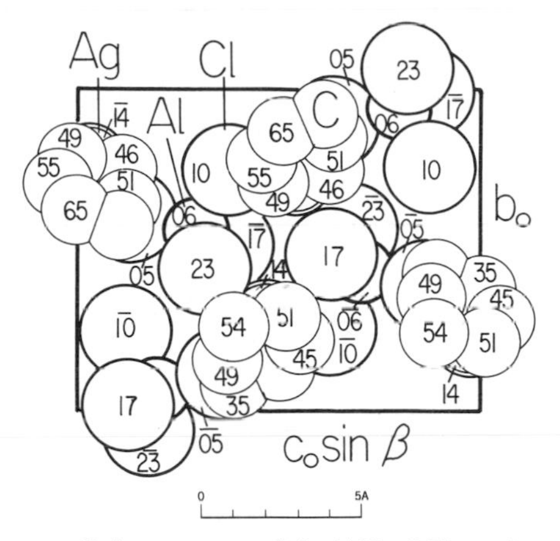

Fig. XVAI,7. The monoclinic structure of $AgAlCl_4 \cdot C_6H_6$ projected along its a_0 axis.

XV,aI,9. The complex *cobalt dimercury hexathiocyanate–benzene*, $CoHg_2(SCN)_6 \cdot C_6H_6$, is triclinic with a unimolecular cell of the dimensions:

$$a_0 = 7.59 \text{ A.}; \quad b_0 = 8.37 \text{ A.}; \quad c_0 = 8.87 \text{ A.}$$
$$\alpha = 98.31°; \quad \beta = 101.01°; \quad \gamma = 95.50°$$

The space group is C_i^1 ($P\bar{1}$). The cobalt atom is in the origin (1*a*) 000 and all other atoms are in (2*i*) $\pm(xyz)$ with the parameters of Table XVAI,5

TABLE XVAI,5

Parameters of Atoms in $CoHg_2(SCN)_6 \cdot C_6H_6$

Atom	x	y	z
Hg	0.4288	0.4559	−0.2423
S(1)	0.4110	0.1617	−0.3305
S(2)	0.4376	0.7094	−0.3445
S(3)	0.2515	0.4872	−0.0007
N(1)	0.1340	0.0745	−0.1785
N(2)	0.7785	0.8700	−0.1612
N(3)	0.1385	0.7955	0.0040
C(1)	0.2535	0.1105	−0.2440
C(2)	0.6485	0.8032	−0.2330
C(3)	0.1830	0.6670	0.0035
C(4)	−0.0718	0.5420	−0.3600
C(5)	−0.0380	0.3830	−0.4080
C(6)	0.0265	0.3420	−0.5555

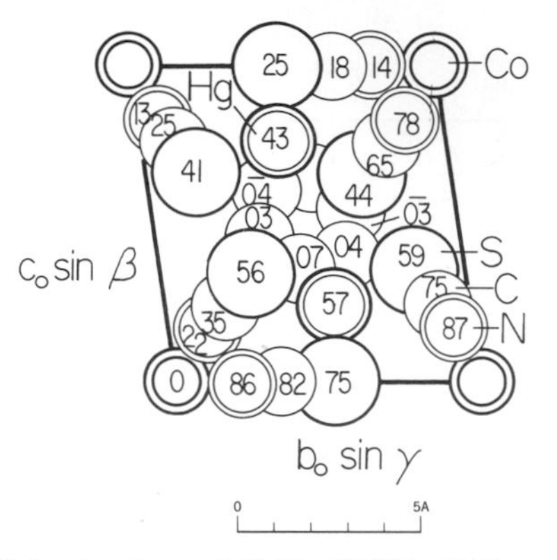

Fig. XVAI,8. The triclinic structure of $CoHg_2(SCN)_6 \cdot C_6H_6$ projected along its a_0 axis.

The structure is shown in Figure XVAI,8. It consists of layers of $CoHg_2(SCN)_6$ in the a_0b_0 direction separated along c_0 by benzene molecules. The thiocyanate radicals are octahedrally distributed about cobalt, with Co–N = 2.08–2.17 A. Each mercury atom has four sulfur neighbors at the corners of a distorted tetrahedron. Pairs of sulfur atoms serve as bridges between neighboring mercury atoms (Hg–S = 2.42, 2.45 and 2.72, 2.85 A.). The benzene carbon atoms are closest to sulfur, with C–S = 1.60–1.76 A.

XV,aI,10. *Nickel cyanide ammonia* unites with *benzene* to form crystals of the composition $Ni(CN)_2 \cdot NH_3 \cdot C_6H_6$. They are tetragonal with a bimolecular cell of the edge lengths:

$$a_0 = 7.242 \text{ A.}, \qquad c_0 = 8.277 \text{ A.}$$

The space group has been chosen as C_{4h}^1 ($P4/m$) and atoms have been placed in the following of its positions:

Ni(1): (1a) 000

Ni(2): (1c) $\frac{1}{2}\,\frac{1}{2}\,0$

NH_3: (2g) $\pm(00u)$ with $u = 0.249$

N: (4j) $\pm(uv0;\ v\bar{u}0)$ with $u = v = 0.209$

C(CN): (4j) with $u = v = 0.328$

For benzene:

C(1): (4i) $\pm(0\ \frac{1}{2}\ u;\ \frac{1}{2}\ 0\ u)$ with $u = 0.330$

C(2): (8l) $\pm(xyz;\ \bar{x}\bar{y}z;\ \bar{y}xz;\ y\bar{x}z)$
with $x = 0.153$, $y = 0.435$, $z = 0.417$

This yields a framework of $Ni(CN)_2 \cdot NH_3$ having large voids that are filled by the C_6H_6 molecules.

II. Compounds containing a Mono-substituted Benzene Ring

XV,aII,1. At −65°C. *fluorobenzene*, C_6H_5F, is tetragonal with a tetramolecular unit of the edge lengths:

$$a_0 = 5.83(5) \text{ A.}, \qquad c_0 = 14.61(7) \text{ A.}$$

According to a preliminary note, the space group appears to be D_4^4 ($P4_12_12$) [or D_4^6 ($P4_22_12$)] with atoms in the positions:

$$(4a) \quad uu0; \quad \bar{u}\,\bar{u}\,{}^1/_2; \quad {}^1/_2-u,u+{}^1/_2,{}^1/_4; \; u+{}^1/_2,{}^1/_2-u,{}^3/_4$$

$$(8b) \quad xyz; \; \bar{x},\bar{y},z+{}^1/_2; \; {}^1/_2-y,x+{}^1/_2,z+{}^1/_4; \; y+{}^1/_2,{}^1/_2-x,z+{}^3/_4;$$
$$yx\bar{z}; \; \bar{y},\bar{x},{}^1/_2-z; \; {}^1/_2-x,y+{}^1/_2,{}^1/_4-z; \; x+{}^1/_2,{}^1/_2-y,{}^3/_4-z$$

The parameters have been given as:

Atom	Position	x	y	z
F	(4*a*)	0.326	0.326	0
C(1)	(4*a*)	0.485	0.485	0
C(2)	(8*b*)	0.455	0.674	0.058
C(3)	(8*b*)	0.627	0.839	0.058
C(4)	(4*a*)	0.820	0.820	0

XV,aII,2. *Nitrobenzene*, $C_6H_5NO_2$, yields monoclinic crystals which, photographed at −30°C., have a tetramolecular unit of the dimensions:

$$a_0 = 3.86 \text{ A.}; \quad b_0 = 11.65 \text{ A.}; \quad c_0 = 13.24 \text{ A.}; \quad \beta = 95°35'$$

The space group is C_{2h}^5 ($P2_1/c$) with atoms in the positions:

$$(4e) \quad \pm(xyz; \; x,{}^1/_2-y,z+{}^1/_2)$$

The determined parameters are those of Table XVAII,1.

TABLE XVAII,1
Parameters of the Atoms in $C_6H_5NO_2$

Atom	x	y	z
C(1)	0.2267	0.3183	0.1300
C(2)	0.1338	0.3415	0.0291
C(3)	−0.0569	0.2553	−0.0258
C(4)	−0.1231	0.1516	0.0178
C(5)	−0.0156	0.1299	0.1161
C(6)	0.1653	0.2181	0.1775
N	0.4216	0.4102	0.1890
O(1)	0.5099	0.3900	0.2779
O(2)	0.4673	0.5002	0.1484

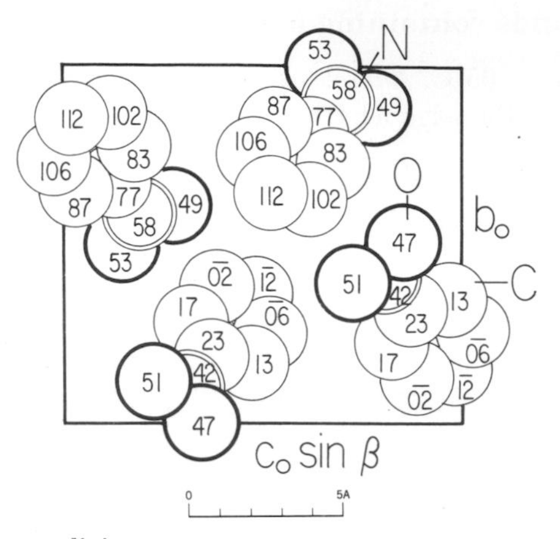

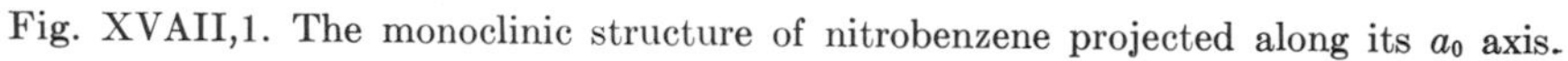

Fig. XVAII,1. The monoclinic structure of nitrobenzene projected along its a_0 axis.

The structure is illustrated by Figure XVAII,1. The entire molecule is planar within the limit of the determination. It has in effect a plane of symmetry and this results in the bond dimensions of Figure XVAII,2. Between molecules the shortest atomic separation is a C–O = 3.46 A.

XV,aII,3. *Benzene dichloroiodide*, $C_6H_5ICl_2$, is monoclinic. The simplest cell contains four molecules and has the dimensions:

$$a_0 = 15.6 \text{ A}; \quad b_0 = 5.45 \text{ A.}; \quad c_0 = 12.3 \text{ A.}; \quad \beta = 128°30'$$

The space group is C_{2h}^5 with, for this cell, the axial sequence $P2_1/a$.

A larger, nonprimitive cell containing eight molecules and centered on the b-face has the dimensions:

$$a_0 = 15.6 \text{ A.}; \quad b_0 = 5.45 \text{ A.}; \quad c_0 = 19.6 \text{ A.}; \quad \beta = 90°30'$$

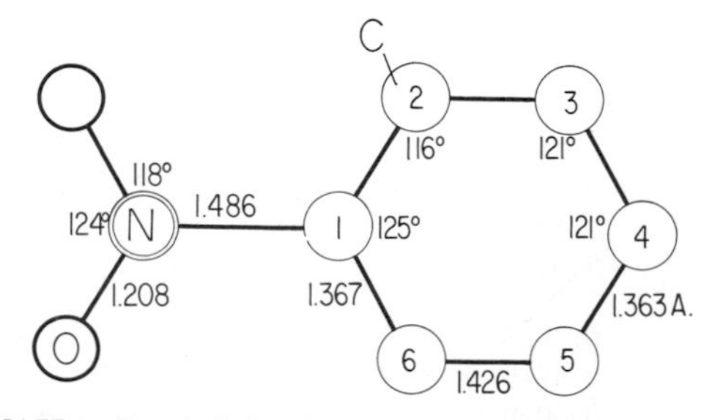

Fig. XVAII,2. Bond dimensions in the molecule of nitrobenzene.

Because it is so nearly orthogonal it was used for the determination of structure. Referred to it atoms were put in the positions:

$$(8e') \quad \pm(xyz;\ x+{}^1/_2,{}^1/_2-y,z;\ x+{}^1/_2,y,z+{}^1/_2;\ x,{}^1/_2-y,z+{}^1/_2)$$

The selected parameters are listed in Table XVAII,2, those for y being based on the assumption that the benzene ring is planar, with C–C = 1.40 A.

TABLE XVAII,2

Parameters of the Atoms in Benzene Dichloroiodide

Atom	x	y	z
I	0.175	0.335	0.077
Cl(1)	0.134	0.011	−0.004
Cl(2)	0.215	0.658	0.160
C(1)	0.058	0.335	0.120
C(2)	0.038	0.171	0.173
C(3)	−0.043	0.171	0.203
C(4)	−0.107	0.335	0.182
C(5)	−0.088	0.498	0.130
C(6)	−0.007	0.498	0.100

This structure is given in Figure XVAII,3. In the ICl_2 group, which makes a large angle with the plane of the ring, I–Cl = 2.45 A. The C–I bond is ca. 2.00 A., and atoms of neighboring molecules are 3.40 A. and more from one another.

XV,aII,4. Crystals of *phenol*, C_6H_5OH, have been the subject of repeated investigation. They were originally described as orthorhombic but x-ray evidence has been adduced indicating that the symmetry is really monoclinic, pseudo-orthorhombic. If this is the case the unit contains six molecules and has the dimensions:

$$a_0 = 6.02(2)\text{ A.};\quad b_0 = 9.04(3)\text{ A.};\quad c_0 = 15.18(4)\text{ A.}$$
$$\beta = 90.0(1)^\circ$$

The space group would be C_2^2 ($P2_1$) with atoms in the positions:

$$(2a) \quad xyz;\ \bar{x},y+{}^1/_2,\bar{z}$$

According to the latest study, the parameters are those of Table XVAII,3. In the preceding investigation (1963: S) parameters close to these were obtained but referred to an origin displaced along the a_0 axis.

TABLE XVAII,3

Parameters of the Atoms in Phenol

Atom	x	y	z
O(1)	0.002(2)	0.627(1)	0.255(0)
C(1)	0.002(3)	0.782(2)	0.255(2)
C(2)	0.183(3)	0.852(2)	0.222(2)
C(3)	0.178(3)	0.009(2)	0.222(2)
C(4)	0.011(4)	0.085(2)	0.253(2)
C(5)	−0.182(4)	0.012(2)	0.287(2)
C(6)	−0.183(2)	0.853(2)	0.287(1)
O(1,1)	0.352(2)	0.478(1)	0.191(1)
C(1,1)	0.339(3)	0.385(2)	0.122(2)
C(1,2)	0.156(3)	0.395(2)	0.068(2)
C(1,3)	0.157(4)	0.288(3)	−0.006(2)
C(1,4)	0.326(4)	0.196(3)	−0.019(2)
C(1,5)	0.516(4)	0.201(3)	0.034(2)
C(1,6)	0.532(3)	0.298(2)	0.109(2)
O(2,1)	−0.350(2)	0.476(1)	0.318(1)
C(2,1)	−0.337(2)	0.386(2)	0.389(1)
C(2,2)	−0.154(3)	0.395(2)	0.443(2)
C(2,3)	−0.159(4)	0.289(3)	0.517(2)
C(2,4)	−0.331(4)	0.195(2)	0.528(2)
C(2,5)	−0.542(4)	0.199(3)	0.477(2)
C(2,6)	−0.519(3)	0.298(2)	0.404(2)

If the symmetry is taken as orthorhombic it has been shown (1967: GP) that practically the same structure can be obtained utilizing the space group D_2^3 ($P2_122_1$). The only essential difference is that in the orthorhombic structure there are two, and in the monoclinic structure three crystallographically different molecules. The orthorhombic parameters are stated in 1967: GP.

The structure is shown in Figure XVAII,4. Its molecules are bound together by hydrogen bonds between the oxygen atoms (O–H–O = 2.63–2.70 A.).

XV,aII,5. A structure has been described for the *hemihydrate* of *phenol*, $C_6H_5OH \cdot {}^1/_2H_2O$. Its orthorhombic unit contains eight molecules and at −5°C. has the edge lengths:

$$a_0 = 13.15 \text{ A.}; \quad b_0 = 10.90 \text{ A.}; \quad c_0 = 7.90 \text{ A.}$$

The space group has been chosen as V_h^{14} ($Pbcn$). Water oxygens are in the positions:

$$(4c) \quad \pm(0\,u\,{}^1/_4;\ {}^1/_2,u+{}^1/_2,{}^1/_4) \qquad \text{with } u = 0.063$$

The other atoms are in the general positions:

$$(8d) \quad \pm(xyz;\ x+{}^1/_2,y+{}^1/_2,{}^1/_2-z;\ {}^1/_2-x,y+{}^1/_2,z;\ x,\bar{y},z+{}^1/_2)$$

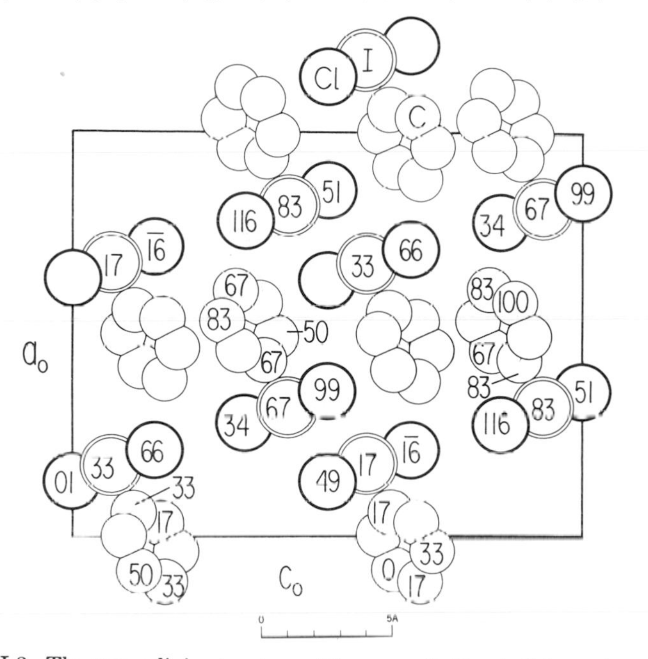

Fig. XVAII,3. The monoclinic structure of benzene dichloroiodide projected along its b_0 axis.

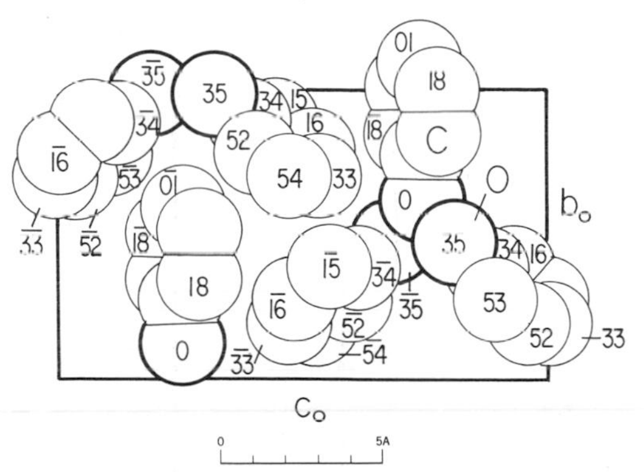

Fig. XVAII,4. The structure of phenol in terms of its monoclinic, pseudo-orthorhombic axes. Projection is along the a_0 axis.

The chosen parameters are those of Table XVAII,4.

TABLE XVAII,4

Parameters of Atoms in $C_6H_5OH \cdot 1/2H_2O$

Atom	x	y	z
O	0.094	0.104	0.55
C(1)	0.133	0.214	0.62
C(2)	0.226	0.210	0.71
C(3)	0.267	0.320	0.78
C(4)	0.216	0.432	0.76
C(5)	0.123	0.435	0.68
C(6)	0.082	0.330	0.61

The resulting structure, shown in Figure XVAII,5, has its molecules tied together by hydrogen bonds, the shortest of which is ca. 2.8 A.

XV,aII,6. *Phenol* forms with gas molecules a series of clathrates similar to those produced with quinol (**XV,aIII,74**). Both are hexagonal. Special attention has been given the SO_2 and H_2S compounds with phenol. For the former:

$$a_0 = 16.21 \text{ A.}, \qquad c_0 = 22.73 \text{ A.}$$

For the latter:

$$a_0 = 16.41 \text{ A.}, \qquad c_0 = 22.33 \text{ A.}$$

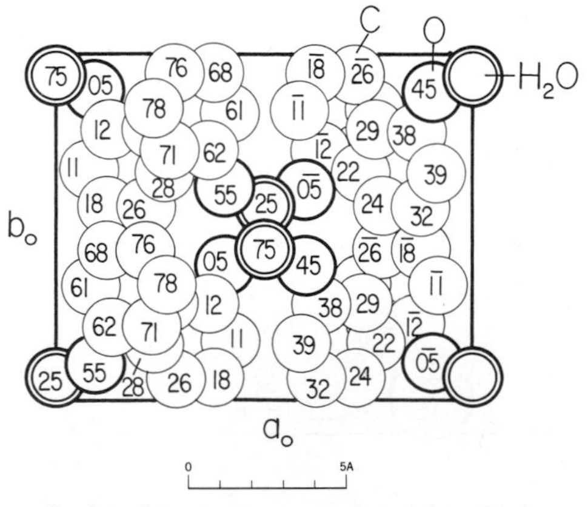

Fig. XVAII,5. The orthorhombic structure of phenol hemihydrate projected along its c_0 axis.

The corresponding unit rhombohedron has for both the approximate dimensions:

$$a_0' = 12.05 \text{ A.}, \qquad \alpha = 85\text{–}86°$$

There are 12 phenol and four guest molecules in this rhombohedron and hence 36 and 12 in the hexagonal cell.

The space group has been chosen as C_{3i}^2 ($R\bar{3}$) with, for the hexagonal cell, all atoms of the phenol molecules in:

$$(18f) \quad \pm(xyz;\ \bar{y},x-y,z;\ y-x,\bar{x},z);\quad \text{rh}$$

The assigned parameters referred to these hexagonal axes are listed in Table XVAII,5.

TABLE XVAII,5

Parameters of Atoms in Phenol Clathrates

Atom	x	y	z
O(1)	0.190	0.100	0.125
O(2)	0.190	0.100	0.875
C(1)	0.262	0.133	0.168
C(2)	0.333	0.109	0.163
C(3)	0.404	0.143	0.206
C(4)	0.404	0.200	0.253
C(5)	0.333	0.224	0.258
C(6)	0.262	0.190	0.215
C(1′–6′)	have the foregoing x and y parameters and $z' = z - 1/4$		

They result in a structure having the phenol molecules stacked along c_0 so that their hydroxyl groups point towards the threefold axes and are tied together by hydrogen bonds to form a regular hexagon (O–H–O = 2.7 A.). The 12 gas molecules per hexagonal unit lie in the hexagonal prismatic voids produced by the stacking. They have been put in the following special positions:

$$(3a) \quad 000;\quad \text{rh}$$
$$(3b) \quad 0\,0\,{}^1/_2;\quad \text{rh}$$
$$(6c) \quad \pm(00u);\quad \text{rh} \qquad \text{with } u = 0.25$$

XV,aII,7. *Benzoic acid*, C_6H_5COOH, forms monoclinic crystals having a tetramolecular unit of the dimensions:

$$a_0 = 5.52 \text{ A.};\quad b_0 = 5.14 \text{ A.};\quad c_0 = 21.90 \text{ A.};\quad \beta = 97°$$

All atoms are in general positions of C_{2h}^5 ($P2_1/c$):

$$(4e) \quad \pm(xyz;\ x,{}^1/_2-y,z+{}^1/_2)$$

with the parameters of Table XVAII,6. The x and z parameters of the hydrogen atoms were determined experimentally; in assigning the y parameters it was assumed they should lie in the plane of the ring.

TABLE XVAII,6

Parameters of the Atoms in Benzoic Acid

Atom	x	y	z
C(1)	0.103	0.278	0.057
C(2)	0.180	0.481	0.104
C(3)	0.383	0.631	0.093
C(4)	0.455	0.823	0.140
C(5)	0.330	0.875	0.190
C(6)	0.133	0.720	0.196
C(7)	0.051	0.516	0.154
O(1)	0.223	0.237	0.013
O(2)	−0.089	0.140	0.064

The structure thus obtained is shown in Figure XVAII,6. The molecules themselves are planar to within 0.01 A. except for the C(1) and O(2) atoms which depart 0.04 and 0.07 A. from the molecular plane. The bond dimensions are those of Figure XVAII,7. In the crystal these molecules are connected by hydrogen bonds between carboxyl oxygen atoms, with O–H–O = 2.64 A. Other intermolecular atomic distances correspond to van der Waals separations, the shortest being an O(1)–C(1) of 3.34 A.

XV,aII,8. *Potassium acid dibenzoate*, $KH(C_6H_5COO)_2$, is monoclinic with a tetramolecular cell of the dimensions:

$$a_0 = 29.53\text{ A.};\quad b_0 = 3.88\text{ A.};\quad c_0 = 11.20\text{ A.};\quad \beta = 95°48'$$

Atoms are in the following positions of the space group C_{2h}^6 ($C2/c$):

$$(4e) \quad \pm(0\,u\,{}^1/_4;\ {}^1/_2,u+{}^1/_2,{}^1/_4)$$

$$(8f) \quad \pm(xyz;\ x,\bar{y},z+{}^1/_2;\ x+{}^1/_2,y+{}^1/_2,z;\ x+{}^1/_2,{}^1/_2-y,z+{}^1/_2)$$

The parameters are listed in Table XVAII,7.

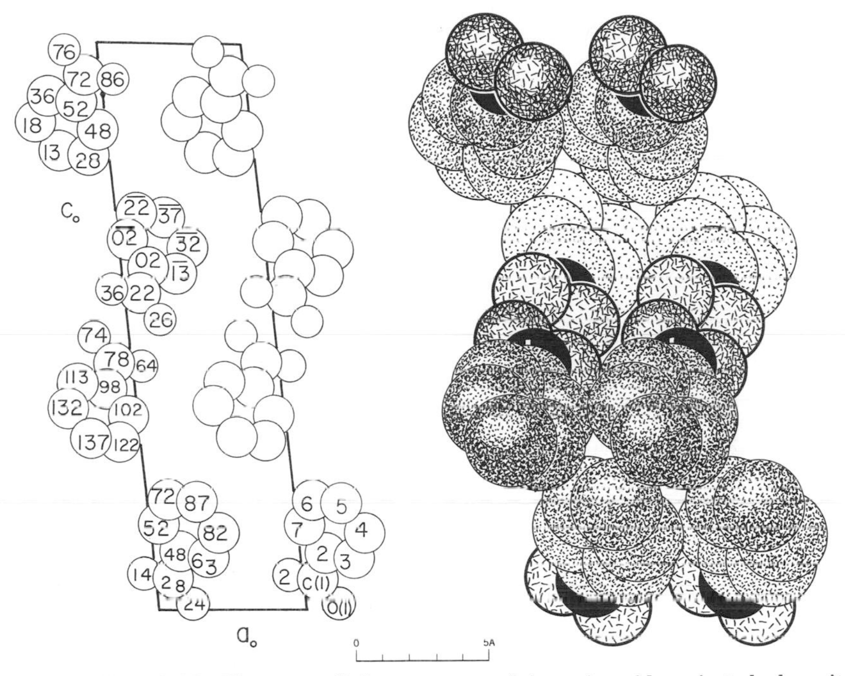

Fig. XVAII,6a (left). The monoclinic structure of benzoic acid projected along its b_0 axis. Left-hand axes.

Fig. XVAII,6b (right). A packing drawing of the monoclinic benzoic acid structure seen along its b_0 axis. The carboxyl carbon atoms are black, the oxygens line shaded and heavily ringed. Benzene carbons are dotted. Left-hand axes.

TABLE XVAII,7

Positions and Parameters of the Atoms in $KH(C_6H_5COO)_2$

Atom	Position	x	y	z
K	(4*e*)	0	0.328	1/4
O(1)	(8*f*)	0.0555	0.165	−0.1259
O(2)	(8*f*)	0.0386	0.063	0.0527
C(1)	(8*f*)	0.0663	0.148	−0.0178
C(2)	(8*f*)	0.1153	0.205	0.0366
C(3)	(8*f*)	0.1462	0.358	−0.0316
C(4)	(8*f*)	0.1908	0.414	0.0170
C(5)	(8*f*)	0.2042	0.321	0.1325
C(6)	(8*f*)	0.1735	0.169	0.2007
C(7)	(8*f*)	0.1293	0.111	0.1521

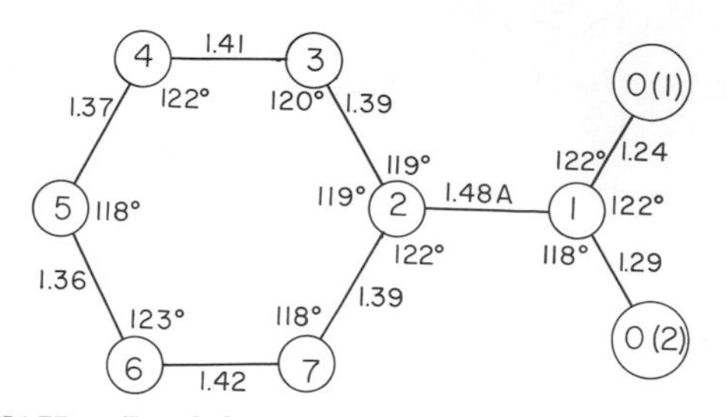

Fig. XVAII,7. Bond dimensions in the benzoic acid molecule.

The structure is that shown in Figure XVAII,8. Within the planar benzoate ions the atomic separations are those of Figure XVAII,9.

XV,aII,9. *Copper benzoate trihydrate*, $Cu(C_6H_5CO_2)_2 \cdot 3H_2O$, forms monoclinic, pseudo-orthorhombic crystals. Their tetramolecular unit has the dimensions:

$$a_0 = 6.98(2) \text{ A.}; \quad b_0 = 34.12(6) \text{ A.}; \quad c_0 = 6.30(2) \text{ A.}; \quad \beta = 90°$$

The space group is C_{2h}^6 ($I2/c$) with copper atoms in the positions:

$$(4a) \quad 000; 0\,0\,{}^1/_2; \quad \text{B.C.}$$

The other atoms are in either:

$$(4e) \quad \pm(0\,u\,{}^1/_4; {}^1/_2, u+{}^1/_2, {}^3/_4)$$

or

$$(8f) \quad \pm(xyz; x,\bar{y},z+{}^1/_2); \quad \text{B.C.}$$

with the parameters of Table XVAII,8.

TABLE XVAII,8
Parameters of Atoms in $Cu(C_6H_5CO_2)_2 \cdot 3H_2O$

Atom	Position	x	y	z
C(1)	(4*e*)	0	0.0709	1/4
C(2)	(4*e*)	0	0.1140	1/4
C(3)	(8*f*)	0.069	0.1356	0.077
C(4)	(8*f*)	0.072	0.1771	0.070
C(5)	(4*e*)	0	0.1964	1/4
C(6)	(4*e*)	0	−0.4041	1/4
C(7)	(4*e*)	0	−0.3602	1/4
C(8)	(8*f*)	0.420	0.1605	0.578
C(9)	(8*f*)	0.420	0.2016	0.585
C(10)	(4*e*)	0	−0.2786	1/4
O(1)	(8*f*)	0.025	0.0539	0.078
O(2)	(8*f*)	0.403	0.0789	0.611
$H_2O(1)$	(8*f*)	0.216	0.0125	0.690
$H_2O(2)$	(4*e*)	0	0.4595	1/4

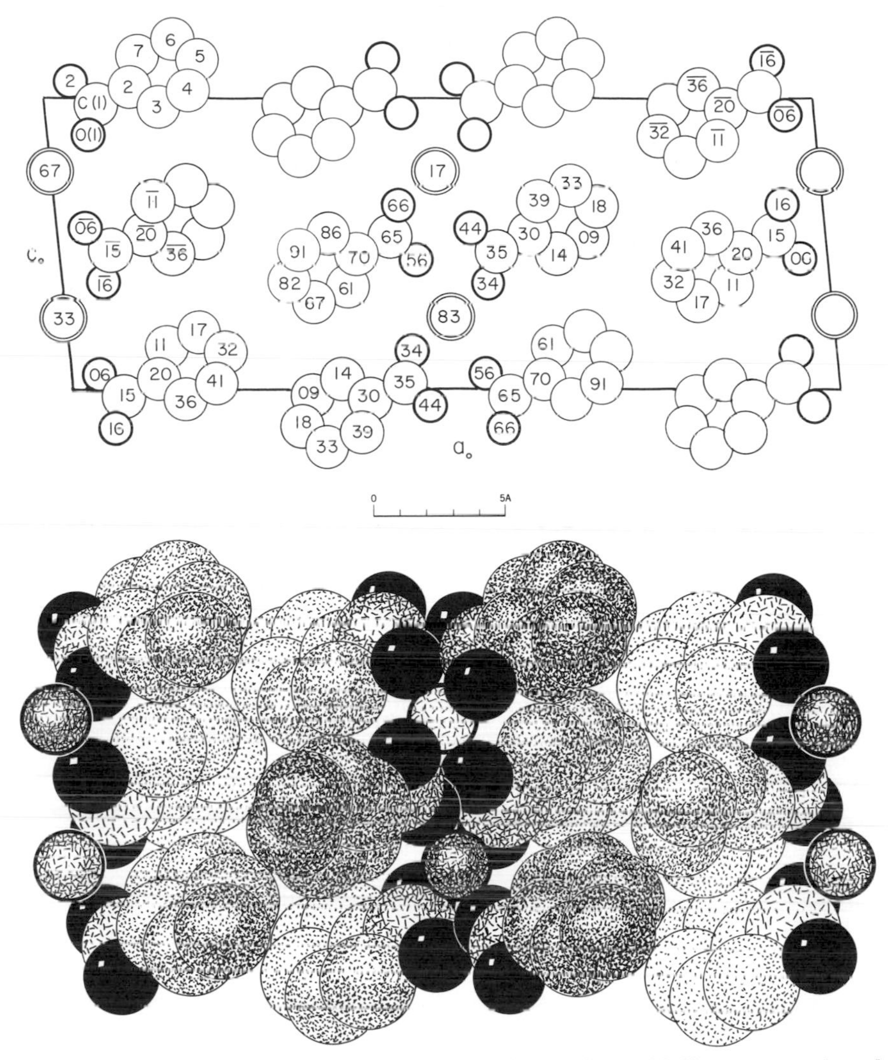

Fig. XVAII,8a (top). The monoclinic structure of potassium acid dibenzoate projected along its b_0 axis. The potassium atoms are doubly ringed. Left-hand axes.

Fig. XVAII,8b (bottom). A packing drawing of the monoclinic potassium acid dibenzoate arrangement viewed along its b_0 axis. The potassium atoms are heavily ringed and line shaded. The carboxyl carbons, also line shaded, are larger, as are the dotted ring carbons. Oxygen atoms are black. Left-hand axes.

The structure, shown in Figure XVAII,10, contains two kinds of benzoate anion. In each the benzene ring is planar, but in one anion the CO_2 group is twisted through 15° and in the other 10° to this plane. The oxygen atoms of the first anion are in contact with the copper atoms; those of the second

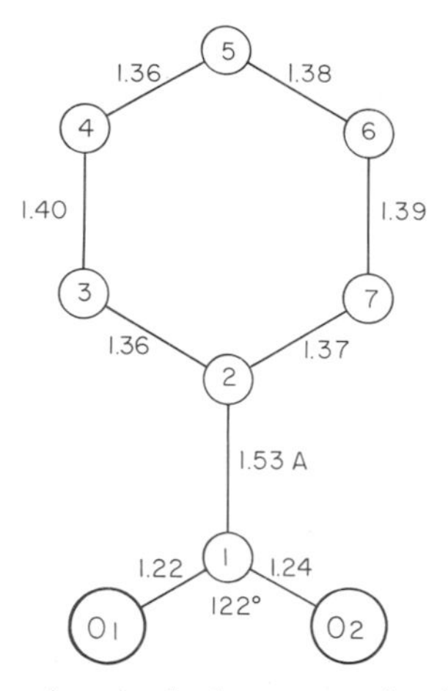

Fig. XVAII,9. Bond dimensions in the benzoate anion in its acid potassium salt.

form hydrogen bonds of length 2.66 and 2.71 A. with water molecules. The cupric ions have their usual square coordination, with 2Cu–O = 1.91 A. and $2Cu–OH_2$ = 1.97 A.; two additional $Cu–OH_2$ = 2.51 A. result in a distorted octahedron. All angles O–Cu–O = 87–91°.

XV,aII,10. The triclinic complex *methyl benzoate–chromium tricarbonyl*, $C_6H_5COOCH_3 \cdot Cr(CO)_3$, has been assigned a tetramolecular non-primitive cell of the dimensions:

$$a_0 = 10.46 \text{ A.}; \quad b_0 = 11.43 \text{ A.}; \quad c_0 = 10.50 \text{ A.}$$
$$\alpha = 108°0'; \quad \beta = 101°11'; \quad \gamma = 101°4'$$

The space group is C_i^1 which, for this four-molecule cell, carries the symbol $C\bar{1}$. Atoms, in the positions (4*i*) $\pm(xyz; x+{}^1/_2,y+{}^1/_2,z)$, have the parameters listed in Table XVAII,9.

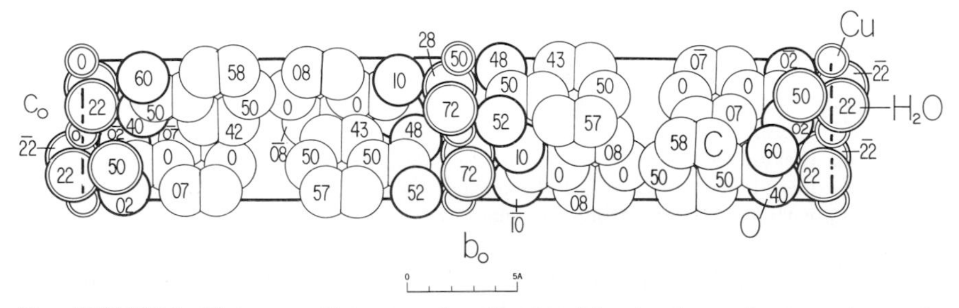

Fig. XVAII,10. The monoclinic, pseudo-orthorhombic structure of copper benzoate trihydrate projected along its a_0 axis.

TABLE XVAII,9

Parameters of the Atoms in $C_6H_5COOCH_3 \cdot Cr(CO)_3$

Atom	x	y	z
C(1)	0.0631	0.2308	0.4251
C(2)	0.1269	0.3443	0.4126
C(3)	0.2146	0.3413	0.3207
C(4)	0.2332	0.2254	0.2427
C(5)	0.1638	0.1108	0.2549
C(6)	0.0824	0.1134	0.3448
C(7)	−0.0266	0.2269	0.5192
O(8)	−0.0870	0.1305	0.5267
O(9)	−0.0280	0.3463	0.5936
C(10)	−0.1045	0.3504	0.6978
C(11)	0.0177	0.2062	0.0250
O(12)	0.0181	0.2040	−0.0828
C(13)	−0.1019	0.3121	0.2023
O(14)	−0.1794	0.3754	0.2023
C(15)	−0.1416	0.0697	0.1202
O(16)	−0.2346	−0.0136	0.0717
Cr	0.0103	0.2097	0.2031

Bond dimensions in the benzoate and carbonyl components are shown in Figure XVAII,11. Viewed normal to the plane of the benzene ring, the chromium atom is above its center with the carbonyl carbons overlying the C(2), C(4), and C(6) atoms of the ring. The distance from chromium to the ring carbons lies between 2.20 and 2.25 A.

XV,aII,11. The clathrate structure described for *tetra-n-butyl ammonium benzoate hydrate*, $(n\text{-}C_4H_9)_4N \cdot C_6H_5CO_2 \cdot 39^1/_2H_2O$, is much the same as that already referred to in paragraph **XIV,c20** for $(n\text{-}C_4H_9)_4NF \cdot 32.8H_2O$. Its crystals are tetragonal with a tetramolecular unit of substantially the same size as for the simpler fluoride:

$$a_0 = 23.57(4) \text{ A.}, \qquad c_0 = 12.45(2) \text{ A.} \quad (-30°\text{C.})$$

The x-ray data are compatible with the high symmetry space group D_{4h}^{14} ($P4_2/mnm$) on the assumption that there is a considerable measure of disorder in the structure itself. Atomic positions and parameters are given in the original paper for such an imperfect structure; in it the organic radicals are embedded in a framework of water molecules suggesting those found in the phosphotungstates.

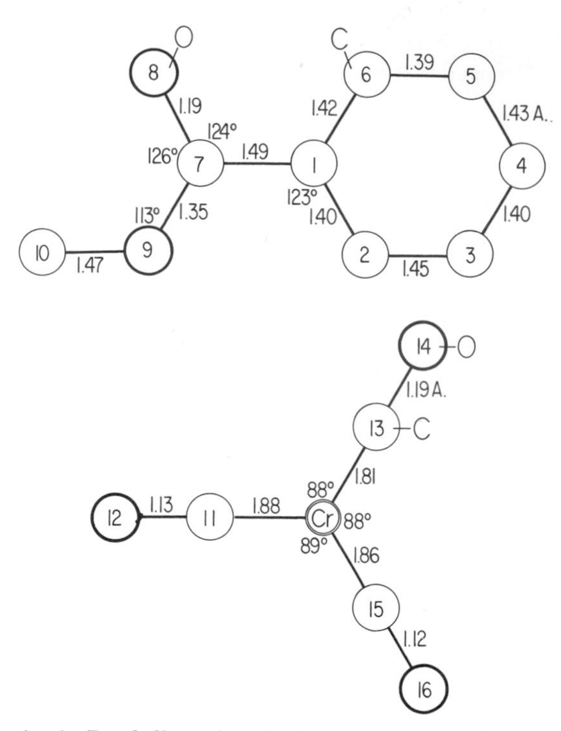

Fig. XVAII,11a (top). Bond dimensions in the organic part of the methyl benzoate–chromium tricarbonyl complex.

Fig. XVAII,11b (bottom). Bond dimensions in the $Cr(CO)_3$ portion of the methyl benzoate–chromium tricarbonyl complex.

XV,aII,12. The complex *aluminum chloride–benzoyl chloride*, $AlCl_3 \cdot C_6H_5COCl$, forms monoclinic crystals whose bimolecular unit has the dimensions:

$$a_0 = 8.681 \text{ A.}; \quad b_0 = 7.286 \text{ A.}; \quad c_0 = 9.330 \text{ A.}; \quad \beta = 104.33°$$

The space group has been chosen as C_{2h}^2 ($P2_1/m$). All atoms except one set of chlorine are in the special positions:

$$(2e) \quad \pm(u\ {}^1/_4\ v)$$

with the parameters of Table XVAII,10. The Cl(1) atoms are in:

$$(4f) \quad \pm(xyz;\ x,{}^1/_2-y,z)$$
$$\text{with } x = 0.0854(4),\ y = 0.5122(4),\ z = 0.7192(4)$$

TABLE XVAII,10

Parameters of the Atoms in $AlCl_3 \cdot C_6H_5COCl$

Atom	u	v
Cl(2)	0.1795(5)	0.1216(5)
Cl(3)	0.5492(6)	0.0498(6)
Al	0.9587(6)	0.1745(5)
O	0.8187(12)	0.9935(12)
C(1)	0.6758(17)	0.9346(17)
C(2)	0.6102(17)	0.7721(17)
C(3)	0.7162(20)	0.6898(18)
C(4)	0.6597(25)	0.5431(21)
C(5)	0.4966(28)	0.4783(20)
C(6)	0.3951(23)	0.5676(23)
C(7)	0.4476(18)	0.7186(20)
H(3)	0.817(20)	0.731(18)
H(4)	0.726(20)	0.486(18)
H(5)	0.439(19)	0.366(18)
H(6)	0.265(20)	0.508(18)
H(7)	0.365(19)	0.777(18)

The structure is shown in Figure XVAII,12. Its molecules are strictly planar except for Cl(1) and have the bond dimensions of Figure XVAII,13. In the crystal the molecules lie in layers. Those in the same layer have a

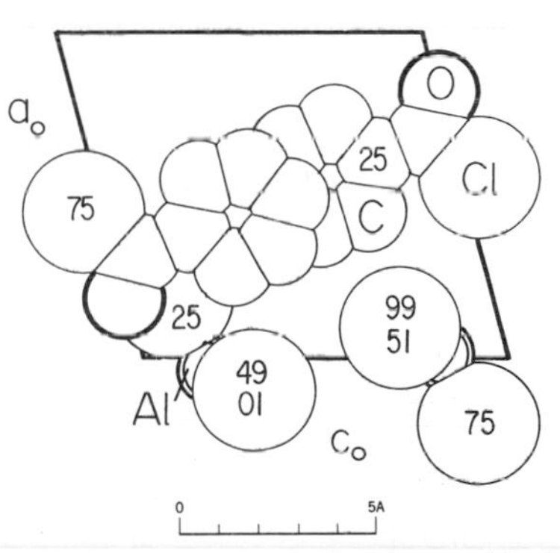

Fig. XVAII,12. The monoclinic structure of aluminum chloride–benzoyl chloride projected along its b_0 axis.

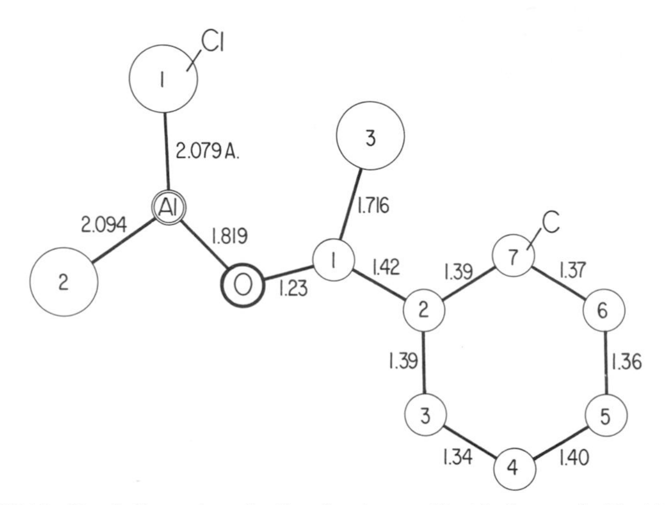

Fig. XVAII,13. Bond dimensions in the aluminum chloride–benzoyl chloride complex.

short Cl–Cl = 3.44 A.; between layers the shortest atomic approach is Cl–Cl = 3.80 A.

XV,aII,13. Crystals of *α-chloroacetophenone*, $C_6H_5COCH_2Cl$, are orthorhombic. Their tetramolecular unit has the edge lengths:

$$a_0 = 19.13(3) \text{ A.}; \quad b_0 = 9.63(2) \text{ A.}; \quad c_0 = 4.13(2) \text{ A.}$$

The space group is V^4 ($P2_12_12_1$) with atoms in the positions:

$$(4a) \quad xyz;\ {}^1/_2-x,\bar{y},z+{}^1/_2;\ x+{}^1/_2,{}^1/_2-y,\bar{z};\ \bar{x},y+{}^1/_2,{}^1/_2-z$$

According to a preliminary note the approximate parameters are those of Table XVAII,11.

TABLE XVAII,11
Parameters of the Atoms in α-Chloroacetophenone

Atom	x	y	z
C(1)	−0.021	0.161	0.452
C(2)	0.031	0.296	0.452
C(3)	0.099	0.251	0.288
C(4)	0.116	0.115	0.166
C(5)	0.182	0.088	0.014
C(6)	0.227	0.196	−0.002
C(7)	0.214	0.339	0.131
C(8)	0.148	0.367	0.276
O	0.019	0.405	0.554
Cl	−0.103	0.226	0.621

XV,aII,14. *Potassium hydrogen bis(phenylacetate)*, $KH(C_6H_5CH_2COO)_2$, and the isostructural rubidium salt are monoclinic with large tetramolecular cells of the dimensions:

For $KH(C_6H_5CH_2COO)_2$:

$$a_0 = 28.4 \text{ A.}; \quad b_0 = 4.50 \text{ A.}; \quad c_0 = 11.97 \text{ A.}; \quad \beta = 90°24'$$

For $RbH(C_6H_5CH_2COO)_2$:

$$a_0 = 29.0 \text{ A.}; \quad b_0 = 4.59 \text{ A.}; \quad c_0 = 12.36 \text{ A.}; \quad \beta = \text{ca. } 90°$$

The alkali cations are in special positions of the space group C_{2h}^6 ($I2/c$):

$$(4e) \quad \pm(0\ u\ {}^1/_4;\ {}^1/_2,u+{}^1/_2,{}^3/_4)$$

All other atoms (except hydrogen) are in general positions:

$$(8f) \quad \pm(xyz;\ x,\bar{y},z+{}^1/_2); \qquad \text{B.C.}$$

The established parameters are listed in Table XVAII,12.

TABLE XVAII,12

Parameters of the Atoms in Potassium Hydrogen Bis(Phenylacetate)

Atom	x	y	z
K	0	0.206	$^1/_4$
O(1)	0.0446	0.290	−0.1430
O(2)	0.0283	0.200	0.0350
C(1)	0.0483	0.318	−0.0458
C(2)	0.0850	0.540	0.0070
C(3)	0.1296	0.373	0.0467
C(4)	0.1354	0.373	0.1667
C(5)	0.1742	0.218	0.2067
C(6)	0.2042	0.062	0.1400
C(7)	0.1983	0.062	0.0208
C(8)	0.1604	0.218	−0.0200

Comparison of the resulting structure, shown in Figure XVAII,14, with that of Figure XVAII,8 brings out its relation to the benzoate.

The interatomic distances within the organic anions of this crystal are given in Figure XVAII,15. The phenyl group is planar but the angle C(1)–C(2)–C(3) = 111°.

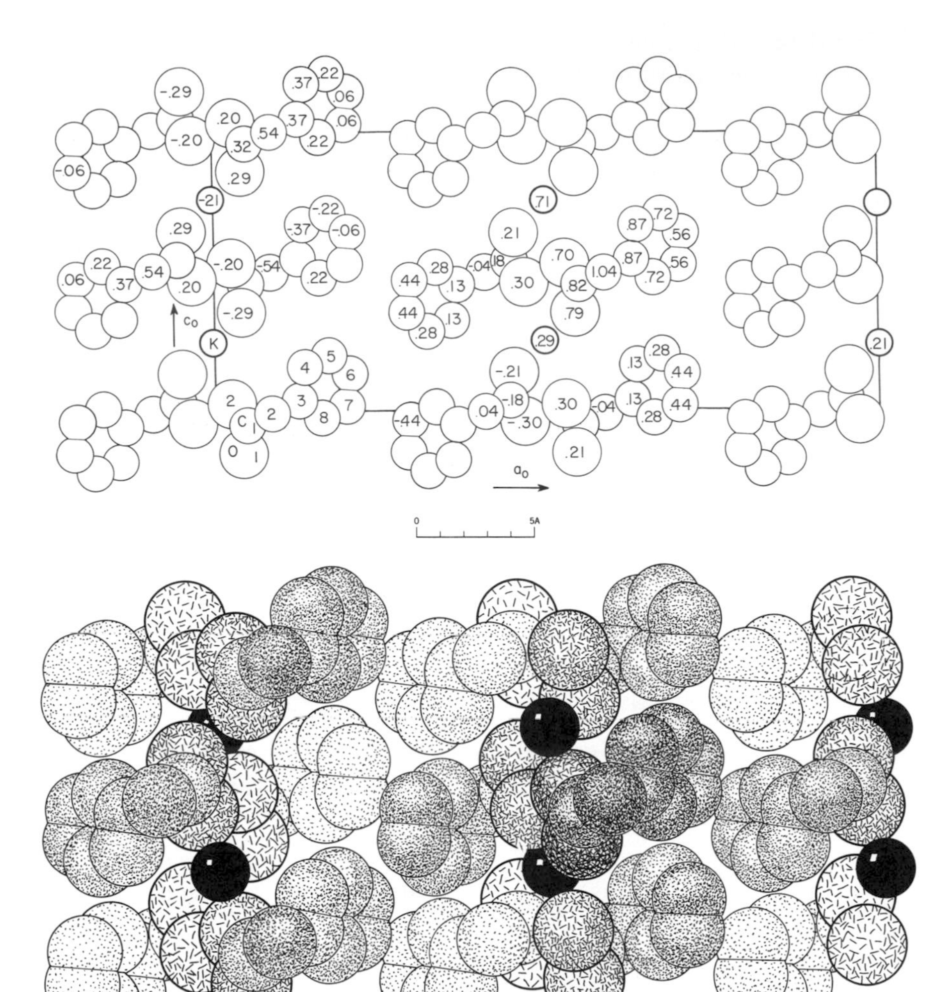

Fig. XVAII,14a (top). The monoclinic structure of potassium acid bis(phenylacetate) projected along its b_0 axis. Left-hand axes.

Fig. XVAII,14b (bottom). A packing drawing of the monoclinic structure of potassium acid bis(phenylacetate) seen along its b_0 axis. The carbon atoms are dotted, the oxygens line shaded. The potassium atoms are black. Left-hand axes.

The acid hydrogen atoms [H(O)] are in symmetry centers:

$$(4a) \quad 000; \ {}^1/_2 \, {}^1/_2 \, 0; \quad \text{B.C.}$$

midway between O(2) atoms of adjacent molecules; the assigned parameters make this O(2)–O(2) distance 2.54 A., which would correspond to a close

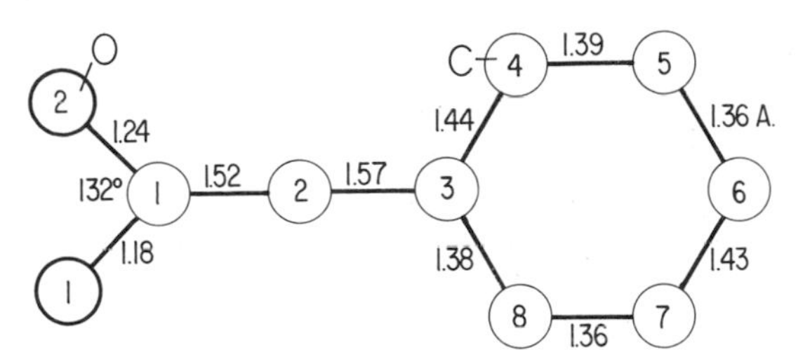

Fig. XVAII,15. Bond dimensions in the phenylacetate anion in its potassium acid salt.

hydrogen bonding. Each potassium ion is surrounded by an octahedron of oxygen atoms with K–O distances of 2.75 and 2.88 A. The closest approach of the carbon atoms of nearby benzene rings is 3.72 A.

More recently neutron diffraction data obtained at 120°K. and at room temperature have confirmed the x-ray parameters for the heavy atoms and shown where the hydrogens are located. The hydrogen x and z parameters were thus found experimentally but the y values were calculated on the assumption that the benzene hydrogens lie in the plane of the ring. Hydrogen parameters established in this way are listed in Table XVAII,13.

TABLE XVAII,13

Neutron Parameters of the Hydrogen Atoms in $KH(C_6H_5CH_2CO_2)_2$

Atom	x	y	z
H(2)	0.098	(0.676)	−0.080
H(2′)	0.071	(0.684)	0.056
H(4)	0.113	(0.502)	0.213
H(5)	0.177	(0.228)	0.301
H(6)	0.234	(−0.071)	0.181
H(7)	0.224	(−0.080)	−0.033
H(8)	0.152	(0.240)	−0.109

XV,aII,15. Crystals of *bis(1-phenyl-1,3-butanedionato)*[*or benzoylacetonato*]*copper*, $Cu[CH_3C(O)CHC(O)C_6H_5]_2$, are monoclinic with a bimolecular unit of the dimensions:

$$a_0 = 4.475(5)\text{ A.};\quad b_0 = 10.640(5)\text{ A.};\quad c_0 = 18.486(10)\text{ A.}$$
$$\beta = 97°40(5)'$$

The space group is C_{2h}^5 ($P2_1/c$). The copper atom is in:

$$(2a)\quad 000;\ 0\ {}^1/_2\ {}^1/_2$$

The other atoms are in:

$$(4e) \quad \pm(xyz;\ x,{}^1/_2-y,z+{}^1/_2)$$

with the parameters of Table XVAII,14.

TABLE XVAII,14

Parameters of the Atoms in $Cu[CH_3C(O)CHC(O)C_6H_5]_2$

Atom	x	y	z
O(1)	0.1951(10)	0.1504(5)	−0.0287(3)
O(2)	0.2528(10)	−0.0094(5)	0.0919(2)
C(1)	0.5502(22)	0.3133(8)	−0.0310(5)
C(2)	0.4218(16)	0.2012(7)	0.0053(4)
C(3)	0.5738(17)	0.1612(7)	0.0729(4)
C(4)	0.6266(16)	0.0380(6)	0.1897(3)
C(5)	0.8193(19)	0.1254(8)	0.2280(4)
C(6)	0.9423(21)	0.1014(9)	0.2987(4)
C(7)	0.4738(16)	0.0637(6)	0.1142(4)
C(8)	0.8797(21)	−0.0095(10)	0.3314(4)
C(9)	0.6921(21)	−0.0974(8)	0.2959(4)
C(10)	0.5613(20)	−0.0734(8)	0.2237(4)

The resulting structure is shown in Figure XVAII,16. Its molecules have the bond dimensions of Figure XVAII,17. Except for C(3) which is 0.05 A.

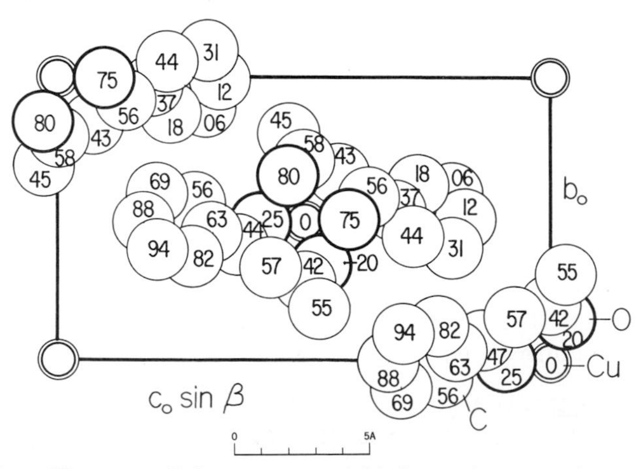

Fig. XVAII,16. The monoclinic structure of bis(benzoylacetonato)copper projected along its short a_0 axis.

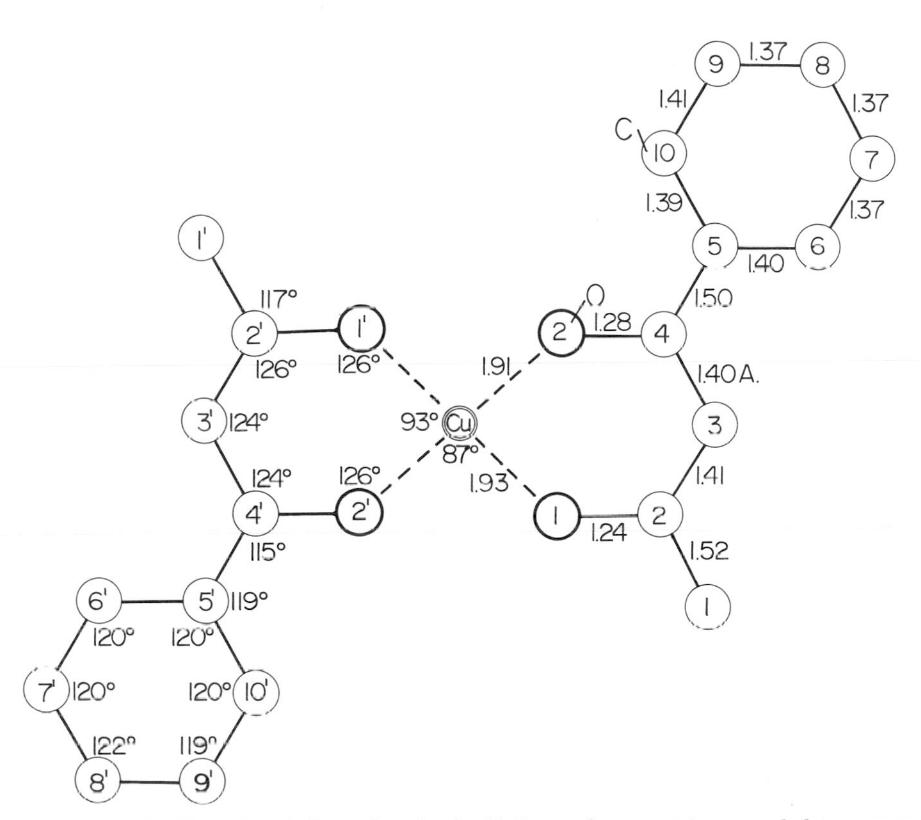

Fig. XVAII,17. Bond dimensions in the bis(benzoylacetonato)copper chelate.

outside, the chelating group of atoms including copper is planar and so is the benzene ring. The two planes are turned through 14° with respect to one another. Approximate parameters for the hydrogen atoms are given in the original article.

XV,aII,16. Crystals of *trans-bis(1-phenyl-1,3-butanedionato)palladium-(II)*, $[C_6H_5C(O)CHC(O)CH_3]_2Pd$, are monoclinic with a bimolecular unit of the dimensions:

$$a_0 = 9.367 \text{ A.}; \quad b_0 = 10.518 \text{ A.}; \quad c_0 = 9.454 \text{ A.}; \quad \beta = 108°0'$$

Atoms of palladium are in the following positions of C_{2h}^5 ($P2_1/c$):

$$(2a) \quad 000; 0\ {}^1/_2\ {}^1/_2$$

The other atoms are in:

$$(4e) \quad \pm(xyz;\ x,{}^1/_2-y,z+{}^1/_2)$$

with the parameters of Table XVAII,15.

TABLE XVAII,15

Parameters of Atoms in *trans*-bis(1-Phenyl-1,3-butanedionato) Palladium

Atom	x	y	z
O(1)	−0.1207(13)	0.1005(14)	−0.1703(13)
O(2)	0.1395(12)	0.1419(13)	0.0818(13)
C(1)	−0.2112(19)	0.2792(20)	−0.3246(19)
C(2)	−0.1018(18)	0.2151(23)	−0.1851(18)
C(3)	0.0120(18)	0.3002(23)	−0.0892(27)
C(4)	0.1291(17)	0.2574(18)	0.0285(19)
C(5)	0.2484(18)	0.3394(18)	0.1137(18)
C(6)	0.2850(20)	0.4462(19)	0.0496(21)
C(7)	0.4081(23)	0.5259(22)	0.1252(24)
C(8)	0.4978(22)	0.4944(24)	0.2695(22)
C(9)	0.4581(23)	0.3880(23)	0.3340(22)
C(10)	0.3333(20)	0.3053(23)	0.2595(21)

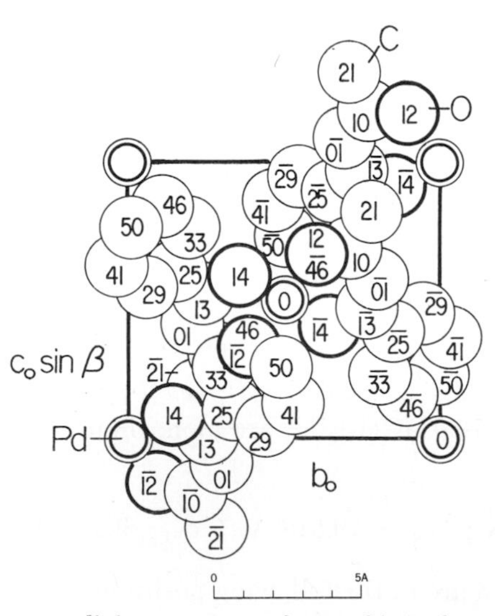

Fig. XVAII,18. The monoclinic structure of *trans*-bis(1-phenyl-1,3-butanedionato)-palladium projected along its a_0 axis.

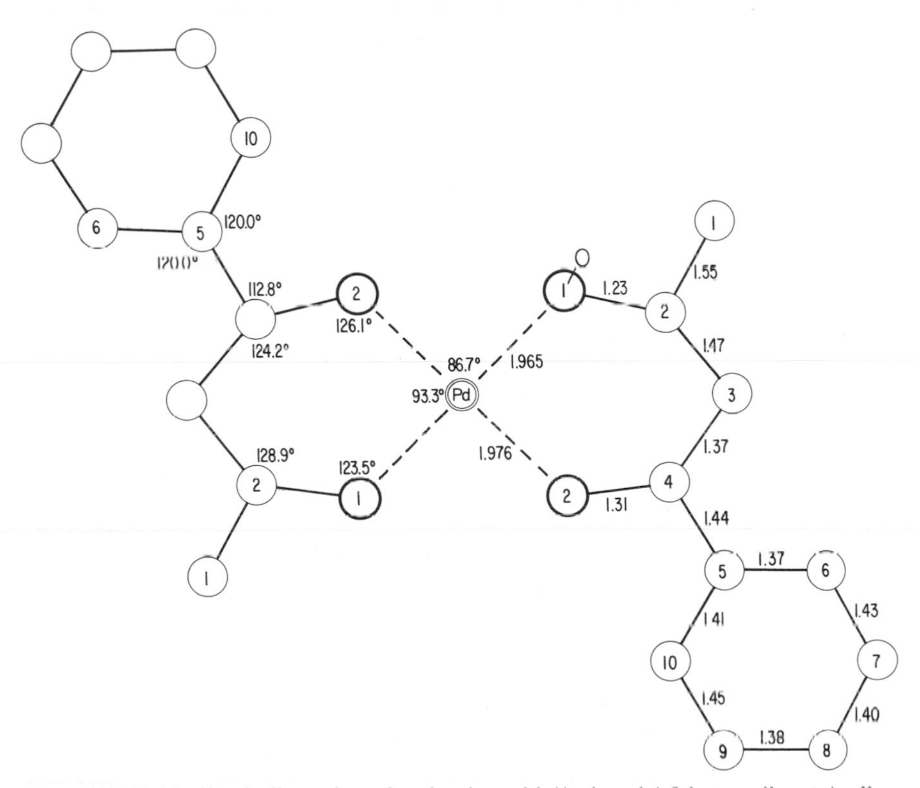

Fig. XVAII,19. Bond dimensions in the *trans*-bis(1-phenyl-1,3-butanedionato)palladium chelate.

The structure, shown in Figure XVAII,18, has molecules with the bond dimensions of Figure XVAII,19. The palladium and the four oxygen atoms in square coordination about it are coplanar. The benzene rings are planar and so are the attached atoms forming the chelating pseudo-ring; between the two the angle is 23°13′.

XV,aII,17. Crystals of *bis(3-phenyl-2,4-pentanedionato)copper* [bis(3-phenylacetylacetonate)copper], $Cu[C_6H_5CH(C(O)CH_3)_2]_2$, are monoclinic with a bimolecular unit of the dimensions:

$$a_0 = 10.250(10) \text{ A.}; \quad b_0 = 6.778(6) \text{ A.}; \quad c_0 = 13.763(13) \text{ A.}$$
$$\beta = 93°33(5)'$$

The space group is C_{2h}^5 ($P2_1/c$). The copper atoms are in:

$$(2a) \quad 000; 0 \, {}^1/_2 \, {}^1/_2$$

The other atoms are in the general positions:

$$(4e) \quad \pm(xyz;\ x,{}^1/_2-y,z+{}^1/_2)$$

with the parameters of Table XVAII,16.

TABLE XVAII,16

Parameters of Atoms in Bis(3-Phenyl-2,4-pentanedionato)copper

Atom	x	y	z
O(1)	0.1634(6)	0.1226(14)	−0.0222(4)
O(2)	−0.0466(6)	0.2105(13)	0.0824(5)
C(1)	0.3548(9)	0.3130(24)	−0.0048(8)
C(2)	0.2168(8)	0.2646(21)	0.0244(6)
C(3)	0.1600(8)	0.3816(20)	0.0980(6)
C(4)	0.0245(9)	0.3447(22)	0.1181(6)
C(5)	−0.0409(9)	0.4899(22)	0.1863(8)
C(6)	0.2365(8)	0.5381(21)	0.1516(6)
C(7)	0.2728(9)	0.7130(22)	0.1081(7)
C(8)	0.3483(10)	0.8553(23)	0.1587(8)
C(9)	0.3833(9)	0.8237(23)	0.2592(8)
C(10)	0.3439(9)	0.6465(22)	0.3033(7)
C(11)	0.2717(8)	0.5036(21)	0.2517(6)

The structure is that of Figure XVAII,20. Its molecules have the bond dimensions of Figure XVAII,21. The chelating part of the molecule is

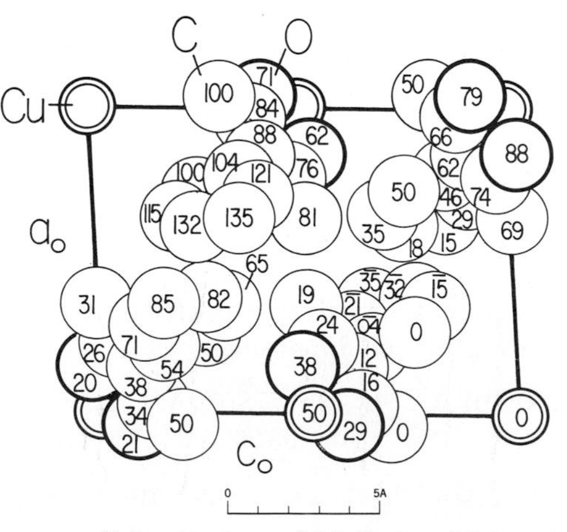

Fig. XVAII,20. The monoclinic structure of bis(3-phenyl-2,4-pentanedionato)copper projected along its b_0 axis.

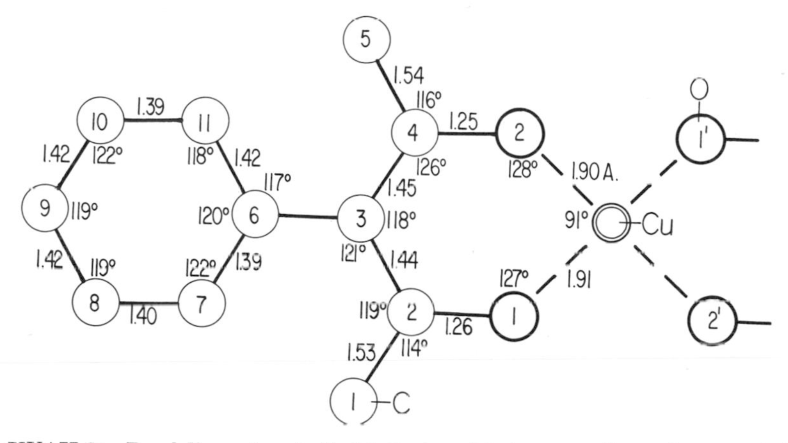

Fig. XVAII,21. Bond dimensions in the bis(3-phenyl-2,4-pentanedionato)copper chelate.

approximately planar and is turned through an angle of about 70° to the plane of the benzene ring. The cupric atom has its usual square coordination.

XV,aII,18. Crystals of the *cis* form of *bis(1-phenyl-1,3-butanedionato)-vanadyl*, $[C_6H_5C(O)CHC(O)CH_3]_2VO$, are monoclinic with a tetramolecular unit of the dimensions:

$$a_0 = 8.130(3) \text{ A.}; \quad b_0 = 22.599(10) \text{ A.}; \quad c_0 = 10.505(4) \text{ A.}$$
$$\beta = 106°47(3)'$$

All atoms are in the general positions of the space group C_{2h}^5 ($P2_1/c$):

$$(4e) \quad \pm(xyz;\ x,{}^1/_2-y,z+{}^1/_2)$$

The determined parameters are listed in Table XVAII,17.

TABLE XVAII,17
Parameters of the Atoms in Bis(1-Phenyl-1,3-butanedionato)vanadyl

Atom	x	y	z
V	0.1865(3)	0.09907(8)	0.12413(19)
O(1)	0.3355(12)	0.1205(3)	0.2538(8)
O(2)	0.0254(11)	0.0513(3)	0.1873(7)
O(3)	0.2769(11)	0.0240(3)	0.0724(7)
O(4)	0.2415(11)	0.1342(3)	−0.0317(7)
O(5)	0.0027(11)	0.1573(3)	0.0969(7)
C(1)	0.2794(20)	0.2022(6)	−0.1951(13)
C(2)	0.1913(19)	0.1839(5)	−0.0904(11)
C(3)	0.0555(18)	0.2192(5)	−0.0695(12)
C(4)	−0.0257(17)	0.2040(5)	0.0228(11)

(continued)

TABLE XVAII,17 (*continued*)

Atom	x	y	z
C(5)	−0.1779(16)	0.2405(4)	0.0389(10)
C(6)	−0.2140(19)	0.2970(5)	−0.0192(11)
C(7)	−0.3588(20)	0.3278(6)	−0.0010(13)
C(8)	−0.4627(21)	0.3050(6)	0.0745(14)
C(9)	−0.4237(20)	0.2496(6)	0.1336(13)
C(10)	−0.2790(20)	0.2176(5)	0.1169(11)
C(11)	0.3776(17)	−0.0753(5)	0.0777(12)
C(12)	0.2622(18)	−0.0288(4)	0.1171(11)
C(13)	0.1495(18)	−0.0433(5)	0.1934(11)
C(14)	0.0348(17)	−0.0035(5)	0.2240(10)
C(15)	−0.0815(17)	−0.0208(5)	0.3020(10)
C(16)	−0.2327(20)	0.0131(5)	0.2894(11)
C(17)	−0.3465(20)	−0.0010(6)	0.3648(13)
C(18)	−0.3046(23)	−0.0501(6)	0.4525(12)
C(19)	−0.1538(22)	−0.0822(6)	0.4667(12)
C(20)	−0.0407(18)	−0.0696(5)	0.3904(11)

The resulting structure is that of Figure XVAII,22. Its molecules, which are nearly planar, have the bond dimensions of Figure XVAII,23. Each

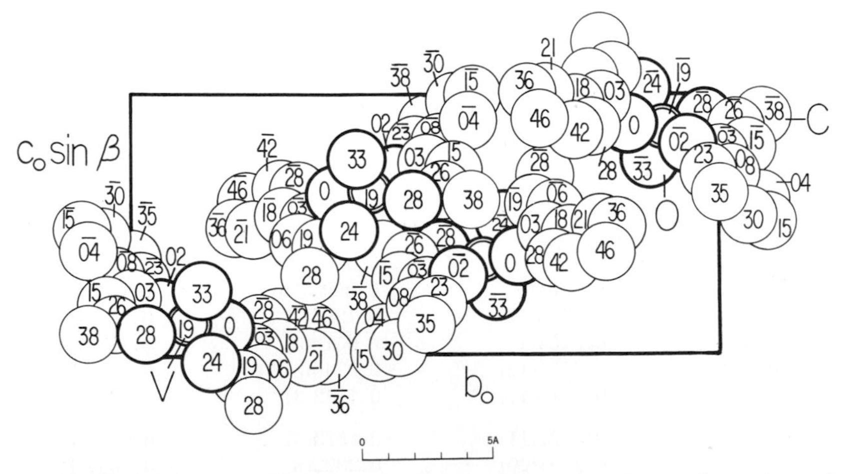

Fig. XVAII,22. The monoclinic structure of bis(1-phenyl-1,3-butanedionato)vanadyl projected along its a_0 axis.

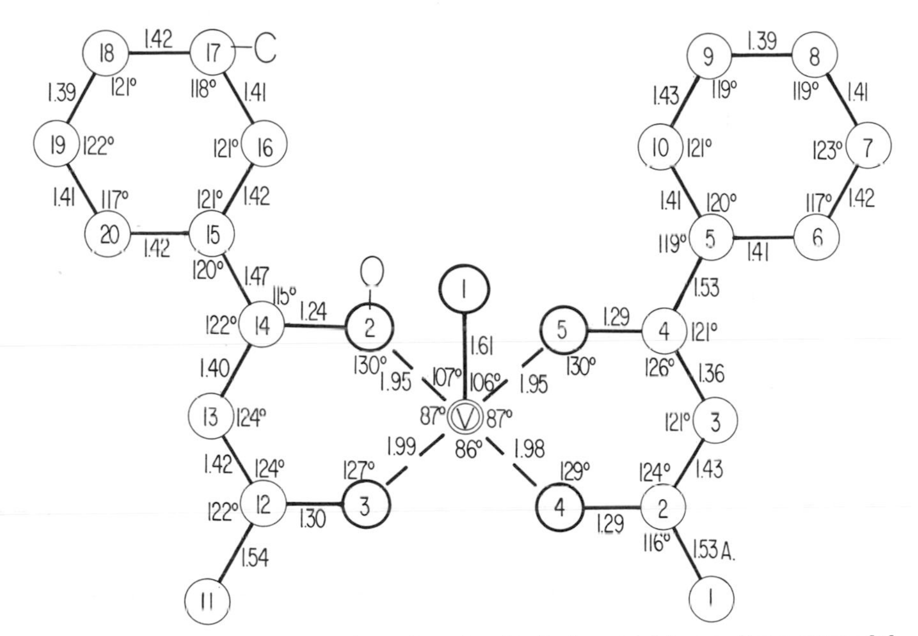

Fig. XVAII,23. Bond dimensions in the bis(1-phenyl-1,3-butanedionato)vanadyl chelate.

vanadium atom is surrounded by an approximately square pyramid of oxygen atoms, that at the apex [O(1)] belonging to the vanadyl group. Intermolecular distances range upwards from a C–C = 3.51 A. and a C–O = 3.55 A.

XV,aII,19. The monoclinic crystals of *styrene palladous chloride*, $C_6H_5CH{=}CH_2 \cdot PdCl_2$, have a tetramolecular unit of the dimensions:

$$a_0 = 13.50 \text{ A.}; \quad b_0 = 4.99 \text{ A.}; \quad c_0 = 13.69 \text{ A.}; \quad \beta = 90°24'$$

The space group is C_{2h}^5 ($P2_1/n$) with atoms in the positions:

$$(4e) \quad \pm(xyz;\ x+{}^1/_2,{}^1/_2-y,z+{}^1/_2)$$

The determined parameters are listed in Table XVAII,18.

TABLE XVAII,18

Parameters of the Atoms in Styrene–$PdCl_2$

Atom	x	y	z
Pd	0.082	0.124	0.086
Cl(1)	0.085	0.828	0.947
Cl(2)	0.248	0.083	0.106
C(1)	0.014	0.860	0.700
C(2)	0.005	0.940	0.357
C(3)	0.087	0.176	0.576
C(4)	0.166	0.027	0.579
C(5)	0.194	0.826	0.636
C(6)	0.107	0.727	0.706
C(7)	0.072	0.243	0.247
C(8)	0.084	0.397	0.170

The structure that results is shown in Figure XVAII,24. The palladium atoms in its dimeric molecules have, approximately, their usual square

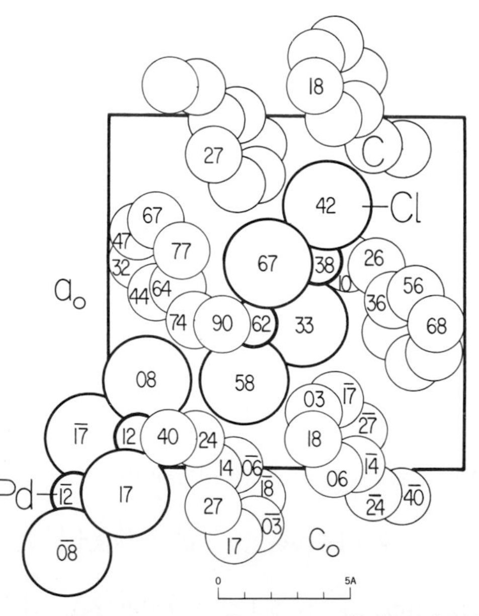

Fig. XVAII,24. The monoclinic structure of styrene–palladous chloride projected along its b_0 axis.

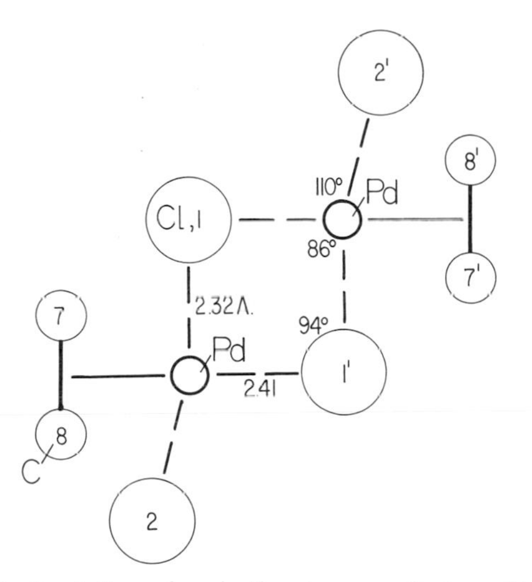

Fig. XVAII,25. Bond dimensions in the styrene–palladous chloride complex.

planar coordination; bond dimensions are shown in Figure XVAII,25. The Cl(2) atoms are 0.13 A. outside the plane of the palladium and Cl(1) atoms; in this plane the palladium–styrene approach is to a line between the C(7) and C(8) atoms. In the styrene part, C(7) lies ca. 0.16 A. out of the plane of the rest and this plane makes 86° with the Pd–Cl(1) line.

XV,aII,20. Crystals of *ammonium hydrogen dicinnamate*, $NH_4H(C_6H_5$-$CH{=}CHCO_2)_2$, are monoclinic with a tetramolecular unit of the dimensions:

$$a_0 = 37.87(12) \text{ A.}; \quad b_0 = 5.84(2) \text{ A.}; \quad c_0 = 7.62(3) \text{ A.}$$
$$\beta = 95°30(3)'$$

The space group is C_{2h}^6 in the body-centered orientation $I2/a$. Nitrogen atoms are in the positions:

$$NH_4: (4e) \quad \pm(0\,u\,{}^1/_4;\ {}^1/_2,{}^1/_2-u,{}^1/_4) \qquad \text{with } u = 0.4747$$

All other atoms are in:

$$(8f) \quad \pm(xyz;\ x+{}^1/_2,{}^1/_2-y,z); \quad \text{B.C.}$$

The parameters are those of Table XVAII,19.

TABLE XVAII,19

Parameters of Atoms in $NH_4H(C_9H_7O_2)_2$

Atom	x	y	z
O(1)	0.0302	0.0772	0.0509
O(2)	0.0466	−0.2849	0.0405
C(1)	0.0531	−0.0837	0.0683
C(2)	0.0921	−0.0404	0.1407
C(3)	0.1034	0.1552	0.1895
C(4)	0.1414	0.2039	0.2616
C(5)	0.1480	0.4055	0.3540
C(6)	0.1821	0.4644	0.4233
C(7)	0.2099	0.3140	0.3995
C(8)	0.2033	0.1155	0.3098
C(9)	0.1694	0.0581	0.2380
H(C,5)	0.127	0.517	0.369
H(C,6)	0.187	0.617	0.492
H(C,7)	0.236	0.355	0.454
H(C,8)	0.224	0.004	0.293
H(C,9)	0.165	−0.098	0.170
H(C,2)	0.109	−0.175	0.146
H(C,3)	0.087	0.282	0.182

The resulting structure is shown in Figure XVAII,26. Its cinnamate anions have the bond dimensions of Figure XVAII,27. In them the benzene ring is planar. The atoms of the ethylenic radical and apparently also of the

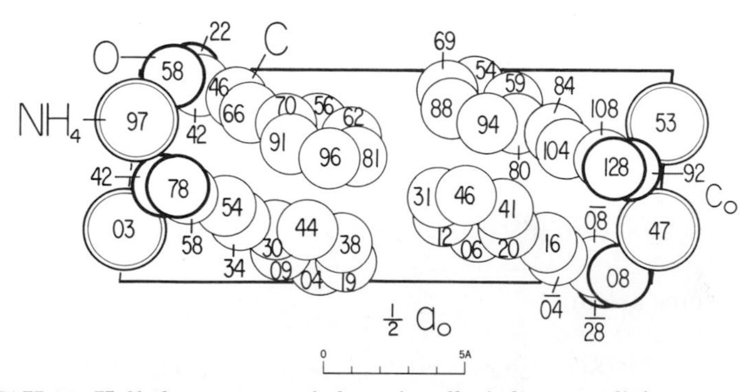

Fig. XVAII,26. Half the contents of the unit cell of the monoclinic structure of ammonium acid dicinnamate projected along its b_0 axis.

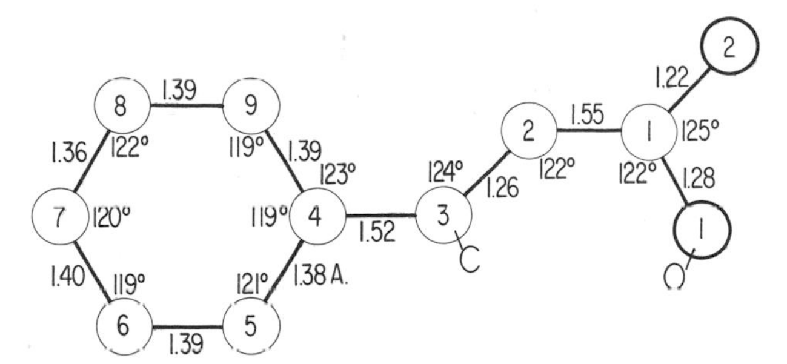

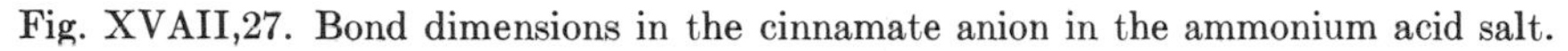
Fig. XVAII,27. Bond dimensions in the cinnamate anion in the ammonium acid salt.

carboxyl group are coplanar, this second plane making an angle of about 17° with the ring. Between anions there is a short O–O bond (presumably hydrogen) of 2.51 A.

Potassium hydrogen dicinnamate, $KH(C_6H_5CH{=}CHCO_2)_2$, is isostructural with a cell of the dimensions:

$$a_0 = 37.7 \text{ A.}; \quad b_0 = 5.74 \text{ A.}; \quad c_0 = 7.65 \text{ A.}; \quad \beta = 93°30'$$

XV,aII,21. *Phenylpropiolic acid,* $C_6H_5C{\equiv}C{\cdot}COOH$, forms orthorhombic crystals whose tetramolecular unit has the edge lengths:

$$a_0 = 5.113(10) \text{ A.}; \quad b_0 = 15.022(15) \text{ A.}; \quad c_0 = 9.963(15) \text{ A.}$$

The space group is V_h^{12} (*Pnnm*) with atoms in the positions:

$$(4g) \quad \pm(uv0;\ u+{}^1/_2,{}^1/_2-v,{}^1/_2)$$
$$(8h) \quad \pm(xyz;\ \bar{x}\bar{y}z;\ {}^1/_2-x,y+{}^1/_2,z+{}^1/_2;\ x+{}^1/_2,{}^1/_2-y,z+{}^1/_2)$$

The parameters are those of Table XVAII,20.

TABLE XVAII,20

Positions and Parameters of the Atoms in Phenylpropiolic Acid

Atom	Position	x	y	z
C(1)	(4g)	0.669	0.2229	0
C(2)	(8h)	0.572	0.1906	0.1217
C(3)	(8h)	0.367	0.1223	0.1217
C(4)	(4g)	0.282	0.0941	0
C(5)	(4g)	0.871	0.2903	0
C(6)	(4g)	0.039	0.3463	0
C(7)	(4g)	0.234	0.4113	0
O	(8h)	0.318	0.4394	0.1105

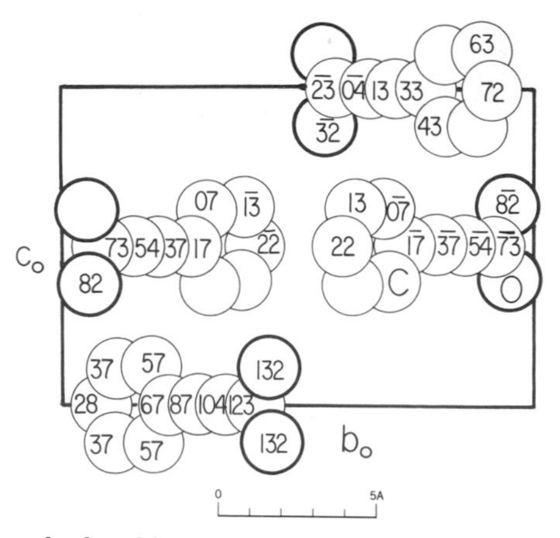

Fig. XVAII,28. The orthorhombic structure of phenylpropiolic acid projected along its a_0 axis.

The structure is shown in Figure XVAII,28. As this indicates, the molecules are hydrogen-bonded to form dimers, with O–H–O = 2.60 A. It is considered that a degree of disorder prevents determination of accurate bond lengths but that the acetylenic C(5)–C(6) = 1.20 A.

XV,aII,22. *Phenylethynyl(trimethylphosphine)silver*, $C_6H_5C{\equiv}C \cdot Ag \cdot P(CH_3)_3$, forms monoclinic crystals which have a unit that contains eight molecules and has the dimensions:

$$a_0 = 11.50(2) \text{ A.}; \quad b_0 = 20.58(3) \text{ A.}; \quad c_0 = 12.12(2) \text{ A.}$$
$$\beta = 123°25(10)'$$

The space group is C_{2h}^6 ($C2/c$). The silver atoms are in the special positions:

$$\text{Ag(1)}: (4a) \quad 000;\ 0\,0\,{}^1/_2;\ {}^1/_2\,{}^1/_2\,0;\ {}^1/_2\,{}^1/_2\,{}^1/_2$$
$$\text{Ag(2)}: (4e) \quad \pm(0\,u\,{}^1/_4;\ {}^1/_2,u+{}^1/_2,{}^1/_4)$$
$$\text{with } u = -0.0065$$

All other atoms are in the general positions:

$$(8f) \quad \pm(xyz;\ x,\bar{y},z+{}^1/_2;\ x+{}^1/_2,y+{}^1/_2,z;\ x+{}^1/_2,{}^1/_2-y,z+{}^1/_2)$$

Their parameters are those of Table XVAII,21.

TABLE XVAII,21

Parameters of Atoms in $C_6H_5C{\equiv}C \cdot Ag \cdot P(CH_3)_3$

Atom	x	y	z
P	0.2203	−0.0685	0.3909
C(1)	0.0958	0.0670	0.1483
C(2)	0.1656	0.1050	0.2369
C(3)	0.2589	0.1489	0.3432
C(4)	0.3349	0.1954	0.3238
C(5)	0.4276	0.2368	0.4304
C(6)	0.4403	0.2329	0.5478
C(7)	0.3653	0.1869	0.5702
C(8)	0.2716	0.1462	0.4625
C(9)	0.3032	−0.0812	0.3004
C(10)	0.3643	−0.0378	0.5521
C(11)	0.2056	−0.1519	0.4363

The bond dimensions in the complicated molecules present in this structure are those of Figure XVAII,29. The coordination about Ag(1) is linear but about Ag(2) it is roughly tetrahedral taking into account the probable π bonding between silver and the C≡C groups.

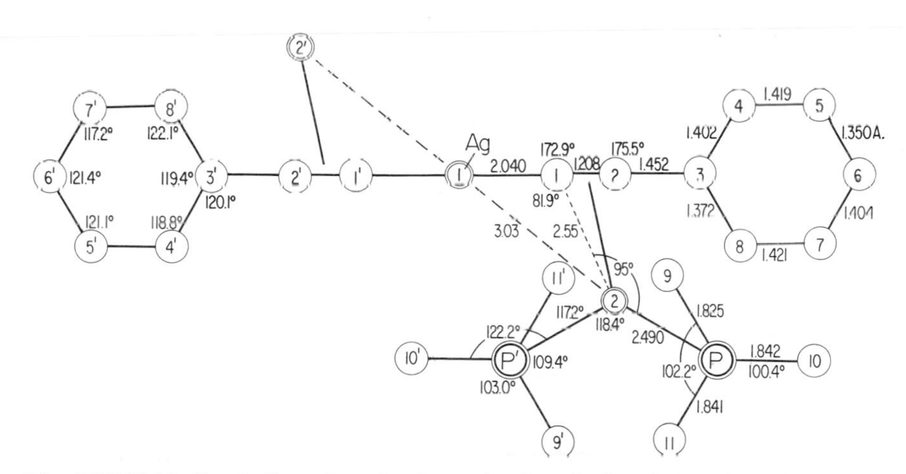

Fig. XVAII,29. Bond dimensions in the molecules of phenylethynyl(trimethylphosphine)silver.

XV,aII,23. *Phenylethynyl(trimethylphosphine)copper(I)*, $C_6H_5C{\equiv}C \cdot Cu \cdot P(CH_3)_3$, forms orthorhombic crystals. Their large 16-molecule unit has the cell edges:

$$a_0 = 12.69(2) \text{ A.}; \quad b_0 = 17.25(3) \text{ A.}; \quad c_0 = 21.39(3) \text{ A.}$$

The space group is V_h^{15} (*Pbca*) with atoms in the positions:

$$(8c) \quad \pm(xyz;\ x+{}^1/_2,{}^1/_2-y,\bar{z};\ \bar{x},y+{}^1/_2,{}^1/_2-z;\ {}^1/_2-x,\bar{y},z+{}^1/_2)$$

The parameters are those of Table XVAII,22.

TABLE XVAII,22

Parameters of the Atoms in Phenylethynyl(trimethyl phosphine)copper(I)

Atom	x	y	z
Cu(1)	−0.0073	0.0514	0.0471
Cu(2)	0.1553	0.1275	0.0401
P(1)	0.2832	0.0676	0.0936
P(2)	0.1751	0.2420	−0.0061
C(1)	0.0221	0.1342	0.1073
C(2)	0.0284	0.1835	0.1479
C(3)	0.0338	0.2426	0.1958
C(4)	−0.0573	0.2602	0.2303
C(5)	−0.0503	0.3171	0.2771
C(6)	0.0411	0.3547	0.2880
C(7)	0.1314	0.3388	0.2534
C(8)	0.1282	0.2818	0.2081
C(9)	0.1061	0.0399	−0.0220
C(10)	0.1449	0.0066	−0.0677
C(11)	0.2263	−0.0053	−0.1157
C(12)	0.2215	−0.0688	−0.1559
C(13)	0.3076	−0.0760	−0.2005
C(14)	0.3875	−0.0239	−0.2018
C(15)	0.3888	0.0394	−0.1617
C(16)	0.3109	0.0478	−0.1169
C(17)	0.2380	−0.0181	0.1353
C(18)	0.3911	0.0291	0.0471
C(19)	0.3541	0.1217	0.1549
C(20)	0.0796	0.3165	0.0343
C(21)	0.1244	0.2475	−0.0796
C(22)	0.2874	0.3027	0.0043
C(23)	0.0707	0.2928	−0.0368
C(24)	0.2580	0.2390	−0.0920
C(25)	0.2887	0.2993	0.0323

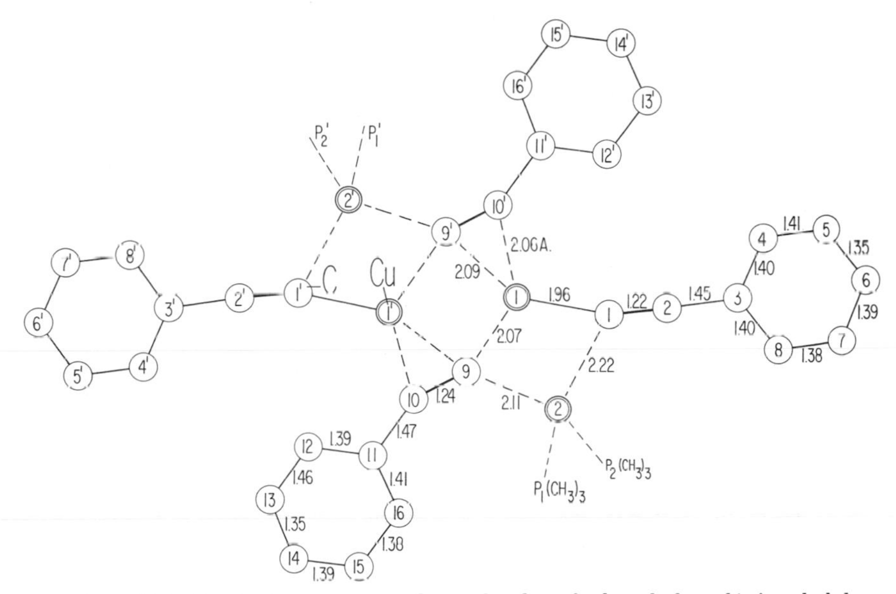

Fig. XVAII,30. Bond dimensions in the molecules of phenylethynyl(trimethylphosphine)copper.

The structure is built up of tetramers which are planar except for the $P(CH_3)_3$ groups; the general shape of these molecules and their bond dimensions are shown in Figure XVAII,30. The Cu(2) atoms are tetrahedrally surrounded by C(9) and C(1) and by P(1) and P(2) at distances of 2.24 and 2.22 A.; the mean distance of the methyl carbons from phosphorus is 1.83 A. The Cu(1) atoms are considered to have a σ bond to C(1) and π bonds to C(9) and C(10).

XV,aII,24. Crystals of *phenylethynyl(isopropylamine)gold*, $i\text{-}C_3H_7NH_2\cdot Au\cdot C{\equiv}CC_6H_5$, are orthorhombic with an eight-molecule cell of the edge lengths:

$$a_0 = 17.92(1)\text{ A.};\quad b_0 = 17.15(2)\text{ A.};\quad c_0 = 7.22(1)\text{ A.}$$

The space group is V_h^{10} (*Pccn*) with atoms in the positions:

$$(8e)\quad \pm(xyz;\ x+{}^1/_2,y+{}^1/_2,\bar{z};\ {}^1/_2-x,y,z+{}^1/_2;\ x,{}^1/_2-y,z+{}^1/_2)$$

The parameters are listed in Table XVAII,23.

TABLE XVAII,23

Parameters of the Atoms in $C_3H_7NH_2 \cdot Au \cdot C{\equiv}CC_6H_5$

Atom	x	y	z
Au	0.2249	0.1582	0.0476
N	0.3377	0.1505	0.0361
C(1)	0.1173	0.1607	0.0687
C(2)	0.0507	0.1572	0.0940
C(3)	−0.0311	0.1539	0.1205
C(4)	−0.0724	0.0966	0.0324
C(5)	−0.1505	0.0955	0.0579
C(6)	−0.1859	0.1480	0.1722
C(7)	−0.1440	0.2030	0.2582
C(8)	−0.0672	0.2083	0.2379
C(9)	0.3736	0.0721	0.0938
C(10)	0.4582	0.0830	0.1105
C(11)	0.3523	0.0116	−0.0443

The structure is given in Figure XVAII,31. Its molecules have the bond dimensions of Figure XVAII,32. Between molecules the shortest atomic separation is an Au–Au = 3.274 A.

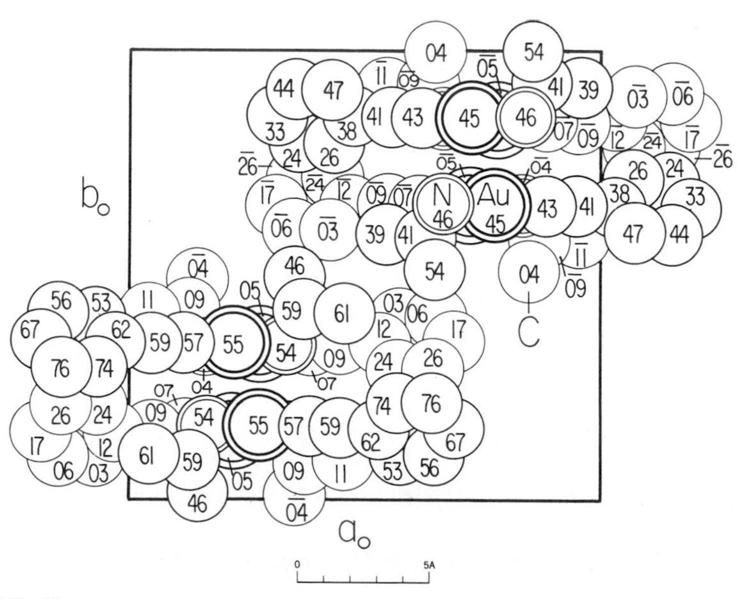

Fig. XVAII,31. The orthorhombic structure of phenylethynyl(isopropylamine)gold projected along its c_0 axis.

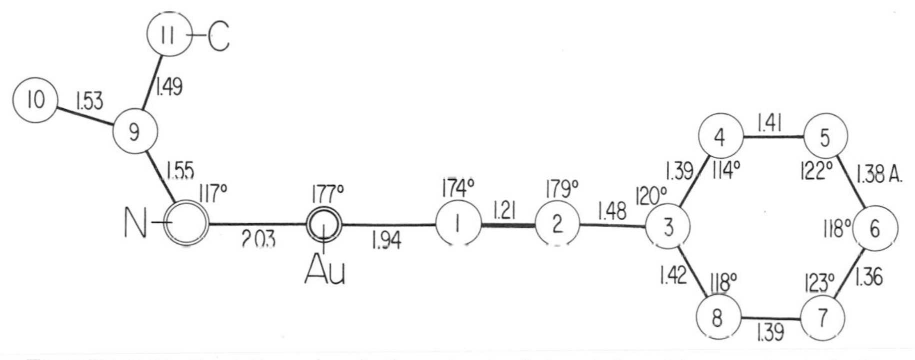

Fig. XVAII,32. Bond dimensions in the molecule of phenylethynyl(isopropylamine)gold.

The following nonyl compound is probably isostructural; its cell dimensions are, for $n\text{-}C_9H_{19}NH_2 \cdot Au \cdot C{\equiv}CC_6H_5$:

$$a_0 = 29.2 \text{ A.}; \quad b_0 = 17.0 \text{ A.}; \quad c_0 = 7.2 \text{ A.}$$

The amyl analogue $n\text{-}C_5H_{11}NH_2 \cdot Au \cdot C{\equiv}CC_6H_5$, though monoclinic, has the similar cell:

$$a_0 = 22.0 \text{ A.}; \quad b_0 = 17.0 \text{ A.}; \quad c_0 = 7.2 \text{ A.}; \quad \beta = 98°$$

XV,aII,25. The crystals of *trans-bis(phenylethynyl)bis(triethylphosphine)nickel(II)*, $[P(C_2H_5)_3]_2Ni[C{\equiv}CC_6H_5]_2$, are monoclinic with a bimolecular unit of the dimensions:

$$a_0 = 15.31(1) \text{ A.}; \quad b_0 = 10.88(1) \text{ A.}; \quad c_0 = 9.08(1) \text{ A.}$$
$$\beta = 105°30(10)'$$

The nickel atoms are in:

$$(2a) \quad 000; \ {}^1/_2\, {}^1/_2\, 0$$

of the space group $C_{2h}^5(P2_1/a)$. All other atoms are in:

$$(4e) \quad \pm(xyz;\ x+{}^1/_2, {}^1/_2-y, z)$$

with the parameters of Table XVAII,21.

TABLE XVAII,24

Parameters of Atoms in $[P(C_2H_5)_3]_2Ni[C{\equiv}CC_6H_5]_2$

Atom	x	y	z
P	0.0931(2)	0.9265(3)	0.8713(4)
C(1)	0.0992(8)	0.0331(12)	0.1680(12)
C(2)	0.1660(8)	0.0535(11)	0.2735(14)
C(3)	0.2490(7)	0.0735(13)	0.3964(12)
C(4)	0.2618(9)	0.1869(14)	0.4748(15)
C(5)	0.3171(8)	0.9816(13)	0.4316(14)
C(6)	0.3425(10)	0.2020(15)	0.5969(17)
C(7)	0.3987(9)	−0.0008(16)	0.5491(18)
C(8)	0.4083(10)	0.1083(17)	0.6320(17)
C(9)	0.0844(10)	0.0145(18)	0.6939(15)
C(10)	0.1060(13)	0.1550(21)	0.7296(24)
C(11)	0.0623(9)	0.7713(14)	0.7981(16)
C(12)	0.0648(13)	0.6780(16)	0.9260(22)
C(13)	0.2185(8)	0.9202(15)	0.9712(17)
C(14)	0.2845(10)	0.8749(19)	0.8793(22)

The structure, as shown in Figure XVAII,33, is built up of molecules having the bond dimensions of Figure XVAII,34. The coordination about the nickel atom is square. It, the phosphorus and the triply bound carbon atoms are all in one plane; in the same plane are the C(3) and C(8) atoms of the benzene rings which are turned through 43.6° with respect to the first plane.

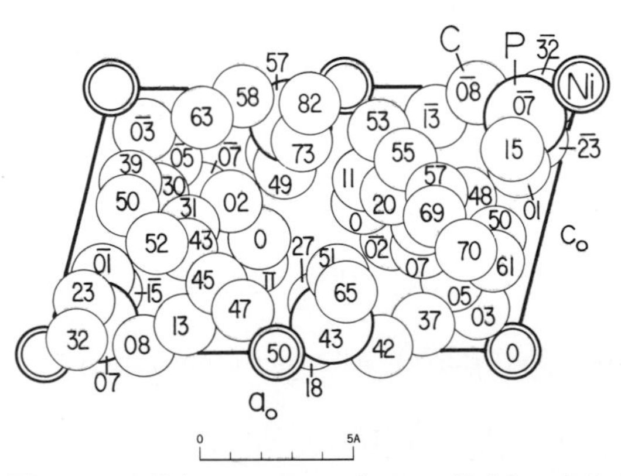

Fig. XVAII,33. The monoclinic structure of *trans*-bis(phenylethynyl)-bis(triethylphosphine)nickel projected along its b_0 axis.

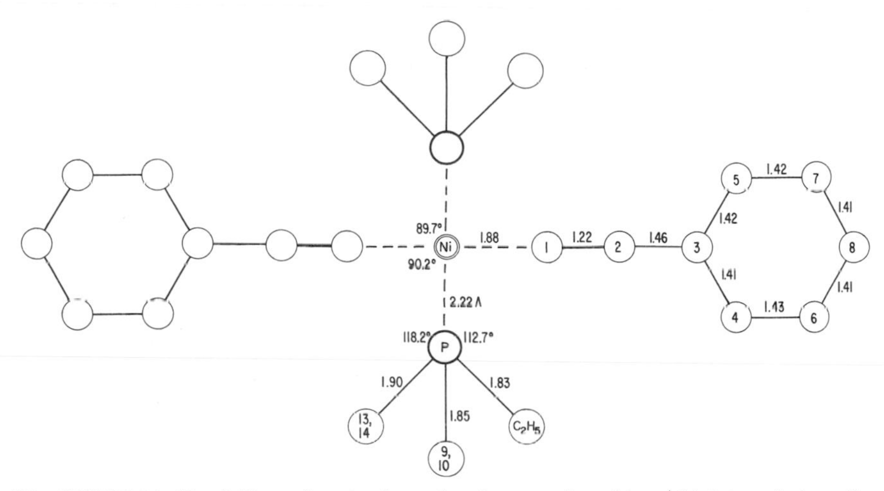

Fig. XVAII,34. Bond dimensions in the molecular complex of *trans*-bis(phenylethynyl)-bis(triethylphosphine)nickel.

XV,aII,26. At room temperature *aniline hydrochloride*, $C_6H_5NH_2 \cdot HCl$, forms monoclinic crystals with a tetramolecular unit of the dimensions:

$$a_0 = 15.84 \text{ A.}; \quad b_0 = 5.33 \text{ A.}; \quad c_0 = 8.11 \text{ A.}; \quad \beta = 98°30'$$

All atoms are in general positions of C_s^4 (*Cc*):

$$(4a) \quad xyz; \; x,\bar{y},z+{}^1/_2; \; x+{}^1/_2,y+{}^1/_2,z; \; x+{}^1/_2,{}^1/_2-y,z+{}^1/_2$$

with the parameters of Table XVAII,25.

TABLE XVAII,25

Parameters of the Atoms in Aniline Hydrochloride

Atom	x	y	z
Cl	0	0.768	0
N	0.430	0.232	0.324
C(1)	0.343	0.232	0.292
C(2)	0.302	0.030	0.210
C(3)	0.212	0.030	0.178
C(4)	0.171	0.232	0.236
C(5)	0.212	0.434	0.320
C(6)	0.302	0.434	0.352

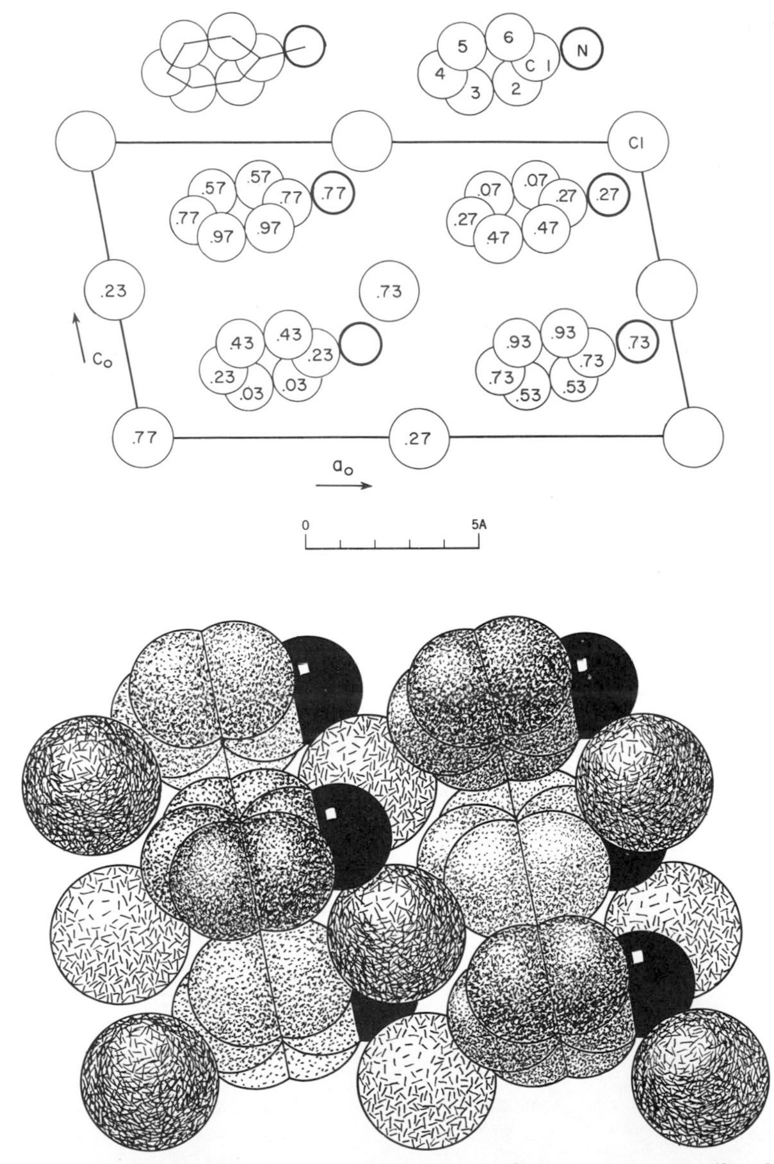

Fig. XVAII,35a (top). The room temperature, monoclinic structure of aniline hydrochloride projected along its b_0 axis. Left-hand axes.

Fig. XVAII,35b (bottom). A packing drawing of the monoclinic structure of aniline hydrochloride viewed along its b_0 axis. The chlorine atoms are large and line shaded, the nitrogens are black. Atoms of carbon are dotted. Left-hand axes.

As Figure XVAII,35 shows, this is a polarized ionic arrangement in which layers of organic cations, all pointed in the same direction, and chloride anions succeed one another normal to the b_0c_0 plane. Each chloride ion has three nearest nitrogen neighbors at distances of 3.16 and 3.18 A. In the planar cations, C–C = 1.38–1.40 A. and C–N = 1.35 A. Between molecules the shortest C–C and the shortest C–Cl are both 3.85 A.

XV,aII,27. Above 27.5°C. *aniline hydrobromide*, $C_6H_5NH_2 \cdot HBr$, is orthorhombic with a tetramolecular unit of the edge lengths:

$$a_0 = 16.77(6) \text{ A.}; \quad b_0 = 6.05(3) \text{ A.}; \quad c_0 = 6.86(3) \text{ A.}$$

Atoms have been placed in the following positions of the space group V_h^{10} (*Pnaa*):

(4c) $\pm(u\ ^1/_4\ ^1/_4;\ u+^1/_2,^1/_4,^1/_4)$
(4d) $\pm(u\ ^3/_4\ ^1/_4;\ u+^1/_2,^3/_4,^1/_4)$
(8e) $\pm(xyz;\ x,^1/_2-y,^1/_2-z;\ ^1/_2-x,y+^1/_2,\bar{z};\ ^1/_2-x,\bar{y},z+^1/_2)$

The determined positions and parameters are those of Table XVAII,26.

TABLE XVAII,26

Positions and Parameters of the Atoms in Orthorhombic Aniline Hydrobromide

Atom	Position	x	y	z
Br	(4c)	0.0230	$^1/_4$	$^1/_4$
N	(4d)	0.4441	$^3/_4$	$^1/_4$
C(1)	(4d)	0.3564	$^3/_4$	$^1/_4$
C(2)	(8e)	0.3152	0.5650	0.1833
C(3)	(8e)	0.2323	0.5650	0.1833
C(4)	(4d)	0.1914	$^3/_4$	$^1/_4$

The structure is shown in Figure XVAII,36. As this indicates, it consists of a succession along the a_0 axis of $C_6H_5NH_3^+$ and Br^- layers. The shortest separations of the bromine and nitrogen atoms of these ions are 3.30 and 3.48 A.; between the cations there are C–C distances of 3.54 A. and greater.

XV,aII,28. *Benzamide*, $C_6H_5CONH_2$, forms monoclinic crystals having a tetramolecular cell of the dimensions:

$$a_0 = 5.59(1) \text{ A.}; \quad b_0 = 5.01(1) \text{ A.}; \quad c_0 = 21.93(5) \text{ A.}$$
$$\beta = 90°45(10)'$$

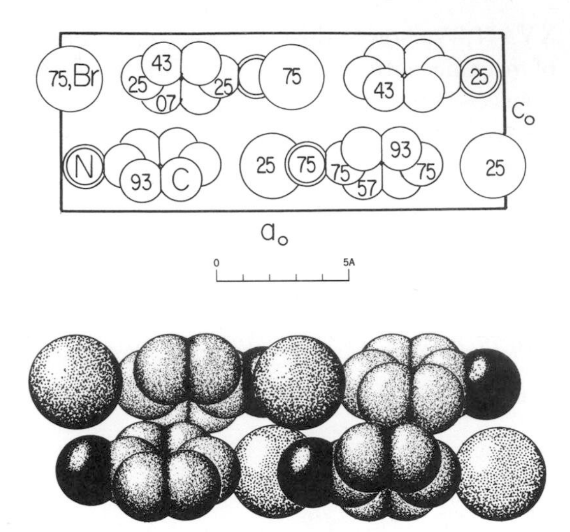

Fig. XVAII,36a (top). The orthorhombic, high temperature structure of aniline hydrobromide projected along its b_0 axis.

Fig. XVAII,36b (bottom). A packing drawing of the orthorhombic aniline hydrobromide structure viewed along its b_0 axis. The bromine atoms are the large, the carbon atoms the smaller dotted circles. Atoms of nitrogen are black.

The space group is C_{2h}^5 ($P2_1/c$) with all atoms in the positions:

$$(4e) \quad \pm(xyz;\ x,{}^1/_2-y,z+{}^1/_2)$$

The chosen parameters are those of Table XVAII,27.

TABLE XVAII,27

Parameters of Atoms in Benzamide

Atom	x	y	z
C(1)	0.292	0.031	0.0608
C(2)	0.478	0.073	0.1086
C(3)	0.644	0.272	0.1038
C(4)	0.822	0.296	0.1482
C(5)	0.823	0.129	0.1976
C(6)	0.647	−0.071	0.2036
C(7)	0.481	−0.097	0.1589
O	0.195	−0.189	0.0530
N	0.230	0.241	0.0284

The structure thus obtained is shown in Figure XVAII,37. It is made up of molecules having the bond dimensions of Figure XVAII,38. The carbon atoms are coplanar but the oxygen and nitrogen atoms are each about 0.5 A. on either side of this plane. The molecules appear to be tied together by N–H–O bonds with lengths of 2.91 and 2.96 A. Hydrogen positions were suggested in the original.

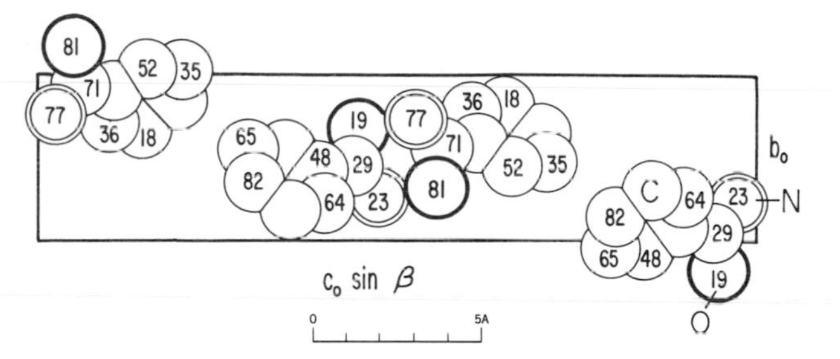

Fig. XVAII,37. The monoclinic structure of benzamide projected along its a_0 axis.

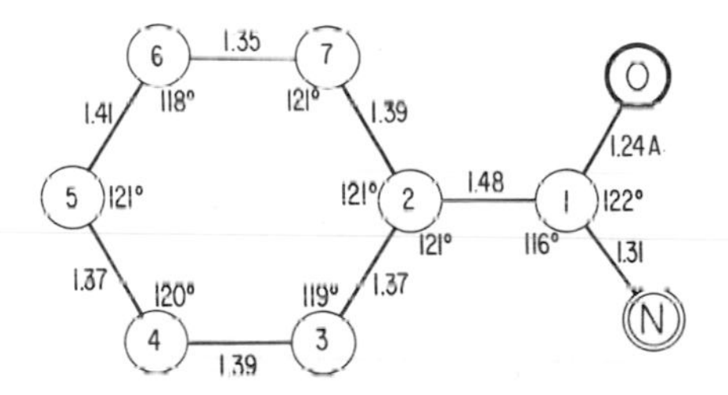

Fig. XVAII,38. Bond dimensions in the molecule of benzamide.

XV,aII,29. An approximate structure has been described for the complex *benzamide-hydrogen triiodide*, $2C_6H_5CONH_2 \cdot HI_3$. Its symmetry is triclinic with a large tetramolecular unit of the dimensions:

$$a_0 = 20.828 \text{ A.}; \quad b_0 = 9.885 \text{ A.}; \quad c_0 = 9.588 \text{ A.}$$
$$\alpha = 95°46'; \quad \beta = 101°54'; \quad \gamma = 94°44'$$

Atoms are in the positions $(2i)$ $\pm(xyz)$ of the space group C_i^1 ($P\bar{1}$). The assigned parameters are listed in Table XVAII,28.

TABLE XVAII,28

Parameters of the Atoms in $2C_6H_5CONH_2 \cdot HI_3$

Atom	x	y	z
I(1)	0.0886(5)	0.1180(6)	0.6543(5)
I(2)	0.1011(4)	0.0950(5)	0.9639(5)
I(3)	0.1074(5)	0.0786(7)	0.2666(5)
I(4)	0.4151(5)	0.5791(6)	0.0739(5)
I(5)	0.4026(4)	0.5497(7)	0.3675(6)
I(6)	0.3918(4)	0.5382(8)	0.6674(6)
C(1,2)[a]	0.07	0.60	0.13,0.63
C(3,4)	0.12	0.61	0.05,0.55
C(5,6)	0.16	0.48	0.02,0.52
C(7,8)	0.14	0.73	0.01,0.51
C(9,10)	0.21	0.50	0.96,0.46
C(11,12)	0.19	0.73	0.93,0.43
C(13,14)	0.22	0.63	0.93,0.43
O,N(1,2)	0.03	0.50	0.15,0.65
O,N(3,4)	0.02	0.72	0.15,0.65
C(15,16)	0.44	0.08	0.60,0.10
C(17,18)	0.38	0.07	0.43,0.93
C(19,20)	0.37	0.93	0.38,0.88
C(21,22)	0.34	0.17	0.35,0.85
C(23,24)	0.32	0.92	0.27,0.77
C(25,26)	0.29	0.16	0.25,0.75
C(27,28)	0.28	0.03	0.18,0.68
O,N(5,6)	0.47	0.21	0.66,0.16
O,N(7,8)	0.47	0.98	0.66,0.16

[a] The two atoms grouped together in a line throughout the rest of the table have parameters that differ only in their z-values.

In the resulting structure, nearly linear I_3^- ions are distributed in pairs between stacks of benzamide molecules. In these anions, I–I = 2.90–2.96 A. and the apparent departure from linearity is about 3°. The channeled structure thus obtained has been thought to be a model for the starch–iodine complex.

XV,aII,30. *Acetanilide,* $C_6H_5NHCOCH_3$, forms orthorhombic crystals whose large, eight-molecule unit has the edge lengths:

$$a_0 = 19.640 \text{ A.}; \quad b_0 = 9.483 \text{ A.}; \quad c_0 = 7.979 \text{ A.}$$

The space group is V_h^{15} ($Pbca$) with all atoms in the positions:

$$(8c) \quad \pm(xyz;\ x+{}^1/_2,{}^1/_2-y,\bar{z};\ \bar{x},y+{}^1/_2,{}^1/_2-z;\ {}^1/_2-x,\bar{y},z+{}^1/_2)$$

The recently refined parameters are those of Table XVAII,29.

TABLE XVAII,29

Parameters of the Atoms in Acetanilide

Atom	x	y	z
C(1)	0.4076	0.0759	0.1294
C(2)	0.3553	−0.0183	0.1691
C(3)	0.2902	0.0052	0.1110
C(4)	0.2761	0.1213	0.0104
C(5)	0.3283	0.2115	−0.0313
C(6)	0.3941	0.1901	0.0251
C(7)	0.5253	0.1367	0.2168
C(8)	0.5864	0.0770	0.3035
N	0.4724	0.0466	0.1981
O	0.5243	0.2580	0.1664
H(1)	0.4736	−0.0593	0.2478
H(2)	0.3660	−0.1097	0.2454
H(3)	0.2498	−0.0671	0.1438
H(4)	0.2250	0.1404	−0.0344
H(5)	0.3176	0.3015	−0.1101
H(6)	0.4345	0.2611	−0.0114
H(7)	0.5898	0.1202	0.4285
H(8)	0.6317	0.1040	0.2340
H(9)	0.5817	−0.0363	0.3108

The resulting structure is shown in Figure XVAII,39. Bond dimensions in its molecules are those of Figure XVAII,40. The benzene ring is planar, with nitrogen 0.046 A. out of this plane. The atoms of the side chain, C(7), C(8), N and O, are also coplanar, with the normals to the two planes making 17.6° with one another.

XV,aII,31. Crystals of *N-methyl acetanilide*, $C_6H_5N(CH_3)\cdot COCH_3$, are orthorhombic with a tetramolecular unit of the edge lengths:

$$a_0 = 17.151 \text{ A.};\quad b_0 = 7.328 \text{ A.};\quad c_0 = 6.779 \text{ A.}$$

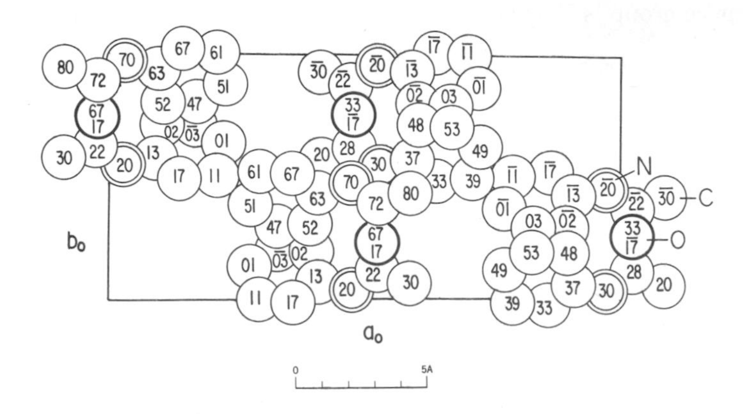

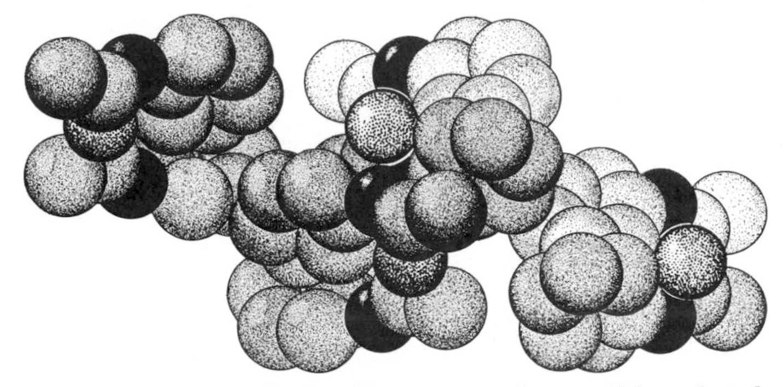

Fig. XVAII,39a (top). The orthorhombic structure of acetanilide projected along its c_0 axis.

Fig. XVAII,39b (bottom). A packing drawing of the orthorhombic structure of acetanilide viewed along its c_0 axis. The nitrogen atoms are black; carbon atoms are finely, oxygen atoms coarsely dotted.

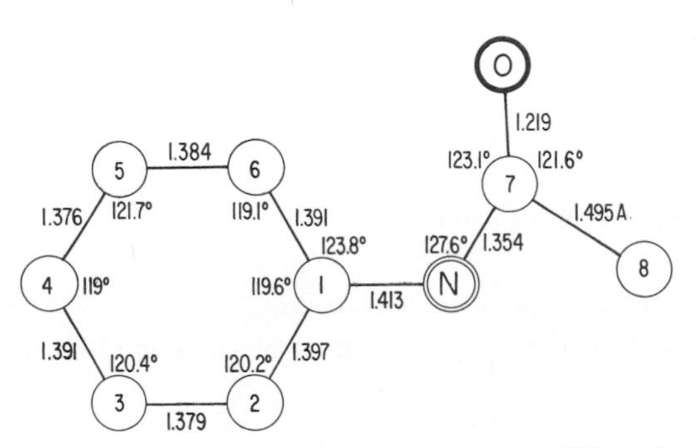

Fig. XVAII,40. Bond dimensions in the acetanilide molecule.

Atoms are in the following positions of the space group V_h^{16} (*Pnma*):

(4*c*) $\pm(u\ ^1/_4\ v;\ u+{}^1/_2,{}^1/_4,{}^1/_2-v)$
(8*d*) $\pm(xyz;\ {}^1/_2-x,y+{}^1/_2,z+{}^1/_2;\ x,{}^1/_2-y,z;\ x+{}^1/_2,y,{}^1/_2-z)$

Atomic parameters, including those for hydrogen, are stated in Table XVAII,30.

TABLE XVAII,30

Positions and Parameters of the Atoms in $C_6H_5N(CH_3)COCH_3$

Atom	Position	x	y	z
O	(4*c*)	−0.0491(3)	1/4	−0.0759(9)
N	(4*c*)	0.0283(4)	1/4	0.1854(11)
C(1)	(4*c*)	0.0191(6)	1/4	−0.0087(15)
C(2)	(4*c*)	−0.0379(5)	1/4	0.3143(15)
C(3)	(4*c*)	0.0898(6)	1/4	−0.1306(15)
C(4)	(8*d*)	0.2107(3)	0.0830(13)	0.4209(8)
C(5)	(8*d*)	0.1406(3)	0.0830(10)	0.3202(7)
C(6)	(4*c*)	0.2110(6)	1/4	0.4680(13)
C(7)	(4*c*)	0.1070(4)	1/4	0.2727(11)
H(1)	(4*c*)	0.292(6)	1/4	0.523(14)
H(2)	(4*c*)	0.082(7)	1/4	−0.255(24)
H(3)	(4*c*)	−0.031(7)	1/4	0.452(20)
H(4)	(8*d*)	0.232(4)	0.557(12)	0.456(13)
H(5)	(8*d*)	0.110(3)	0.557(11)	0.284(9)
H(6)	(8*d*)	−0.081(3)	0.173(7)	0.303(9)
H(7)	(8*d*)	0.138(4)	0.173(10)	−0.113(12)

The structure is shown in Figure XVAII,41. Its molecules have the bond dimensions of Figure XVAII,42. Both the phenyl and the acetamido groups are planar, lying at right angles to one another.

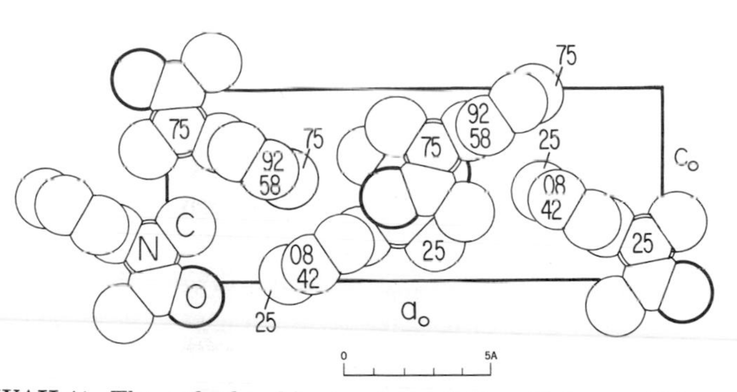

Fig. XVAII,41. The orthorhombic structure of *N*-methyl acetanilide projected along its b_0 axis.

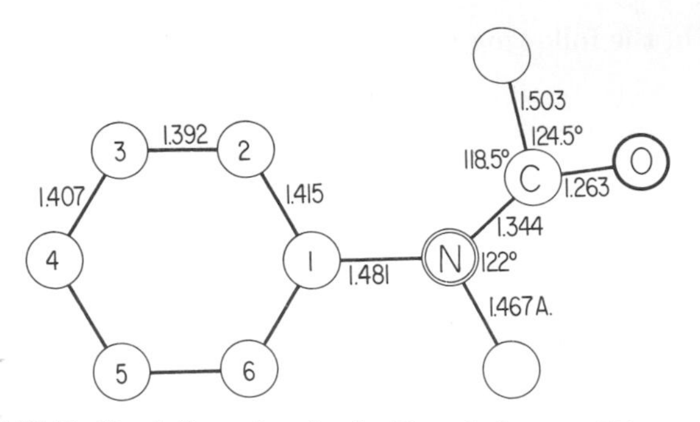

Fig. XVAII,42. Bond dimensions in the *N*-methyl acetanilide molecule.

XV,aII,32. The orthorhombic crystals of *benzenediazonium chloride*, $C_6H_5N_2Cl$, have a tetramolecular cell of the edge lengths:

$$a_0 = 15.152(6) \text{ A.}; \quad b_0 = 4.928(2) \text{ A.}; \quad c_0 = 9.044(3) \text{ A.}$$

The space group is V^5 ($C222_1$) with atoms in the positions:

(4*a*) $u00;\ \bar{u}\,0\,{}^1/_2;\ u+{}^1/_2,{}^1/_2,0;\ {}^1/_2-u,{}^1/_2,{}^1/_2$

(4*b*) $0\,u\,{}^1/_4;\ 0\,\bar{u}\,{}^3/_4;\ {}^1/_2,u+{}^1/_2,{}^1/_4;\ {}^1/_2,{}^1/_2-u,{}^3/_4$

(8*c*) $xyz;\quad x\bar{y}\bar{z};\quad \bar{x},\bar{y},z+{}^1/_2;\quad \bar{x},y,{}^1/_2-z;$
$x+{}^1/_2,y+{}^1/_2,z;\ x+{}^1/_2,{}^1/_2-y,\bar{z};\ {}^1/_2-x,{}^1/_2-y,z+{}^1/_2;$
${}^1/_2-x,y+{}^1/_2,{}^1/_2-z$

The determined positions and parameters are listed in Table XVAII,31. Parameters assumed for the hydrogen atoms were given in the original paper.

TABLE XVAII,31

Positions and Parameters of the Atoms in $C_6H_5N_2Cl$

Atom	Position	x	y	z
Cl	(4*b*)	0	0.0450	1/4
N(1)	(4*a*)	0.4614	0	0
N(2)	(4*a*)	0.5339	0	0
C(1)	(4*a*)	0.3700	0	0
C(2)	(8*c*)	0.1720	0.673	0.4037
C(3)	(8*c*)	0.2633	0.678	0.4034
C(4)	(4*a*)	0.1925	0	0

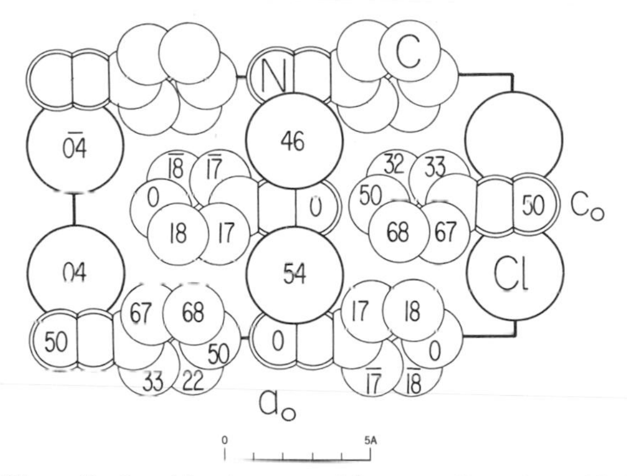

Fig. XVAII,43. The orthorhombic structure of benzene diazonium chloride projected along its b_0 axis.

The structure, as shown in Figure XVAII,43, is ionic with $(C_6H_5N_2)^+$ cations in which N–N = 1.10 A., N–C = C–C = 1.38 A. The ions lie in planes normal to the a_0 axis. In such planes the N(2)–N(1) bonds intersect squares of Cl^- anions. These intersections are 3.18 or 3.51 A. distant from chlorine while Cl–N(2) = 3.225 or 3.548 A. and Cl N(1) – 3.237 or 3.559 A.

XV,aII,33. The monoclinic crystals of *benzenediazonium tribromide*, $C_6H_5N_2Br_3$, have a tetramolecular unit of the dimensions:

$$a_0 = 6.73 \text{ A.};\quad b_0 = 14.10 \text{ A.};\quad c_0 = 11.00 \text{ A.};\quad \beta = 106°48'$$

The space group is C_{2h}^6 ($C2/c$) with atoms in the positions:

Br(1): (4*a*) 000; $^1/_2\,^1/_2\,0$; $0\,0\,^1/_2$; $^1/_2\,^1/_2\,^1/_2$
N(1): (4*e*) $\pm(0\,u\,^1/_4;\ ^1/_2,u+^1/_2,^1/_4)$
with u = 0.762
N(2): (4*e*) with u = 0.683
C(1): (4*e*) with u = 0.583
C(4): (4*e*) with u = 0.395

The other atoms are in the general positions:

$$(8f)\quad \pm(xyz;\ x,\bar{y},z+{}^1/_2;\ x+{}^1/_2,y+{}^1/_2,z;\ x+{}^1/_2,{}^1/_2-y,z+{}^1/_2)$$

For them the parameters are:

Br(2): x = 0.0049, y = 0.1775, z = 0.0418
C(2): x = 0.106, y – 0.543, z = 0.174
C(3): x = 0.104, y = 0.440, z = 0.175

The structure shown in Figure XVAII,44 is composed of $(C_6H_5N_2)^+$ and Br_3^- ions. In the cations, N–N = 1.11 A. and C–N = 1.41 A. The C–C bonds of its benzene ring are not equal; instead, C(2)–C(3) = 1.45 A. while the other two are 1.38 A. Corresponding to this distortion, C(1)–C(2)–C(3) = 114° and C(2)–C(1)–C(6) = 131°. In the symmetrical Br_3^- ion, Br–Br = 2.543(4) A. The shortest anion–cation distances are N(1)–Br(2) = 3.46 A. and N(2)–Br(3) = 3.31 A.

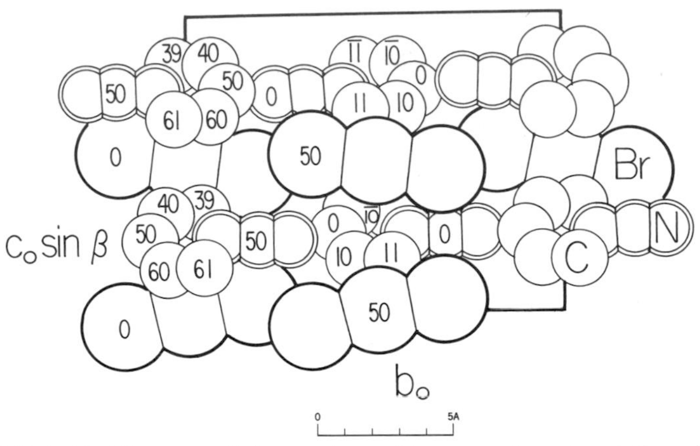

Fig. XVAII,44. The monoclinic structure of benzene diazonium tribromide projected along its a_0 axis.

XV,aII,34. *Bis(trimethylbenzyl ammonium)copper(II) tetrachloride*, $[C_6H_5CH_2N(CH_3)_3]_2CuCl_4$, possesses a monoclinic modification, for which the tetramolecular cell has the dimensions:

$$a_0 = 9.584(10) \text{ A.}; \quad b_0 = 9.104(5) \text{ A.}; \quad c_0 = 28.434(10) \text{ A.}$$
$$\beta = 92°50'$$

The space group is C_{2h}^5 ($P2_1/n$) with atoms in the positions:

$$(4e) \quad \pm(xyz;\ x+{}^1/_2,{}^1/_2-y,z+{}^1/_2)$$

The chosen parameters for all the atoms, are given in Table XVAII,32.

TABLE XVAII,32

Parameters of Atoms in $[C_6H_5CH_2N(CH_3)_3]_2CuCl_4$

Atom	x	y	z
Cu	0.2278(1)	0.1697(1)	0.3882(1)
Cl(1)	0.1663(3)	0.3259(2)	0.4437(1)
Cl(2)	0.2857(2)	0.3298(2)	0.3317(1)
Cl(3)	0.4112(2)	0.0144(2)	0.3844(1)
Cl(4)	0.0522(2)	0.0041(2)	0.3924(1)
N(1)	0.4279(6)	0.1868(6)	0.2091(2)
N(2)	0.7348(6)	0.1967(6)	0.4829(2)
C(1)	0.9339(8)	0.2117(9)	0.2702(3)
C(2)	0.8990(8)	0.2472(12)	0.2237(3)
C(3)	0.7630(8)	0.2851(10)	0.2100(3)
C(4)	0.6596(7)	0.2857(8)	0.2417(2)
C(5)	0.6951(8)	0.2548(10)	0.2890(3)
C(6)	0.8337(9)	0.2197(11)	0.3026(3)
C(7)	0.5141(8)	0.3227(7)	0.2263(3)
C(8)	0.4869(8)	0.1236(9)	0.1656(3)
C(9)	0.2817(7)	0.2382(9)	0.1977(3)
C(10)	0.4243(9)	0.0715(8)	0.2473(3)
C(11)	0.6433(9)	0.7036(9)	0.4083(3)
C(12)	0.5416(8)	0.5938(11)	0.4008(3)
C(13)	0.5776(8)	0.4492(9)	0.4091(3)
C(14)	0.7142(7)	0.4111(7)	0.4250(2)
C(15)	0.8124(7)	0.5199(8)	0.4318(3)
C(16)	0.7772(8)	0.6665(8)	0.4237(3)
C(17)	0.7491(8)	0.2492(8)	0.4327(2)
C(18)	0.5873(10)	0.2158(12)	0.4970(3)
C(19)	0.7706(10)	0.0376(8)	0.4836(3)
C(20)	0.8328(11)	0.2792(10)	0.5169(3)

Note: Parameters for the hydrogen atoms are given in the original.

The structure is seen in Figure XVAII,45. Its two crystallographically different cations have much the same dimensions: the benzene ring and C(7), for instance, are coplanar and the angle C(4)–C(7)–N(1) is roughly tetrahedral, as is the distribution about nitrogen. The coordination around the copper atom of the anion is also distortedly tetrahedral, with Cu–Cl = 2.229–2.268 A.

XV,aII,35. According to a preliminary description, crystals of *trans-dichloro(cis-2-butene)Sα-phenethylamine platinum(II)*, $PtCl_2 \cdot NH_2CH(CH_3)C_6H_5 \cdot C_2H_2(CH_3)_2$, are orthorhombic with a tetramolecular unit of the edge lengths:

$$a_0 = 15.64(5) \text{ A.}; \quad b_0 = 10.42(3) \text{ A.}; \quad c_0 = 9.16(3) \text{ A.}$$

Atoms are in the positions:

$$(4a) \quad xyz;\ {}^1/_2-x,\bar{y},z+{}^1/_2;\ x+{}^1/_2,{}^1/_2-y,\bar{z};\ \bar{x},y+{}^1/_2,{}^1/_2-z$$

of the space group V^4 ($P2_12_12_1$). The approximate parameters are those of Table XVAII,33.

TABLE XVAII,33

Parameters of the Atoms in $PtCl_2 \cdot NH_2CH(CH_3)_2C_6H_5 \cdot C_2H_2(CH_3)_2$

Atom	x	y	z
Pt	0.315	0.171	0.722
Cl(1)	0.241	0.203	0.498
Cl(2)	0.381	0.127	0.943
N	0.297	−0.020	0.720
C(1)	0.240	0.439	0.663
C(2)	0.305	0.367	0.764
C(3)	0.385	0.341	0.712
C(4)	0.409	0.384	0.557
C(5)	0.384	−0.077	0.682
C(6)	0.379	−0.216	0.614
C(7)	0.428	−0.026	0.546
C(8)	0.497	0.070	0.549
C(9)	0.535	0.114	0.437
C(10)	0.507	0.070	0.300
C(11)	0.448	−0.019	0.284
C(12)	0.408	−0.067	0.407

In this structure the platinum has its usual square coordination, with 2Pt–Cl = 2.34 A. The nitrogen atom is at a third corner of the square, while the fourth reaches to the center of the C=C bond.

XV,aII,36. Crystals of *(+)-2-benzylglutamic acid hydrobromide dihydrate*, $C_6H_5CH_2C(NH_2)(COOH) \cdot (CH_2)_2COOH \cdot HBr \cdot 2H_2O$, are orthorhombic with a tetramolecular unit of the edge lengths:

$$a_0 = 8.28(2) \text{ A.}; \quad b_0 = 10.54(2) \text{ A.}; \quad c_0 = 17.68(2) \text{ A.}$$

Atoms are in general positions of the space group V^4 ($P2_12_12_1$):

$$(4a) \quad xyz;\ {}^1/_2-x,\bar{y},z+{}^1/_2;\ x+{}^1/_2,{}^1/_2-y,\bar{z};\ \bar{x},y+{}^1/_2,{}^1/_2-z$$

The parameters are listed in Table XVAII,34.

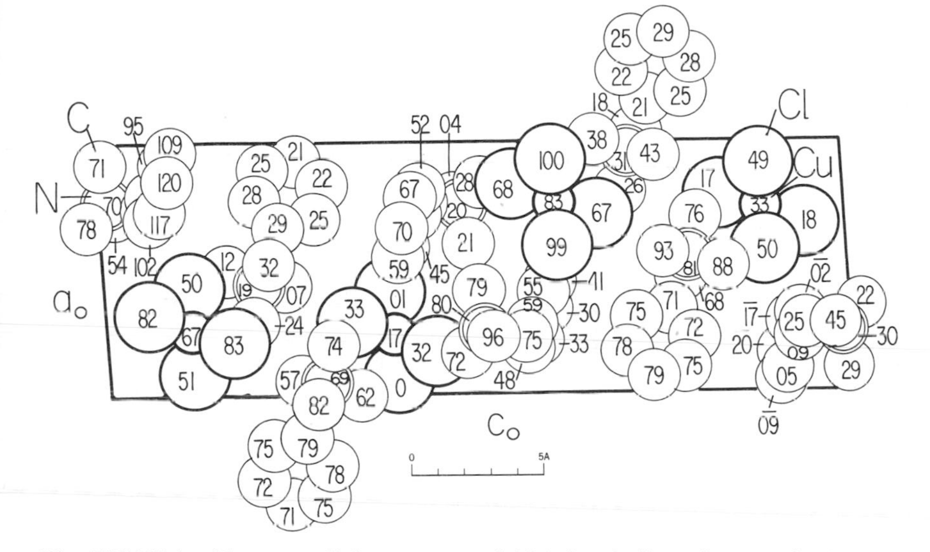

Fig. XVAII,45. The monoclinic structure of bis(trimethylbenzyl ammonium)copper tetrachloride projected along its b_0 axis.

TABLE XVAII,34

Parameters of the Atoms in $C_6H_5CH_2C(NH_2)(COOH)\cdot(CH_2)_2COOH\cdot HBr\cdot 2H_2O$

Atom	x	y	z
Br	0.5337	0.6500	0.0522
O(1)	0.1964	0.0839	0.1724
O(2)	0.4133	0.0278	0.2388
O(3)	0.0447	0.4867	0.1679
O(4)	0.2032	0.6425	0.2078
$H_2O(1)$	0.1660	0.8610	0.1245
$H_2O(2)$	0.9212	0.6229	0.0638
N	0.4945	0.2635	0.2760
C(1)	0.3226	0.1056	0.2128
C(2)	0.3504	0.2506	0.2243
C(3)	0.1997	0.3058	0.2628
C(4)	0.2048	0.4478	0.2785
C(5)	0.1502	0.5338	0.2139
C(6)	0.3946	0.3082	0.1459
C(7)	0.5441	0.2500	0.1105
C(8)	0.6924	0.3051	0.1232
C(9)	0.8329	0.2481	0.0926
C(10)	0.8218	0.1356	0.0512
C(11)	0.6702	0.0809	0.0385
C(12)	0.5321	0.1394	0.0675

The structure, shown in Figure XVAII,46, is composed of molecules which have the bond dimensions of Figure XVAII,47. Two of the hydrogen bonds connecting these molecules, involving carboxyl and water oxygens, have lengths as short as O–H–O = 2.51 and 2.55 A. Other hydrogen bonds are described between the nitrogen atoms and either bromide anions (3.27 A.) or oxygens.

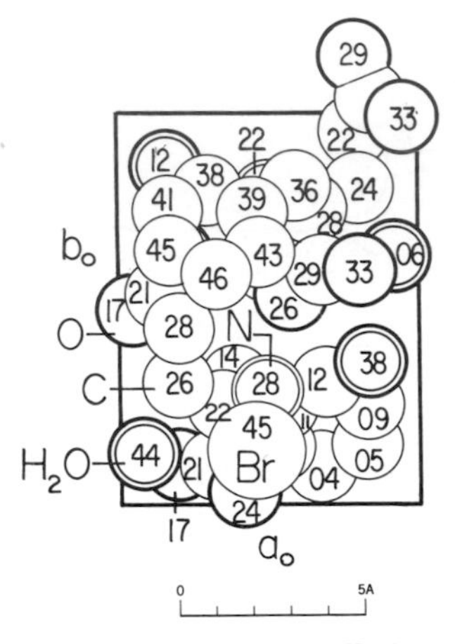

Fig. XVAII,46. Half the molecules in the unit cell of the orthorhombic 2-benzylglutamic acid hydrobromide dihydrate.

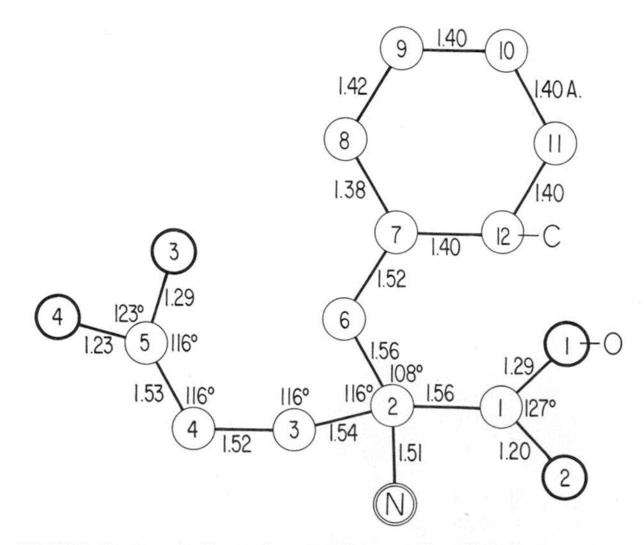

Fig. XVAII,47. Bond dimensions in the complex 2-benzylglutamate cation.

XV,aII,37. Crystals of *bis(α-hydroxy-α-phenylbutyramidine)copper(II) dihydrate*, $[C_6H_5C(O)(C_2H_5)C(NH)NH_2]_2Cu \cdot 2H_2O$, are monoclinic with a bimolecular unit of the dimensions:

$$a_0 = 12.46(1) \text{ A.}; \quad b_0 = 8.50(1) \text{ A.}; \quad c_0 = 11.47(1) \text{ A.}$$
$$\beta = 120.0(1)°$$

The space group has been chosen as C_2^2 ($P2_1$) with atoms in the positions:

$$(2a) \quad xyz;\ \bar{x},y+{}^1/_2,\bar{z}$$

Parameters are listed in Table XVAII,35. Some of the hydrogen atoms were located experimentally, others were assigned positions determined by their expected bond lengths and angles; these parameters are stated in the original.

TABLE XVAII,35

Parameters of Atoms in $Cu[C_{10}H_{13}N_2O]_2 \cdot 2H_2O$

Atom	x	y	z
Cu	0.0522	0.0000	0.1626
O(1)	0.2174	0.0808	0.2084
O(2)	0.1984	0.6018	0.1673
O(1′)	−0.1107	0.0751	0.1117
O(2′)	−0.1105	0.3919	0.0761
N(1)	0.1317	0.2045	0.1923
N(2)	0.3273	0.3208	0.2912
N(1′)	−0.0267	−0.2046	0.1410
N(2′)	−0.1955	−0.3323	0.1362
C(1)	0.2544	0.1904	0.2555
C(2)	0.3068	0.0269	0.2987
C(3)	0.4261	0.0016	0.2979
C(4)	0.4194	0.0319	0.1588
C(5)	0.3340	−0.0076	0.4465
C(6)	0.4237	0.0913	0.5516
C(7)	0.4426	0.0683	0.6812
C(8)	0.3804	−0.0439	0.7060
C(9)	0.2940	−0.1352	0.6002
C(10)	0.2712	−0.1173	0.4692
C(1′)	−0.1290	−0.2002	0.1421

(continued)

TABLE XVAIV,35 (*continued*)

Atom	x	y	z
C(2′)	−0.1764	−0.0358	0.1474
C(3′)	−0.3165	−0.0215	0.0520
C(4′)	−0.3670	0.1552	0.0461
C(5′)	−0.1506	−0.0089	0.2954
C(6′)	−0.2239	−0.0897	0.3384
C(7′)	−0.1936	−0.0900	0.4758
C(8′)	−0.0917	0.0132	0.5613
C(9′)	−0.0184	0.0843	0.5199
C(10′)	−0.0527	0.0783	0.3794

Note: Parameters for the hydrogen atoms, some established experimentally, are given in the original.

The structure is shown in Figure XVAII,48. Its crystallographically different molecules have the same, rather than opposite rotations. Their averaged bond dimensions are given in Figure XVAII,49. The coordination about copper is square, the neighboring nitrogen and oxygen atoms being

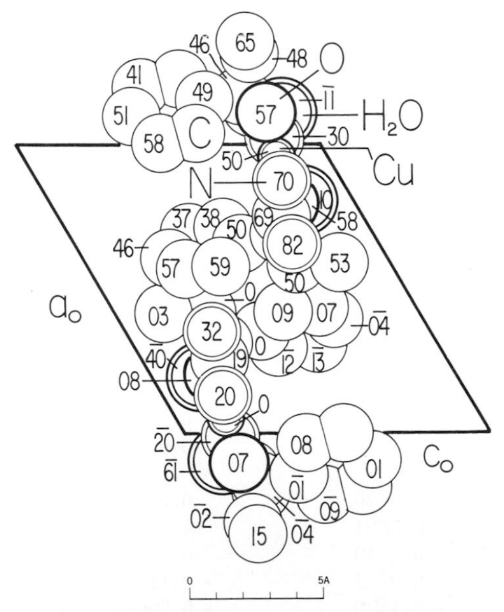

Fig. XVAII,48. The monoclinic structure of bis(α-hydroxy-α-phenylbutyramidine) copper dihydrate projected along its b_0 axis.

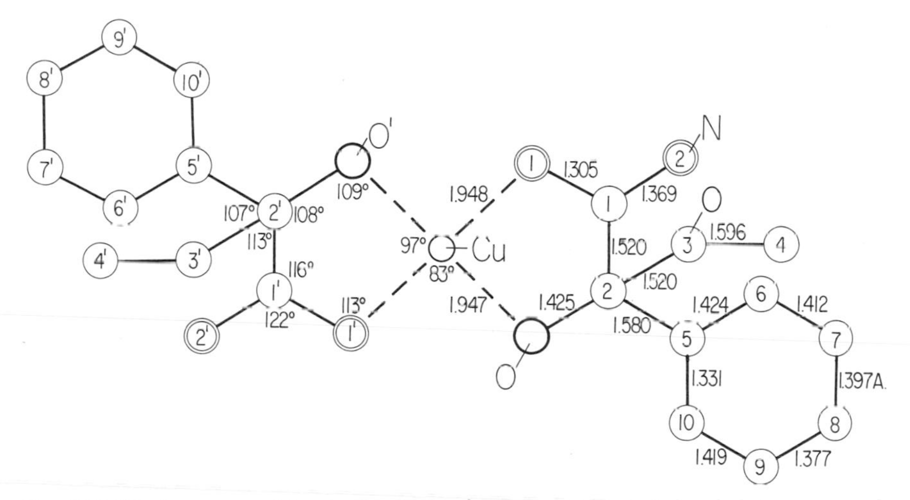

Fig. XVAII,49. Averaged bond dimensions in the molecules of the chelate compound bis(α-hydroxy-α-phenylbutyramidine) copper dihydrate.

no more than 0.04 A. from the best plane through the five atoms. Water molecules are nearest to nitrogen and oxygen atoms of adjacent molecules along the screw axis, with N(2)–OH_2 = 2.814 A. and O(1)–OH_2 = 2.727 A. The shortest intermolecular distances are N–N = 3.452 A. and C–C = 3.646 A.

XV,aII,38. Crystals of *bis(nitrosophenyl hydroxylaminate)copper(II)* [cupferronate], $(C_6H_5N_2O)_2Cu$, are monoclinic. Their bimolecular unit has the dimensions:

$$a_0 = 8.39(2)\text{ A.};\quad b_0 = 5.70(2)\text{ A.};\quad c_0 = 13.28(2)\text{ A.}$$
$$\beta = 94°$$

The space group is C_{2h}^5 ($P2_1/c$). The copper atoms are in:

$$(2a)\quad 000;\ 0\ {}^1/_2\ {}^1/_2$$

All other atoms are in:

$$(4e)\quad \pm(xyz;\ x,{}^1/_2-y,z+{}^1/_2)$$

with the parameters of Table XVAII,36.

TABLE XVAII,36

Parameters of Atoms in bis(Nitrosophenyl hydroxylaminate)copper

Atom	x	y	z
O(1)	0.048	0.259	0.083
O(2)	0.170	0.142	−0.064
N(1)	0.174	0.377	0.047
N(2)	0.246	0.323	−0.031
C(1)	0.221	0.565	0.113
C(2)	0.332	0.716	0.072
C(3)	0.402	0.902	0.127
C(4)	0.358	0.938	0.225
C(5)	0.247	0.792	0.270
C(6)	0.183	0.610	0.212

The atomic arrangement is that of Figure XVAII,50. Its molecules have the bond dimensions of Figure XVAII,51. The benzene ring is planar, as is also the N_2O_2 group with the copper atom, the angle between the two planes thus defined being 12°.

This structure differs from most copper chelates composed of ligands containing both nitrogen and oxygen: here only oxygens are bound to the metal atom.

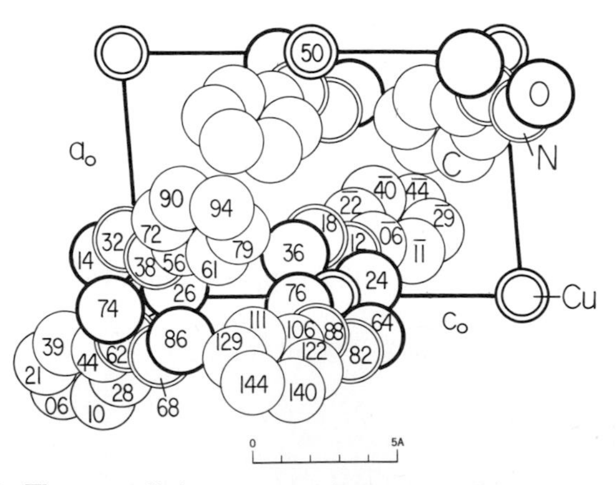

Fig. XVAII,50. The monoclinic structure of copper cupferronate projected along its b_0 axis.

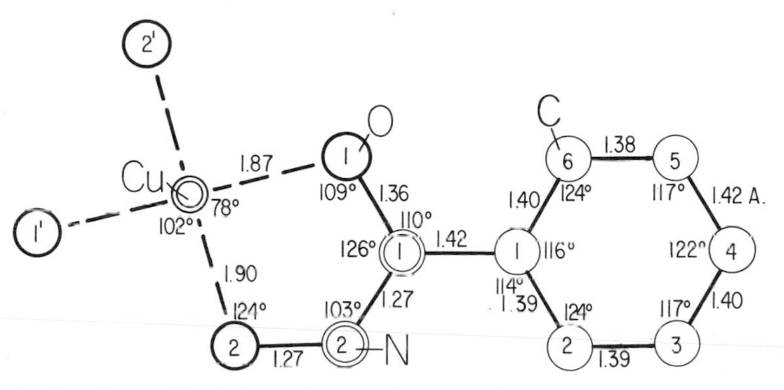

Fig. XVAII,51. Bond dimensions in the molecule of copper cupferronate.

XV,aII,39. Crystals of *tris(nitrosophenyl hydroxylaminate)iron(III)* [cupferronate], $(C_6H_5N_2O)_3Fe$, are monoclinic with a tetramolecular cell of the dimensions:

$$a_0 = 12.50(1) \text{ A.}; \quad b_0 = 17.45(5) \text{ A.}; \quad c_0 = 11.15(4) \text{ A.}$$
$$\beta = 122°19(5)'$$

Atoms are in the general positions of C_{2h}^5 ($P2_1/a$):

$$(4e) \quad \pm(xyz;\ x+{}^1/_2,{}^1/_2-y,z)$$

The parameters are listed in Table XVAII,37.

TABLE XVAII,37

Parameters of the Atoms in Iron Cupferronate

Atom	x	y	z
Fe	0.079	0.236	0.241
O(1)	0.034	0.202	0.049
O(2)	0.209	0.169	0.397
O(3)	0.939	0.312	0.145
O(4)	0.216	0.280	0.219
O(5)	0.980	0.146	0.243
O(6)	0.102	0.305	0.396
N(1)	0.101	0.238	0.005
N(2)	0.159	0.113	0.430
N(3)	0.934	0.300	0.232

(continued)

TABLE XVAIV,37 (*continued*)

Atom	x	y	z
N(4)	0.197	0.281	0.091
N(5)	0.038	0.097	0.350
N(6)	0.021	0.360	0.366
C(1)	0.070	0.229	0.863
C(2)	0.150	0.255	0.821
C(3)	0.114	0.246	0.681
C(4)	0.998	0.213	0.585
C(5)	0.917	0.191	0.623
C(6)	0.954	0.196	0.769
C(7)	0.235	0.067	0.553
C(8)	0.186	0.004	0.580
C(9)	0.262	0.963	0.706
C(10)	0.383	0.988	0.800
C(11)	0.431	0.049	0.773
C(12)	0.358	0.093	0.647
C(13)	0.839	0.420	0.179
C(14)	0.825	0.464	0.276
C(15)	0.732	0.522	0.222
C(16)	0.662	0.534	0.073
C(17)	0.678	0.491	0.983
C(18)	0.769	0.431	0.035

A portion of the resulting structure is shown in Figure XVAII,52. The molecules it contains have the bond dimensions of Figure XVAII,53. The benzene rings are planar and their nitrosohydroxylamino attachments are nearly coplanar with a maximum departure [of the N(3) atom] of 0.04 A. The angle between the two planes is ca. 10°. The sixfold coordination of oxygen atoms around the chelated iron atoms is that of a badly distorted octahedron that has practically equal Fe–O distances (2.00 A.). Between molecules the shortest inter-atomic distances are an O(3)–N(4) = 3.19 A. and O(6)–C(3) = 3.27 A.

XV,aII,40. The crystals of *ephedrine hydrochloride*, $C_6H_5 \cdot CH(OH) \cdot CH(CH_3) \cdot NHCH_3 \cdot HCl$, are monoclinic with the bimolecular unit:

$$a_0 = 12.65 \text{ A.}; \quad b_0 = 6.09 \text{ A.}; \quad c_0 = 7.32 \text{ A.}; \quad \beta = 102°15'$$

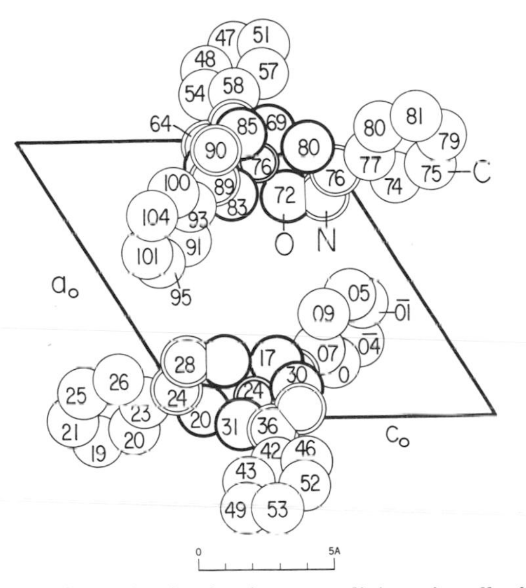

Fig. XVAII,52. Half the molecules in the monoclinic unit cell of iron cupferronate projected along its b_0 axis. The iron atoms are the smaller doubly ringed circles.

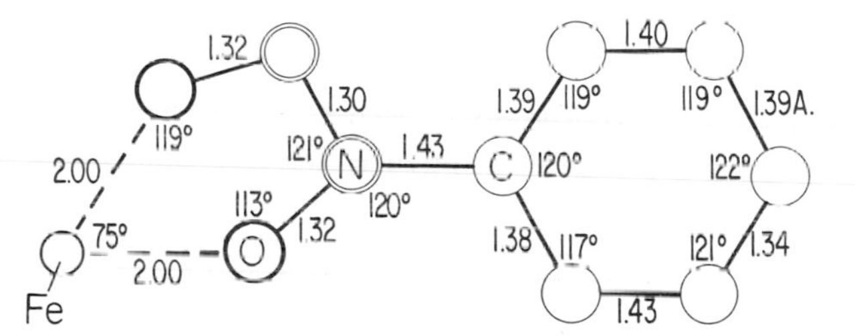

Fig. XVAII,53. The mean dimensions of the bonds in the cupferronate molecule in the iron compound.

All atoms are in the general positions of the space group C_2^2 ($P2_1$):

$$(2a) \quad xyz;\ \bar{x},y+{}^1/_2,\bar{z}$$

The determined parameters are those of Table XVAII,38, the x values of this table being the simple average of the two sets given in the original paper.

TABLE XVAII,38

Parameters of the Atoms in Ephedrine Hydrochloride

Atom	x	y	z
C(1)	0.3984	0.3242	0.7863
C(2)	0.5151	0.2856	0.8297
C(3)	0.5789	0.4450	0.7830
C(4)	0.5364	0.6288	0.6390
C(5)	0.4197	0.6688	0.6420
C(6)	0.3577	0.5110	0.6970
C(7)	0.2342	0.5480	0.6418
C(8)	0.1965	0.7040	0.7766
C(9)	0.1967	0.6070	0.9672
C(10)	0.0474	0.9608	0.8060
O	0.1700	0.3582	0.6252
N	0.0845	0.8006	0.6974
Cl	0.1245	0.0000	0.3257

The molecules that result have the bond angles and lengths of Figure XVAII,54. In the crystal they are packed as in Figure XVAII,55. Between molecules each chlorine ion has two nitrogen neighbors at 3.12 and 3.20 A. and one close oxygen contact with Cl–O = 3.06 A.

XV,aII,41. The orthorhombic crystals of *β-phenylethylamine hydrochloride*, $C_6H_5(CH_2)_2NH_2 \cdot HCl$, have a tetramolecular unit of the edge lengths:

$$a_0 = 32.30(4) \text{ A.}; \quad b_0 = 5.92(2) \text{ A.}; \quad c_0 = 4.61(2) \text{ A.}$$

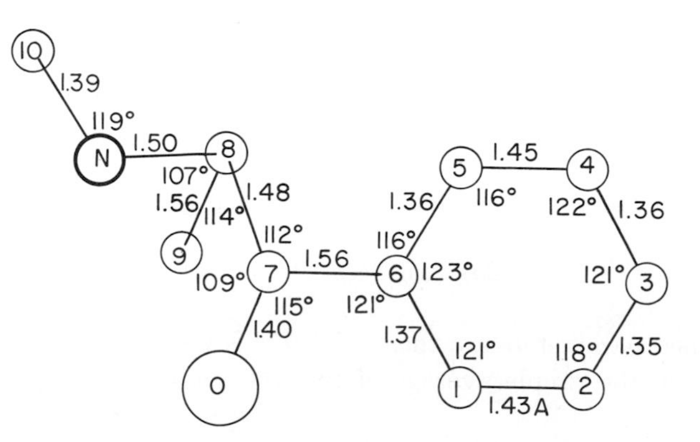

Fig. XVAII,54. Bond dimensions in the molecule of ephedrine hydrochloride.

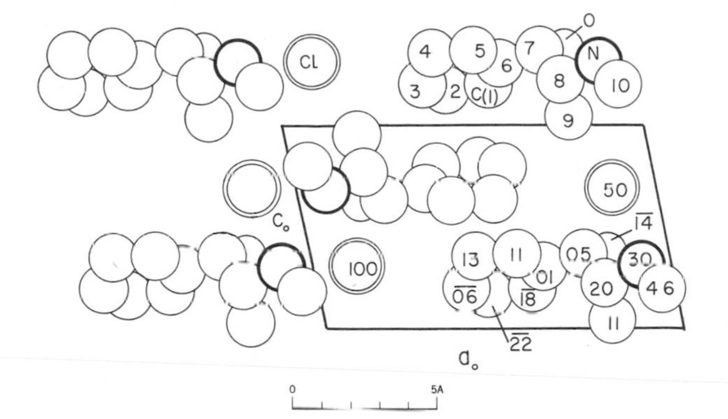

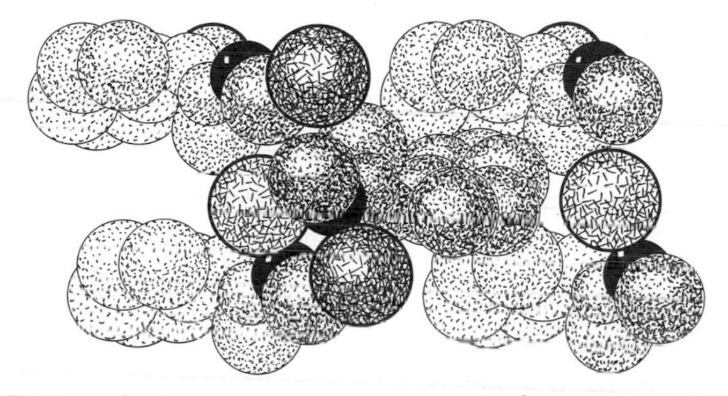

Fig. XVAII,55a (top). The monoclinic structure of ephedrine hydrochloride projected along its b_0 axis. Left-hand axes.

Fig. XVAII,55b (bottom). A packing drawing of the monoclinic ephedrine hydrochloride arrangement seen along its b_0 axis. The nitrogen atoms are black, the oxygens are heavily outlined and dotted. The larger heavily outlined, line shaded circles are chlorine. The dotted carbon atoms of the benzene ring are larger than the others. Left-hand axes.

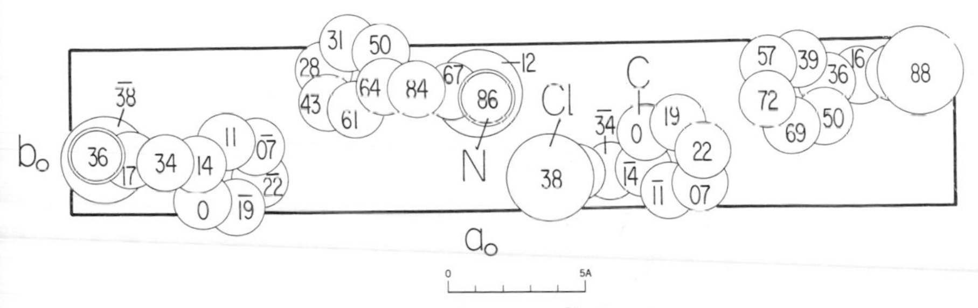

Fig. XVAII,56. The elongated, orthorhombic cell of β-phenylethylamine hydrochloride projected along its c_0 axis.

The space group is V^4 ($P2_12_12_1$) with atoms in the positions:

$$(4a) \quad xyz;\ {}^1/_2-x,\bar{y},z+{}^1/_2;\ x+{}^1/_2,{}^1/_2-y,\bar{z};\ \bar{x},y+{}^1/_2,{}^1/_2-z$$

The parameters are those of Table XVAII,39.

TABLE XVAII,39

Parameters of the Atoms in $C_6H_5(CH_2)_2NH_2 \cdot HCl$

Atom	x	y	z
Cl	0.5386	0.180	0.379
N	0.532	0.177	0.642
C(1)	0.714	0.334	0.217
C(2)	0.710	0.124	0.070
C(3)	0.675	0.093	0.888
C(4)	0.645	0.251	0.856
C(5)	0.650	0.460	0.996
C(6)	0.684	0.499	0.187
C(7)	0.609	0.209	0.659
C(8)	0.569	0.184	0.833

The structure is shown in Figure XVAII,56. In the cation, atoms C(1) through C(7) are coplanar. Around the nitrogen atom, at the positive end of the ion, there are three Cl^- anions at distances of 3.17, 3.17, and 3.23 A.; together with C(8) they form a tetrahedron about this atom.

The hydrobromide is isostructural with a unit of the dimensions:

$$a_0 = 32.01(4) \text{ A.};\quad b_0 = 6.16(2) \text{ A.};\quad c_0 = 4.70(2) \text{ A.}$$

Atomic parameters were not established.

XV,aII,42. *Phenylalanine hydrochloride,* $C_6H_5CH_2CH(NH_2)COOH \cdot HCl$, is orthorhombic with a tetramolecular unit of the edge lengths:

$$a_0 = 27.68 \text{ A.};\quad b_0 = 6.98 \text{ A.};\quad c_0 = 5.34 \text{ A.}$$

The space group has been chosen as V^4 ($P2_12_12_1$) with atoms in the positions:

$$(4a) \quad xyz;\ {}^1/_2-x,\bar{y},z+{}^1/_2;\ x+{}^1/_2,{}^1/_2-y,\bar{z};\ \bar{x},y+{}^1/_2,{}^1/_2-z$$

The parameters are listed in Table XVAII,40.

TABLE XVAII,40

Parameters of the Atoms in Phenylalanine·HCl

Atom	x	y	z
Cl	0.0561	0.1511	0.9768
C(1)	0.0645	0.7020	0.3188
C(2)	0.0686	0.5819	0.5509
C(3)	0.1206	0.5449	0.6465
C(4)	0.1534	0.4398	0.4506
C(5)	0.1765	0.5404	0.2678
C(6)	0.2075	0.4543	0.0984
C(7)	0.2134	0.2588	0.1134
C(8)	0.1888	0.1544	0.3054
C(9)	0.1587	0.2470	0.4724
O(1)	0.0860	0.8711	0.3588
O(2)	0.0457	0.6591	0.1316
N	0.0424	0.4027	0.4956
H(1)	0.065	0.917	0.777
H(2)	0.056	0.347	0.623
H(3)	0.041	0.332	0.359
H(4)	0.011	0.440	0.543
H(5)	0.048	0.641	0.297
H(6)	0.140	0.660	0.322
H(7)	0.114	0.585	0.144
H(8)	0.170	0.695	0.745
H(9)	0.228	0.535	0.043
H(10)	0.236	0.185	0.014
H(11)	0.194	0.005	0.682
H(12)	0.138	0.165	0.386

The molecular cations in the resulting structure (Fig. XVAII,57) have the bond dimensions of Figure XVAII,58. The atoms of the benzene ring and its attached C(3) are coplanar, as are the atoms of the carboxyl group and C(2). The chloride ion is considered to be involved in an elaborate system of hydrogen bonds ranging upwards in length from an N–H . . . Cl = 3.13 A. and an O–H . . . Cl = 2.94 A.

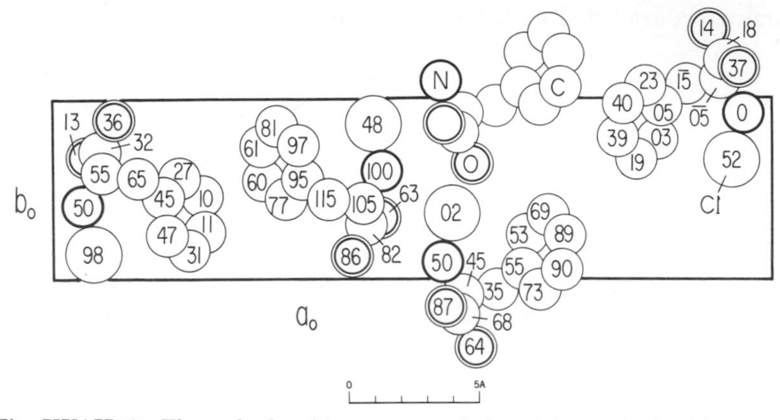

Fig. XVAII,57. The orthorhombic structure of phenylalanine hydrochloride projected along its c_0 axis.

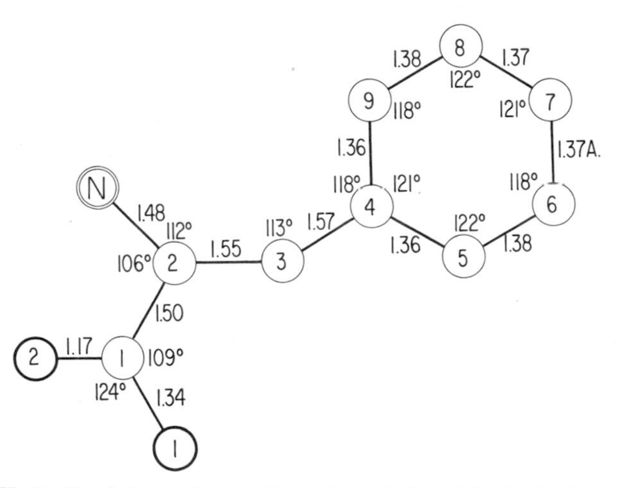

Fig. XVAII,58. Bond dimensions in the cation of phenylalanine hydrochloride.

XV,aII,43. The tripeptid *glycyl phenylalanylglycine hemihydrate*, $C_6H_5CH_2CH[NHC(O)CH_2NH_2]\ [C(O)NHCH_2COOH]\cdot{}^1/_2H_2O$, forms orthorhombic crystals. Their tetramolecular unit has the edge lengths:

$$a_0 = 29.72 \text{ A.};\quad b_0 = 9.98 \text{ A.};\quad c_0 = 4.90 \text{ A.},\qquad \text{all } \pm 0.5\%$$

The space group is V^4 ($P2_12_12_1$) with atoms in the positions:

$$(4a)\quad xyz;\ {}^1/_2-x,\bar{y},z+{}^1/_2;\ x+{}^1/_2,{}^1/_2-y,\bar{z};\ \bar{x},y+{}^1/_2,{}^1/_2-z$$

Parameters are listed in Table XVAII,41; calculated hydrogen positions are given in the original article.

TABLE XVAII,41

Parameters of the Atoms in Glycylphenylalanyl Glycine Hemihydrate

Atom	x	y	z
C(1)	0.1690	0.5447	0.850
C(2)	0.1254	0.6153	0.869
C(3)	0.1366	0.8492	0.755
C(4)	0.1389	0.9941	0.866
C(5)	0.2100	0.1126	0.931
C(6)	0.2480	0.1810	0.772
C(7)	0.0989	0.0730	0.760
C(8)	0.0543	0.0319	0.877
C(9)	0.0353	0.1084	0.074
C(10)	−0.0072	0.0757	0.191
C(11)	−0.0300	0.9674	0.097
C(12)	−0.0126	0.8940	0.892
C(13)	0.0309	0.9236	0.779
N(1)	0.1286	0.7560	0.946
N(2)	0.1794	0.0553	0.752
N(3)	0.2532	0.3192	0.865
O(1)	0.2051	0.6071	0.881
O(2)	0.1678	0.4228	0.787
O(3)	0.1432	0.8240	0.512
O(4)	0.2064	0.1181	0.176
O(H_2O)	0.1276	0.3115	0.323

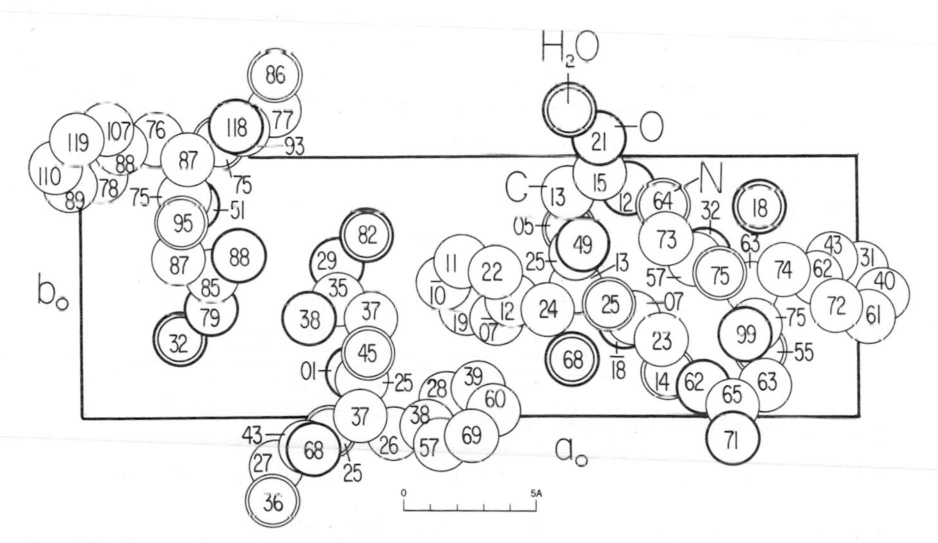

Fig. XVAII,59. The orthorhombic structure of glycyl phenylalanylglycine hemihydrate projected along its c_0 axis.

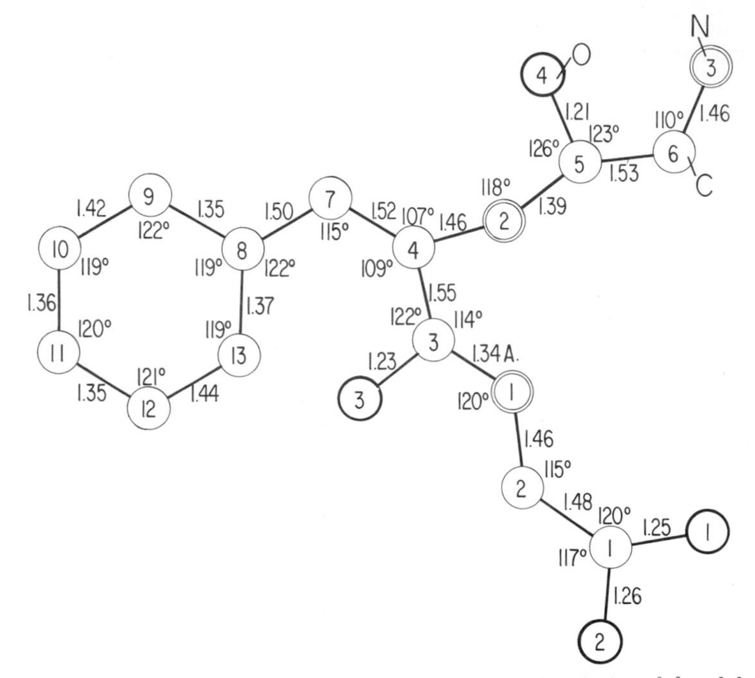

Fig. XVAII,60. Bond dimensions in the molecule of glycyl phenylalanylglycine.

The structure, as shown in Figure XVAII,59, is made up of molecules having the bond dimensions of Figure XVAII,60. Four planes of atoms are discernible in this large molecule: (*a*) the carboxyl group and its attached C(2), (*b*) the benzene ring and its attached C(7), and the two peptid groups (*c*) [C(2,3,4), N(1) and O(3)] and (*d*) [C(4,5,6), N(2) and O(4)]. The departure from planarity in none exceeds about 0.035 A. As in other peptides and in the amino acids, the molecules are tied together by hydrogen bonds (N–H–O = 2.77–3.00 A.). There are also some bonds between the water and other oxygen atoms (O–H–O = 2.80–3.12 A.).

XV,aII,44. *Magnesium benzenesulfonate hexahydrate*, $Mg(C_6H_5SO_3)_2 \cdot 6H_2O$, is monoclinic with a bimolecular unit of the dimensions:

$$a_0 = 22.6 \text{ A.}; \quad b_0 = 6.32 \text{ A.}; \quad c_0 = 6.94 \text{ A.}; \quad \beta = 93°36'$$

The isostructural *zinc* salt has:

$$a_0 = 22.5 \text{ A.}; \quad b_0 = 6.32 \text{ A.}; \quad c_0 = 6.98 \text{ A.}; \quad \beta = 93°36'$$

The space group is C_{2h}^5 ($P2_1/n$). Metallic atoms are in:

$$(2a) \quad 000; \ {}^1/_2\, {}^1/_2\, {}^1/_2$$

All other atoms are in:

$$(4e) \quad \pm(xyz;\ x+{}^1/_2,{}^1/_2-y,z+{}^1/_2)$$

The determined parameters are those of Table XVAII,42, values for the zinc compound being in parentheses. The parameters for y and z are stated to be less accurate than those for x.

TABLE XVAII,42

Parameters of the Atoms in $M(C_6H_5SO_3)_2 \cdot 6H_2O$, where M = Mg or Zn

Atom	x	y	z
Mg[Zn]	0	0	0
$H_2O(1)$	0.043 (0.043)	0.277 (0.272)	0.019 (0.011)
$H_2O(2)$	0.055 (0.060)	0.886 (0.886)	0.242 (0.232)
$H_2O(3)$	0.057 (0.057)	0.886 (0.886)	0.800 (0.782)
C(1)	0.303 (0.304)	0.514 (0.519)	0.509 (0.495)
C(2)	0.285 (0.287)	0.312 (0.283)	0.566 (0.566)
C(3)	0.227 (0.225)	0.285 (0.272)	0.566 (0.564)
C(4)	0.187 (0.184)	0.443 (0.474)	0.501 (0.494)
C(5)	0.198 (0.204)	0.635 (0.679)	0.436 (0.438)
C(6)	0.265 (0.266)	0.680 (0.679)	0.437 (0.446)
S	0.105 (0.106)	0.386 (0.386)	0.519 (0.522)
O(1)	0.102 (0.102)	0.155 (0.161)	0.519 (0.522)
O(2)	0.093 (0.085)	0.480 (0.480)	0.699 (0.687)
O(3)	0.077 (0.077)	0.480 (0.480)	0.354 (0.356)

Note: The parameters for the Zn compound are in parentheses.

In this structure (Fig. XVAII,61) the metallic atoms are surrounded by an octahedron of water molecules which approach closest to sulfonate oxygen atoms. Significant atomic separations are Mg(or Zn) OH_2 = 2.08 A., S–O = ca. 1.40 A., S–C = 1.82 A., H_2O–O(sulfonate) = 2.72–2.86 A.

XV,aII,45. *Methylphenyl sulfone*, $C_6H_5SO_2CH_3$, forms monoclinic crystals whose tetramolecular unit has the dimensions:

$$a_0 = 8.35 \text{ A.};\quad b_0 = 9.20 \text{ A.};\quad c_0 = 10.98 \text{ A.};\quad \beta = 112^\circ$$

Atoms are in the following positions of C_{2h}^5 ($P2_1/c$):

$$(4e) \quad \pm(xyz;\ x,{}^1/_2-y,z+{}^1/_2)$$

The parameters are listed in Table XVAII,43.

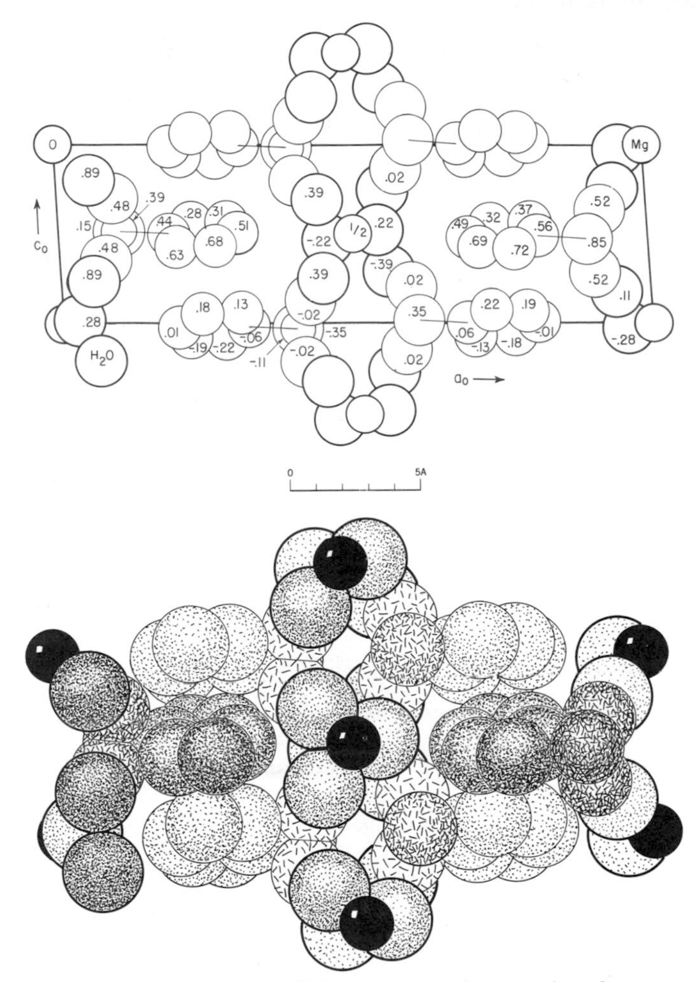

Fig. XVAII,61a (top). The monoclinic structure of magnesium benzene sulfonate hexahydrate projected along its b_0 axis. The sulfonate oxygen atoms are the large light circles, water molecules are large and more heavily ringed. The smallest circles are sulfur if light and magnesium if heavy. Left-hand axes.

Fig. XVAII,61b (bottom). A packing drawing of the monoclinic magnesium benzene sulfonate hexahydrate arrangement seen along its b_0 axis. The magnesium atoms are black, the oxygen atoms are the line shaded light circles; they hide the sulfur atoms. Water molecules are the heavily outlined, dotted circles. Left-hand axes.

TABLE XVAII,43
Parameters of the Atoms in $C_6H_5SO_2CH_3$

Atom	x	y	z
S	0.369	0.137	0.169
O(1)	0.320	0.011	0.083
O(2)	0.456	0.109	0.310
C(1)	0.177	0.245	0.151
C(2)	0.022	0.228	0.039
C(3)	0.876	0.311	0.036
C(4)	0.896	0.423	0.130
C(5)	0.065	0.440	0.230
C(6)	0.210	0.365	0.237
CH_3	0.515	0.231	0.116

The structure (Fig. XVAII,62) is composed of molecules having the bond dimensions of Figure XVAII,63. Distribution about its sulfur atoms is, as would be expected, that of a distorted tetrahedron.

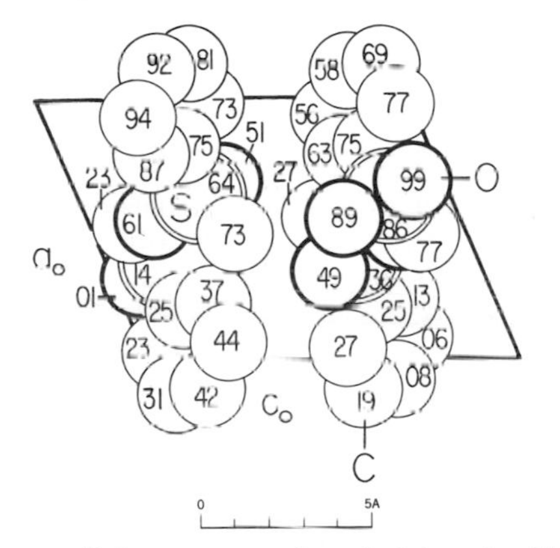

Fig. XVAII,62. The monoclinic structure of methylphenyl sulfone projected along its b_0 axis.

XV,aII,46. Crystals of *potassium α-hydroxybenzyl sulfonate* (benzaldehyde-potassium bisulfite), $KCH(C_6H_5)(OH)SO_3$, are monoclinic with a tetramolecular unit of the dimensions:

$$a_0 = 16.37(2) \text{ A.}; \quad b_0 = 6.07(2) \text{ A.}; \quad c_0 = 9.12(2) \text{ A.}$$
$$\beta = 103°24(6)'$$

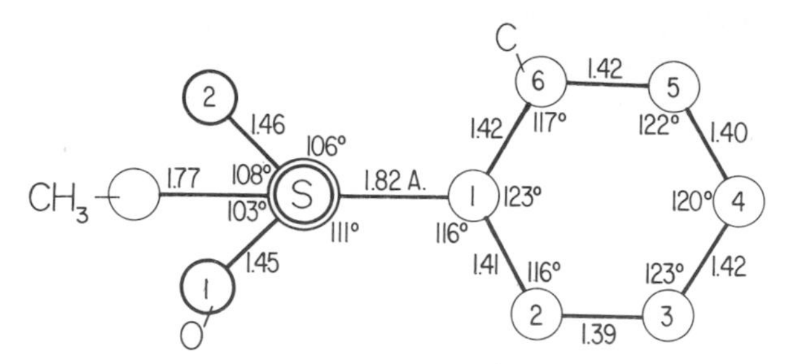

Fig. XVAII,63. Bond dimensions in the molecule of methylphenyl sulfone.

All atoms are in the positions of the space group C_{2h}^5 ($P2_1/c$)

$$(4e) \quad \pm(xyz;\ x,{}^1/_2-y,z+{}^1/_2)$$

The parameters are listed in Table XVAII,44.

TABLE XVAII,44

Parameters of the Atoms in $KCH(C_6H_5)(OH)SO_3$

Atom	x	y	z
K	0.0609	0.2105	0.1143
S	0.1249	0.7064	0.2804
O(1)	0.0579	0.7076	0.1432
O(2)	0.1113	0.5371	0.3810
O(3)	0.1372	0.9228	0.3470
O(4)	0.2080	0.4542	0.1330
C(1)	0.2218	0.6440	0.2214
C(2)	0.2929	0.6107	0.3586
C(3)	0.3033	0.4101	0.4296
C(4)	0.3696	0.3789	0.5564
C(5)	0.4252	0.5489	0.6047
C(6)	0.4145	0.7491	0.5353
C(7)	0.3471	0.7840	0.4079

The structure, shown in Figure XVAII,64, has anions with the bond dimensions of Figure XVAII,65. In them the benzene ring and its attached C(1) are planar. Atoms O(1), S, C(1) and C(2) also roughly define a plane that makes a dihedral angle of 82° with that of benzene. The potassium atoms have nine oxygen neighbors at distances between 2.73 and 3.27 A., all but one of these being sulfonate oxygens. The anions are tied together

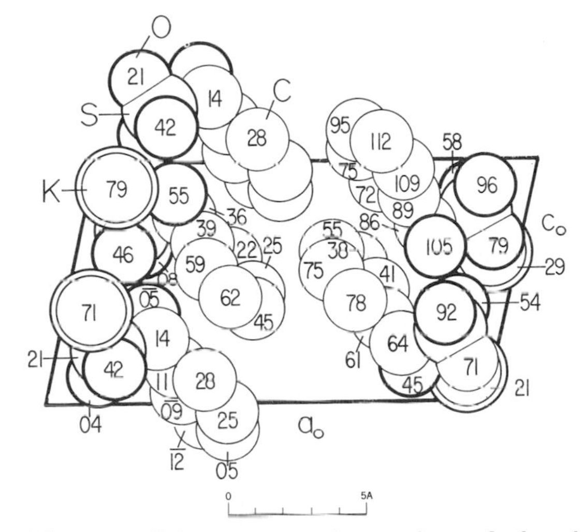

Fig. XVAII,64. The monoclinic structure of potassium α-hydroxybenzyl sulfonate projected along its b_0 axis.

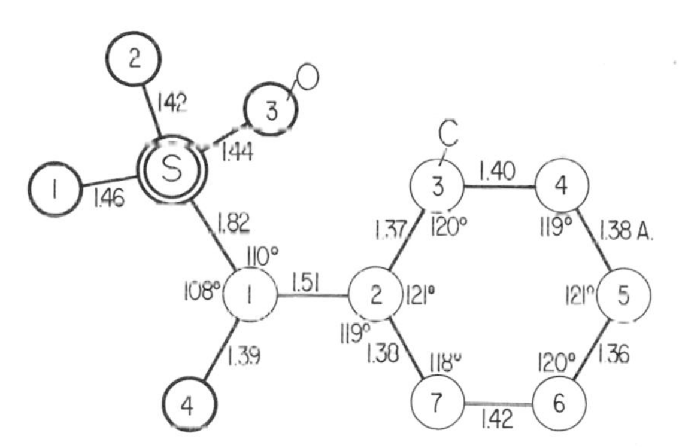

Fig. XVAII,65. Bond dimensions in the anion of potassium α-hydroxybenzyl sulfonate.

by a hydrogen bond of length 2.70 A. between the hydroxyl O(4) and an O(3) atom.

XV,aII,47. Crystals of *N-methyl-2-methylsulfonyl-2-phenylsulfonyl vinylidineamine*, $(CH_3SO_2)(C_6H_5SO_2)C{=}C{=}NCH_3$, are monoclinic with a tetramolecular unit of the dimensions:

$$a_0 = 15.248(5) \text{ A.}; \quad b_0 = 13.030(5) \text{ A.}; \quad c_0 = 10.660(5) \text{ A.}$$
$$\beta - 144°27'$$

The space group is C_{2h}^5 which, for this choice of cell axes, has the orientation $P2_1/a$. Atoms are in the positions:

$$(4e) \quad \pm(xyz; x+1/2, 1/2-y, z)$$

The parameters are those of Table XVAII,45.

TABLE XVAII,45

Parameters of the Atoms in $(CH_3SO_2)(C_6H_5SO_2)C{=}C{=}NCH_3$

Atom	x	y	z
S(1)	0.0159	0.0430	−0.2138
S(2)	−0.2343	−0.1128	0.4602
O(1)	0.0864	0.1180	−0.2097
O(2)	0.1012	−0.0432	−0.0714
O(3)	−0.3689	−0.1159	0.2277
O(4)	−0.2439	−0.1120	−0.4156
N	−0.2276	0.0958	0.2351
C(1)	−0.0552	0.1065	−0.1632
C(2)	−0.1376	−0.0035	−0.4736
C(3)	−0.1897	0.0500	0.3657
C(4)	−0.2897	0.1586	0.0721
C(5)	−0.1221	−0.2199	−0.4516
C(6)	−0.0568	−0.2788	−0.2817
C(7)	0.0263	−0.3656	−0.2180
C(8)	0.0387	−0.3914	−0.3320
C(9)	−0.0271	−0.3301	−0.4991
C(10)	−0.1111	−0.2447	0.4362

The structure, shown in Figure XVAII,66, is built up of molecules having the bond dimensions of Figure XVAII,67. Such molecules closely resemble those seen in the analogous $(CH_3SO_2)_2C{=}C{=}NCH_3$ [**XIV,b27**].

XV,aII,48. *Dimethylphenylsulfonium perchlorate*, $(CH_3)_2S(C_6H_5)ClO_4$, forms monoclinic crystals whose tetramolecular unit has the dimensions:

$$a_0 = 12.81(3) \text{ A.}; \quad b_0 = 10.06(3) \text{ A.}; \quad c_0 = 9.33(3) \text{ A.}$$
$$\beta = 110.7(3)^\circ$$

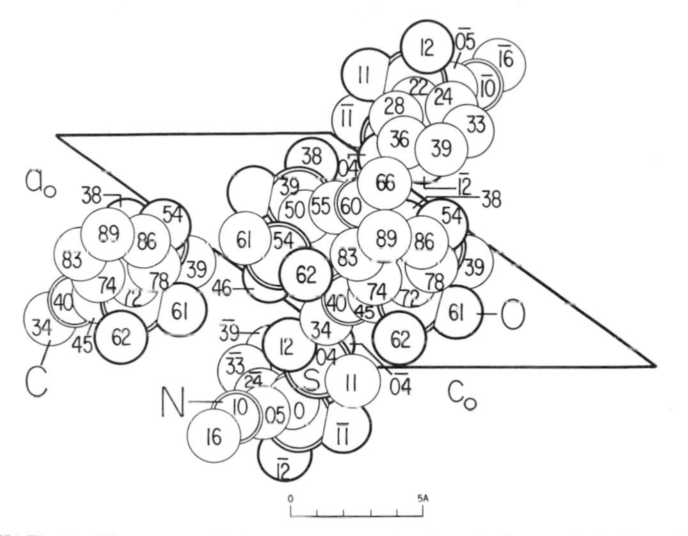

Fig. XVAII,66. The monoclinic structure of *N*-methyl-2-methylsulfonyl-2-phenylsulfonyl vinylidineamine projected along its b_0 axis.

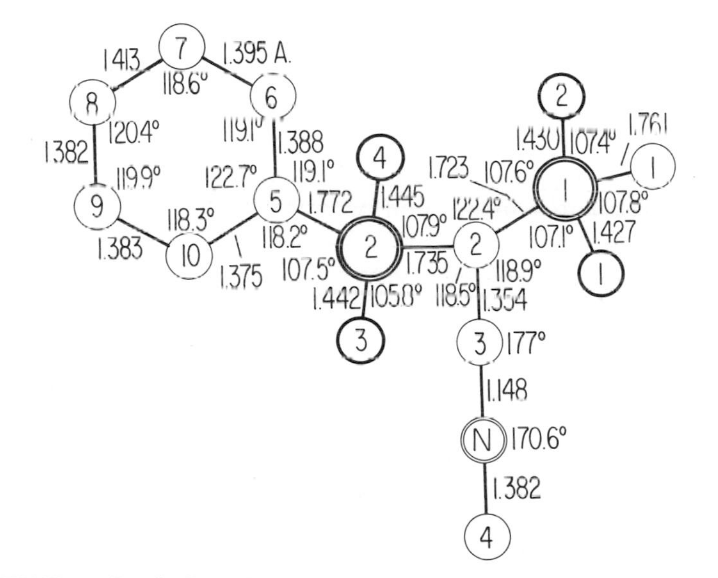

Fig. XVAII,67. Bond dimensions in the molecule of *N*-methyl-2-methylsulfonyl-2-phenylsulfonyl vinylidineamine.

The space group is C_{2h}^5 ($P2_1/c$) with atoms in the positions:

$$(4e)\quad \pm(xyz;\ x,{}^1/_2-y,z+{}^1/_2)$$

The determined parameters are listed in Table XVAII,46; calculated positions for the atoms of hydrogen are to be found in the original article.

TABLE XVAII,46

Parameters of the Atoms in $(CH_3)_2S(C_6H_5)ClO_4$

Atom	x	y	z
Cl	0.1873	0.0669	−0.0848
S	0.1641	−0.3831	−0.1539
O(1)	0.1522	−0.0414	−0.1980
O(2)	0.0903	0.1369	−0.0761
O(3)	0.2386	−0.0151	0.0631
O(4)	0.2688	0.1497	−0.0859
C(1)	0.3089	−0.3767	−0.0259
C(2)	0.3494	−0.4565	0.1030
C(3)	0.4680	−0.4483	0.1949
C(4)	0.5305	−0.3542	0.1609
C(5)	0.4797	−0.2780	0.0281
C(6)	0.3726	−0.2883	−0.0656
C(7)	0.0886	−0.3311	−0.0342
C(8)	0.1301	−0.5588	−0.1808

The resulting structure, shown in Figure XVAII,68, consists of $[(CH_3)_2(C_6H_5)S]^+$ and ClO_4^- ions. In the cations the three S–C bonds are pyramidally distributed about sulfur, with S–C = 1.81–1.83 A.; the angles C–S–S = 102–105°. The ClO_4^- anion has its usual tetrahedral shape but the Cl–O distances are very unequal: 1.38–1.60 A.

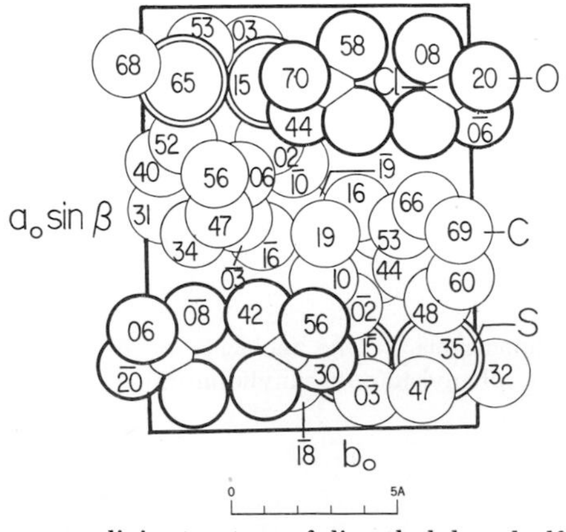

Fig. XVAII,68. The monoclinic structure of dimethylphenylsulfonium perchlorate projected along its c_0 axis.

XV,aII,49. The monoclinic crystals of *benzene seleninic acid*, $C_6H_5SeO_2H$ have a tetramolecular unit of the dimensions:

$$a_0 = 10.29 \text{ A.}; \quad b_0 = 5.14 \text{ A.}; \quad c_0 = 12.63 \text{ A.}; \quad \beta = 99°15'$$

All atoms are in the general positions of C_{2h}^5 ($P2_1/c$):

$$(4e) \quad \pm(xyz;\ x,{}^1/_2-y,z+{}^1/_2)$$

with the parameters of Table XVAII,47.

TABLE XVAII,47

Parameters of the Atoms in $C_6H_5SeO_2H$

Atom	x	y	z
Se	0.0966	0.9979	0.1614
O(1)	0.094	0.041	0.295
O(2)	0.949	0.144	0.099
C(1)	0.215	0.270	0.140
C(2)	0.282	0.406	0.227
C(3)	0.371	0.507	0.209
C(4)	0.388	0.673	0.103
C(5)	0.317	0.548	0.017
C(6)	0.228	0.345	0.036

The molecule thus defined has the bond dimensions shown in Figure XVAII,69. Its phenyl group and the attached selenium atom are substantially planar, with the O(1) atom 0.325 A. outside this plane. In the crystal these molecules pack as shown in Figure XVAII,70. They can be thought of as bound together in chains parallel to the b_0 axis by hydrogen bonds between oxygen atoms, with O(2)–H O(1) = 2.520 A.

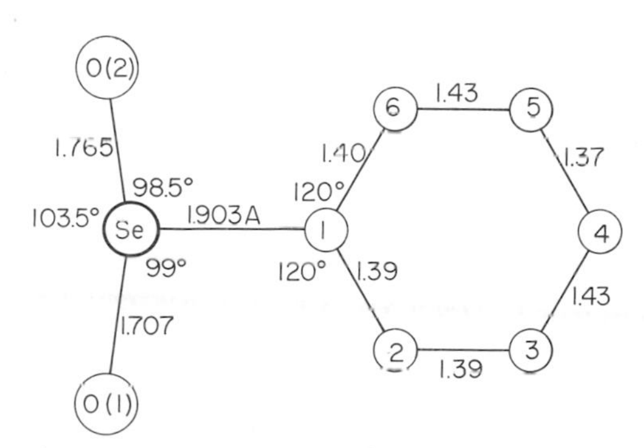

Fig. XVAII,69. Bond dimensions in the molecule of benzene seleninic acid.

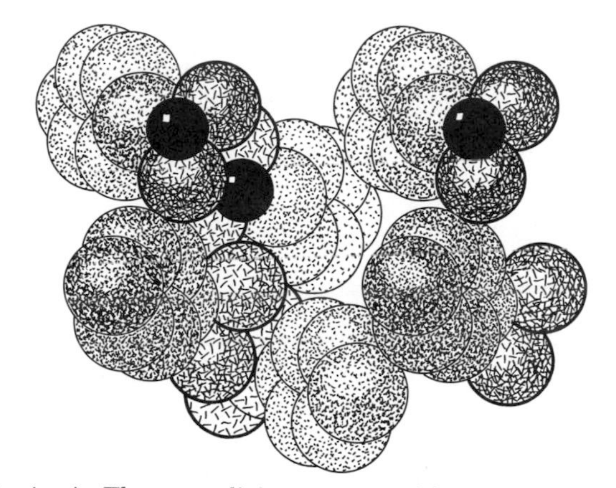

Fig. XVAII,70a (top). The monoclinic structure of benzene seleninic acid projected along its b_0 axis. Left-hand axes.

Fig. XVAII,70b (bottom). A packing drawing of the monoclinic structure of benzene seleninic acid seen along its b_0 axis. The selenium atoms are black, the oxygens are heavily outlined and line shaded. Atoms of carbon are dotted. Left-hand axes.

XV,aII,50. *Phenyl bis(thiourea) tellurium(II) chloride*, $C_6H_5TeCl \cdot 2CS(NH_2)_2$, forms orthorhombic crystals having a tetramolecular unit of the edge lengths:

$$a_0 = 11.98 \text{ A.}; \quad b_0 = 20.80 \text{ A.}; \quad c_0 = 5.79 \text{ A.}$$

The space group is V^4 $(P2_12_12_1)$ with atoms in the positions:

$$(4a) \quad xyz;\ x+{}^1/_2,{}^1/_2-y,\bar{z};\ {}^1/_2-x,\bar{y},z+{}^1/_2;\ \bar{x},y+{}^1/_2,{}^1/_2-z$$

The determined parameters are listed in Table XVAII,48.

TABLE XVAII,48

Parameters of the Atoms in $C_6H_5TeCl \cdot 2CS(NH_2)_2$

Atom	x	y	z
Te	0.4270	0.1181	0.360
Cl	0.1489	0.1804	0.440
S(1)	0.4079	0.0098	0.585
C(1)	0.409	−0.050	0.366
N(1)	0.391	−0.037	0.144
N(2)	0.426	−0.111	0.428
S(2)	0.4775	0.2313	0.141
C(2)	0.380	0.241	−0.092
N(3)	0.416	0.264	−0.293
N(4)	0.273	0.226	−0.060
C(3)	0.601	0.101	0.367
C(4)	0.654	0.072	0.177
C(5)	0.768	0.061	0.181
C(6)	0.828	0.078	0.376
C(7)	0.775	0.107	0.566
C(8)	0.661	0.118	0.562

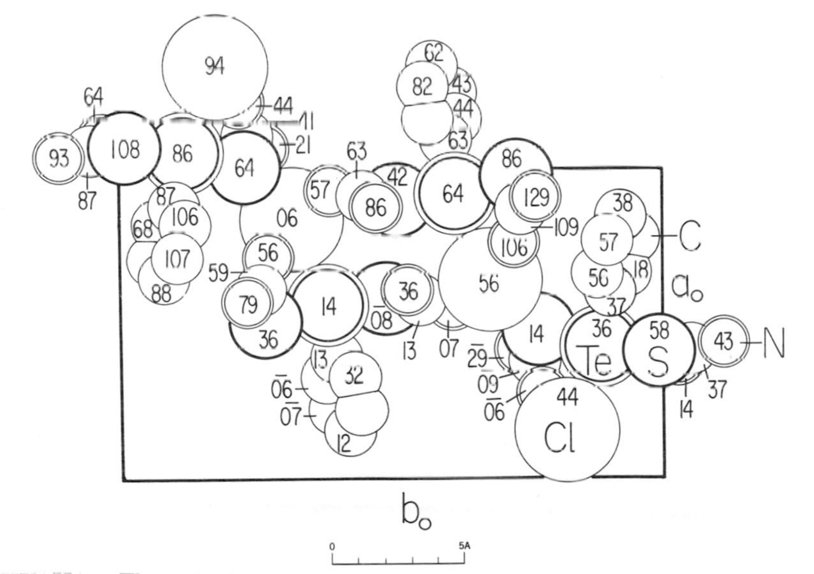

Fig. XVAII,71. The orthorhombic structure of phenyl bis(thiourea) tellurium chloride projected along its c_0 axis.

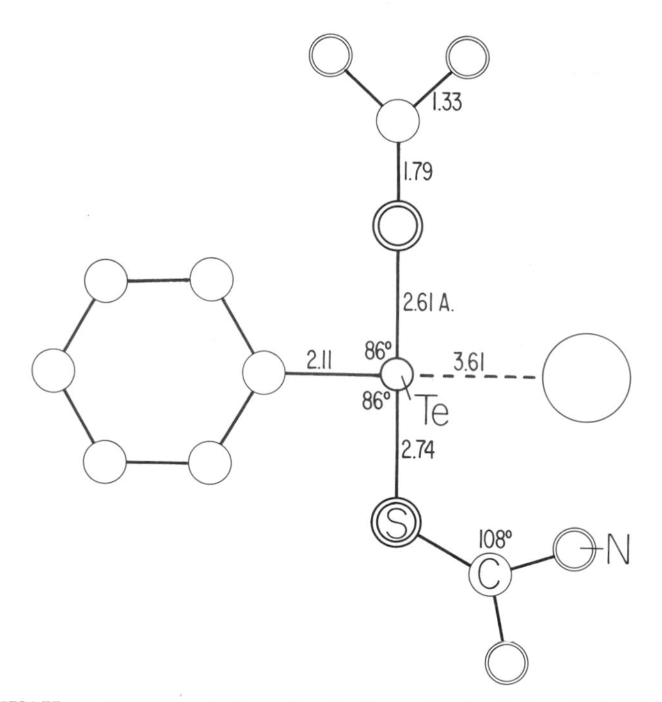

Fig. XVAII,72. Some bond dimensions in the molecule of phenyl bis(thiourea) tellurium chloride.

The structure is shown in Figure XVAII,71. Bond dimensions within it are indicated in Figure XVAII,72. The center of the benzene ring, the two sulfur atoms and tellurium are essentially planar, with the Cl^- lying 0.78 A. outside this plane. The plane through the ring and the tellurium atom makes 85° with this central plane while the planar thioureas make 89 and 46° with it. It is thought that the N–Cl separations of 3.14–3.41 A. correspond to hydrogen bonds.

XV,aII,51. Two determinations have been published of the structure of *phenylarsonic acid*, $C_6H_5AsO_3H_2$. Its symmetry is orthorhombic with a tetramolecular unit of the edge lengths:

$$a_0 = 10.42(2) \text{ A.}; \quad b_0 = 14.92(2) \text{ A.}; \quad c_0 = 4.70(2) \text{ A.}$$

Atoms are in the positions:

$$(4a) \quad xyz;\ {}^1/_2-x,\bar{y},z+{}^1/_2;\ x+{}^1/_2,{}^1/_2-y,\bar{z};\ \bar{x},y+{}^1/_2,{}^1/_2-z$$

of the space group V^4 ($P2_12_12_1$). The parameters found in one of these determinations are listed in Table XVAII,49. For the other study (1960: Struchkov) the origin was differently chosen but if it is displaced by $^{1}/_{4}a_0$ with respect to the above, there is acceptable agreement for the x and y parameters (Struchkov in parentheses); there is, however, rather poor agreement along c_0.

TABLE XVAII,49

Parameters of the Atoms in $C_6H_5AsO_3H_2$

Atom	x	y	z
As	0.3344 (0.0848)	0.1265 (0.1262)	0.2112 (0.1419)
O(1)	0.457 (0.203)	0.111 (0.111)	−0.032 (−0.120)
O(2)	0.192 (−0.063)	0.109 (0.109)	0.079 (−0.032)
O(3)	0.372 (0.121)	0.060 (0.057)	0.499 (0.409)
C(1)	0.338 (0.084)	0.249 (0.247)	0.320 (0.323)
C(2)	0.429 (0.175)	0.313 (0.311)	0.218 (0.243)
C(3)	0.424 (0.174)	0.400 (0.397)	0.332 (0.370)
C(4)	0.336 (0.082)	0.423 (0.419)	0.531 (0.576)
C(5)	0.249 (−0.010)	0.360 (0.355)	0.633 (0.656)
C(6)	0.242 (−0.009)	0.271 (0.269)	0.520 (0.529)

The structure, shown in Figure XVAII,73, is built up of molecules having the bond dimensions of Figure XVAII,74. Around the arsenic atoms the distribution is tetrahedral. The molecules are held together in

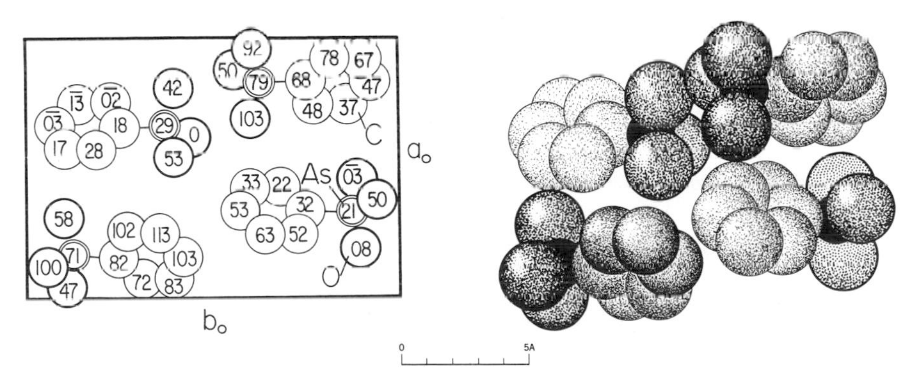

Fig. XVAII,73a (left). The orthorhombic structure of phenylarsonic acid projected along its c_0 axis.

Fig. XVAII,73b (right). A packing drawing of the orthorhombic structure of phenylarsonic acid viewed along its c_0 axis. The arsenic atoms are black, the oxygens are heavily outlined and coarsely dot shaded. The carbon atoms are finely dotted.

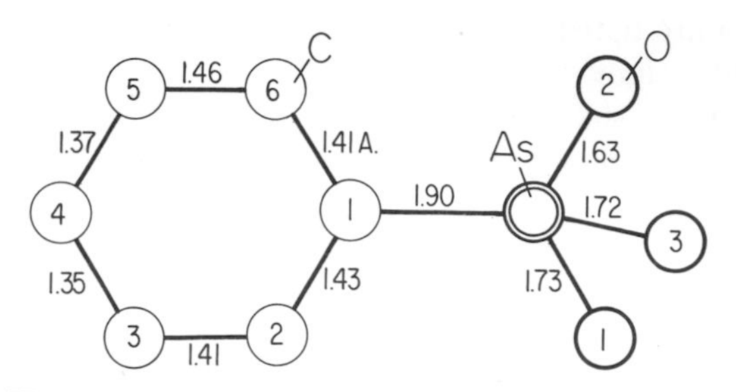

Fig. XVAII,74. Bond dimensions in the molecule of phenylarsonic acid.

pairs by hydrogen bonds of lengths O–H–O = 2.49 and 2.64 A. The shortest intermolecular distances other than these are C–O = 3.29 and 3.40 A. and O–O = 3.39 A.

XV,aII,52. *Phenylarsenic bis(diethyldithiocarbamate)*, $C_6H_5As[S_2CN(C_2H_5)_2]_2$, forms monoclinic crystals whose tetramolecular cells have the edge lengths:

$$a_0 = 14.82(2) \text{ A.}; \quad b_0 = 9.48(2) \text{ A.}; \quad c_0 = 16.63(2) \text{ A.}$$
$$\beta = 112°40(20)'$$

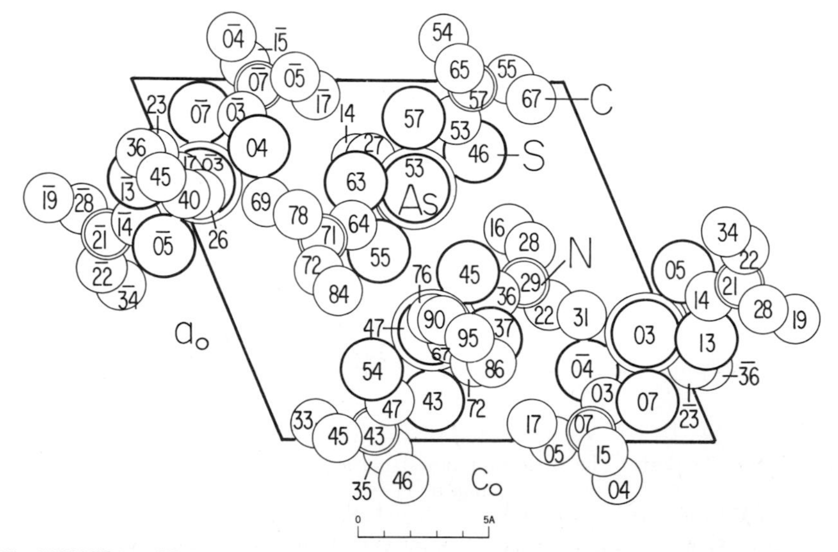

Fig. XVAII,75. The monoclinic structure of phenylarsenic bis(diethyldithiocarbamate).

The space group is C_{2h}^5 ($P2_1/c$) with atoms in the positions:

$$(4c) \quad \pm(xyz;\ x,{}^1/_2-y,z+{}^1/_2)$$

The determined parameters are listed in Table XVAII,50.

TABLE XVAII,50

Parameters of the Atoms in $C_6H_5As[S_2CN(C_2H_5)_2]_2$

Atom	x	y	z
As	0.2841(1)	0.4725(1)	0.4570(1)
S(1)	0.2015(2)	0.5450(4)	0.2726(2)
S(2)	0.4719(2)	0.4482(3)	0.5925(2)
S(3)	0.1163(2)	0.4299(4)	0.3936(2)
S(4)	0.2737(2)	0.3706(3)	0.5802(2)
N(1)	0.0247(7)	0.4278(12)	0.2214(6)
N(2)	0.4349(6)	0.2926(10)	0.7115(5)
C(1)	0.1080(8)	0.4689(13)	0.2861(7)
C(2)	0.4012(7)	0.3636(12)	0.6359(6)
C(3)	0.0099(9)	0.4534(16)	0.1297(9)
C(4)	0.3703(8)	0.2172(14)	0.7473(7)
C(5)	0.0457(11)	0.3260(19)	0.0912(10)
C(6)	0.3530(11)	0.3077(18)	0.8172(10)
C(7)	−0.0571(10)	0.3539(17)	0.2349(9)
C(8)	0.5416(8)	0.2875(14)	0.7617(8)
C(9)	−0.1300(13)	0.4599(21)	0.2440(12)
C(10)	0.5862(10)	0.1634(17)	0.7305(9)
C(11)	0.2797(8)	0.6703(13)	0.4889(7)
C(12)	0.2204(9)	0.7201(16)	0.5325(8)
C(13)	0.2268(10)	0.8645(17)	0.5517(9)
C(14)	0.2866(11)	0.9529(18)	0.5311(10)
C(15)	0.3461(11)	0.9021(19)	0.4889(10)
C(16)	0.3404(10)	0.7593(17)	0.4686(9)

The structure is shown in Figure XVAII,75. Its molecules have the bond dimensions of Figure XVAII,76. In them the plane of the benzene ring is nearly normal (91.5°) to the plane through the four sulfur atoms. The atoms $S_2CN\langle^{C}_{C}$ are coplanar and the two planes they define make an angle of 150° with one another. Arsenic does not, however, lie in their intersection.

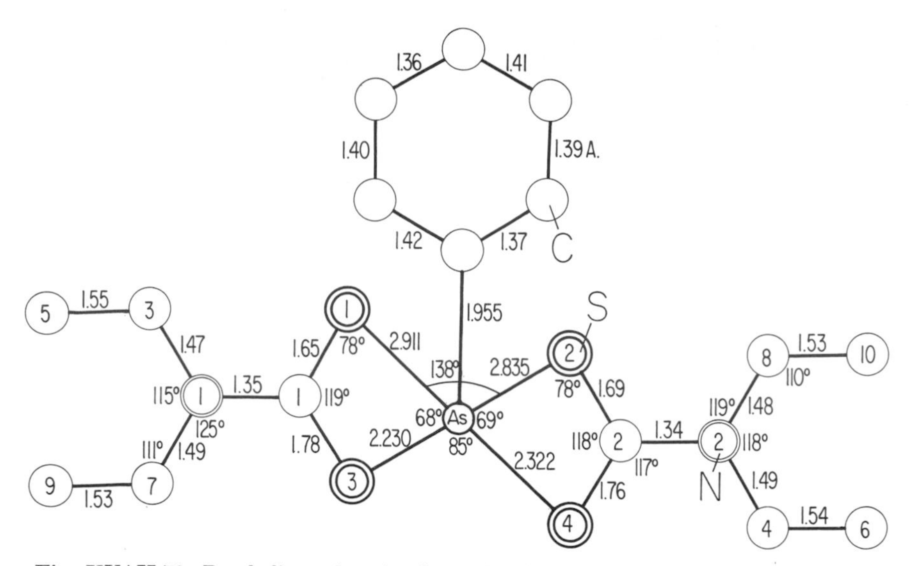

Fig. XVAII,76. Bond dimensions in the molecule of phenylarsenic bis(diethyldithiocarbamate).

III. Compounds containing a Bi-substituted Benzene Ring

XV,aIII,1. At room temperature crystals of *p-dichlorobenzene*, $p\text{-}C_6H_4Cl_2$, and of its *bromine* analogue are monoclinic. Their elongated bimolecular units have the dimensions:

$$\text{For } p\text{-}C_6H_4Cl_2\text{: } a_0 = 14.83 \text{ A.}; \quad b_0 = 5.88 \text{ A.}; \quad c_0 = 4.10 \text{ A.} \quad \beta = 112°30'$$

$$\text{For } p\text{-}C_6H_4Br_2\text{: } a_0 = 15.36 \text{ A.}; \quad b_0 = 5.75 \text{ A.}; \quad c_0 = 4.10 \text{ A.} \quad \beta = 112°30'$$

The space group is $C_{2h}^5(P2_1/a)$ with all atoms in the positions:

$$(4e) \quad \pm(xyz;\ x+{}^1/_2,{}^1/_2-y,z)$$

According to the latest study the parameters for the chloro compound are those given below; the similar values for $p\text{-}C_6H_4Br_2$ are in parentheses.

Atom	x	y	z
Cl(Br)[Cl,Br]	0.162(0.169)[0.169]	0.180(0.170)[0.173]	0.977(0.976)[0.966]
C(1)	0.072(0.074)[0.081]	0.345(0.355)[0.356]	0.979(0.992)[0.982]
C(2)	0.095(0.097)[0.095]	0.570(0.571)[0.533]	0.145(0.197)[0.166]
C(3)	0.030(0.021)[0.013]	0.711(0.716)[0.676]	0.203(0.203)[0.178]

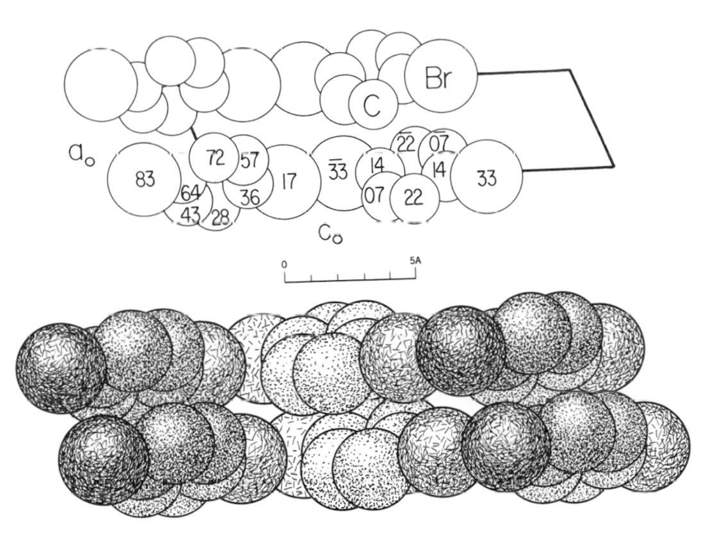

Fig. XVAIII,1a (top). The monoclinic, room temperature, structure of *p*-dibromobenzene projected along its b_0 axis.

Fig. XVAIII,1b (bottom). A packing drawing of the monoclinic structure of *p*-dibromobenzene viewed along its b_0 axis. The carbon atoms are dotted; the atoms of bromine are line shaded.

The resulting structure, which places the long axes of the molecules nearly in the a_0b_0 plane, is shown in Figure XVAIII,1.

The mixed compound *p*-C_6H_4ClBr is isostructural with a cell of the dimensions:

$$a_0 = 15.2 \text{ A.}; \quad b_0 = 5.86 \text{ A.}; \quad c_0 = 4.11 \text{ A.}; \quad \beta = 113°$$

The x-ray data are very similar to those for the compounds just discussed but they do not define separate positions for the chlorine and bromine atoms. Therefore there appears to be disorder in the way these halogen atoms are oriented along the a_0 axis. Parameters are given in square brackets above.

The structure of the chloro compound remains unchanged on cooling to -140°C. Its cell dimensions then are:

$$a_0 = 14.63(6) \text{ A.}; \quad b_0 = 5.66(3) \text{ A.}; \quad c_0 = 3.89(2) \text{ A.}$$
$$\beta = 111°45(20)'$$

At this temperature the determined parameters, including those assigned the hydrogen atoms, are as follows:

Atom	x	y	z
Cl	0.167	0.172	0.973
C(1)	0.074	0.357	0.991
C(2)	0.098	0.559	0.187
C(3)	0.023	0.712	0.191
H(2)	0.155	0.594	0.346
H(3)	0.134	0.843	0.337

XV,aIII,2. There is a triclinic modification of *p-dichlorobenzene*, p-$C_6H_4Cl_2$, which has a unimolecular cell of the dimensions:

$$a_0 = 7.32(1) \text{ A.}; \quad b_0 = 5.95(1) \text{ A.}; \quad c_0 = 3.98(1) \text{ A.}$$
$$\alpha = 93°10'; \qquad \beta = 113°35'; \qquad \gamma = 93°30'$$

Its space group is C_i^1 ($P\bar{1}$) with all atoms in the positions (2i) $\pm(xyz)$. The parameters have been given as the following:

Atom	x	y	z
Cl	0.671	0.305	0.036
C(1)	0.857	0.132	0.010
C(2)	0.051	0.208	0.185
C(3)	0.200	0.070	0.160

The resulting very simple structure is shown in Figure XVAIII,2. With the foregoing parameters the centrosymmetric molecule has the bond

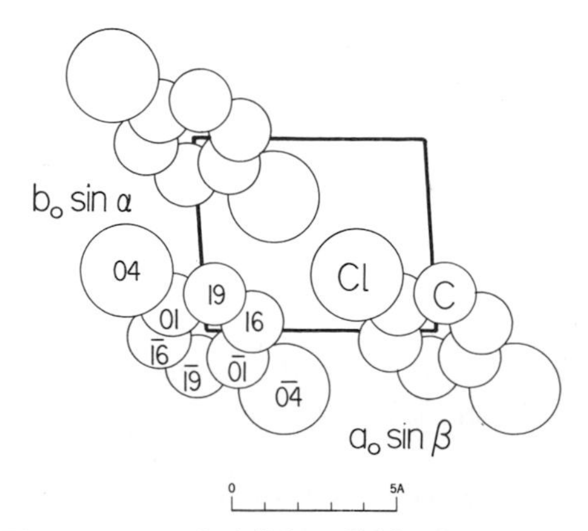

Fig. XVAIII,2. The structure of triclinic p-dichlorobenzene projected along its c_0 axis.

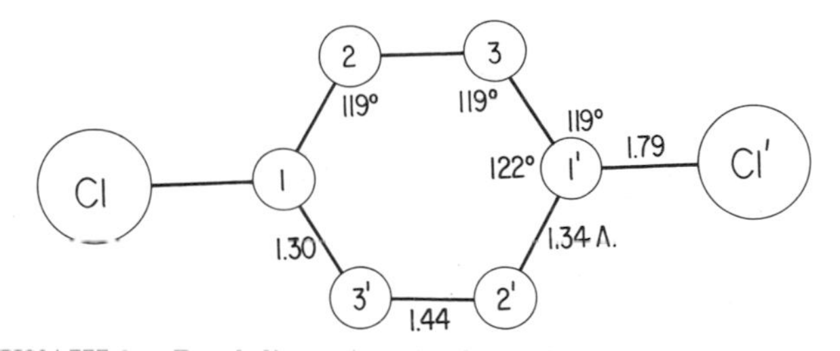

Fig. XVAIII,3. Bond dimensions in the molecule of the triclinic form of *p*-dichlorobenzene.

dimensions of Figure XVAIII,3. It is not strictly planar, the departures of the ring bonds from the best plane through the molecule being about 4°.

XV,aIII,3. Crystals of *p-diiodobenzene*, $C_6H_4I_2$, are orthorhombic with a tetramolecular unit of the edge lengths:

$$a_0 = 17.008(2) \text{ A.}; \quad b_0 = 7.321(2) \text{ A.}; \quad c_0 = 5.949(2) \text{ A.}$$

Atoms are in the positions

$$(8c) \quad \pm(xyz;\ x+{}^1/_2,{}^1/_2-y,\bar{z};\ \bar{x},y+{}^1/_2,{}^1/_2-z;\ {}^1/_2-x,\bar{y},z+{}^1/_2)$$

of the space group V_h^{15} (*Pbca*). The determined parameters are as follows:

Atom	x	y	z
I	0.1708	0.0396	0.2998
C(1)	0.070	0.016	0.123
C(2)	0.072	−0.048	−0.099
C(3)	0.002	−0.065	−0.221

These parameters correspond to a C–C = 1.40 A. and a C–I = 2.02 A. Possible positions have been calculated for the hydrogen atoms on the assumption that C–H = 1.08 A. and H–C–C = 120°. The structure is shown in Figure XVAIII,4.

XV,aIII,4. The monoclinic crystals of *p-iodobenzonitrile*, IC_6H_4CN, have a tetramolecular unit of the dimensions:

$$a_0 = 10.36(5) \text{ A.}; \quad b_0 = 10.63(5) \text{ A.}; \quad c_0 = 9.10(5) \text{ A.}$$
$$\beta = 133°6(12)'$$

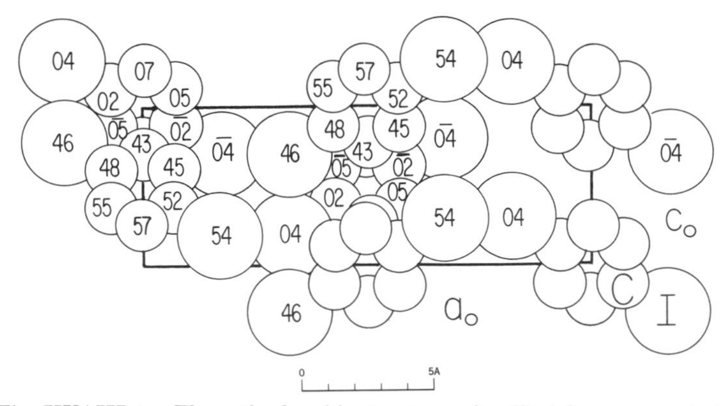

Fig. XVAIII,4. The orthorhombic structure of *p*-diiodobenzene projected along its b_0 axis.

The space group was chosen as C_{2h}^6 ($C2/c$) with atoms in the positions:

$$\begin{array}{rll} \text{I}: (4e) & \pm(0\,u\,{}^1/_4;\ {}^1/_2,u+{}^1/_2,{}^1/_4) & \text{with } u = 0.0265 \\ \text{C}(1): (4e) & \text{with } u = 0.8324 & \\ \text{C}(4): (4e) & \text{with } u = 0.5675 & \\ \text{C}(5): (4e) & \text{with } u = 0.4386 & \\ \text{N}: (4e) & \text{with } u = 0.3258 & \end{array}$$

The other carbon atoms are in:

$$(8f)\quad \pm(xyz;\ x,\bar{y},z+{}^1/_2;\ x+{}^1/_2,y+{}^1/_2,z;\ x+{}^1/_2,{}^1/_2-y,z+{}^1/_2)$$

with the parameters:

$$\begin{array}{l} \text{C}(2): x = 0.0926,\ y = 0.7642,\ z = 0.2141 \\ \text{C}(3): x = 0.0945,\ y = 0.6339,\ z = 0.2162 \end{array}$$

The structure, as shown in Figure XVAIII,5, is built of planar molecules having the bond dimensions of Figure XVAIII,6.

XV,aIII,5. Crystals of *p-iodonitrosobenzene*, $C_6H_4I(NO)$, are monoclinic with a tetramolecular unit of the dimensions:

$$a_0 = 7.86\text{ A.};\quad b_0 = 9.99\text{ A.};\quad c_0 = 10.52\text{ A.};\quad \beta = 122°$$

The chosen space group, C_s^4 (Aa), places atoms in the positions:

$$(4a)\quad xyz;\ x+{}^1/_2,\bar{y},z;\ x,y+{}^1/_2,z+{}^1/_2;\ x+{}^1/_2,{}^1/_2-y,z+{}^1/_2$$

The parameters are listed in Table XVAIII,1.

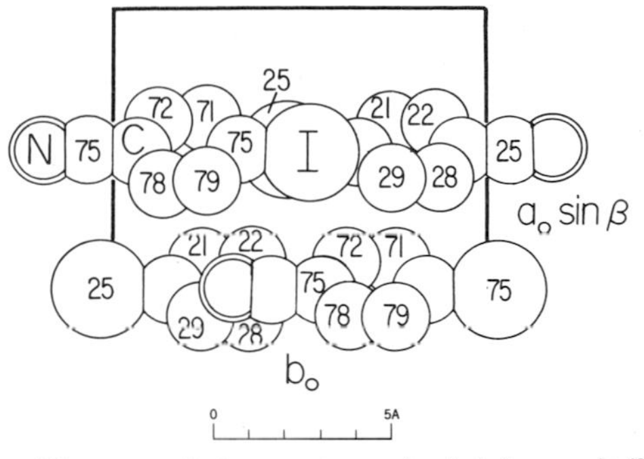

Fig. XVAIII,5. The monoclinic structure of *p*-iodobenzonitrile projected along its c_0 axis.

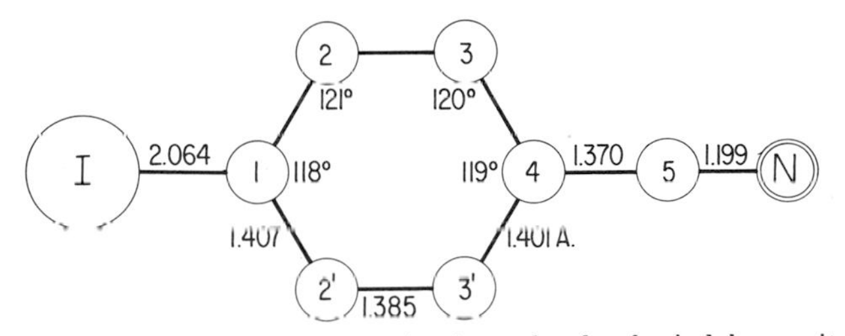

Fig. XVAIII,6. Bond dimensions in the molecule of *p*-iodobenzonitrile.

TABLE XVAIII,1

Parameters of the Atoms in $C_6H_4I(NO)$

Atom	x	y	z
I	0.000	0.203	0.000
N	0.020	0.804	0.053
O	−0.006	0.888	−0.043
C(1)	0.000	0.403	0.010
C(2)	−0.027	0.480	−0.107
C(3)	−0.024	0.620	−0.097
C(4)	0.007	0.677	0.033
C(5)	0.030	0.600	0.150
C(6)	0.028	0.460	0.140

Molecules of the resulting structure are planar, with I–C = 2.00 A. Between them there is a short I–O = 3.2 A.

XV,aIII,6. The monoclinic crystals of *p-chloroiodoxybenzene*, ClC_6H_4-(IO_2), possess a tetramolecular unit of the dimensions:

$$a_0 = 14.4 \text{ A.}; \quad b_0 = 6.50 \text{ A.}; \quad c_0 = 8.11 \text{ A.}; \quad \beta = 98°30'$$

The space group is C_{2h}^5 $(P2_1/a)$ with atoms in the positions:

$$(4e) \quad \pm(xyz;\ x+{}^1/_2,{}^1/_2-y,z)$$

The parameters, determined some years ago, are those of Table XVAIII,2.

TABLE XVAIII,2

Parameters of the Atoms in *p*-Chloroiodoxybenzene

Atom	x	y	z
C(1)	0.174	0.322	0.388
C(2)	0.178	0.522	0.455
C(3)	0.260	0.637	0.451
C(4)	0.337	0.552	0.383
C(5)	0.333	0.352	0.316
C(6)	0.251	0.237	0.320
Cl	0.439	0.710	0.368
I	0.052	0.208	0.400
O(1)	0.092	0.023	0.520
O(2)	0.042	0.098	0.213

The resulting structure is shown in Figure XVAIII,7. The molecule has the plane of its IO_2 radical nearly normal to that of the benzene ring. Within this molecule, C–I = 1.93 A., I–O(1) = 1.60 A., I–O(2) = 1.65 A., C–Cl = 1.80 A. In the crystal the IO_2 groups of adjacent molecules are in contact, with O(1)–O(1) = 2.62 A. and I–O = 2.87 A. Similarly, the chlorine atoms at the other ends of the molecules approach one another closely, with Cl–Cl = 3.75 A. The shortest carbon–carbon distance between neighboring molecules is about 3.9 A.

The analogous *p*-methyliodoxybenzene, $CH_3C_6H_4(IO_2)$, is said to give x-ray diffraction data indistinguishable from those of the chloride and therefore to have a cell of the same dimensions.

XV,aIII,7. A structure has been described for *p-bromophenyl boric acid*, $BrC_6H_4B(OH)_2$. Its hexagonal crystals have a large 36-molecule unit of the edge lengths:

$$a_0 = 28.73 \text{ A.}, \qquad c_0 = 9.74 \text{ A.}$$

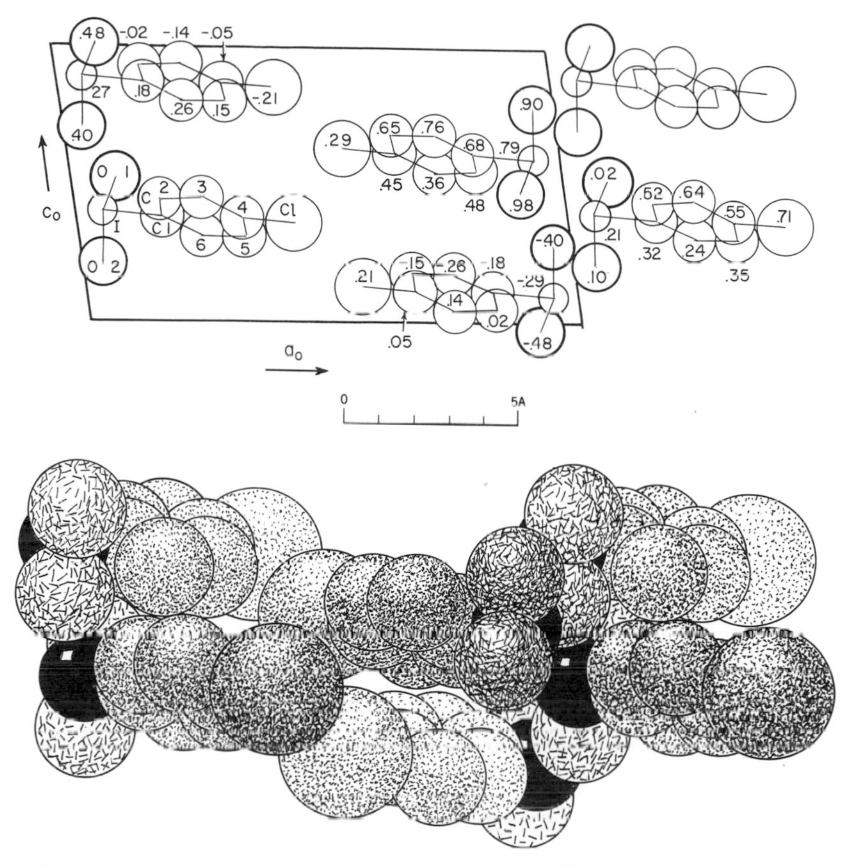

Fig. XVAIII,7a (top). The monoclinic structure of *p*-chloroiodoxybenzene projected along its b_0 axis. Left hand axes.

Fig. XVAIII,7b (bottom). A packing drawing of the monoclinic structure of *p*-chloroiodoxybenzene viewed along its b_0 axis. The iodine atoms are black; the oxygens are line shaded. Atoms of chlorine are the large circles; carbon atoms are the small circles.

The space group was chosen as D_{6h}^2 ($P6/mcc$) with atoms in the positions:

$$(12l)\quad \pm(uv0;\ \bar{v},u-v,0;\quad u+v,\bar{u},0;\ v\ u\ {}^1/_2;\ \bar{u},v-u,{}^1/_2;\ u-v,\bar{v},{}^1/_2)$$

$$(24m)\quad \pm(xyz;\quad \bar{y},x-y,z;\qquad y-x,\bar{x},z;\ y,x,z+{}^1/_2;\ \bar{x},y-x,z+{}^1/_2;\ x-y,\bar{y},z+{}^1/_2;\ \bar{x}\bar{y}z;\qquad y,y-x,z;\qquad x-y,x,z;\ \bar{y},\bar{x},z+{}^1/_2;\ x,x-y,z+{}^1/_2;\ y-x,y,z+{}^1/_2)$$

The parameters are those of Table XVAIII,3.

TABLE XVAIII,3

Positions and Parameters of the Atoms in *p*-Bromophenyl Boric Acid

Atom	Position	x	y	z
Br	(12*l*)	0.106	0.978	0
C(1)	(12*l*)	0.179	0.027	0
C(2)	(24*m*)	0.206	0.045	0.123
C(3)	(24*m*)	0.261	0.082	0.123
C(4)	(12*l*)	0.287	0.099	0
B	(12*l*)	0.347	0.140	0
O(1)	(24*m*)	0.373	0.157	0.122
Br′	(12*l*)	0.966	0.356	0
C(1′)	(12*l*)	0.024	0.42	0
C(2′)	(24*m*)	0.046	0.450	0.123
C(3′)	(24*m*)	0.090	0.503	0.123
C(4′)	(12*l*)	0.111	0.528	0
B′	(12*l*)	0.160	0.586	0
O(1′)	(24*m*)	0.181	0.611	0.122
Br″	(12*l*)	0.275	0.580	0
C(1″)	(12*l*)	0.218	0.511	0
C(2″)	(24*m*)	0.197	0.485	0.123
C(3″)	(24*m*)	0.153	0.432	0.123
C(4″)	(12*l*)	0.133	0.407	0
B″	(12*l*)	0.085	0.349	0
O″	(24*m*)	0.065	0.324	0.122

In the molecules, which are planar, Br–C = 1.85 A., C–C = 1.37–1.41 A., C–B = 1.54 A. and B–O = 1.36 A. These molecules stand normal to the *c*-plane, with a closest intermolecular O–O = 2.76 A.

XV,aIII,8. Crystals of *syn-p-chlorobenzaldoxime*, ClC_6H_4CHNHO, are monoclinic with a tetramolecular unit of the dimensions:

$$a_0 = 6.06 \text{ A.}; \quad b_0 = 4.73 \text{ A.}; \quad c_0 = 25.06 \text{ A.}; \quad \beta = 93°24'$$

Atoms are in the positions

$$(4e) \quad \pm(xyz;\ x,{}^1/_2-y,z+{}^1/_2)$$

of the space group C_{2h}^5 ($P2_1/c$). The parameters are those listed in Table XVAIII,4.

TABLE XVAIII,4
Parameters of the Atoms in *syn-p*-Chlorobenzaldoxime

Atom	x	y	z
Cl(1)	0.0921(2)	−0.1305(4)	0.2014(1)
C(2)	0.2524(9)	0.1014(12)	0.1671(2)
C(3)	0.4584(10)	0.1679(11)	0.1886(2)
C(4)	0.5900(10)	0.3542(13)	0.1612(2)
C(5)	0.5082(9)	0.4681(13)	0.1126(2)
C(6)	0.3042(10)	0.3938(14)	0.0909(2)
C(7)	0.1737(10)	0.2076(13)	0.1176(2)
C(8)	0.6544(11)	0.6700(15)	0.0861(3)
N(9)	0.5839(9)	0.8095(12)	0.0458(2)
O(10)	0.7505(8)	0.9864(13)	0.0276(2)

The structure, shown in Figure XVAIII,8, yields molecules having the bond dimensions of Figure XVAIII,9. Though this and the *anti*-compound of paragraph **XV,aIII,9** possess units of similar shapes and dimensions, their molecular distributions are quite different.

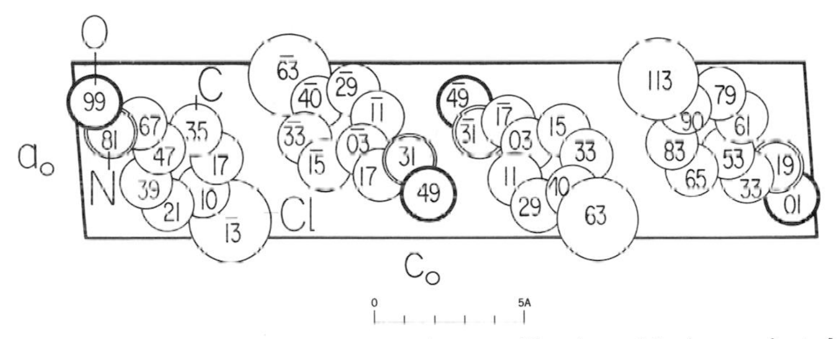

Fig. XVAIII,8. The monoclinic structure of *syn-p*-chlorobenzaldoxime projected along its b_0 axis.

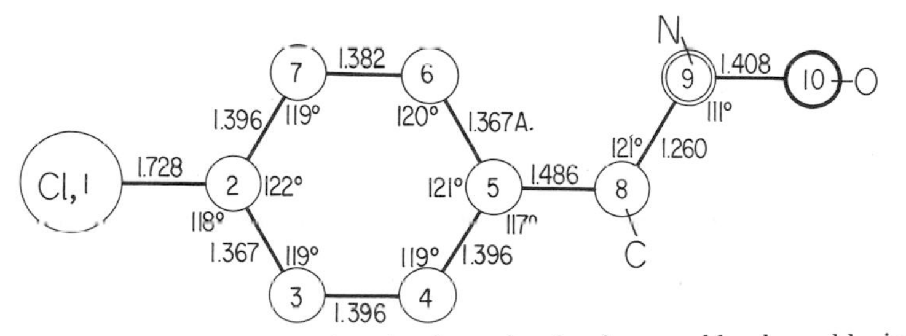

Fig. XVAIII,9. Bond dimensions in the molecule of *syn-p*-chlorobenzaldoxime.

XV,aIII,9. The *anti-p-chlorobenzaldoxime*, ClC_6H_4CHNHO, forms orthorhombic crystals having a tetramolecular unit of the edge lengths:

$$a_0 = 6.60 \text{ A.}; \quad b_0 = 4.67 \text{ A.}; \quad c_0 = 23.52 \text{ A.}$$

The space group is V^4 ($P2_12_12_1$) with atoms in the positions:

$$(4a) \quad xyz; \ ^1/_2-x,\bar{y},z+^1/_2; \ x+^1/_2,^1/_2-y,\bar{z}; \ \bar{x},y+^1/_2,^1/_2-z$$

Parameters are those of Table XVAIII,5.

TABLE XVAIII,5

Parameters of the Atoms in *anti-p*-Chlorobenzaldoxime

Atom	x	y	z
Cl(1)	0.0002(12)	0.3207(18)	0.0528(4)
C(2)	0.7782(36)	0.4466(49)	0.0898(9)
C(3)	0.6730(57)	0.6506(53)	0.0627(9)
C(4)	0.4924(45)	0.7493(47)	0.0905(9)
C(5)	0.4471(31)	0.6398(46)	0.1457(7)
C(6)	0.5555(41)	0.4286(52)	0.1692(8)
C(7)	0.7497(59)	0.3304(57)	0.1434(9)
C(8)	0.2412(48)	0.7594(46)	0.1723(9)
N(9)	0.1432(34)	0.6570(42)	0.2111(9)
O(10)	0.2027(34)	0.3948(37)	0.2338(7)

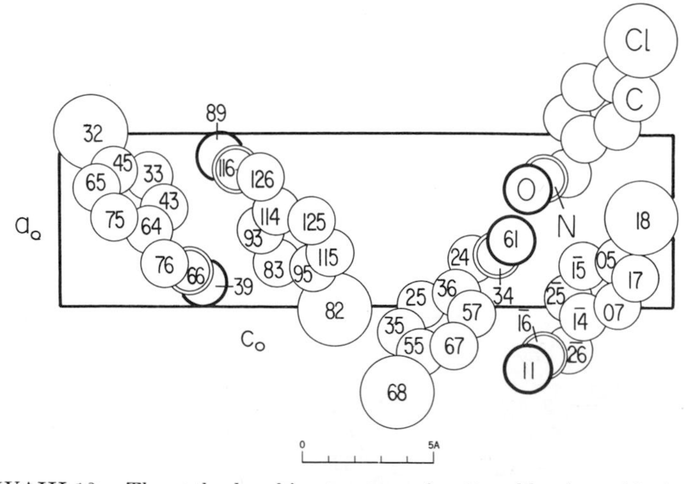

Fig. XVAIII,10. The orthorhombic structure of *anti-p*-chlorobenzaldoxime projected along its b_0 axis.

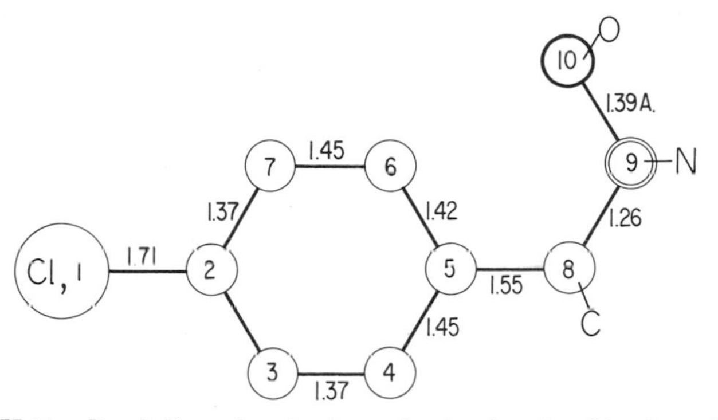

Fig. XVAIII,11. Bond dimensions in the molecule of *anti-p*-chlorobenzaldoxime.

The resulting structure, as shown in Figure XVAIII,10, is built of molecules having the bond dimensions of Figure XVAIII,11. Unlike the methyl-substituted compound of **XV,aIII,10,** the aldoxime part of the molecule departs considerably from the plane of the rest; the plane of C(8), N(9) and O(10), in fact, makes an angle of 19° to that of the ring.

XV,aIII,10. Crystals of *N-methyl-p-chlorobenzaldoxime,* $ClC_6H_4CHN(CH_3)O$, are monoclinic with a tetramolecular unit of the dimensions:

$$a_0 = 7.50(2) \text{ A.}; \quad b_0 = 9.91(2) \text{ A.}; \quad c_0 = 11.677(5) \text{ A.}$$
$$\beta = 108°1(8)'$$

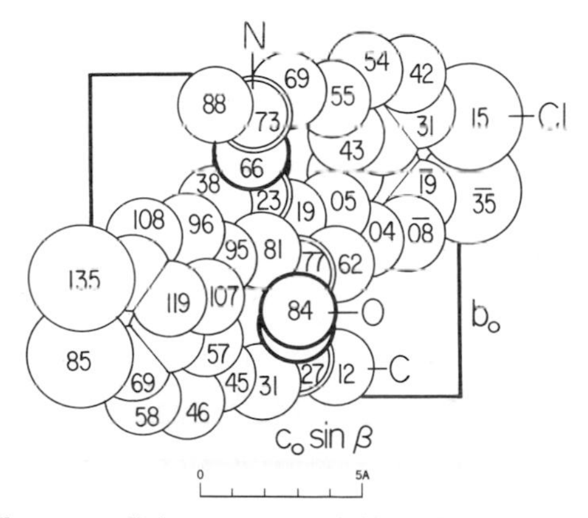

Fig. XVAIII,12. The monoclinic structure of *N*-methyl-*p*-chlorobenzaldoxime projected along its a_0 axis.

All atoms are in the positions

$$(4e) \quad \pm(xyz;\, x+{}^1/_2,{}^1/_2-y,z)$$

of the space group C_{2h}^5 ($P2_1/a$). The parameters are those of Table XVAIII,6.

TABLE XVAIII,6
Parameters of the Atoms in *N*-Methyl-*p*-chlorobenzaldoxime

Atom	x	y	z
Cl(1)	0.8464(4)	0.1341(2)	0.0410(2)
C(2)	0.6914(10)	0.1122(6)	0.8936(5)
C(3)	0.5834(11)	−0.0011(6)	0.8675(6)
C(4)	0.4617(10)	−0.0198(6)	0.7521(5)
C(5)	0.4499(8)	0.0780(5)	0.6621(5)
C(6)	0.5675(9)	0.1907(6)	0.6930(5)
C(7)	0.6883(10)	0.2081(6)	0.8093(6)
C(8)	0.3133(9)	0.0468(6)	0.5452(5)
N(9)	0.2685(8)	0.1280(4)	0.4522(4)
O(10)	0.3439(7)	0.2434(4)	0.4480(4)
C(11)	0.1196(11)	0.0849(7)	0.3397(6)
H(3)	0.604	−0.065	0.934
H(4)	0.361	−0.097	0.724
H(6)	0.553	0.257	0.629
H(7)	0.777	0.285	0.827
H(8)	0.244	−0.053	0.536
H(11)	0.075	−0.013	0.337
H(11′)	0.175	0.100	0.268
H(11″)	0.016	0.155	0.327

Molecules in the structure that results (Fig. XVAIII,12) have the bond dimensions of Figure XVAIII,13. They are nearly planar, the C(8) atom

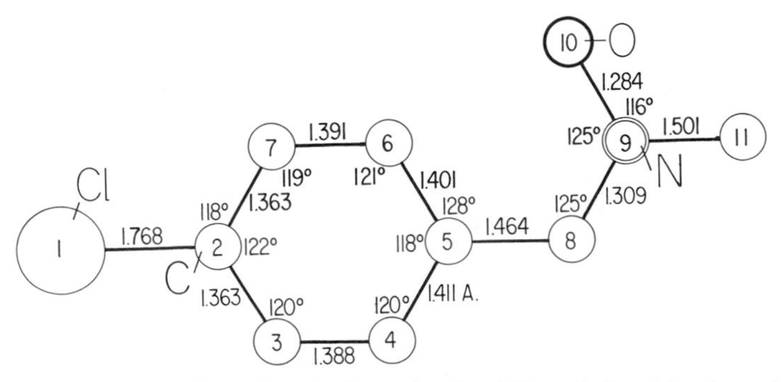

Fig. XVAIII,13. Bond dimensions in the molecule of *N*-methyl-*p*-chlorobenzaldoxime.

being only 0.04 A. and the N(9), O(10), and C(11) atoms about 0.2 A. out of the plane of the rest.

XV,aIII,11. The symmetry of *α,5(2′-chloroethoxy)o-quinone-2-oxime*, $ClC_2H_4OC_6H_4(O)NO$, is monoclinic. Its tetramolecular unit has the dimensions:

$$a_0 = 4.01(1) \text{ A.}; \quad b_0 = 10.87(3) \text{ A.}; \quad c_0 = 20.08(5) \text{ A.}$$
$$\beta = 103°36(10)' \qquad (-180°\text{C.})$$
$$a_0 = 4.12(12) \text{ A.}; \quad b_0 = 10.83(2) \text{ A.}; \quad c_0 = 20.35(4) \text{ A.}$$
$$\beta = 104°48(10)' \qquad (20°\text{C.})$$

The space group is C_{2h}^5 ($P2_1/c$) with all atoms in the positions:

$$(4e) \quad \pm(xyz;\ x,{}^1/_2-y,z+{}^1/_2)$$

The parameters determined at the lower temperature are stated in Table XVAIII,7.

TABLE XVAIII,7

Parameters of the Atoms in $ClC_2H_4OC_6H_4(O)NO$

Atom	x	y	z
C(1)	−0.1138	−0.0564	0.1292
C(2)	0.1423	−0.0540	0.1871
C(3)	0.2736	0.0607	0.2171
C(4)	0.1156	0.1757	0.1819
C(5)	−0.1598	0.1664	0.1218
C(6)	−0.2749	0.0555	0.0969
C(7)	−0.0953	−0.2746	0.1218
C(8)	−0.2556	−0.3741	0.0729
O(1)	−0.2536	−0.1584	0.0967
O(2)	0.1311	0.3795	0.1718
O(3)	0.5086	0.0673	0.2706
N	0.2680	0.2772	0.2074
Cl	−0.1812	−0.3491	−0.0107
H(1)	−0.268	0.245	0.102
H(2)	−0.475	0.044	0.055
H(3)	0.218	−0.128	0.206
H(4)	0.172	−0.274	0.128
H(5)	−0.147	−0.295	0.166
H(6)	−0.525	−0.378	0.063
H(7)	−0.166	−0.464	0.091
H(8)	0.303	0.428	0.192

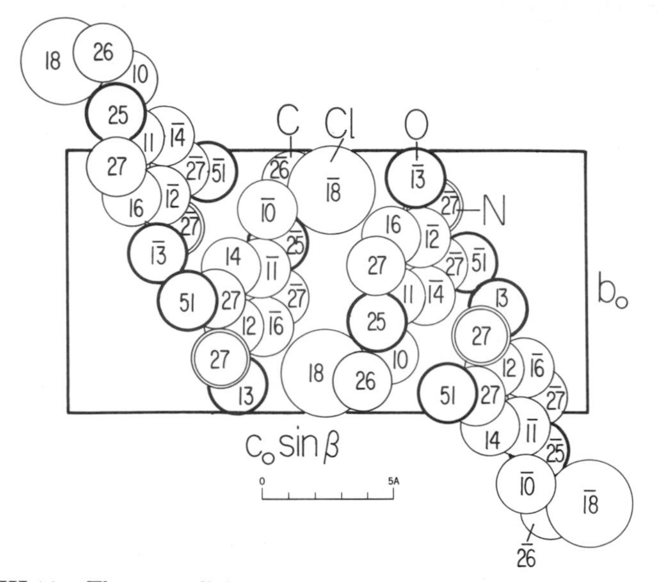

Fig. XVAIII,14. The monoclinic structure of α 5(2′-chloroethoxy)*o*-quinone-2-oxime projected along its a_0 axis.

The resulting structure is shown in Figure XVAIII,14. Its molecules have the bond dimensions of Figure XVAIII,15. Except for the NO and C_2H_4Cl radicals they are planar. For these non-coplanar atoms, N is 0.08 A. and O(2) is −0.24 A. outside the plane; on the other side of the molecule, C(7) is −0.15 A., C(8), − 0.34 A., and Cl, 1.79 A. away from the plane. There is a short O(2)–O(3) = 2.61 A. between molecules; all other intermolecular distances are 3.12 A. or more.

XV,aIII,12. Crystals of *p-chloronitrobenzene*, $ClC_6H_4NO_2$, have been given a partially disordered structure. The bimolecular monoclinic unit has the dimensions:

$$a_0 = 3.84(1) \text{ A.}; \quad b_0 = 6.80(1) \text{ A.}; \quad c_0 = 13.35(2) \text{ A.}$$
$$\beta = 97°31(5)'$$

The space group is C_{2h}^5 ($P2_1/c$) with atoms in the positions:

$$(4e) \quad \pm(xyz;\ x,{}^1/_2-y,z+{}^1/_2)$$

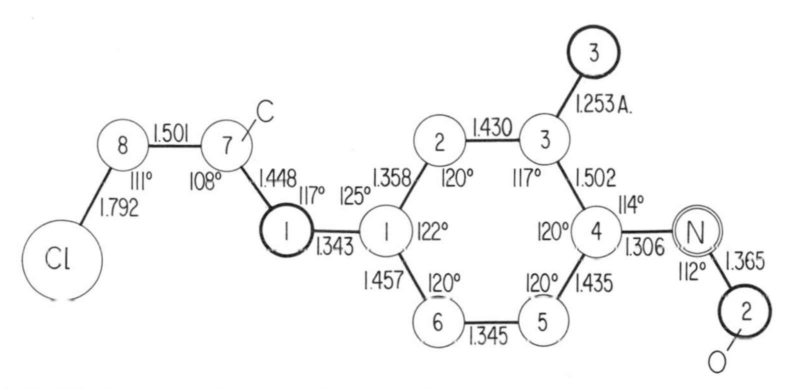

Fig. XVAIII,15. Bond dimensions in the molecule of α,5(2′-chloroethoxy)*o*-quinone-2-oxime.

It is considered that the data point to an arrangement in which the carbon atoms have the parameters:

Atom	x	y	z
C(1)	0.167	0.153	0.064
C(2)	0.110	0.169	−0.038
C(3)	−0.065	0.017	0.102

The ring thus defined has a center of symmetry but there is statistical disorder in the attachment of chlorine and NO_2 to C(1) or C(1′). The half-atomic positions that express this disorientation have the parameters:

Atom	x	y	z
$^1/_2$Cl	0.376	0.340	0.142
$^1/_2$N	0.343	0.312	0.130
$^1/_2$O(1)	0.404	0.277	0.213
$^1/_2$O(2)	0.425	0.444	0.097

This is the type of disorder first found many years ago for *p*-ClC_6H_4Br (**XV,aIII,1**).

XV,aIII,13. Crystals of *m-bromonitrobenzene,* $BrC_6H_4NO_2$, are orthorhombic with a tetramolecular cell of the edge lengths:

$$a_0 = 5.92 \text{ A.}; \quad b_0 = 21.52 \text{ A.}; \quad c_0 = 5.34 \text{ A.}$$

The space group has been chosen as C_{2v}^9 in the orientation $Pbn2_1$. Atoms therefore are in the positions:

$$(4a) \quad xyz;\ \bar{x},\bar{y},z+1/2;\ x+1/2,1/2-y,z+1/2;\ 1/2-x,y+1/2,z$$

with, according to a brief note, the parameters of Table XVAIII,8.

TABLE XVAIII,8

Parameters of the Atoms in m-$BrC_6H_4NO_2$

Atom	x	y	z
Br	0.0412	0.2370	0.0000
C(1)	0.1872	0.1707	0.1625
C(2)	0.0760	0.1440	0.3659
C(3)	0.1812	0.0937	0.4766
C(4)	0.3972	0.0720	0.4137
C(5)	0.5034	0.1004	0.2081
C(6)	0.3988	0.1499	0.0985
N	0.0650	0.0651	0.6875
O(1)	−0.1207	0.0828	0.7607
O(2)	0.1462	0.0230	0.8086

The structure is shown in Figure XVAIII,16. Its molecules are approximately planar. In them Br–C = 1.88(5) A.; the other atomic separations agree with expectations. Between molecules there are relatively short Br–C distances of 3.35 and 3.40 A.

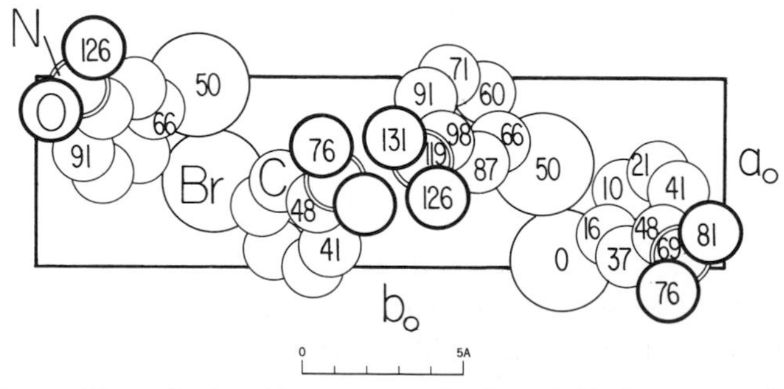

Fig. XVAIII,16. The orthorhombic structure of m-bromonitrobenzene projected along its c_0 axis

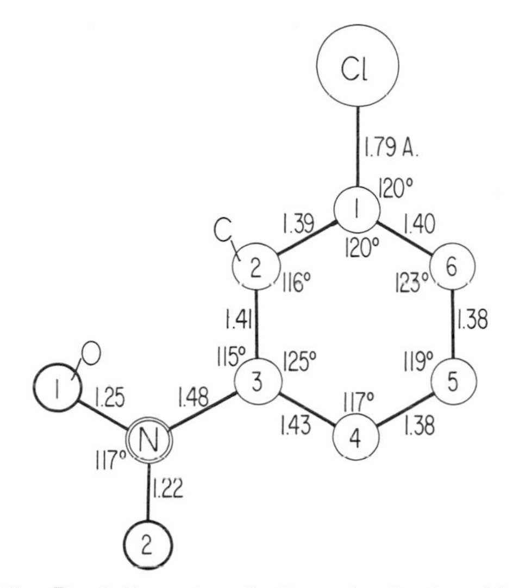

Fig. XVAIII,17. Bond dimensions in the molecule of *m*-chloronitrobenzene.

In the isostructural *m-chloronitrobenzene,* $ClC_6H_4NO_2$, the unit has the edge lengths:

$$a_0 = 6.00 \text{ A.}; \quad b_0 = 21.40 \text{ A.}; \quad c_0 = 5.35 \text{ A.}$$

Determined parameters are those of Table XVAIII,9. Molecules of this crystal, which are planar to within 0.05 A., have the bond dimensions of Figure XVAIII,17. The shortest intermolecular distances are N–O = 3.02 A., O–O = 3.32 and 3.36 A., C–Cl = 3.37 A., and C–O = 3.38 A.

TABLE XVAIII,9

Parameters of the Atoms in $m\text{-}ClC_6H_4NO_2$

Atom	x	y	z
Cl	0.070	0.237	0.000
C(1)	0.187	0.172	0.163
C(2)	0.076	0.146	0.366
C(3)	0.181	0.094	0.477
C(4)	0.398	0.072	0.414
C(5)	0.498	0.099	0.208
C(6)	0.401	0.151	0.099
N	0.064	0.064	0.688
O(1)	−0.120	0.084	0.761
O(2)	0.150	0.022	0.809

XV,aIII,14. Several studies have been made of the structure of *m-dinitrobenzene*, $C_6H_4(NO_2)_2$. Its symmetry is orthorhombic with a tetramolecular cell of the edge lengths:

$$a_0 = 13.257 \text{ A.}; \quad b_0 = 14.048 \text{ A.}; \quad c_0 = 3.806 \text{ A.}$$

The space group is C_{2v}^9 in the orientation $Pbn2_1$. All atoms are in the positions:

$$(4a) \quad xyz;\ \bar{x},\bar{y},z+{}^1/_2;\ x+{}^1/_2,{}^1/_2-y,z+{}^1/_2;\ {}^1/_2-x,y+{}^1/_2,z$$

As most recently determined, the parameters are those listed in Table XVAIII,10; for hydrogen they are approximate.

TABLE XVAIII,10

Parameters of the Atoms in *m*-Dinitrobenzene

Atom	x	y	z
C(1)	0.1399	0.3652	−0.0298
C(2)	0.1800	0.4493	−0.1473
C(3)	0.2828	0.4576	−0.1198
C(4)	0.3450	0.3860	−0.0037
C(5)	0.3000	0.3042	0.1257
C(6)	0.1971	0.2934	0.1101
N(1)	0.0288	0.3531	−0.0416
N(2)	0.3297	0.5469	−0.2506
O(1)	−0.0067	0.2850	0.1268
O(2)	−0.0200	0.4108	−0.1986
O(3)	0.2756	0.6009	−0.4133
O(4)	0.4161	0.5627	−0.1512
H(2)	0.139	0.489	−0.360
H(4)	0.409	0.399	0.029
H(5)	0.326	0.246	0.080
H(6)	0.164	0.239	0.141

The resulting arrangement (Fig. XVAIII,18) is composed of molecules having the bond dimensions of Figure XVAIII,19. The planes of their NO_2 groups are each turned through 13° with respect to the plane of the benzene ring; consequently, within the limit of accuracy of the determination, the molecule has a plane of symmetry, through C(2) and C(5), normal to the plane of the ring.

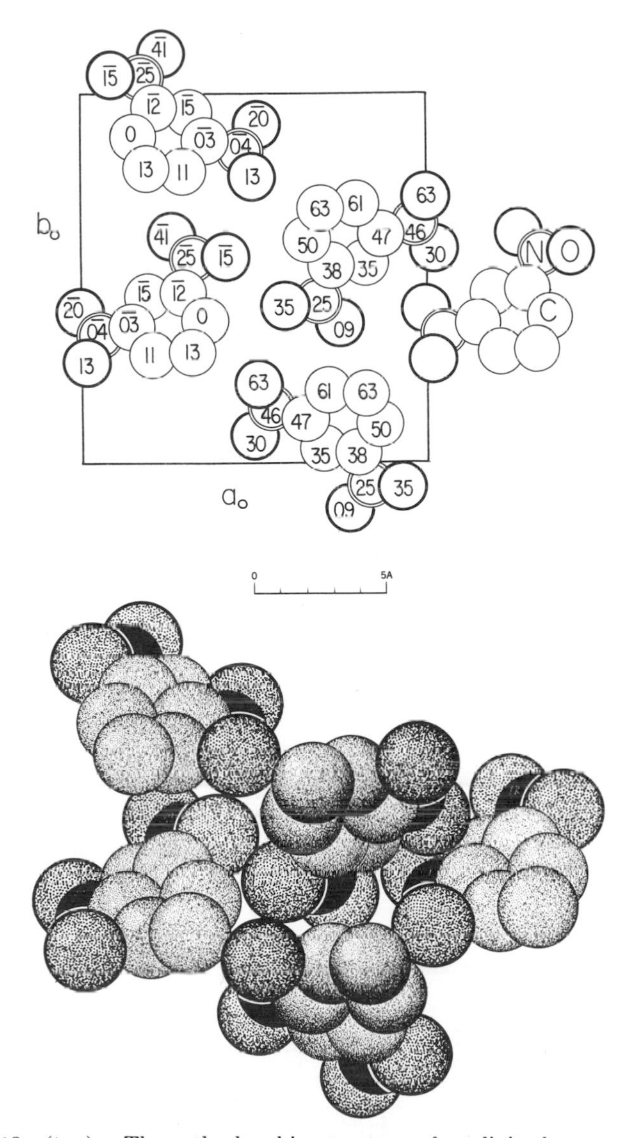

Fig. XVAIII,18a (top). The orthorhombic structure of *m*-dinitrobenzene projected along its c_0 axis.

Fig. XVAIII,18b (bottom). A packing drawing of the orthorhombic structure of *m*-dinitrobenzene viewed along its c_0 axis. The nitrogen atoms are black; the oxygens are coarsely dot shaded and heavily outlined. The atoms of carbon are finely dotted.

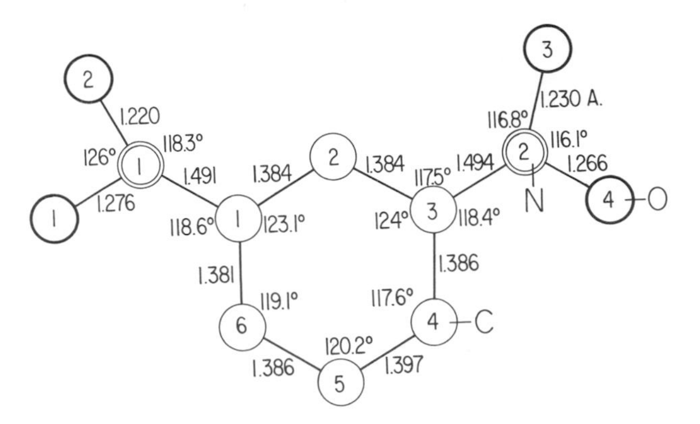

Fig. XVAIII,19. Bond dimensions in the molecule of *m*-dinitrobenzene.

XV,aIII,15. The monoclinic crystals of *p-dinitrobenzene,* $C_6H_4(NO_2)_2$, have a bimolecular unit of the dimensions:

$$a_0 = 11.05 \text{ A.}; \quad b_0 = 5.42 \text{ A.}; \quad c_0 = 5.65 \text{ A.}; \quad \beta = 92°18'$$

The space group is C_{2h}^5 ($P2_1/n$) with atoms in the positions:

$$(4e) \quad \pm(xyz;\ x+1/2, 1/2-y, z+1/2)$$

A recent reworking of the original data has yielded the parameters of Table XVAIII,11.

TABLE XVAIII,11

Parameters of the Atoms in *p*-Dinitrobenzene

Atom	x	y	z
C(1)	0.0688	0.1819	0.0885
C(2)	0.0317	−0.1177	−0.2042
C(3)	0.1006	0.0725	−0.1165
N	0.1440	0.3903	0.1858
O(1)	0.2171	0.4719	0.0587
O(2)	0.1231	0.4624	0.3811

The structure, shown in Figure XVAIII,20, has molecules with the bond dimensions of Figure XVAIII,21. In them the planes of the NO_2 radicals are twisted by about 9° with respect to the plane of the benzene ring.

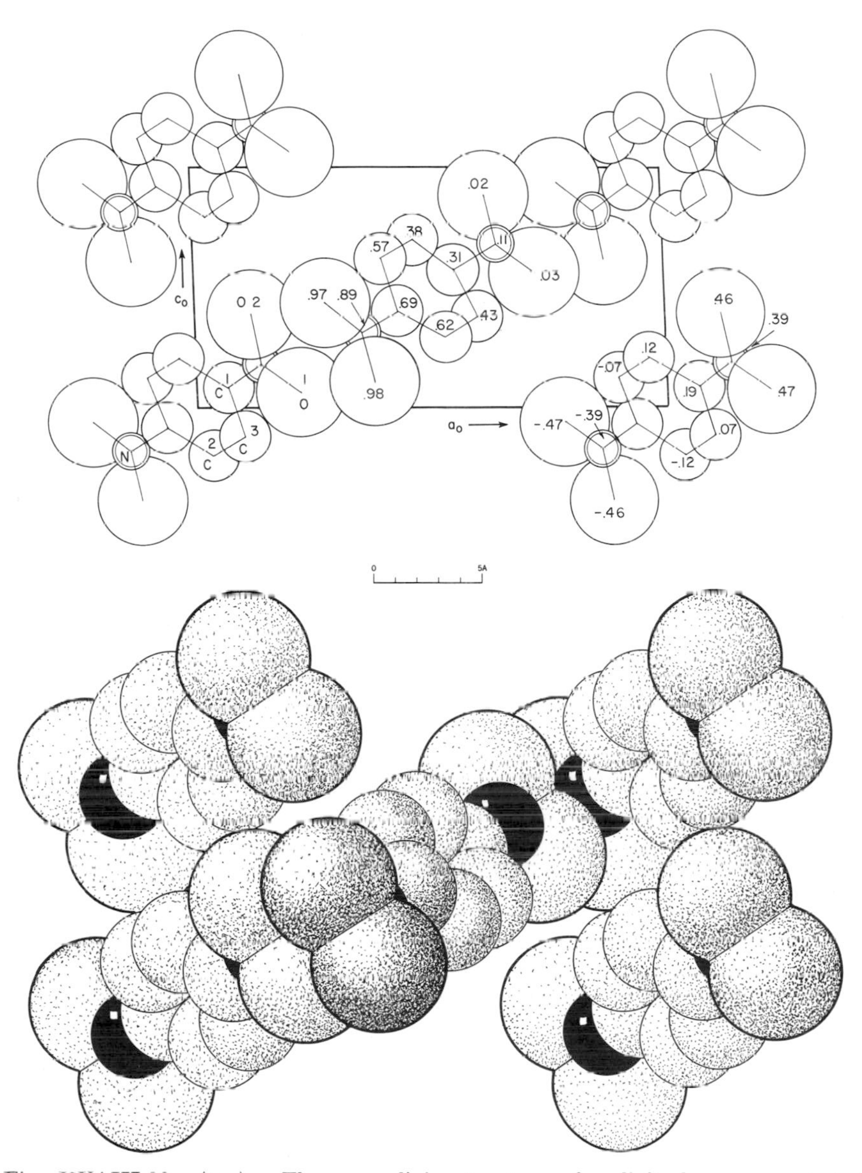

Fig. XVAIII,20a (top). The monoclinic structure of *p*-dinitrobenzene projected along its b_0 axis. Left hand axes.

Fig. XVAIII,20b (bottom). A packing drawing of the monoclinic structure of *p*-dinitrobenzene seen along its b_0 axis. The nitrogen atoms are black, the atoms of oxygen are larger than those of carbon and heavily outlined: both are dotted.

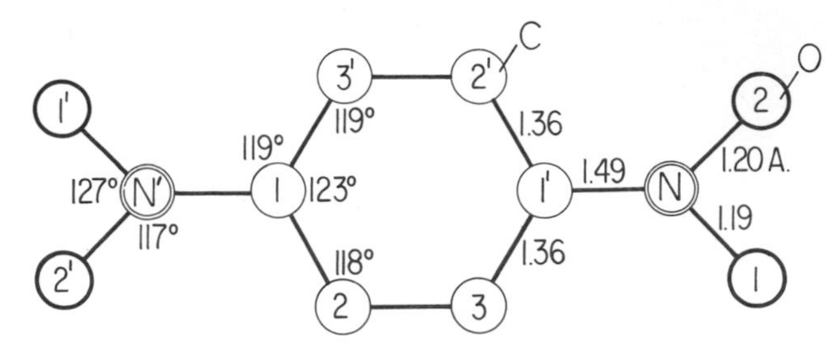

Fig. XVAIII,21. Bond dimensions in the molecule of *p*-dinitrobenzene.

XV,aIII,16. The orthorhombic crystals of *p-nitrophenyl trinitride*, $C_6H_4(NO_2)(N_3)$, possess a tetramolecular cell of the edge lengths:

$$a_0 = 18.05(2) \text{ A.}; \quad b_0 = 10.29(1) \text{ A.}; \quad c_0 = 3.73(1) \text{ A.}$$

The space group is V^4 ($P2_12_12_1$) with all atoms in the positions:

$$(4a) \quad xyz;\ {}^1/_2-x,\bar{y},z+{}^1/_2;\ x+{}^1/_2,{}^1/_2-y,\bar{z};\ \bar{x},y+{}^1/_2,{}^1/_2-z$$

The chosen parameters are stated in Table XVAIII,12.

TABLE XVAIII,12

Parameters of the Atoms in *p*-Nitrophenyl Trinitride

Atom	x	y	z
C(1)	0.4659(8)	0.1729(13)	0.5377(49)
C(2)	0.4199(8)	0.0647(13)	0.5190(50)
C(3)	0.3487(7)	0.0776(11)	0.6566(47)
C(4)	0.3247(7)	0.1949(12)	0.7886(46)
C(5)	0.3715(7)	0.3016(12)	0.7950(47)
C(6)	0.4429(7)	0.2907(12)	0.6681(51)
N(1)	0.5397(6)	0.1614(10)	0.3893(42)
N(2)	0.2507(6)	0.1959(10)	0.9153(40)
N(3)	0.2308(6)	0.3006(10)	0.0675(41)
N(4)	0.2068(7)	0.3878(11)	0.2085(45)
O(1)	0.5607(5)	0.0565(10)	0.2688(37)
O(2)	0.5801(5)	0.2601(9)	0.3767(37)
H(1)	0.441	−0.016	0.424
H(2)	0.316	0.002	0.647
H(3)	0.349	0.392	0.907
H(4)	0.478	0.377	0.679

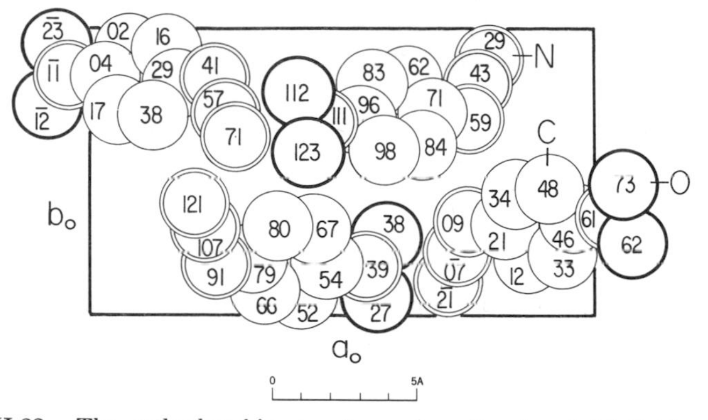

Fig. XVAIII,22. The orthorhombic structure of *p*-nitrophenyl trinitride projected along its c_0 axis.

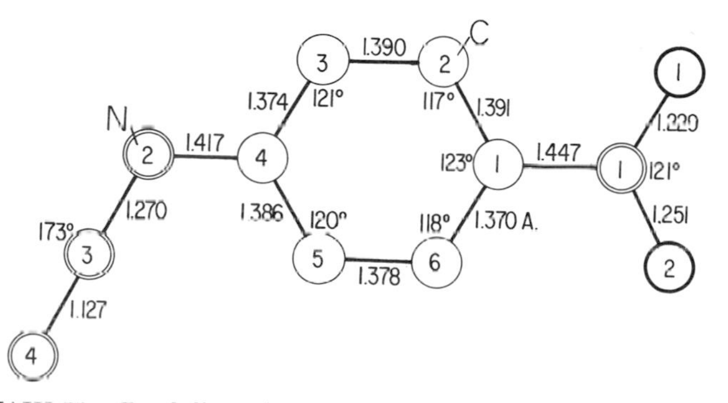

Fig. XVAIII,23. Bond dimensions in the molecule of *p*-nitrophenyl trinitride.

In the resulting structure (Fig. XVAIII,22) molecules have the bond dimensions of Figure XVAIII,23. In contrast to the inorganic trinitrides (Chapter VI) the N–N distances in the trinitride radical as determined here are not equal and the atoms are not exactly collinear. There is a short C O between molecules of 3.18 A. and N–O separations of 3.23 A. upwards.

XV,aIII,17. Crystals of the stable modification of *o-nitrobenzaldehyde*, $C_6H_4(NO_2)CHO$, have been examined through both x-ray and neutron diffraction. They are monoclinic with a bimolecular unit of the dimensions:

$$a_0 = 11.37 \text{ A.};\quad b_0 = 3.960 \text{ A.};\quad c_0 = 7.57 \text{ A.};\quad \beta = 90°11'$$

Atoms are in the positions

$$(2a) \quad xyz;\ \bar{x},y+{}^1/_2,\bar{z}$$

of the space group C_2^2 ($P2_1$). Parameters found with x-rays are given in Table XVAIII,13. The neutron data, providing accurate parameters for the hydrogen atoms, are also stated in the table.

TABLE XVAIII,13

Parameters of the Atoms in $C_6H_4(NO_2)CHO$

Atom	x	y	z
		X-ray data	
O(1)	0.4452(5)	0.659(2)	0.1179(8)
O(2)	0.5147(4)	0.303(3)	0.3119(7)
O(3)	0.1793(4)	0.565(2)	−0.1466(6)
N	0.4336(5)	0.433(2)	0.2274(7)
C(1)	0.3150(5)	0.301(2)	0.2594(8)
C(2)	0.2304(4)	0.326(2)	0.1285(6)
C(3)	0.1190(4)	0.213(2)	0.1681(6)
C(4)	0.0931(5)	0.081(2)	0.3336(10)
C(5)	0.1788(6)	0.060(−)	0.4605(10)
C(6)	0.2918(5)	0.174(3)	0.4254(9)
C(7)	0.2533(6)	0.433(3)	−0.0569(9)
		Neutron data	
H(3)	0.0541(14)	0.210(14)	0.0651(20)
H(4)	0.0066(16)	−0.028(17)	0.3606(25)
H(5)	0.1637(13)	−0.031(20)	0.5919(21)
H(6)	0.3596(16)	0.168(18)	0.5276(24)
H(7)	0.3419(16)	0.366(18)	−0.1090(21)

The crystal structure is shown in Figure XVAIII,24. Bond dimensions of its molecule are those of Figure XVAIII,25. The benzene part is planar but the nitro and aldehyde groups are each turned from this plane by about 30°. There is no evidence that in this compound the aldehyde part is involved in internal hydrogen bonding.

XV,aIII,18. *Potassium p-nitrophenyl dicyanomethide,* $KC_6H_4(NO_2)$-$C(CN)_2$, is orthorhombic with a tetramolecular unit of the dimensions:

$$a_0 = 21.749(1)\text{ A.};\quad b_0 = 11.416(1)\text{ A.};\quad c_0 = 3.771\text{ A.}$$

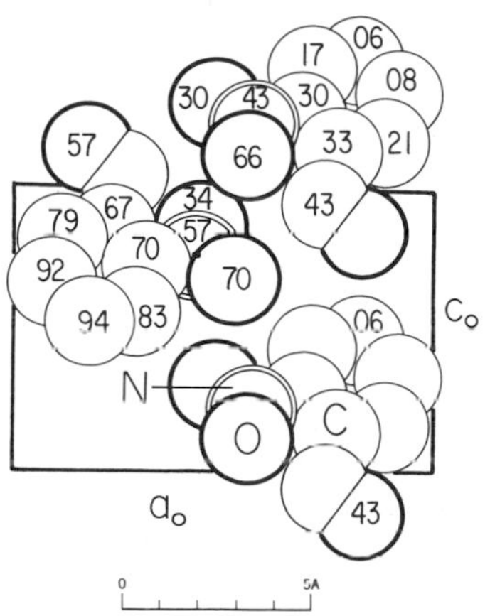

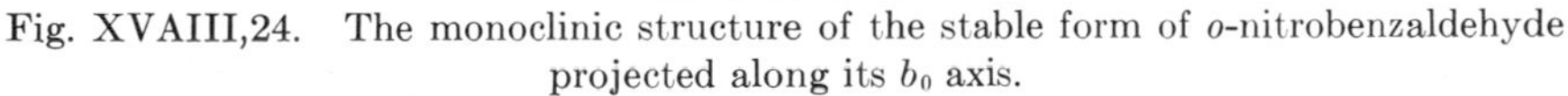

Fig. XVAIII,24. The monoclinic structure of the stable form of *o*-nitrobenzaldehyde projected along its b_0 axis.

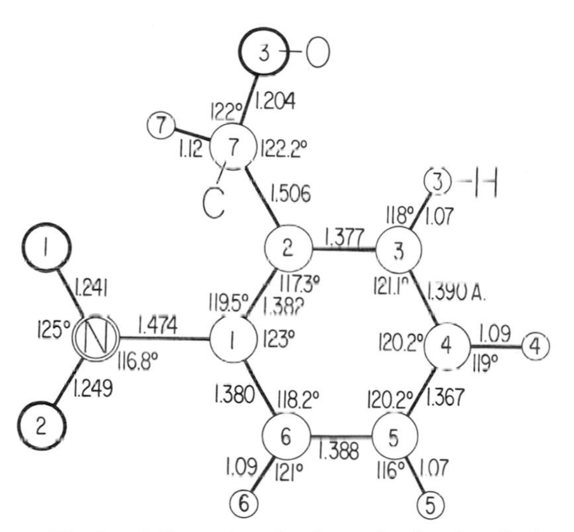

Fig. XVAIII,25. The bond dimensions in the molecule of *o*-nitrobenzaldehyde.

Atoms are in the positions

$$(4a) \quad xyz;\ {}^1/_2-x,\bar{y},z+{}^1/_2;\ x+{}^1/_2,{}^1/_2-y,\bar{z};\ \bar{x},y+{}^1/_2,{}^1/_2-z$$

of the space group V^4 ($P2_12_12_1$). The determined parameters are listed in Table XVAIII,14.

TABLE XVAIII,14

Parameters of the Atoms in $KC_6H_4(NO_2)C(CN)_2$

Atom	x	y	z
O(1)	0.7083(1)	0.3962(2)	−0.2488(8)
O(2)	0.7796(1)	0.2871(2)	−0.0216(10)
N(1)	0.7238(1)	0.3039(3)	−0.0987(11)
N(2)	0.4362(1)	−0.0168(3)	0.0497(10)
N(3)	0.5843(1)	−0.2390(2)	0.4830(13)
C(1)	0.6803(2)	0.2177(3)	−0.0117(12)
C(2)	0.6183(2)	0.2370(3)	−0.0983(12)
C(3)	0.5765(6)	0.1508(3)	−0.0197(13)
C(4)	0.5935(2)	0.0441(3)	0.1340(10)
C(5)	0.6569(2)	0.0281(3)	0.2132(12)
C(6)	0.6990(2)	0.1143(3)	0.1458(10)
C(7)	0.5500(2)	−0.0459(3)	0.2083(10)
C(8)	0.4868(2)	−0.0314(3)	0.1235(11)
C(9)	0.5677(2)	−0.1538(3) [a]	0.3612(10)
K	0.3563(<1)	0.0995(1)	0.5412(2)
H(1)	0.602(1)	0.312(2)	−0.195(8)
H(2)	0.535(1)	0.169(2)	−0.057(10)
H(3)	0.670(2)	0.042(3)	0.316(10)
H(4)	0.744(1)	0.111(3)	0.231(9)

[a] Apparently this y parameter should have the negative value shown here rather than that in the original article.

The structure is shown in Figure XVAIII,26. Its anions, having the bond dimensions of Figure XVAIII,27, are planar except for O(1) and hydrogen atoms which are as much as 0.12 A. outside the plane of the rest. The potassium cations are surrounded by three oxygen and four cyanide nitrogen atoms at distances between 2.783 and 2.995 A.

XV,aIII,19. More recent studies have altered the structure initially given *p-nitroaniline*, $NH_2C_6H_4NO_2$. The symmetry is monoclinic with a tetramolecular unit of the dimensions:

$$a_0 = 12.336(8) \text{ A.}; \quad b_0 = 6.07(2) \text{ A.}; \quad c_0 = 8.592(5) \text{ A.}$$
$$\beta = 91.45(5)^\circ$$

The space group is C_{2h}^5 in the orientation $P2_1/n$. Atoms therefore are in the positions:

$$(4e) \quad \pm(xyz;\ x+1/2,1/2-y,z+1/2)$$

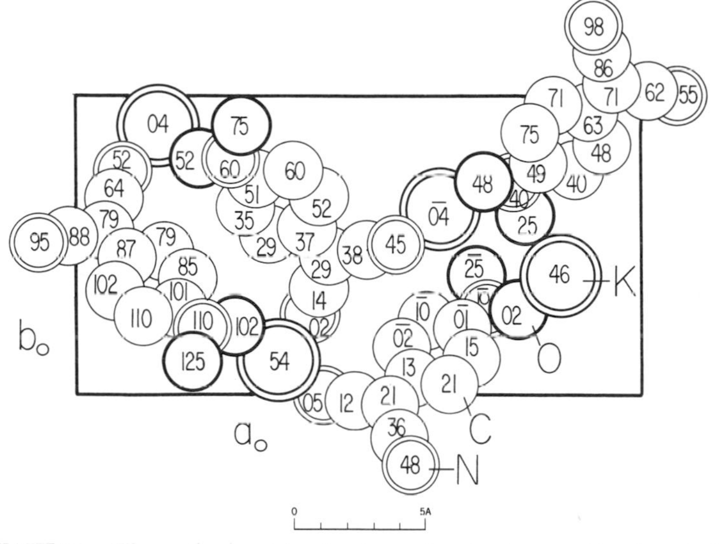

Fig. XVAIII,26. The orthorhombic structure of potassium *p*-nitrophenyl dicyanomethide projected along its c_0 axis.

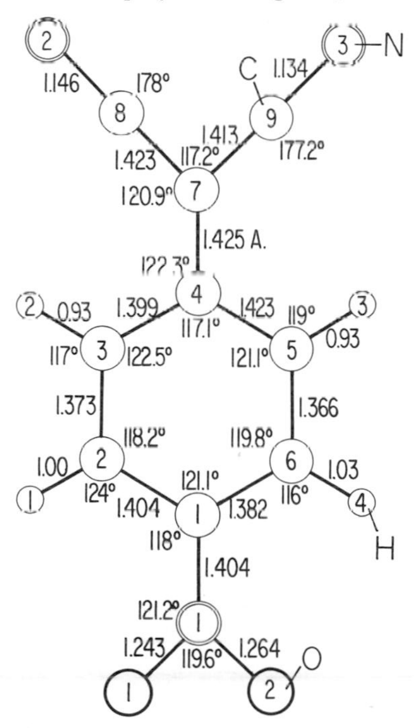

Fig. XVAIII,27. Bond dimensions in the molecular anion of potassium *p*-nitrophenyl dicyanomethide.

The parameters, as refined, are those of Table XVAIII,15.

TABLE XVAIII,15

Parameters of the Atoms in *p*-Nitroaniline

Atom	x	y	z
C(1)	0.7293(3)	0.4158(9)	0.5251(5)
C(2)	0.6781(3)	0.3063(8)	0.6472(5)
C(3)	0.5946(3)	0.4044(8)	0.7243(5)
C(4)	0.5608(3)	0.6136(8)	0.6800(4)
C(5)	0.6087(3)	0.7240(8)	0.5575(4)
C(6)	0.6927(3)	0.6279(8)	0.4813(5)
N(1)	0.8127(3)	0.3219(8)	0.4502(5)
N(2)	0.4730(3)	0.7217(7)	0.7600(4)
O(1)	0.4295(3)	0.6199(6)	0.8649(4)
O(2)	0.4466(3)	0.9093(6)	0.7209(4)
H(1)	0.695(5)	0.158(10)	0.673(7)
H(2)	0.564(5)	0.353(10)	0.798(7)
H(3)	0.591(5)	0.853(10)	0.535(7)
H(4)	0.724(5)	0.695(10)	0.411(7)
H(5)	0.825(6)	0.168(12)	0.466(8)
H(6)	0.855(6)	0.374(12)	0.401(8)

The structure is shown in Figure XVAIII,28. Its molecules may be thought of as tied, by N–H–O bonds, into strings in planes diagonal to the a_0 and c_0 axes; the lengths of these bonds are 3.07 and 3.14 A. Bond dimensions within the molecules are given in Figure XVAIII,29. Their benzene rings are strictly planar, with the nitrogen atoms 0.02 A. and one of the oxygen atoms 0.06 A. outside this plane.

XV,aIII,20. Crystals of *N,N-dimethyl-p-nitroaniline*, $N(CH_3)_2C_6H_4$-(NO_2), are monoclinic with a bimolecular unit of the dimensions:

$$a_0 = 9.73(1) \text{ A.}; \quad b_0 = 10.56(1) \text{ A.}; \quad c_0 = 3.964(5) \text{ A.}$$
$$\beta = 91°28(5)'$$

Atoms are in the positions

$$(2a) \quad xyz; \bar{x},y+{}^1/_2,\bar{z}$$

of the space group C_2^2 ($P2_1$). Parameters are listed in Table XVAIII,16.

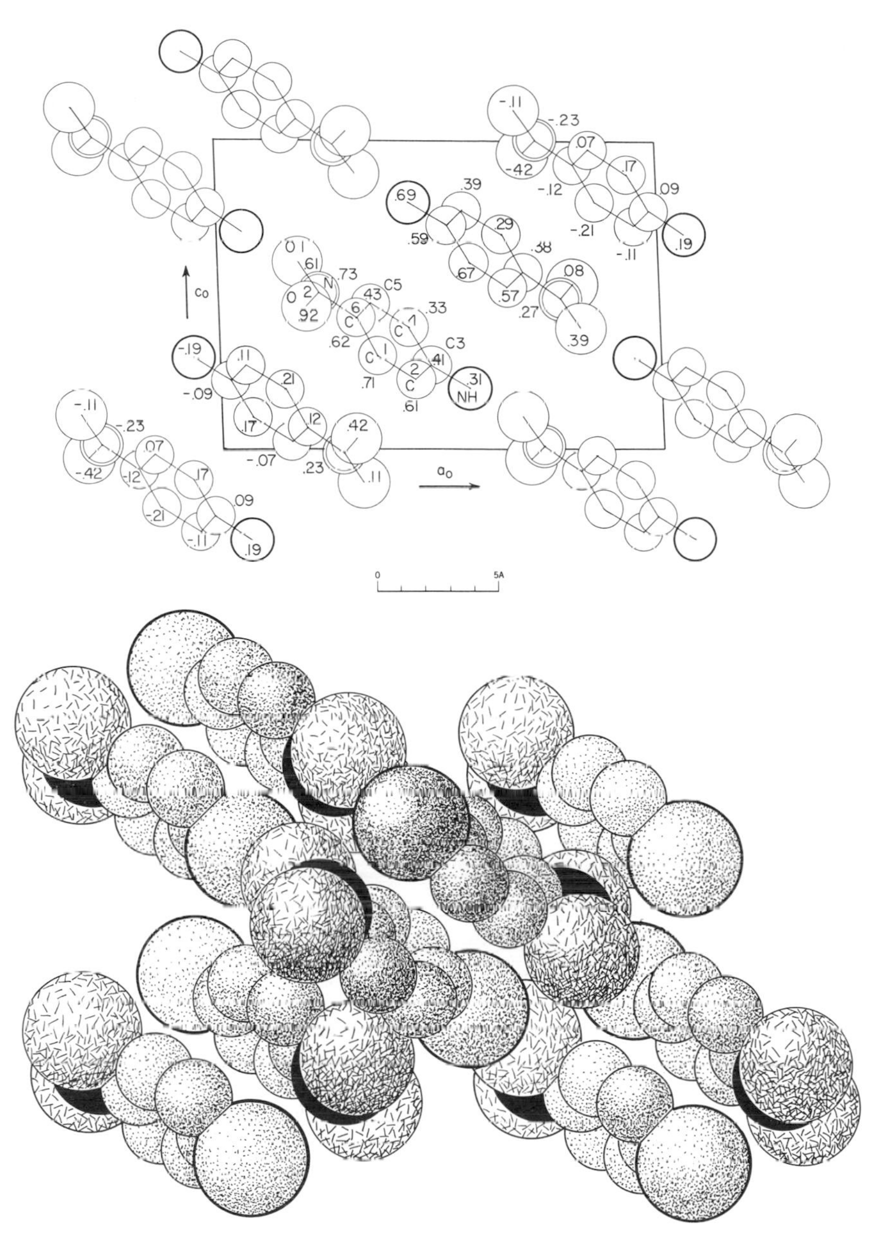

Fig. XVAIII,28a (top). The monoclinic structure of *p*-nitroaniline projected along its b_0 axis. Left hand axes.

Fig. XVAIII, 28b (bottom). A packing drawing of the monoclinic *p*-nitroaniline structure viewed along its b_0 axis. The nitro nitrogens are black; the amino nitrogens are large, dotted and heavily outlined. Oxygens are large and line shaded; the carbon atoms are small and dotted.

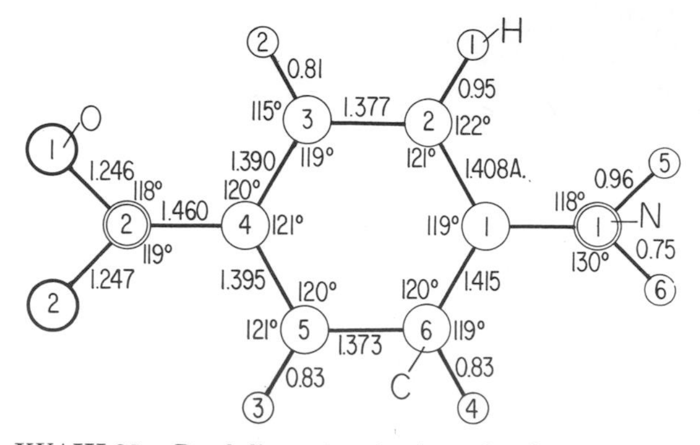

Fig. XVAIII,29. Bond dimensions in the molecule of *p*-nitroaniline.

TABLE XVAIII,16

Parameters of the Atoms in $N(CH_3)_2C_6H_4(NO_2)$

Atom	x	y	z
C(1)	0.3159	0.2490	0.1321
C(2)	0.3744	0.1463	0.3106
C(3)	0.5095	0.1466	0.4127
C(4)	0.5944	0.2473	0.3351
C(5)	0.5429	0.3535	0.1580
C(6)	0.4056	0.3558	0.0625
C(7)	0.0944	0.1354	0.0886
C(8)	0.1153	0.3644	−0.1057
N(9)	0.1815	0.2527	0.0369
N(10)	0.7324	0.2517	0.4429
O(11)	0.7827	0.1562	0.5924
O(12)	0.8014	0.3431	0.3853
H(13)	0.302	0.054	0.332
H(14)	0.541	0.053	0.563
H(15)	0.584	0.428	0.060
H(16)	0.368	0.447	−0.070

The structure, shown in Figure XVAIII,30, is composed of molecules having the bond dimensions of Figure XVAIII,31. They are planar except for the $N(CH_3)_2$ and NO_2 groups which are twisted about their C–N attachments by 7.3 and 2.8°. The amino nitrogen is in the plane of the ring but the nitro nitrogen lies about 0.04 A. to one side of it. Intermolecular separations are normal.

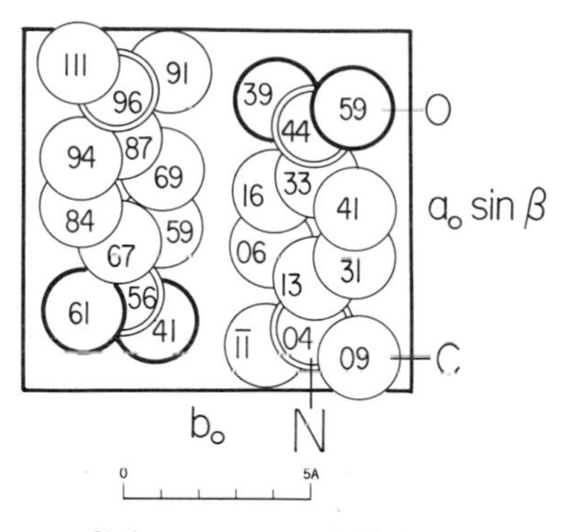

Fig. XVAIII,30. The monoclinic structure of N,N-dimethyl-p-nitroaniline projected along its c_0 axis.

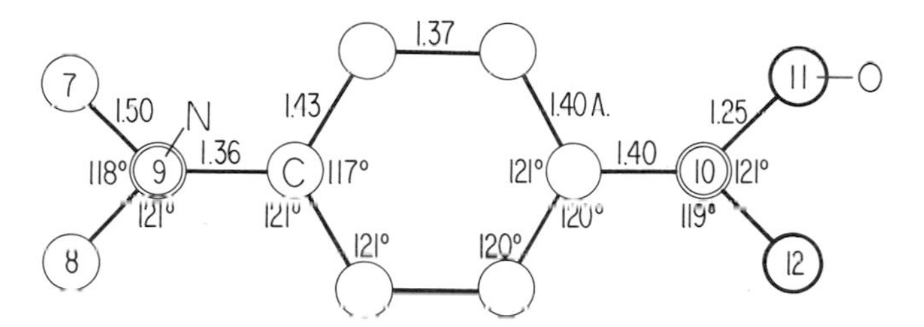

Fig. XVAIII,31. Bond dimensions in the molecule of N,N-dimethyl-p-nitroaniline.

XV,aIII,21. Crystals of what proved to be *bis(p-iodo-N,N-dimethylanilino) hydrochlorido triiodido*, $[IC_6H_4N(CH_3)_2]_2HCl \cdot HI_3$, have monoclinic symmetry. Their tetramolecular unit has the dimensions:

$$a_0 = 16.253(4) \text{ A.}; \quad b_0 = 20.607(3) \text{ A.}; \quad c_0 = 8.283(2) \text{ A.}$$
$$\beta = 109.28(2)°$$

Atoms were put in the following positions of C_{2h}^6 $(C2/c)$:

$$\text{I(3)}: (4c) \quad {}^1/_4\, {}^1/_4\, 0;\ {}^3/_4\, {}^1/_4\, {}^1/_2;\ {}^3/_4\, {}^3/_4\, 0;\ {}^1/_4\, {}^3/_4\, {}^1/_2$$
$$\text{Cl}: (4e) \quad \pm(0\, u\, {}^1/_4;\ {}^1/_2, u+{}^1/_2, {}^1/_4) \qquad \text{with } u = 0.4516$$

The other atoms, in

$$(8f) \quad \pm(xyz;\ x,\bar{y},z+{}^1/_2;\ x+{}^1/_2,y+{}^1/_2,z;\ x+{}^1/_2,{}^1/_2-y,z+{}^1/_2)$$

have the parameters of Table XVAIII,17.

TABLE XVAIII,17

Parameters of the Atoms in $[IC_6H_4N(CH_3)_2]_2HCl \cdot HI_3$

Atom	x	y	z
I(1)	0.0896(2)	0.2460(2)	0.0849(5)
I(2)	0.0831(2)	0.1179(2)	0.5899(5)
C(1)	0.176(3)	0.047(3)	0.719(7)
C(2)	0.253(4)	0.064(3)	0.822(8)
C(3)	0.319(3)	0.014(3)	0.893(7)
C(4)	0.293(4)	0.047(3)	0.355(7)
C(5)	0.212(4)	0.066(3)	0.252(8)
C(6)	0.148(4)	0.021(3)	0.165(8)
C(7)	0.415(3)	0.114(2)	0.304(7)
C(8)	0.326(3)	0.169(2)	0.496(8)
N	0.359(3)	0.100(2)	0.429(6)

This structure (Fig. XVAIII,32) appears as an assemblage of organic cations and Cl^- and I_3^- anions. Choosing C_{2h}^6 as space group makes the

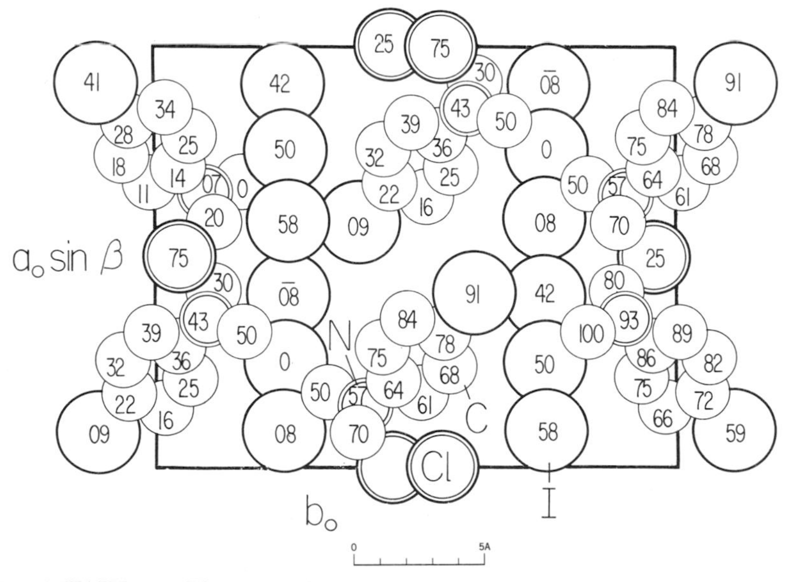

Fig. XVAIII,32. The monoclinic structure of bis(p-iodo-N,N-dimethylaniline) hydrochloride triiodide projected along its c_0 axis. Only six of the eight organic cations per cell are shown; the other two would be located in the center of the drawing, more or less completely obscuring the two that appear there.

I_3^- anion necessarily centrosymmetric (I–I = 2.917 A.). If a group of lower symmetry were selected, an asymmetric anion would be possible but calculations on that basis did not lead to better general agreement with the data.

XV,aIII,22. Three independent determinations have recently been made of the structure of *p-chloroaniline*, $ClC_6H_4NH_2$. Expressing the results in terms of the same set of axes, the determined atomic positions are in close agreement. The symmetry is orthorhombic. When the axial sequence is chosen as

$$a_0 = 8.666 \text{ A.}; \quad b_0 = 7.397 \text{ A.}; \quad c_0 = 9.281 \text{ A.}$$

the space group is V_h^{16} in the orientation *Pnma*. The atoms of the four molecules in this cell are in the positions:

(4*c*) $\pm(u\ {}^1/_4\ v;\ u+{}^1/_2,{}^1/_4,{}^1/_2-v)$

(8*d*) $\pm(xyz;\ x+{}^1/_2,{}^1/_2-y,{}^1/_2-z;\ x,{}^1/_2-y,z;\ x+{}^1/_2,y,{}^1/_2-z)$

The three sets of parameters, referred to this cell, are listed for comparison in Table XVAIII,18. Assuming, as stated in this table, that z(N) in 1966: T,W&Z should be − 0.2049 (instead of the 0.0249 of the original), the agreement is good.

TABLE XVAIII,18

Parameters of the Atoms in *p*-Chloroaniline

Atom	Position	x P	x T,W&Z	x S	y	z P	z T,W&Z	z S
C(1)	(4*c*)	0.6197	0.6168	0.6200	$^1/_4$	0.8941	0.8952	0.8867
C(2)[a]	(8*d*)	0.6828	0.6835	0.6839	—[a]	0.9438	0.9439	0.9441
C(3)[b]	(8*d*)	0.8090	0.8068	0.8066	—[b]	0.0399	0.0390	0.0411
C(4)	(4*c*)	0.8674	0.8676	0.8728	$^1/_4$	0.0860	0.0871	0.0862
N	(4*c*)	0.4984	0.4990	0.4916	$^1/_4$	0.7946	0.7951	0.7975
Cl	(4*c*)	0.0254	0.0256	0.0262	$^1/_4$	0.2037	0.2034	0.2033
H(3)	(8*d*)	—	−0.159	—	0.571	—	0.052	—
H(2)	(8*d*)	—	−0.353	—	0.547	—	−0.089	—
N	(8*d*)	—	−0.535	—	0.368	—	0.775	—

[a] For C(2): y = 0.4134, 0.4139, 0.4141.
[b] For C(3): y = 0.4122, 0.4137, 0.4147.

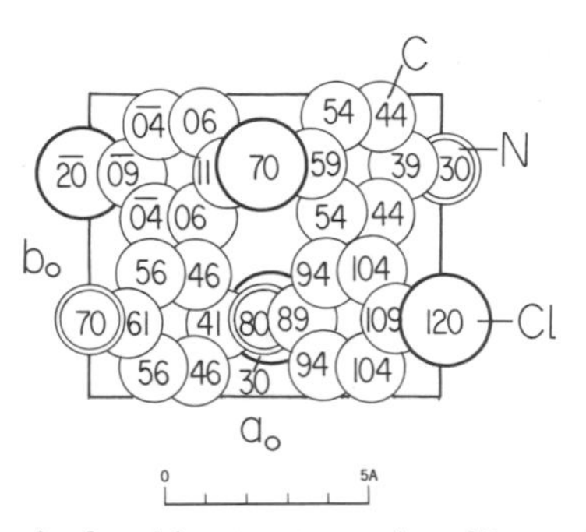

Fig. XVAIII,33. The orthorhombic structure of *p*-chloroaniline projected along its c_0 axis.

The structure is illustrated in Figure XVAIII,33. Bond dimensions are normal though, because it would be difficult to select the most accurate of the three sets of parameters, no specific values are stated.

XV,aIII,23. According to a preliminary note, the structure originally given *p-toluidine*, $CH_3C_6H_4NH_2$, is in error. Its orthorhombic crystals have a unit which contains eight molecules in a cell of the edge lengths:

$$a_0 = 9.02 \text{ A.}; \quad b_0 = 6.00 \text{ A.}; \quad c_0 = 23.31 \text{ A.}$$

The crystals are piezoelectric and the space group is C_{2v}^9 ($Pna2_1$). Atoms therefore are in the positions:

$$(4a) \quad xyz;\ \bar{x},\bar{y},z+{}^1/_2;\ {}^1/_2-x,y+{}^1/_2,z+{}^1/_2;\ x+{}^1/_2,{}^1/_2-y,z$$

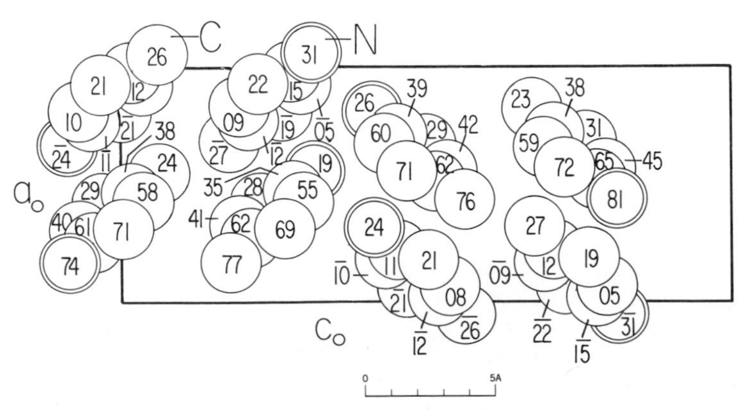

Fig. XVAIII,34. The orthorhombic structure of *p*-toluidine projected along its b_0 axis.

The assigned parameters are those of Table XVAIII,19.

TABLE XVAIII,19

Parameters of the Atoms in *p*-Toluidine

	x		y		z	
Atom	Mol I	Mol II	Mol I	Mol II	Mol I	Mol II
C(1)	−0.223	−0.215	−0.114	−0.121	−0.052	0.208
C(2)	−0.182	−0.169	−0.206	−0.190	0.000	0.263
C(3)	−0.087	−0.074	−0.084	−0.048	0.035	0.295
C(4)	−0.038	−0.024	0.124	0.154	0.018	0.277
C(5)	−0.086	−0.074	0.210	0.224	−0.035	0.223
C(6)	−0.188	−0.169	0.098	0.094	−0.069	0.191
N	−0.322	0.062	−0.237	0.306	−0.088	0.313
C(8)	0.055	−0.321	0.256	−0.273	0.060	0.175

According to this new determination, the structure is that shown in Figure XVAIII,34. It contains two crystallographically different molecules in which, in contrast to the single molecule of the original determination, the C–N bonds have their usual length (1.43 A.).

XV,aIII,24. The structure described for *cobaltous chloride di p-toluidine*, $CoCl_2 \cdot 2NH_2C_6H_4CH_3$, is monoclinic with a tetramolecular unit of the dimensions:

$$a_0 = 12.32(5) \text{ A.}; \quad b_0 = 4.60(1) \text{ A.}; \quad c_0 = 26.15(10) \text{ A.}$$
$$\beta = 93°45'$$

The cobalt atoms are in

$$(4e) \quad \pm(^1/_4\ u\ 0;\ ^3/_4,u+^1/_2,^1/_2) \qquad \text{with } u = 0.384$$

of C_{2h}^6 in the orientation $I2/a$. The other atoms, in

$$(8f) \quad \pm(xyz;\ x+^1/_2,\bar{y},z); \quad \text{B.C.}$$

have the parameters of Table XVAIII,20.

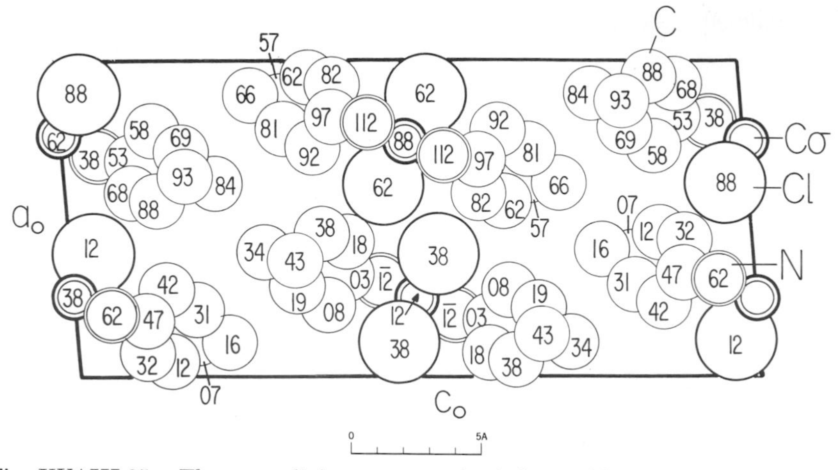

Fig. XVAIII,35. The monoclinic structure of cobaltous chloride di-p-toluidine projected along its b_0 axis.

TABLE XVAIII,20

Parameters of Atoms in $CoCl_2 \cdot 2NH_2C_6H_4CH_3$

Atom	x	y	z
Cl	0.386	0.116	0.034
N	0.184	0.616	0.054
C(1)	0.170	0.470	0.103
C(2)	0.260	0.419	0.133
C(3)	0.205	0.306	0.175
C(4)	0.125	0.066	0.180
C(5)	0.042	0.116	0.135
C(6)	0.073	0.320	0.100
C(7)	0.095	0.155	0.220

In this structure (Fig. XVAIII,35) the cobalt atoms are tetrahedrally surrounded by two chlorines (Co–Cl = 2.26 A.) and two nitrogens (Co–N = 1.95 A.).

XV,aIII,25. The monoclinic crystals of *bis(p-toluidinium)hexachlororhenate(IV)*, $(CH_3C_6H_4NH_3)_2ReCl_6$, have a tetramolecular unit of the dimensions:

$$a_0 = 7.01(2) \text{ A.}; \quad b_0 = 25.04(5) \text{ A.}; \quad c_0 = 11.54(2) \text{ A.}$$
$$\beta = 90°0(6)'$$

All atoms are in the positions

$$(4e) \quad \pm(xyz;\ x,{}^1/_2-y,z+{}^1/_2)$$

of the space group C_{2h}^5 ($P2_1/c$). Parameters are stated in Table XVAIII,21.

TABLE XVAIII,21

Parameters of the Atoms in $(CH_3C_6H_4NH_3)_2ReCl_6$

Atom	x	y	z
Re	0.2736(2)	0.02819(4)	0.2532(1)
Cl(1)	0.2737(16)	−0.0339(3)	0.0999(7)
Cl(2)	0.2625(14)	0.0880(3)	0.4096(6)
Cl(3)	0.0204(15)	−0.0198(3)	0.3411(7)
Cl(4)	0.5306(16)	0.0748(3)	0.1729(7)
Cl(5)	0.4961(13)	−0.0252(3)	0.3526(6)
Cl(6)	0.0537(14)	0.0796(3)	0.1491(6)
N(1)	−0.252(4)	0.078(1)	0.431(2)
N(2)	0.250(4)	0.437(1)	0.385(2)
C(1)	−0.229(5)	0.135(1)	0.460(3)
C(2)	−0.157(6)	0.171(1)	0.378(3)
C(3)	−0.133(6)	0.223(1)	0.400(3)
C(4)	−0.179(6)	0.246(1)	0.503(3)
C(5)	−0.268(7)	0.208(2)	0.592(4)
C(6)	−0.270(6)	0.153(2)	0.567(3)
C(7)	−0.160(7)	0.303(2)	0.538(4)
C(8)	0.266(5)	0.379(1)	0.349(3)
C(9)	0.334(5)	0.344(1)	0.430(3)
C(10)	0.362(6)	0.293(1)	0.395(3)
C(11)	0.307(6)	0.275(2)	0.287(3)
C(12)	0.242(7)	0.312(2)	0.206(4)
C(13)	0.219(5)	0.364(1)	0.239(3)
C(14)	0.338(6)	0.217(2)	0.249(3)

In this ionic structure (Fig. XVAIII,36) there are two crystallographically independent toluidinium cations. These essentially planar cations have their expected ring dimensions as well as CH_3–C = 1.47 and 1.54 A., and C–NH_3 = 1.48 and 1.52 A. The $ReCl_6^{2-}$ anions are octahedra in which the angles Cl–Re–Cl lie between 88.1 and 91.8° and Re–Cl = 2.34–2.37 A.

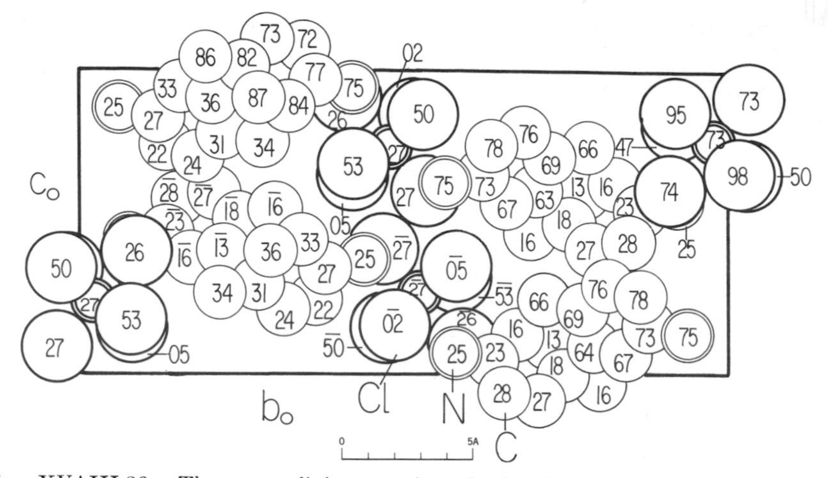

Fig. XVAIII,36. The monoclinic, pseudo-orthorhombic structure of bis(p-toluidinium)hexachlororhenate projected along its a_0 axis. Atoms of rhenium are the small doubly ringed circles.

XV,aIII,26. The monoclinic complex *tricarbonyl chromium-o-toluidine*, $Cr(CO)_3 \cdot C_6H_4(CH_3)NH_2$, has an eight-molecule unit of the dimensions:

$$a_0 = 19.25 \text{ A.}; \quad b_0 = 8.32 \text{ A.}; \quad c_0 = 13.80 \text{ A.}; \quad \beta = 107°0'$$

Its space group is C_{2h}^6 ($C2/c$) with atoms in the positions:

$$(8f) \quad \pm(xyz;\ x,\bar{y},z+{}^1/_2;\ x+{}^1/_2,y+{}^1/_2,z;\ x+{}^1/_2,{}^1/_2-y,z+{}^1/_2)$$

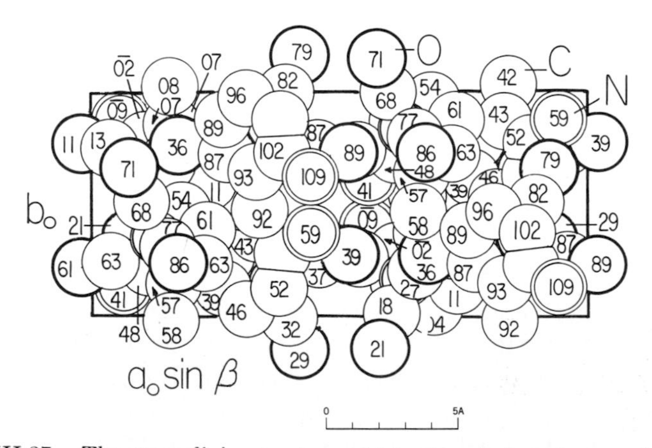

Fig. XVAIII,37. The monoclinic structure of tricarbonyl chromium-*o*-toluidine projected along its c_0 axis. The atoms of chromium scarcely show.

Parameters are listed in Table XVAIII,22.

TABLE XVAIII,22

Parameters of the Atoms in $Cr(CO)_3 \cdot C_6H_4(CH_3)NH_2$

Atom	x	y	z
C(1)	0.3824	0.2740	0.5175
C(2)	0.3324	0.3590	0.4347
C(3)	0.2729	0.2719	0.3748
C(4)	0.2626	0.1110	0.3850
C(5)	0.3137	0.0236	0.4568
C(6)	0.3743	0.1091	0.5200
N(7)	0.4402	0.3566	0.5855
C(8)	0.3412	0.5343	0.4251
C(9)	0.3984	−0.0326	0.3169
O(10)	0.4185	−0.1576	0.2934
C(11)	0.4595	0.2287	0.3726
O(12)	0.5185	0.2740	0.3858
C(13)	0.3431	0.2088	0.2285
O(14)	0.3247	0.2498	0.1411
Cr	0.3703	0.1499	0.3585

The toluidine molecule in the structure (Fig. XVAIII,37) has its expected dimensions. The $Cr(CO)_3$ portion is the usual triangular pyramid with chromium at the apex and Cr–O = 1.76–1.80 A. and C–O = 1.16 1.20 A.; opposite the oxygens the chromium is approximately centered over the benzene ring of toluidine.

XV,aIII,27. Crystals of *aniline p-thiocyanate*, $NH_2C_6H_4SCN$, are monoclinic with a tetramolecular unit of the dimensions:

$$a_0 = 12.35 \text{ A.}; \quad b_0 = 4.40 \text{ A.}; \quad c_0 = 18.94 \text{ A.}$$
$$\beta = 135°8'$$

Atoms are in the positions

$$(4e) \quad \pm(xyz;\ x,{}^1/_2-y,z+{}^1/_2)$$

of C_{2h}^5 ($P2_1/c$). The parameters are given in Table XVAIII,23.

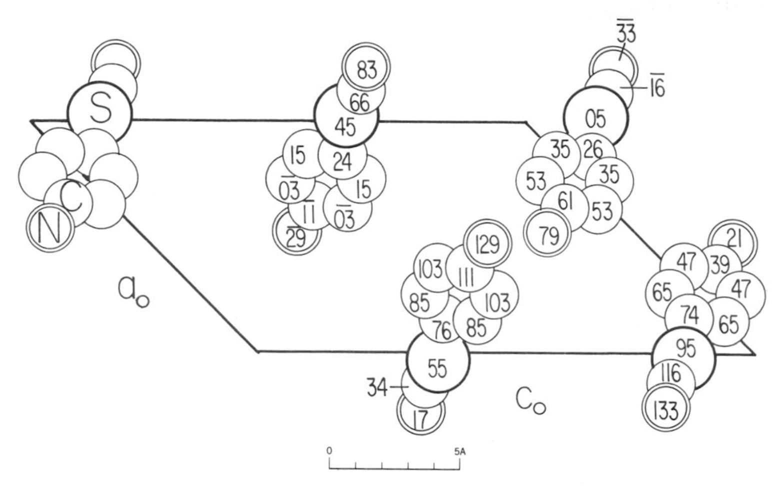

Fig. XVAIII,38. The monoclinic arrangement found for crystals of aniline-*p*-thiocyanate projected along its b_0 axis.

TABLE XVAIII,23

Parameters of the Atoms in *p*-$NH_2C_6H_4SCN$

Atom	x	y	z
S	0.995	0.55	0.341
C(1)	0.130	0.76	0.434
C(2)	0.132	0.85	0.505
C(3)	0.247	0.03	0.587
C(4)	0.359	0.11	0.594
C(5)	0.365	0.03	0.528
C(6)	0.252	0.85	0.447
C(7)	0.850	0.34	0.267
N(1)	0.738	0.17	0.209
N(2)	0.472	0.29	0.674

The structure thus defined is illustrated in Figure XVAIII,38.

XV,aIII,28. According to a preliminary announcement, crystals of *p-aminobenzamide*, $C_6H_4(NH_2)CONH_2$, are monoclinic with a bimolecular unit of the dimensions:

$$a_0 = 8.42(1) \text{ A.}; \quad b_0 = 5.29(1) \text{ A.}; \quad c_0 = 7.91(1) \text{ A.}$$
$$\beta = 108°0(10)'$$

Atoms are in the positions

$$(2a) \quad xyz;\ \bar{x},y+{}^1/_2,\bar{z}$$

of the low symmetry C_2^2 ($P2_1$). The parameters are listed in Table XVAIII,24.

TABLE XVAIII,24

Parameters of the Atoms in *p*-Aminobenzamide

Atom	*x*	*y*	*z*
C(1)	0.570	0.205	0.278[a]
C(2)	0.510	0.000	0.165
C(3)	0.340	−0.055	0.110
C(4)	0.225	0.100	0.170
C(5)	0.290	0.305	0.275
C(6)	0.460	0.365	0.335
C(7)	0.755	0.260	0.345
N(1)	0.810	0.490	0.395
N(2)	0.055	0.045	0.115
O	0.850	0.075	0.360

[a] The correct parameters, stated here, have been provided through the courtesy of Professor Wyart.

The structure, shown in Figure XVAIII,39, is composed of molecules bound together along the a_0 direction by hydrogen bonds involving the oxygen and both the nitrogen atoms.

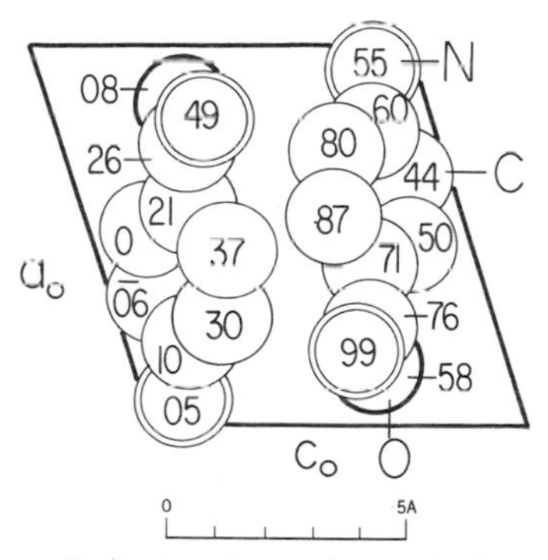

Fig. XVAIII,39. The monoclinic structure of *p*-aminobenzamide projected along its b_0 axis.

XV,aIII,29. The monoclinic crystals of *m-methyl benzamide*, $C_6H_4(CH_3)$-$CONH_2$, have a tetramolecular unit of the dimensions:

$$a_0 = 8.93(3) \text{ A.}; \quad b_0 = 16.02(3) \text{ A.}; \quad c_0 = 5.12(2) \text{ A.}$$
$$\beta = 95°0(36)'$$

The space group is C_{2h}^5 ($P2_1/a$) with atoms in the positions:

$$(4e) \quad \pm(xyz; x+{}^1/_2,{}^1/_2-y,z)$$

The determined parameters are listed in Table XVAIII,25; values assigned to the hydrogen atoms can be found in the original paper.

TABLE XVAIII,25

Parameters of the Atoms in $C_6H_4(CH_3)CONH_2$

Atom	x	y	z
C(1)	0.354	0.207	0.423
C(2)	0.418	0.266	0.595
C(3)	0.387	0.350	0.565
C(4)	0.285	0.375	0.360
C(5)	0.219	0.316	0.195
C(6)	0.254	0.233	0.224
C(7)	0.392	0.116	0.458
C(8)	0.464	0.416	0.745
N	0.386	0.068	0.254
O	0.433	0.089	0.681

Fig. XVAIII,40. The monoclinic structure of m-methyl benzamide projected along its c_0 axis.

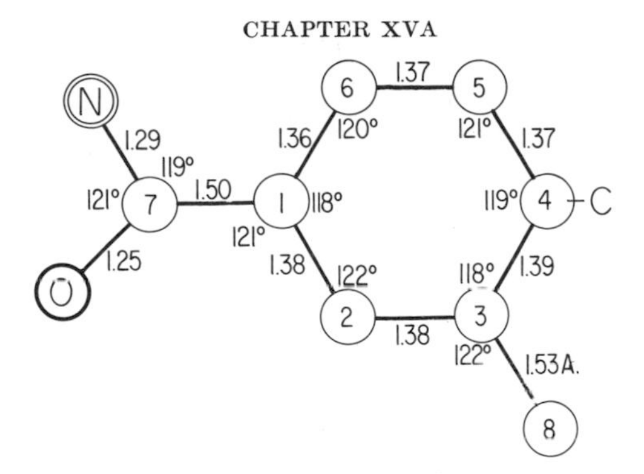

Fig. XVAIII,41. Bond dimensions in the molecule of *m*-methyl benzamide.

The molecules in this structure (Fig. XVAIII,40) have the bond dimensions of Figure XVAIII,41. In them the methyl C(8) atom is 0.06 A. outside the plane set by the benzene ring and C(7). Between this plane and that through the amide group there is an angle of 28.7°. The molecules themselves are connected by N–H–O hydrogen bonds of lengths 2.97 and 3.22 A.

XV,aIII,30. The orthorhombic crystals of *p-chloroacetanilide*, ClC_6H_4-$NH \cdot COCH_3$, possess a tetramolecular unit of the edge lengths:

$$a_0 = 9.73 \text{ A.}; \quad b_0 = 12.88 \text{ A.}; \quad c_0 = 6.56 \text{ A.}$$

Atoms are in the positions

$$(4a) \quad xyz;\ x,\bar{y},z+1/2;\ 1/2-x,y+1/2,z+1/2;\ x+1/2,1/2-y,z$$

of the space group C_{2v}^9 ($Pna2_1$). Parameters are listed in Table XVAIII,26.

TABLE XVAIII,26

Parameters of the Atoms in *p*-Chloroacetanilide

Atom	x	y	z
Cl	0.2586	0.5391	−0.2500
O	0.4392	0.2596	0.5788
N	0.2151	0.2922	0.4913
C(1)	0.2508	0.4666	−0.0340
C(2)	0.1197	0.4340	0.0271
C(3)	0.1119	0.3752	0.2001
C(4)	0.2341	0.3514	0.3259
C(5)	0.3655	0.3816	0.2387
C(6)	0.3736	0.4403	0.0576
C(7)	0.3180	0.2472	0.6172
C(8)	0.2575	0.1858	0.8090

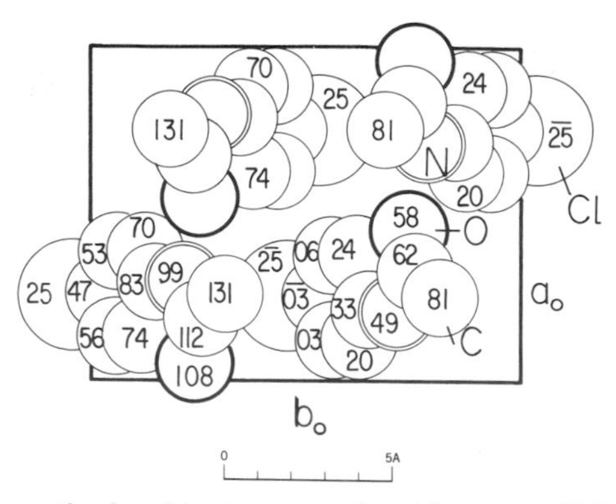

Fig. XVAIII,42. The orthorhombic structure of *p*-chloroacetanilide projected along its c_0 axis.

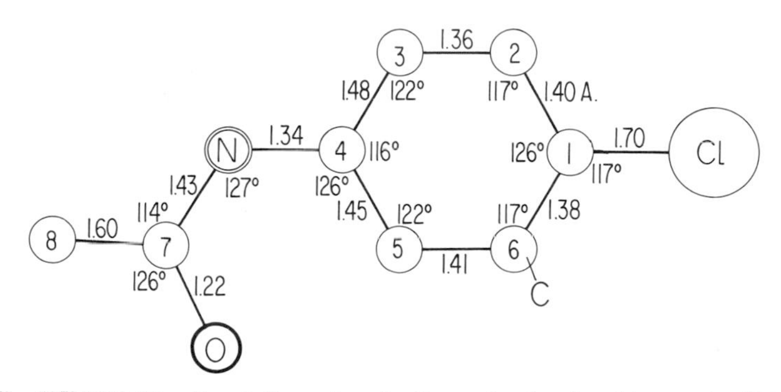

Fig. XVAIII,43. Bond dimensions in the molecule of *p*-chloroacetanilide.

The structure, shown in Figure XVAIII,42, is composed of molecules having the bond dimensions of Figure XVAIII,43. All atoms except those of the acetyl group are coplanar to within about 0.06 A. This group and its attached nitrogen atom also are planar, the angle between it and the chlorobenzyl part of the molecule being only 5°. Molecules are tied together in strings by N–H . . . O bonds of length 2.83 A.

XV,aIII,31. Crystals of *7,7,8,8-tetracyanoquinodimethane*, C_6H_4-$[C(CN)_2]_2$, are monoclinic with a tetramolecular unit of the dimensions:

$$a_0 = 8.906(6) \text{ A.}; \quad b_0 = 7.060(4) \text{ A.}; \quad c_0 = 16.395(5) \text{ A.}$$
$$\beta = 98°32(2)'$$

All atoms are in the positions

$$(8f) \quad \pm(xyz;\ x,\bar{y},z+{}^1/_2;\ x+{}^1/_2,y+{}^1/_2,z;\ x+{}^1/_2,{}^1/_2-y,z+{}^1/_2)$$

of the space group C_{2h}^6 ($C2/c$). Their parameters are stated in Table XVAIII,27.

TABLE XVAIII,27

Parameters of the Atoms in $C_6H_4[C(CN)_2]_2$

Atom	x	y	z
C(1)	−0.0341(2)	0.0778(3)	0.0741(1)
C(2)	0.1497(2)	−0.0253(3)	−0.0162(1)
C(3)	0.1214(2)	0.0550(2)	0.0611(1)
C(4)	0.2393(2)	0.1077(3)	0.1204(1)
C(5)	0.2148(2)	0.1877(3)	0.1979(1)
C(6)	0.3950(2)	0.0840(3)	0.1091(1)
N(1)	0.1960(2)	0.2524(4)	0.2592(1)
N(2)	0.5179(2)	0.0648(3)	0.0996(1)
H(1)	−0.0577(34)	0.1312(45)	0.1216(20)
H(2)	0.2447(31)	−0.0295(37)	−0.0253(16)

The structure (Fig. XVAIII,44) is made up of molecules that are planar to within 0.01 A. and have the bond dimensions of Figure XVAIII,45. The shortest intermolecular distances are N–C = 3.18 A. and N–N = 3.35 A.

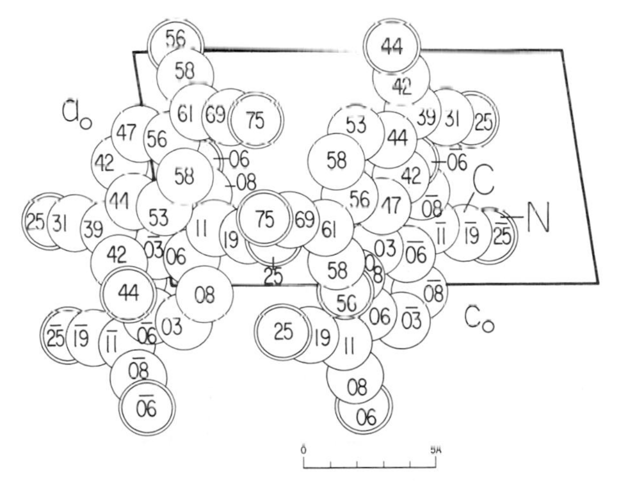

Fig. XVAIII,44. The monoclinic structure of 7,7,8,8 tetracyanoquinodimethane projected along its b_0 axis.

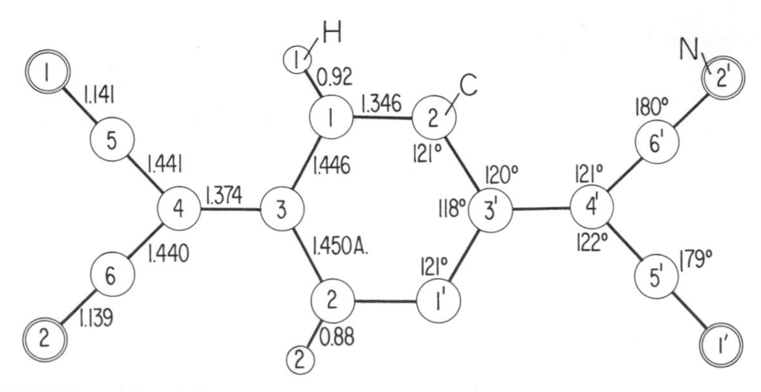

Fig. XVAIII,45. Bond dimensions in the molecule of 7,7,8,8-tetracyanoquinodimethane.

XV,aIII,32. The 1:1 addition compound of *7,7,8,8-tetracyanoquinodimethane* and *N,N,N′,N′-tetramethyl-p-phenylenediamine*, $C_6H_4[C(CN)_2]_2 \cdot C_6H_4[N(CH_3)_2]_2$, forms monoclinic crystals. Their bimolecular unit has the dimensions:

$$a_0 = 9.88(3) \text{ A.}; \quad b_0 = 12.71(4) \text{ A.}; \quad c_0 = 7.72(3) \text{ A.}$$
$$\beta = 97°20(3)'$$

The space group was chosen as C_{2h}^3 ($C2/m$) with atoms in the positions:

$$(4i) \quad \pm(u\,0\,v;\ u+{}^1/_2,{}^1/_2,v)$$
$$(8j) \quad \pm(xyz;\ x\bar{y}z;\ x+{}^1/_2,y+{}^1/_2,z;\ x+{}^1/_2,{}^1/_2-y,z)$$

Parameters are given in Table XVAIII,28.

TABLE XVAIII,28

Parameters of the Atoms in $C_6H_4[C(CN)_2]_2 \cdot C_6H_4[N(CH_3)_2]_2$

Atom	Position	x	y	z
C(1)	(8*j*)	0.0590	0.0952	−0.0383
C(2)	(4*i*)	0.1224	0	−0.0779
C(3)	(4*i*)	0.2448	0	−0.1529
C(4)	(8*j*)	0.3118	0.0941	−0.1890
N(5)	(8*j*)	0.3683	0.1705	−0.2199
H(1)[a]	(8*j*)	0.101	0.164	−0.065
C(6)	(8*j*)	0.0593	0.0951	0.4622
C(7)	(4*i*)	0.1229	0	0.4214
N(8)	(4*i*)	0.2407	0	0.3463
C(9)	(8*j*)	0.3127	0.0987	0.3168
H(6)[a]	(8*j*)	0.103	0.161	0.435

[a] Positions for the hydrogen atoms were not refined.

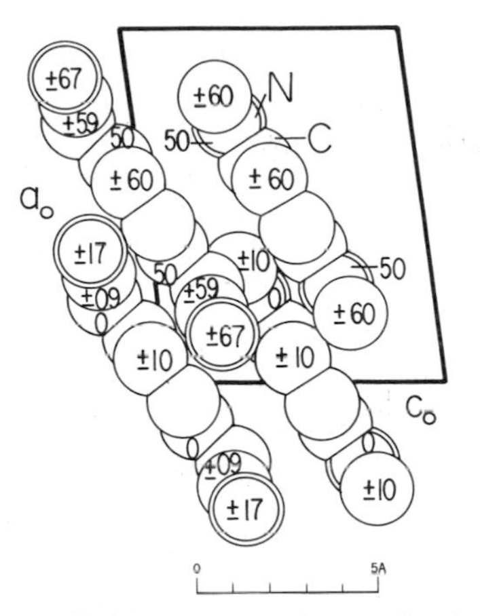

Fig. XVAIII,46. The monoclinic structure of crystals of the addition compound of 7,7,8,8-tetracyanoquinodimethane and *N,N,N′,N′*-tetramethyl-*p*-phenylenediamine projected along its b_0 axis.

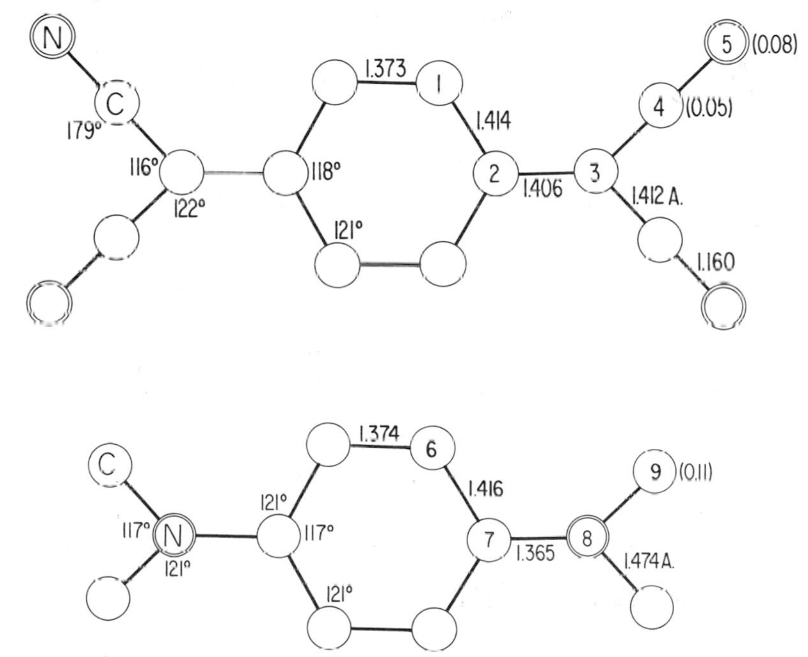

Fig. XVAIII,47a (top). Bond dimensions of the $C_6H_4[C(CN)_2]_2$ part of the addition compound of 7,7,8,8-tetracyanoquinodimethane with *N,N,N′,N′*-tetramethyl-*p*-phenylenediamine.

Fig. XVAIII,47b (bottom). Bond dimensions of the $C_6H_4[N(CH_3)_2]_2$ portion of the addition compound of 7,7,8,8-tetracyanoquinodimethane with *N,N,N′,N′*-tetramethyl-*p*-phenylenediamine.

The structure as a whole is shown in Figure XVAIII,46. Intense thermal motion in the crystal is considered to limit the accuracy of the bond dimensions (Fig. XVAIII,47) of the two kinds of molecule in the complex. Each of these molecules is planar except for the terminal atoms which depart from the planes of the rest by the distances in parentheses in the figure.

XV,aIII,33. Crystals of *p-bis(tertiarybutyl)benzene*, $C_6H_4[C(CH_3)_3]_2$, are monoclinic with a bimolecular cell of the dimensions:

$$a_0 = 9.89 \text{ A.}; \quad b_0 = 10.13 \text{ A.}; \quad c_0 = 6.35 \text{ A.}$$
$$\beta = 94°48'$$

The space group is C_{2h}^5 in the orientation $P2_1/n$. Atoms, therefore, are in the positions:

$$(4e) \quad \pm(xyz;\ x+{}^1/_2,{}^1/_2-y,z+{}^1/_2)$$

Approximate parameters determined many years ago are listed in Table XVAIII,29.

TABLE XVAIII,29

Parameters of the Atoms in *p*-bis(Tertiarybutyl) Benzene

Atom	x	y	z
C(1)	0.015	0.055	0.200
C(2)	0.088	0.035	−0.149
C(3)	0.104	0.090	0.051
C(4)	0.218	0.190	0.108
C(5)	0.230	0.215	0.348
C(6)	0.185	0.321	0.009
C(7)	0.353	0.136	0.042

The structure is shown in Figure XVAIII,48.

XV,aIII,34. Crystals of the complex *carbon tetrabromide-p-xylene*, $CBr_4 \cdot C_6H_4(CH_3)_2$, are orthorhombic with a tetramolecular unit of the edge lengths:

$$a_0 = 8.48(3) \text{ A.}; \quad b_0 = 8.89(3) \text{ A.}; \quad c_0 = 17.46(5) \text{ A.}$$

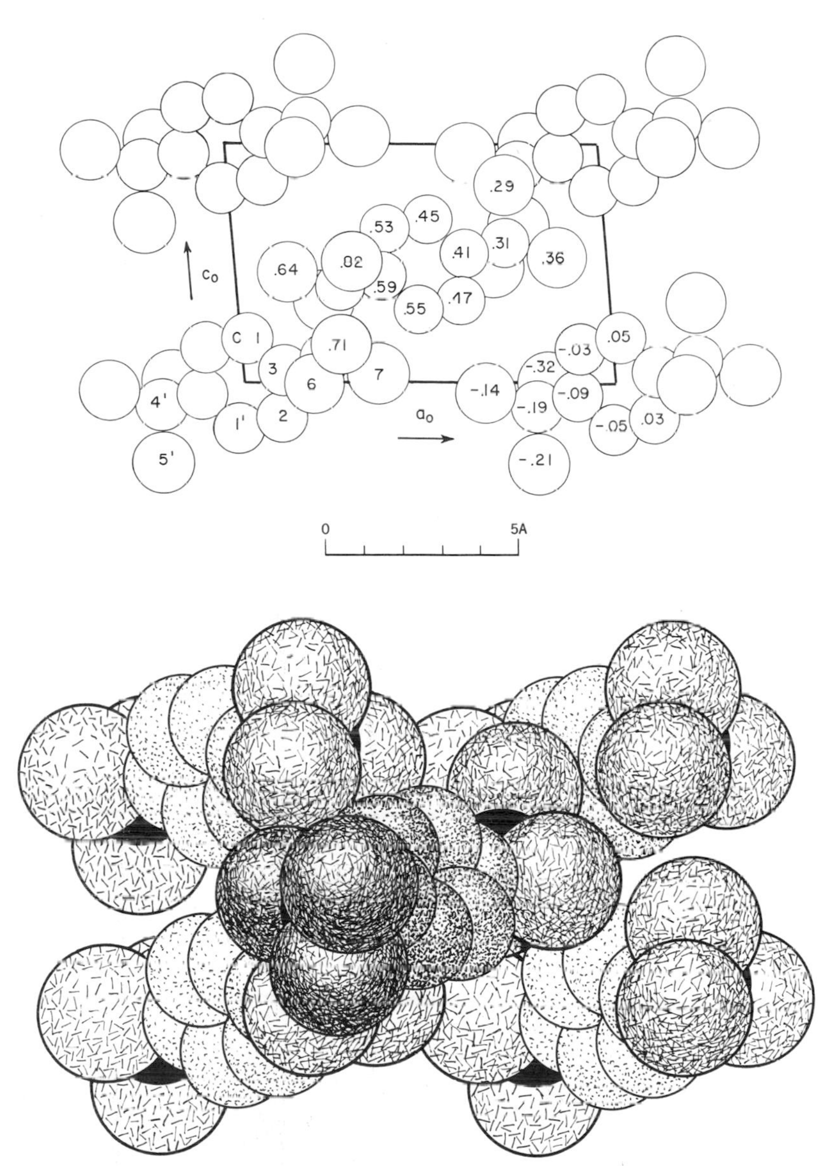

Fig. XVAIII,48a (top). The monoclinic structure of *p*-bis(tertiarybutyl)benzene projected along its b_0 axis. Left hand axes.

Fig. XVAIII,48b (bottom). A packing drawing of the monoclinic *p*-bis(tertiarybutyl)benzene seen along its b_0 axis. The central carbon atom of the butyl radical is black; the other carbon atoms of the group are line shaded. Benzene carbons are smaller and dotted.

Their atoms are in the following positions of the space group V_h^{17} (*Cmcm*):

C(1): (4*c*) $\pm(0\,u\,{}^1/_4;\ {}^1/_2,u+{}^1/_2,{}^1/_4)$
with $u = 0.331(6)$

Br(1): (8*f*) $\pm(0uv;\ 0,u,{}^1/_2-v;\ {}^1/_2,u+{}^1/_2,v;\ {}^1/_2,u+{}^1/_2,{}^1/_2-v)$
with $u = 0.2034(5)$, $v = 0.1608(2)$

Br(2): (8*g*) $\pm(u\,v\,{}^1/_4;\ \bar{u}\,v\,{}^1/_4;\ u+{}^1/_2,v+{}^1/_2,{}^1/_4;\ {}^1/_2-u,v+{}^1/_2,{}^1/_4)$
with $u = 0.1840(5)$, $v = 0.4506(5)$

The carbon atoms of the xylene molecule are in:

C(2): (8*f*) with $u = 0.872(4)$, $v = 0.0497(10)$
C(3): (8*f*) with $u = 0.732(5)$, $v = 0.0970(20)$
C(4): (16*h*) $\pm(xyz;\ x\bar{y}\bar{z};\ x,y,{}^1/_2-z;\ x,\bar{y},z+{}^1/_2;$
$x+{}^1/_2,y+{}^1/_2,z;\ x+{}^1/_2,{}^1/_2-y,\bar{z};\ x+{}^1/_2,y+{}^1/_2,{}^1/_2-z;$
$x+{}^1/_2,{}^1/_2-y,z+{}^1/_2$
with $x = 0.139(3)$, $y = 0.932(3)$, $z = 0.0193(10)$

The resulting structure is shown in Figure XVAIII,49. In the tetrahedral CBr_4, the average C–Br = 1.93 A.

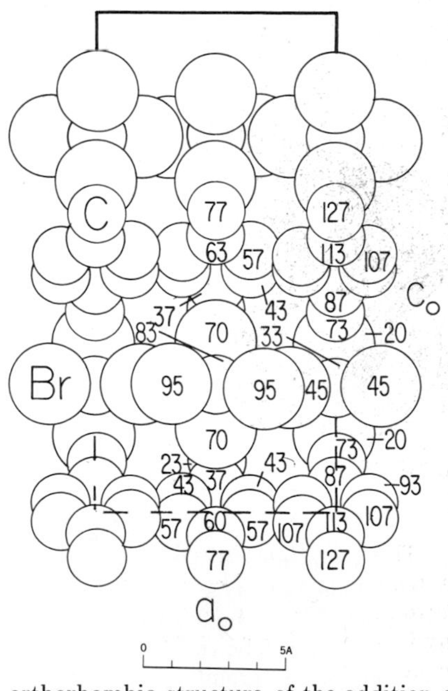

Fig. XVAIII,49. The orthorhombic structure of the addition compound $CBr_4 \cdot C_6H_4$-$(CH_3)_2$ projected along its b_0 axis.

XV,aIII,35. The quasi-racemate of (+)-*m-methoxyphenoxypropionic acid* and (−)-*m-bromophenoxypropionic acid,* $(CH_3O)C_6H_4OC_2H_5COOH \cdot BrC_6H_4OC_2H_5COOH$, is monoclinic with a tetramolecular unit of the dimensions:

$$a_0 = 33.48(6) \text{ A.}; \quad b_0 = 5.15(1) \text{ A.}; \quad c_0 = 11.43(2) \text{ A.}$$
$$\beta = 90°36(15)'$$

Atoms are in the positions

$$(4c) \quad xyz;\ \bar{x}y\bar{z};\ x+{}^1/_2,y+{}^1/_2,z;\ {}^1/_2-x,y+{}^1/_2,\bar{z}$$

of C_2^3 ($C2$). The parameters are stated in Table XVAIII,30.

TABLE XVAIII,30

Parameters of the Atoms in the Quasi-racemate of Methoxy- and Bromo-phenoxypropionic Acids

Atom	x	y	z
Br	0.2205(1)	0.1390(1)	0.0711(12)
C(1)	0.0467(7)	0.380(8)	0.1563(19)
C(2)	0.0795(7)	0.267(7)	0.0826(19)
C(3)	0.0601(6)	0.147(13)	−0.0280(18)
C(4)	0.1311(5)	0.599(9)	0.1181(15)
C(5)	0.1558(6)	0.769(7)	0.0666(17)
C(6)	0.1850(7)	0.889(8)	0.1386(20)
C(7)	0.1848(7)	0.863(9)	0.2624(21)
C(8)	0.1595(7)	0.681(14)	0.3087(20)
C(9)	0.1292(6)	0.554(8)	0.2392(18)
O(1)	0.0301(5)	0.221(5)	0.2313(13)
O(2)	0.0324(5)	0.606(9)	0.1430(14)
O(3)	0.1038(4)	0.479(5)	0.0396(11)
C(11)	0.9557(6)	0.617(12)	0.3398(17)
C(12)	0.9227(6)	0.734(7)	0.4129(17)
C(13)	0.9423(6)	0.860(8)	0.5227(18)
C(14)	0.8719(7)	0.387(8)	0.3896(19)
C(15)	0.8493(6)	0.223(7)	0.4428(16)
C(16)	0.8195(6)	0.089(9)	0.3761(17)
C(17)	0.8180(6)	0.116(11)	0.2555(18)
C(18)	0.8406(7)	0.294(8)	0.2032(19)
C(19)	0.8712(7)	0.444(9)	0.2633(21)
C(20)	0.7665(7)	−0.250(7)	0.3756(19)
O(11)	0.9712(5)	0.783(5)	0.2621(14)
O(12)	0.9705(5)	0.405(6)	0.3512(16)
O(13)	0.9000(4)	0.527(4)	0.4607(11)
O(14)	0.7981(4)	−0.091(5)	0.4396(12)

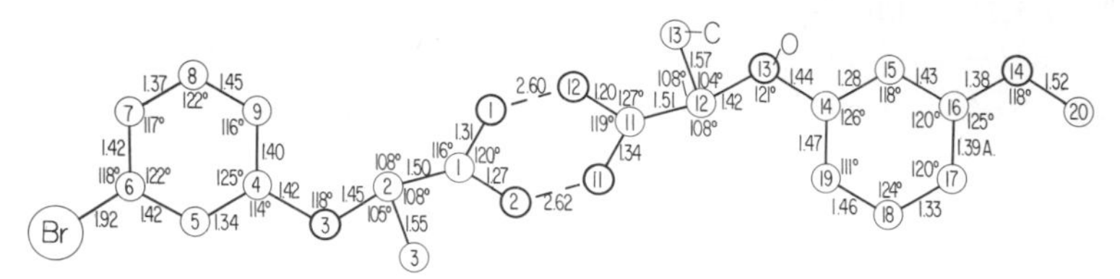

Fig. XVAIII,50. Bond dimensions in the two components of the quasi-racemate of (+)-*m*-methoxyphenoxypropionic acid and (−)-*m*-bromophenoxypropionic acid.

The very similar molecules of the two components in this structure have the bond dimensions of Figure XVAIII,50. They are so arranged that there is a pseudo-center of symmetry between them and they are tied together by a pair of hydrogen bonds of lengths 2.60 and 2.62 A.

XV,aIII,36. Crystals of *1,4-dimethoxy benzene*, $C_6H_4(OCH_3)_2$, are orthorhombic with a tetramolecular unit of the edge lengths:

$$a_0 = 7.29 \text{ A.}; \quad b_0 = 6.30 \text{ A.}; \quad c_0 = 16.55 \text{ A.}$$

The space group is V_h^{15} in the orientation *Pbca*. All atoms are thus in the positions:

$$(8c) \quad \pm(xyz;\ x+{}^1/_2,{}^1/_2-y,\bar{z};\ \bar{x},y+{}^1/_2,{}^1/_2-z;\ {}^1/_2-x,\bar{y},z+{}^1/_2)$$

The parameters, determined some years ago, are those of Table XVAIII,31. The hydrogen parameters of this table are based on the assumptions that C–H = 1.09 A. and H–C–H = 109° in the methyl groups, while in the ring, C–H = 1.08 A. and C–C–H = 120°.

TABLE XVAIII,31

Parameters of the Atoms in 1,4-Dimethoxybenzene

Atom	x	y	z
C(1)	0.0055	0.0797	0.0777
CH(2)	0.0927	−0.1066	0.0612
CH(3)	0.0977	−0.1866	−0.0156
O	0.0104	0.1581	0.1541
$CH_3(4)$	−0.0763	0.3415	0.1730
	Proposed hydrogen positions		
$H(CH_2)$	0.165	−0.189	0.108
$H(CH_3)$	0.173	−0.332	−0.028
$H(1)[CH_3]$	−0.016	0.473	0.137
$H(2)[CH_3]$	−0.058	0.375	0.237
$H(3)[CH_3]$	−0.222	0.327	0.159

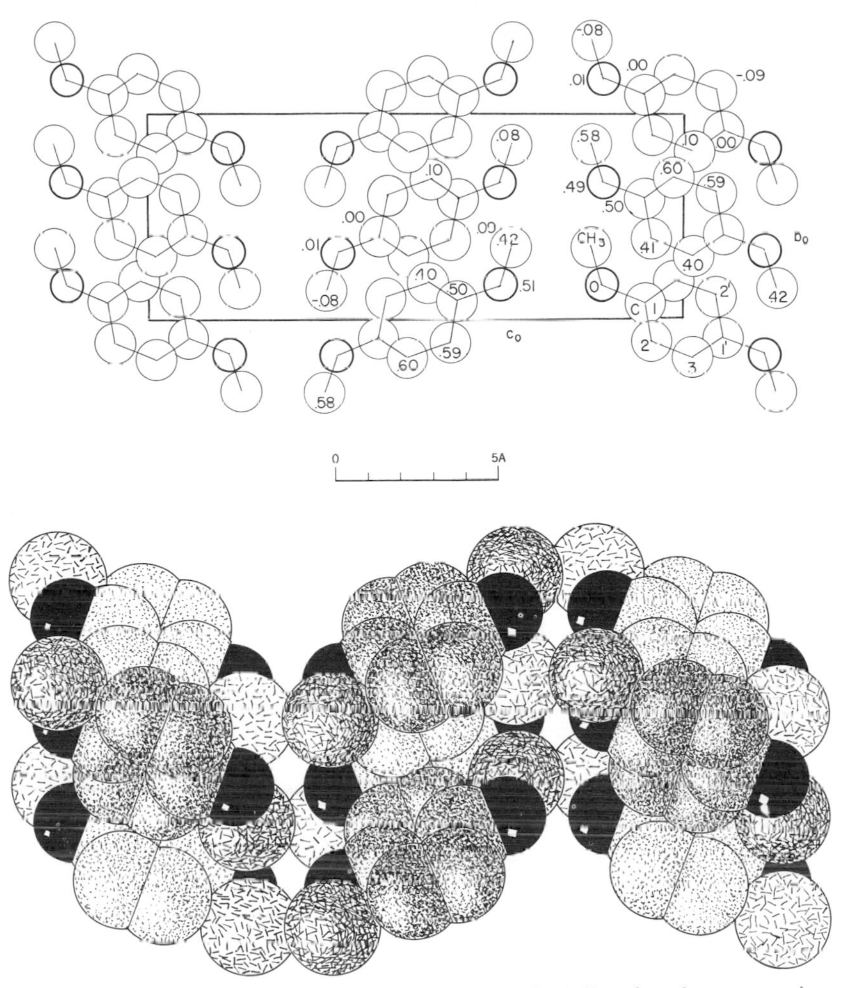

Fig. XVAIII,51a (top). The orthorhombic structure of 1,4-dimethoxybenzene projected along its a_0 axis.

Fig. XVAIII,51b (bottom). A packing drawing of the orthorhombic 1,4-dimethoxybenzene arrangement seen along its a_0 axis. The oxygen atoms are black; the methyl carbons are line shaded. Benzene carbon atoms are dotted.

The structure is illustrated in Figure XVAIII,51, which brings out the pronounced layering of the molecules. Bond dimensions are given in Figure XVAIII,52. Between molecules the shortest atomic distances are O–C(CH_3) = 3.52 A. and C–C = 3.54 A.

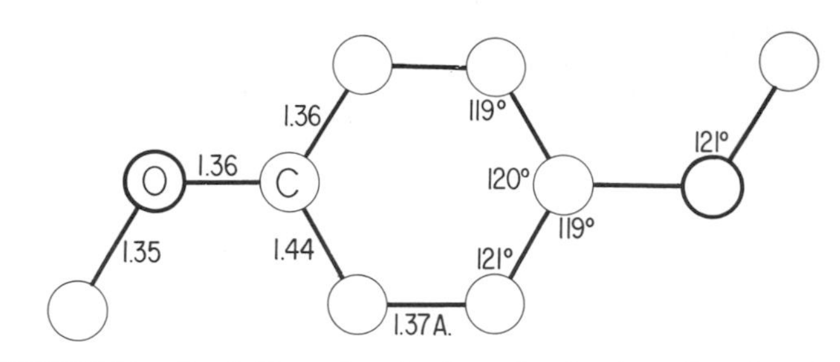

Fig. XVAIII,52. Bond dimensions in the molecule of 1,4-dimethoxybenzene.

XV,aIII,37. The double salt *ferric chloride·o-methoxyphenyldiazonium chloride*, *o*-$CH_3OC_6H_4N_2 \cdot FeCl_4$, is orthorhombic with a tetramolecular unit of the edge lengths:

$$a_0 = 11.16(2) \text{ A.}; \quad b_0 = 16.36(3) \text{ A.}; \quad c_0 = 7.18(2) \text{ A.}$$

The space group, chosen as V_h^{16} (*Pbnm*), places all atoms except one set of chlorines in the special positions:

$$(4c) \quad \pm(u\,v\,{}^1/_4;\ {}^1/_2-u,v+{}^1/_2,{}^1/_4)$$

The remaining atoms of chlorine are in the general positions:

$$(8d) \quad \pm(xyz;\ x+{}^1/_2,{}^1/_2-y,z+{}^1/_2;\ x,y,{}^1/_2-z;\ {}^1/_2-x,y+{}^1/_2,z)$$

Selected positions and parameters are those of Table XVAIII,32.

TABLE XVAIII,32

Positions and Parameters of the Atoms in *o*-$CH_3OC_6H_4N_2 \cdot FeCl_4$

Atom	Position	x	y	z
Fe	(4*c*)	0.152	0.235	$^1/_4$
Cl(1)	(8*d*)	0.038	0.232	0.495
Cl(2)	(4*c*)	0.273	0.132	$^1/_4$
Cl(3)	(4*c*)	0.247	0.355	$^1/_4$
O	(4*c*)	0.592	0.537	$^1/_4$
N(1)	(4*c*)	0.749	0.662	$^1/_4$
N(2)	(4*c*)	0.714	0.725	$^1/_4$
C(1)	(4*c*)	0.793	0.577	$^1/_4$
C(2)	(4*c*)	0.710	0.515	$^1/_4$
C(3)	(4*c*)	0.751	0.434	$^1/_4$
C(4)	(4*c*)	0.877	0.427	$^1/_4$
C(5)	(4*c*)	0.960	0.492	$^1/_4$
C(6)	(4*c*)	0.926	0.580	$^1/_4$
C(7)	(4*c*)	0.500	0.474	$^1/_4$

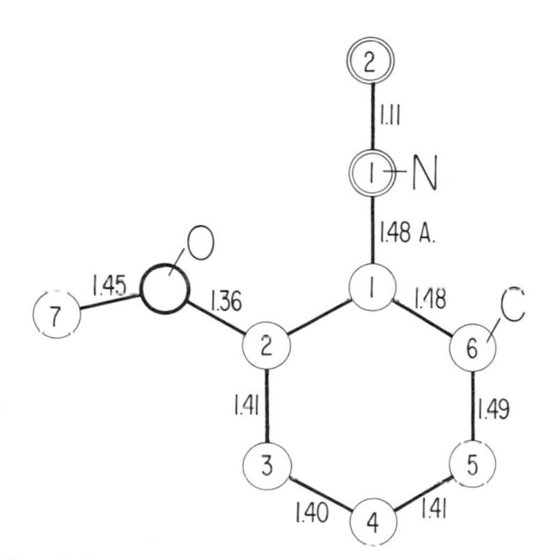

Fig. XVAIII,53. Bond dimensions in the complex cation present in crystals of ferric chloride-*o*-methoxyphenyldiazonium chloride.

This structure is unusual in that all atoms of the organic part are required by symmetry to be exactly in a plane. This portion is a complex cation (Fig. XVAIII,53), the compensating anion $FeCl_4^-$ being tetrahedral in shape, with Fe–Cl = 2.165–2.230 A.

XV,aIII,38. *Benzoquinone*, $C_6H_4O_2$, is monoclinic with a bimolecular unit of the dimensions:

$$a_0 = 7.055(4) \text{ A.}; \quad b_0 = 6.795(4) \text{ A.}; \quad c_0 = 5.767(4) \text{ A.}$$
$$\beta = 101°28(2)'$$

Its atoms are in the positions

$$(4e) \quad \pm(xyz;\ x+{}^1/_2,{}^1/_2-y,z)$$

of the space group C_{2h}^5 ($P2_1/a$). A recent redetermination has yielded the following parameters:

Atom	x	y	z
C(1)	0.1069	−0.0272	0.2334
C(2)	0.0638	0.1703	0.1341
C(3)	−0.0459	0.1848	−0.1095
O	0.1182	0.3168	0.2502

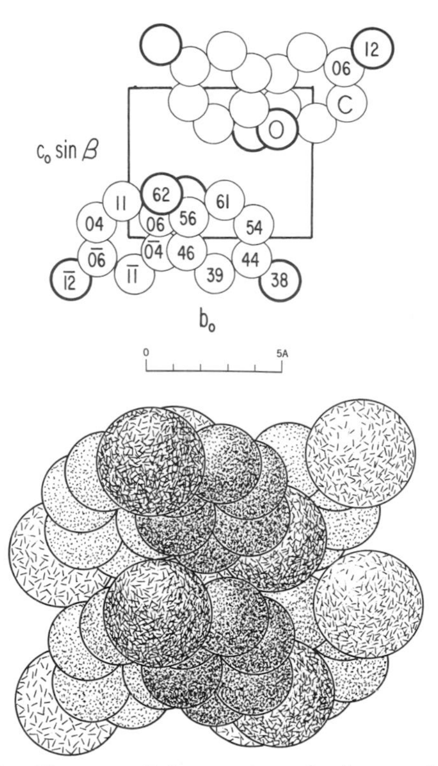

Fig. XVAIII,54a (top). The monoclinic structure of *p*-benzoquinone projected along its a_0 axis.

Fig. XVAIII,54b (bottom). A packing drawing of the monoclinic *p*-benzoquinone structure seen along its a_0 axis. The oxygen atoms are line shaded; the carbon atoms are dotted.

The structure, shown in Figure XVAIII,54, is built up of molecules which are strictly planar and have a center of symmetry. Their bond dimensions are given in Figure XVAIII,55. Between them in the crystal the shortest separations are: C–O = 3.34, 3.39, and 3.45 A., O–O = 3.64 A., and C–C = 3.47 A.

XV,aIII,39. Crystals of *o-fluorobenzoic acid*, C_6H_4FCOOH, are monoclinic with a tetramolecular unit of the dimensions:

$$a_0 = 6.71(2) \text{ A.}; \quad b_0 = 3.81(1) \text{ A.}; \quad c_0 = 25.05(3) \text{ A.}$$
$$\beta = 99°54(6)'$$

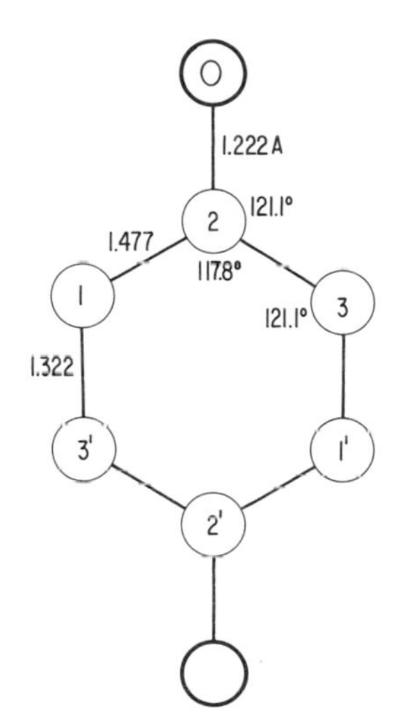

Fig. XVAIII,55. Bond dimensions in the molecule of benzoquinone.

All atoms are in the positions

$$(4e) \quad \pm(xyz;\ x,{}^1/_2-y,z+{}^1/_2)$$

of the space group C_{2h}^5 ($P2_1/c$). The selected parameters are stated in Table XVAIII,33.

TABLE XVAIII,33

Parameters of the Atoms in C_6H_4FCOOH

Atom	x	y	z
F	0.048	0.110	0.1702
O(1)	0.034	0.010	0.0675
O(2)	0.775	0.220	0.0095
C(1)	0.763	0.255	0.1025
C(2)	0.855	0.235	0.1560
C(3)	0.759	0.350	0.1965
C(4)	0.564	0.490	0.1855
C(5)	0.469	0.510	0.1315
C(6)	0.568	0.395	0.0910
C(7)	0.865	0.160	0.0580

The structure is shown in Figure XVAIII,56. The benzene ring and attached fluorine atom are strictly planar, with the carboxyl carbon less

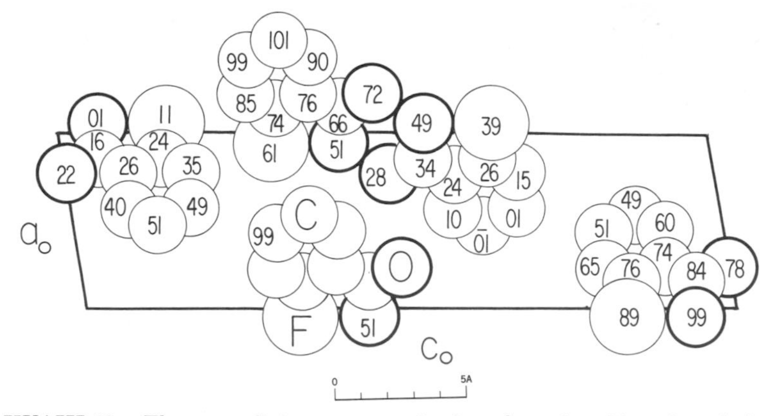

Fig. XVAIII,56. The monoclinic structure of *o*-fluorobenzoic acid projected along its b_0 axis.

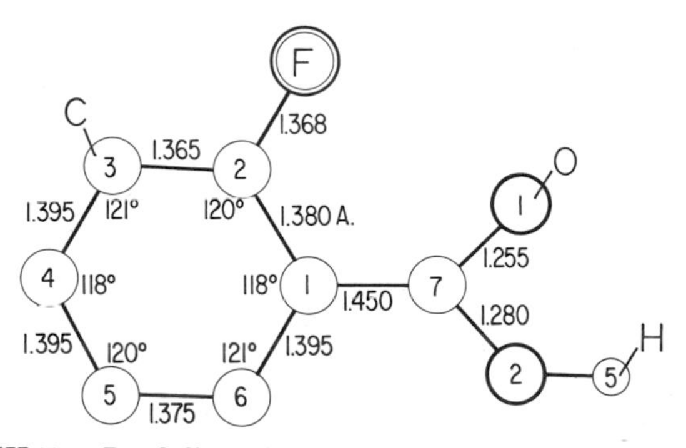

Fig. XVAIII,57. Bond dimensions in the molecule of *o*-fluorobenzoic acid.

than 0.1 A. out of the plane. The carboxyl group is, however, twisted by 21° from it. The bond dimensions of the molecule thus constituted are those of Figure XVAIII,57. Adjacent molecules are tied together by pairs of carboxylic hydrogen bonds, with O–H . . . O = 2.64 A.

XV,aIII,40. Crystals of *o-chlorobenzoic acid*, ClC_6H_4COOH, are monoclinic with a unit that contains eight molecules and has the dimensions:

$$a_0 = 14.73(3) \text{ A.}; \quad b_0 = 3.90(2) \text{ A.}; \quad c_0 = 25.50(5) \text{ A.}$$
$$\beta = 112°40(20)'$$

All atoms are in the positions

$$(8f) \quad \pm(xyz;\ x,\bar{y},z+{}^1/_2;\ x+{}^1/_2,y+{}^1/_2,z;\ x+{}^1/_2,{}^1/_2-y,z+{}^1/_2)$$

of the space group C_{2h}^6 ($C2/c$). The determined parameters are those of Table XVAIII,34.

TABLE XVAIII,34

Parameters of the Atoms in o-ClC_6H_4COOH and o-BrC_6H_4COOH (in parentheses)

Atom	x	y	z
C(1)	0.1961 (0.2029)	0.3430 (0.3511)	0.0550 (0.0567)
C(2)	0.1480 (0.1583)	0.4213 (0.4335)	0.0969 (0.0958)
C(3)	0.1997 (0.2117)	0.5710 (0.5871)	0.1503 (0.1484)
C(4)	0.1498 (0.1636)	0.6447 (0.6701)	0.1852 (0.1830)
C(5)	0.0501 (0.0611)	0.5751 (0.5978)	0.1676 (0.1629)
C(6)	0.0009 (0.0078)	0.4282 (0.4434)	0.1159 (0.1104)
C(7)	0.0501 (0.0560)	0.3488 (0.3571)	0.0798 (0.0753)
O(1)	0.2760 (0.2801)	0.4551 (0.4633)	0.0610 (0.0596)
O(2)	0.1427 (0.1489)	0.1498 (0.1268)	0.0132 (0.0156)
Cl(Br)	0.3247 (0.3523)	0.6642 (0.6911)	0.1775 (0.1833)

The structure, shown in Figure XVAIII,58, has molecules with the bond dimensions of Figure XVAIII,59. In them the benzene ring is strictly planar but its attached carboxyl carbon is displaced by 0.058 A. to one side and its chlorine atom by 0.036 A. to the other side of this plane. The plane defined by the carboxyl group is twisted by 13.7° about C(1)–C(2) with reference to the benzene plane. A short O(1)–Cl = 2.892 A. suggests some strain within the molecule as a whole. Hydrogen bonds (O–H . . . O = 2.632 A.) between carboxyl oxygens tie these molecules together in pairs. Other atomic separations between molecules involve carbon–oxygen distances ranging upwards from 3.32 A.

The corresponding bromine compound *o-bromobenzoic acid*, BrC_6H_4-$COOH$, is essentially isostructural. For it:

$$a_0 = 14.82(4) \text{ A.}; \quad b_0 = 4.10(2) \text{ A.}; \quad c_0 = 25.90(5) \text{ A.}$$
$$\beta = 118°15'$$

Its parameters are given in parentheses in the table.

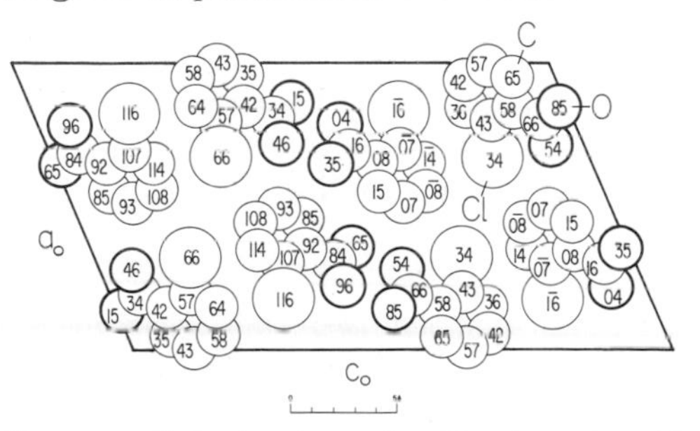

Fig. XVAIII,58. The monoclinic structure of o-chlorobenzoic acid projected along its b_0 axis.

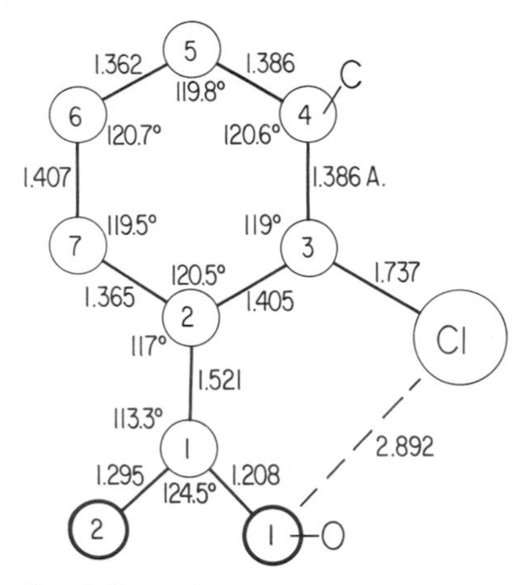

Fig. XVAIII,59. Bond dimensions in the molecule of *o*-chlorobenzoic acid.

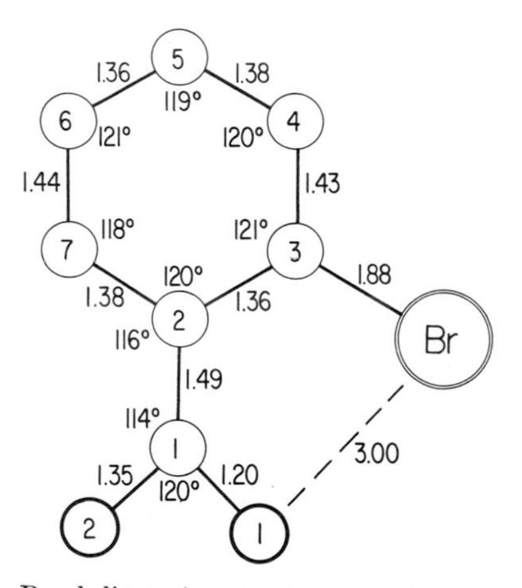

Fig. XVAIII, 60. Bond dimensions in the molecule of *o*-bromobenzoic acid.

The molecule has the bond dimensions of Figure XVAIII,60. The benzene ring is strictly planar with the carboxyl carbon and bromine atoms 0.06 A. on opposite sides of this plane. The twist of the carboxyl group (18.3°) provides room for the atom of bromine but leaves a short O(1)–Br = 3.004 A.

XV,aIII,41. *Potassium hydrogen di-p-chlorobenzoate*, $KH(ClC_6H_4CO_2)_2$, forms monoclinic crystals whose large tetramolecular unit has the dimensions:

$$a_0 = 33.205(6)\text{ A.};\quad b_0 = 3.846(20)\text{ A.};\quad c_0 = 11.212(3)\text{ A.}$$
$$\beta = 89.91(2)^\circ$$

Potassium atoms are in the positions

$$(4e)\quad \pm(0\,u\,{}^1/_4;\ {}^1/_2,u+{}^1/_2,{}^1/_4)\qquad \text{with } u = 0.6417$$

of the space group C_{2h}^6 ($C2/c$). All other atoms are in the general positions:

$$(8f)\quad \pm(xyz;\ x,\bar{y},z+{}^1/_2;\ x+{}^1/_2,y+{}^1/_2,z;\ x+{}^1/_2,{}^1/_2-y,z+{}^1/_2)$$

with the parameters of Table XVAIII,35.

TABLE XVAIII,35

Parameters of Atoms in $KH(ClC_6H_4CO_2)_2$

Atom	x	y	z
O(1)	0.0335	0.0437	−0.0446
O(2)	0.0492	0.1851	0.1420
C(1)	0.0593	0.1452	0.0379
C(2)	0.1011	0.2050	−0.0056
C(3)	0.1288	0.3564	0.0730
C(4)	0.1677	0.4198	0.0335
C(5)	0.1785	0.3302	−0.0808
C(6)	0.1517	0.1693	−0.1573
C(7)	0.1125	0.1054	−0.1184
Cl	0.2262	0.4324	−0.1352
H(C,3)	0.120	0.422	0.159
H(C,4)	0.189	0.540	0.092
H(C,6)	0.160	0.111	−0.244
H(C,7)	0.091	−0.032	−0.177

The structure is shown in Figure XVAIII,61. In its anions (Fig. XVAIII,62) the benzene ring is planar but the chlorine atom is 0.13 A. outside this plane. The plane of the carboxyl group makes an angle of 9° with that of the ring. Between anions there are short O–O = 2.46, 2.48, and 2.55 A., presumably involving hydrogen bonds. Each potassium cation is surrounded by six oxygen atoms belonging to different anions, with K–O = 2.68, 2.83, and 2.91 A.

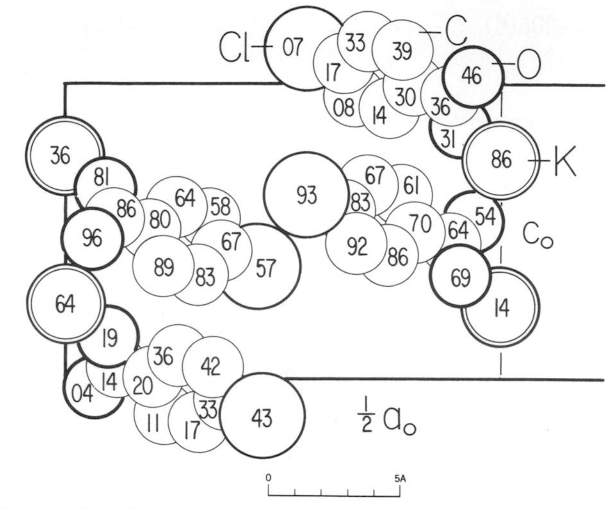

Fig. XVAIII,61. Half the contents of the unit cell of monoclinic $KH(ClC_6H_4CO_2)_2$ projected along its b_0 axis.

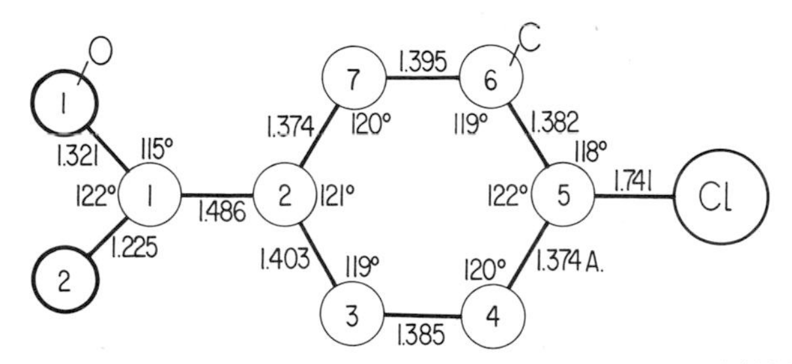

Fig. XVAIII,62. Bond dimensions in the molecular anion of $KH(ClC_6H_4CO_2)_2$.

The corresponding ammonium and rubidium salts are isostructural, with units of the dimensions:

For $NH_4H(ClC_6H_4CO_2)_2$: $a_0 = 32.8(2)$ A.; $b_0 = 3.88(2)$ A.; $c_0 = 11.44(4)$ A.
$\beta = 93°$

For $RbH(ClC_6H_4CO_2)_2$: $a_0 = 33.2(2)$ A.; $b_0 = 3.89(4)$ A.; $c_0 = 11.47(10)$ A.
$\beta = 92.1°$

Values of x and z but not of y are given in the original article.

XV,aIII,42. Crystals of *S-methylthiouronium-p-chlorobenzoate*, CH_3-$SCN_2H_3 \cdot C_6H_4(Cl)CO_2H$, are monoclinic with a tetramolecular unit of the dimensions:

$$a_0 = 9.505(5) \text{ A.}; \quad b_0 = 5.61(1) \text{ A.}; \quad c_0 = 22.176(10) \text{ A.}$$
$$\beta = 103.22(5)°$$

All atoms are in the positions

$$(4e) \quad \pm(xyz;\ x,{}^1/_2-y,z+{}^1/_2)$$

of C_{2h}^5 ($P2_1/c$). The parameters are listed in Table XVAIII,36.

TABLE XVAIII,36

Parameters of the Atoms is *S*-Methylthiouronium-*p*-chlorobenzoate

Atom	x	y	z
Cl	0.2291	0.2507	0.4676
S	0.2259	0.6848	0.8517
O(1)	0.3568	0.0805	0.6878
O(2)	0.1418	0.9583	0.6987
N(1)	0.3545	0.4125	0.7741
N(2)	0.1245	0.3336	0.7779
C(1)	0.2351	0.4558	0.5273
C(2)	0.3292	0.6344	0.5331
C(3)	0.3345	0.7933	0.5820
C(4)	0.2412	0.7701	0.6224
C(5)	0.1474	0.5851	0.6144
C(6)	0.1411	0.4197	0.5667
C(7)	0.2470	0.9497	0.6727
C(8)	0.2393	0.4761	0.7960
C(9)	0.3768	0.8766	0.8484

The structure, shown in Figure XVAIII,63, is composed of benzoate anions and thiouronium cations having the bond dimensions of Figure XVAIII,64. They are united by hydrogen bonds between oxygen and nitrogen to form sheets held together only by van der Waals forces. Both ions are planar except for C(9) which lies 0.11 A. outside the plane of the rest of the cation.

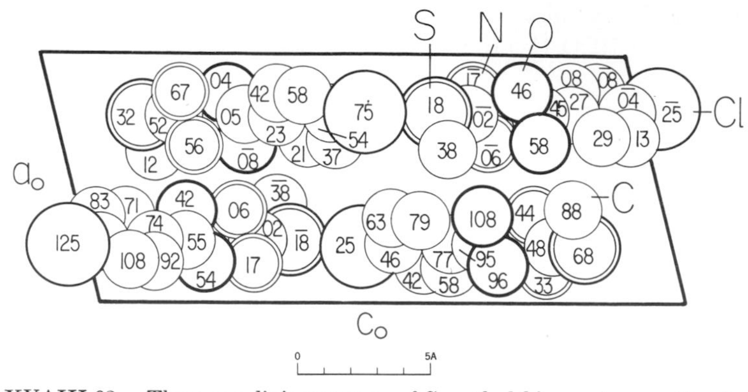

Fig. XVAIII,63. The monoclinic structure of *S*-methylthiouronium-*p*-chlorobenzoate projected along its b_0 axis.

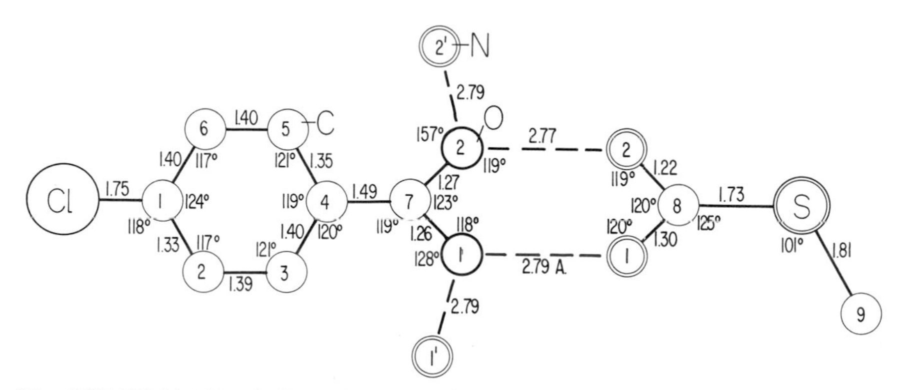

Fig. XVAIII,64. Bond dimensions prevailing in crystals of *S*-methylthiouronium-*p*-chlorobenzoate.

XV,aIII,43. Crystals of *o-nitrobenzoic acid*, $NO_2C_6H_4COOH$, are triclinic with a bimolecular unit of the dimensions:

$$a_0 = 7.55 \text{ A.}; \quad b_0 = 4.99 \text{ A.}; \quad c_0 = 12.50 \text{ A.}$$
$$\alpha = 122°30'; \quad \beta = 95°18'; \quad \gamma = 108°54'$$

All atoms are in the positions $(2i)$ $\pm(xyz)$ of C_i^1 $(P\bar{1})$. The determined parameters are stated in Table XVAIII,37.

TABLE XVAIII,37

Parameters of the Atoms in *o*-Nitrobenzoic Acid

Atom	x	y	z
C(1)	−0.2812(9)	−0.0628(14)	0.1928(6)
C(2)	−0.2308(9)	−0.0972(14)	0.2931(6)
C(3)	−0.3639(9)	−0.1913(14)	0.3521(6)
C(4)	−0.5533(9)	−0.2327(14)	0.3130(6)
C(5)	−0.6118(9)	−0.1928(14)	0.2181(6)
C(6)	−0.4758(9)	−0.1073(14)	0.1568(6)
C(7)	−0.1553(9)	−0.0234(14)	0.1125(6)
N	−0.0243(8)	−0.0158(12)	0.3470(5)
O(1)	−0.0187(7)	−0.1066(10)	0.1047(4)
O(2)	−0.1984(7)	0.0952(10)	0.0486(4)
O(3)	0.1089(7)	0.2759(10)	0.3909(4)
O(4)	0.0045(7)	−0.2355(10)	0.3502(4)

The structure, shown in Figure XVAIII,65, is built of molecules having the bond dimensions of Figure XVAIII,66. Their benzene rings are planar but both the carboxyl and nitro groups are twisted from this plane. The angle to it made by the NO_2 group is 54.7° and by the carboxyl group 23.4°.

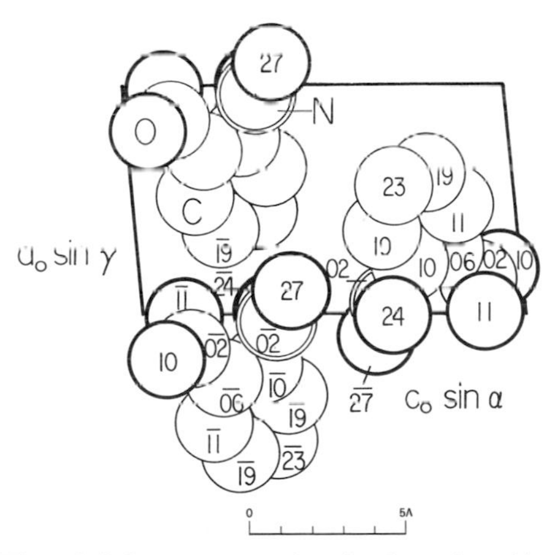

Fig. XVAIII,65. The triclinic structure of *o*-nitrobenzoic acid projected along its b_0 axis.

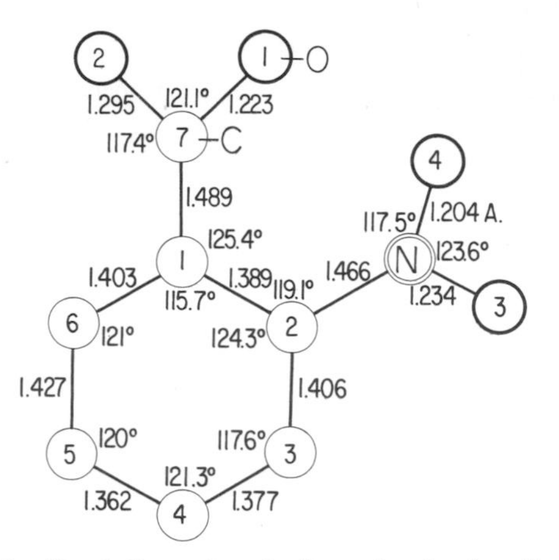

Fig. XVAIII,66. Bond dimensions in the molecule of *o*-nitrobenzoic acid.

A preliminary note utilizing an entirely different unit (1967: K,F&S) has also just been published.

XV,aIII,44. *Rubidium hydrogen di-o-nitrobenzoate*, $RbH(NO_2C_6H_4CO_2)_2$, forms triclinic crystals whose unimolecular cell has the dimensions:

$$a_0 = 13.80 \text{ A.}; \quad b_0 = 4.74 \text{ A.}; \quad c_0 = 6.08 \text{ A.}$$
$$\alpha = 100.5°; \quad \beta = 99.4°; \quad \gamma = 95.0°$$

The space group is C_i^1 ($P\bar{1}$) with the rubidium atom in (1*a*) 000 and the other non-hydrogen atoms in (2*i*) $\pm(xyz)$. Their parameters have been determined to be those of Table XVAIII,38.

TABLE XVAIII,38
Parameters of the Atoms in $RbH(NO_2C_6H_4CO_2)_2$

Atom	x	y	z
O(1)	0.0680	0.587	0.273
O(2)	0.0780	0.925	0.580
O(3)	0.2190	0.027	0.232
O(4)	0.3730	0.987	0.241
N	0.2916	0.945	0.300
C(1)	0.1127	0.740	0.447
C(2)	0.2150	0.679	0.553
C(3)	0.2950	0.787	0.480
C(4)	0.3850	0.732	0.581
C(5)	0.3900	0.580	0.742
C(6)	0.3050	0.467	0.820
C(7)	0.2200	0.520	0.720

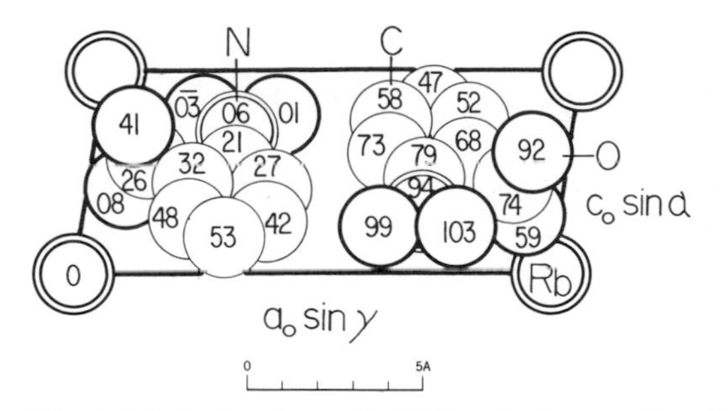

Fig. XVAIII,67. The triclinic structure of rubidium hydrogen di-*o*-nitrobenzoate projected along its b_0 axis.

The structure is shown in Figure XVAIII,67. In its anions the benzene ring and nitro group are coplanar but the carboxyl group is twisted through 70° with respect to this plane. These ions are joined into pairs by a short hydrogen bond of length 2.43 A. between O(2) atoms. The Rb–O distances lie between 2.85 and 2.95 A.

XV,aIII,45. Crystals of *p-nitrobenzoic acid*, $NO_2C_6H_4COOH$, are monoclinic with a large eight-molecule unit having the dimensions:

$$a_0 = 12.97 \text{ A.}; \quad b_0 = 5.07 \text{ A.}; \quad c_0 = 21.43 \text{ A.}; \quad \beta = 96°24'$$

All atoms have the coordinates

$$(8f) \quad \pm(xyz;\ x+{}^1/_2,\bar{y},z;\ x,y+{}^1/_2,z+{}^1/_2;\ x+{}^1/_2,{}^1/_2-y,z+{}^1/_2)$$

of C_{2h}^6 in the orientation $A2/a$. Recently redetermined parameters are given in Table XVAIII,39.

TABLE XVAIII,39
Parameters of the Atoms in *p*-Nitrobenzoic Acid

Atom	x	y	z
C(1)	0.4157(4)	0.4986(11)	0.0897(3)
C(2)	0.4899(4)	0.6398(11)	0.1299(3)
C(3)	0.4584(4)	0.8352(11)	0.1696(3)
C(4)	0.3530(4)	0.8894(11)	0.1659(3)
C(5)	0.2787(4)	0.7479(11)	0.1272(3)
C(6)	0.3101(4)	0.5507(11)	0.0885(3)
C(7)	0.4538(4)	0.2883(11)	0.0486(3)
N	0.3185(4)	0.0898(9)	0.2093(2)
O(1)	0.5464(3)	0.2422(8)	0.0492(2)
O(2)	0.3802(3)	0.1559(8)	0.0144(2)
O(3)	0.3823(3)	0.1767(8)	0.2508(2)
O(4)	0.2310(3)	0.1699(8)	0.1987(2)
H(1)	0.4115(26)	0.0382(65)	−0.0076(15)

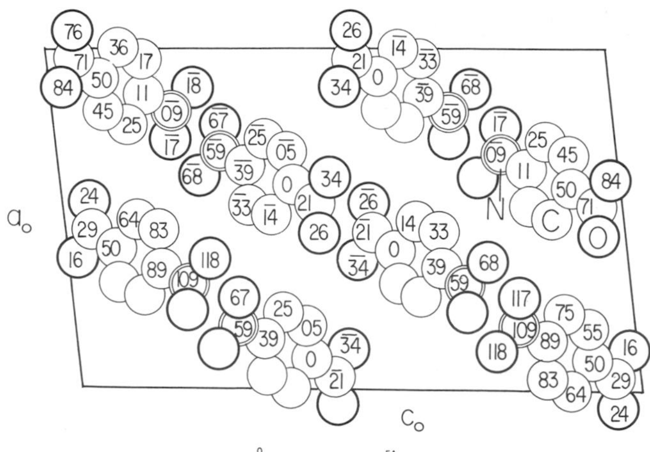

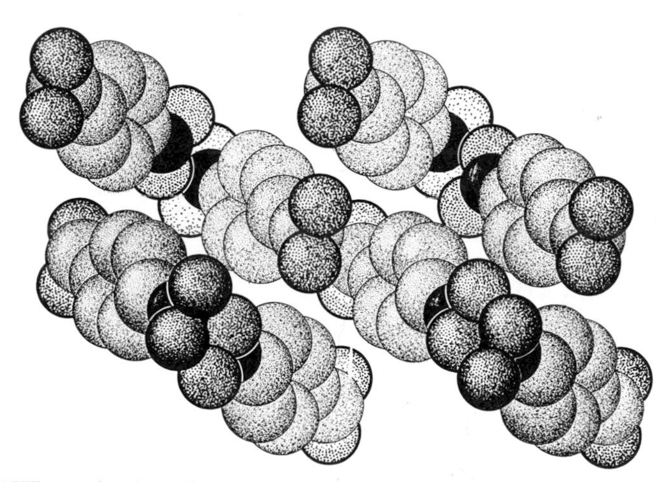

Fig. XVAIII,68a (top). The monoclinic structure of *p*-nitrobenzoic acid projected along its b_0 axis.

Fig. XVAIII,68b (bottom). A packing drawing of the monoclinic structure of *p*-nitrobenzoic acid viewed along its b_0 axis. The nitrogen atoms are black; the atoms of oxygen are heavily outlined and heavily dotted. Atoms of carbon are lightly dot shaded.

The structure (Fig. XVAIII,68) contains molecules with the bond dimensions of Figure XVAIII,69. In these molecules the benzene ring and attached N and C(7) atoms are coplanar but the plane of C(7) and its two oxygen atoms is turned 3.3° with respect to the benzene plane. The plane of the NO_2 radical is rotated through 13.7° with respect to that of the ring.

The carboxyl groups of adjacent molecules face one another to yield hydrogen bonds of length 2.65 A.

XV,aIII,46. *Potassium hydrogen di-p-nitrobenzoate*, $KH(NO_2C_6H_4CO_2)_2$, is triclinic with a bimolecular cell of the dimensions:

$$a_0 = 17.20 \text{ A.}; \quad b_0 = 4.05 \text{ A.}; \quad c_0 = 11.44 \text{ A.}$$
$$\alpha = 93.8°; \quad \beta = 104.1°; \quad \gamma = 90.5°$$

All atoms are in the positions $(2i)$ $\pm(xyz)$ of C_i^1 $(P\bar{1})$. The determined parameters, more accurate than those for the rubidium orthonitrobenzoate (**XV,aIII,44**), are given in Table XVAIII,40; possible values for the hydrogen atoms are stated in the original paper.

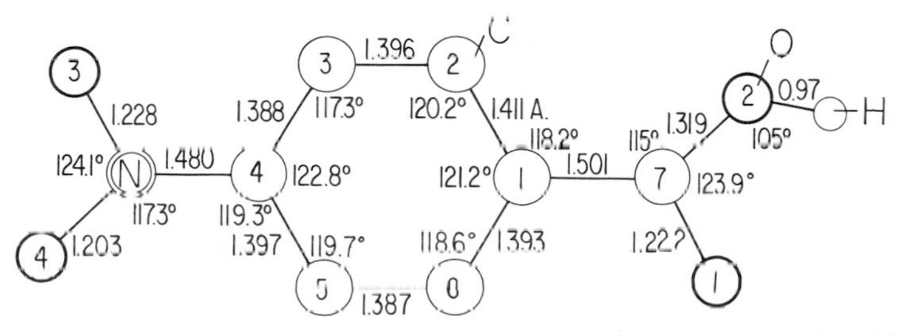

Fig. XVAIII,69. Bond dimensions in the molecule of *p*-nitrobenzoic acid.

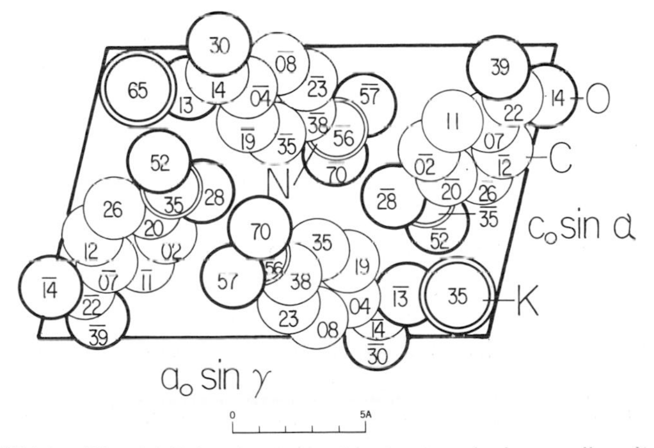

Fig. XVAIII,70. The triclinic structure of potassium hydrogen di-*p*-nitrobenzoate projected along its b_0 axis.

TABLE XVAIII,40

Parameters of the Atoms in $KH(NO_2C_6H_4CO_2)_2$

Atom	x	y	z
K	0.09075	0.34914	0.14252
O(1)	0.20249	−0.13449	0.14498
O(2)	0.24954	−0.29750	−0.01429
O(3)	0.59816	0.57259	0.21544
O(4)	0.56426	0.69715	0.38155
O(5)	0.11834	0.39474	−0.07826
O(6)	0.00551	0.14065	−0.17499
O(7)	0.17253	−0.52368	−0.61787
O(8)	0.28551	−0.28577	−0.53461
N(1)	0.55283	0.55892	0.28326
N(2)	0.21729	−0.35078	−0.53672
C(1)	0.25553	−0.13531	0.09116
C(2)	0.33435	0.04238	0.13938
C(3)	0.38818	0.08298	0.06542
C(4)	0.46070	0.23061	0.11710
C(5)	0.47530	0.37792	0.23075
C(6)	0.42284	0.34705	0.30373
C(7)	0.34986	0.18937	0.25634
C(8)	0.07646	0.22000	−0.16724
C(9)	0.11485	0.07500	−0.26249
C(10)	0.06697	−0.11652	−0.36086
C(11)	0.10039	−0.25701	−0.45330
C(12)	0.18103	−0.19924	−0.44415
C(13)	0.23040	−0.01721	−0.34710
C(14)	0.19601	0.11568	−0.25640

The structure is shown in Figure XVAIII,70. Its nitrobenzoate groups are of two sorts. One makes three close oxygen contacts [O(5), O(6), and O(8)] with potassium and is considered to be the anion. The other makes only one such contact [through O(1)] and is thought to be a neutral molecule. Both groups are planar except for twists of 5–10° for the NO_2 and CO_2 radicals; their bond dimensions are those of Figure XVAIII,71. In the crystal they are joined by short hydrogen bonds of length 2.492 A. Each

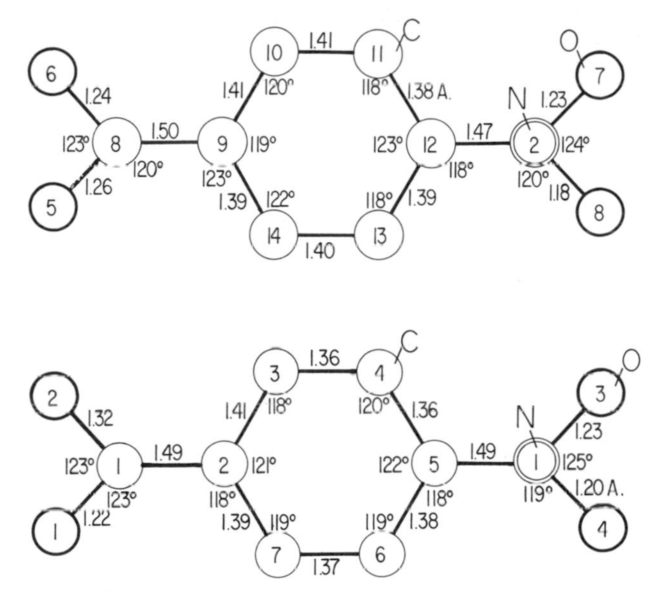

Fig. XVAIII,71a (top). Bond dimensions in the anionic molecular group in potassium hydrogen di-*p*-nitrobenzoate.

Fig. XVAIII,71b (bottom). Bond dimensions in the neutral molecular group in crystals of potassium hydrogen di-*p*-nitrobenzoate.

potassium cation is octahedrally surrounded by six oxygens (one belonging to an NO_2 group) at distances which for five of the six lie between 2.70 and 2.82 A.

XV,aIII,47. The monoclinic crystals of *o-nitroperoxybenzoic acid*, $C_6H_4(NO_2)C(O)O_2H$, have a tetramolecular unit of the dimensions:

$$a_0 = 13.84(2) \text{ A.}; \quad b_0 = 8.03\ (2) \text{ A.}; \quad c_0 = 7.51(2) \text{ A.}$$
$$\beta = 112°0(30)' \quad \text{at } 25°\text{C.}$$
$$a_0 = 13.75(2) \text{ A.}; \quad b_0 = 7.95(1) \text{ A.}; \quad c_0 = 7.47(1) \text{ A.}$$
$$\beta = 112°40(10)' \quad \text{at } -15°\text{C.}$$

The space group is C_{2h}^5 ($P2_1/c$) with all atoms in the positions:

$$(4e) \quad \pm(xyz;\ x,{}^1/_2-y,z+{}^1/_2)$$

Determined parameters are listed in Table XVAIII,41.

TABLE XVAIII,41

Parameters of the Atoms in $C_6H_4(NO_2)C(O)O_2H$

Atom	x	y	z
C(1)	0.1783	−0.3022	0.8096
C(2)	0.2451	−0.4381	0.8219
C(3)	0.3454	−0.4130	0.8291
C(4)	0.3839	−0.2502	0.8259
C(5)	0.3175	−0.1152	0.8183
C(6)	0.2145	−0.1377	0.8056
C(7)	0.1383	0.0023	0.7826
O(1)	0.0912	0.0272	0.8872
O(2)	0.1233	0.0881	0.6198
O(3)	0.0475	0.2257	0.5994
O(4)	0.3253	0.1651	0.9061
O(5)	0.4364	0.0759	0.7845
N	0.3628	0.0550	0.8348
H(1)	0.105	−0.307	0.790
H(2)	0.215	−0.552	0.840
H(3)	0.395	−0.505	0.815
H(4)	0.460	−0.225	0.830
H(5)	0.065	0.340	0.530

Note: The parameters given here are those found in the least squares refinement.

The structure, shown in Figure XVAIII,72, has molecules with the bond dimensions of Figure XVAIII,73. The plane of the NO_2 group makes 28° and that of the carboxyl group 58° with the benzene ring. Between the COO and OOH planes there is a dihedral angle of 146°.

XV,aIII,48. Crystals of *p-aminobenzoic acid,* $NH_2C_6H_4COOH$, are monoclinic with a unit that contains eight molecules and has the dimensions:

$$a_0 = 18.551(2) \text{ A.}; \quad b_0 = 3.860(10) \text{ A.}; \quad c_0 = 18.642(3) \text{ A.}$$
$$\beta = 93.56(2)°$$

Atoms are in the positions

$$(4e) \quad \pm(xyz;\ x+{}^1/_2,{}^1/_2-y,z+{}^1/_2)$$

of the space group C_{2h}^5 ($P2_1/n$). This large unit requires that there be two kinds of molecule in the structure: the parameters of their atoms are listed in Table XVAIII,42.

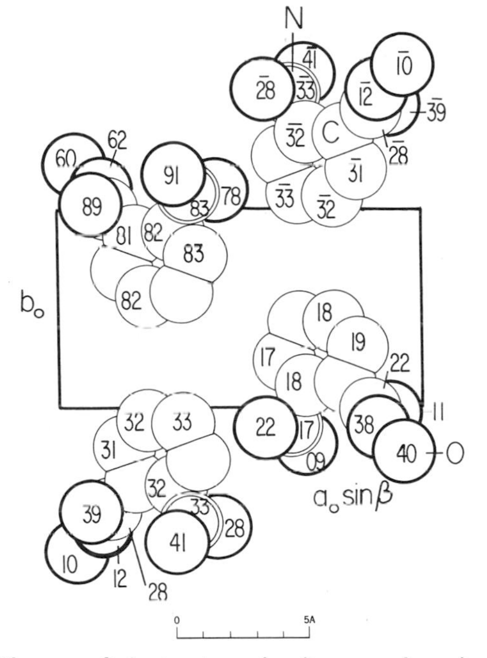

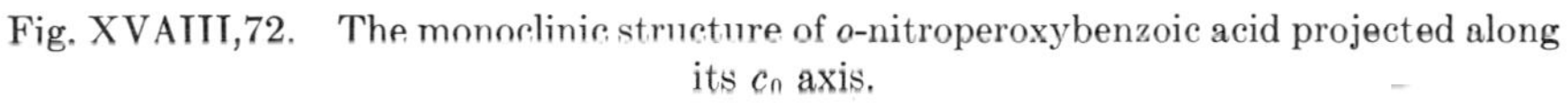
Fig. XVAIII,72. The monoclinic structure of *o*-nitroperoxybenzoic acid projected along its c_0 axis.

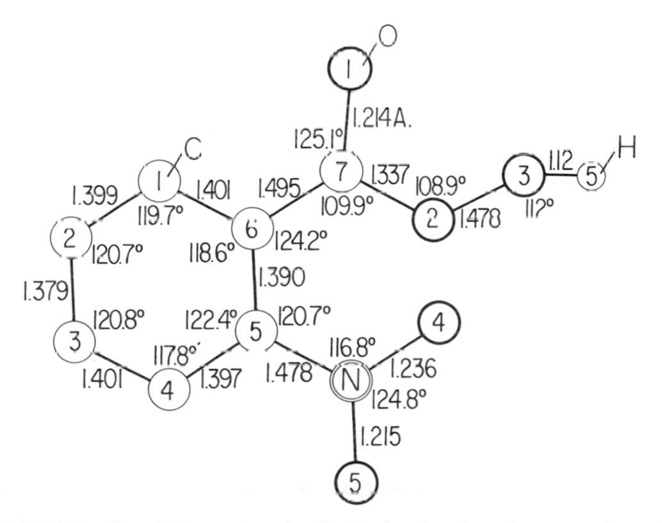

Fig. XVAIII,73. Bond dimensions in the molecule of *o*-nitroperoxybenzoic acid.

TABLE XVAIII,42

Parameters of the Atoms in *p*-Aminobenzoic Acid

Atom	x	y	z
	Molecule A		
N	0.3461(1)	0.5079(8)	−0.1666(2)
C(1)	0.1542(1)	0.2733(7)	−0.0711(1)
C(2)	0.2214(2)	0.1898(8)	−0.0372(2)
C(3)	0.2842(2)	0.2738(8)	−0.0675(2)
C(4)	0.2829(2)	0.4379(8)	−0.1343(2)
C(5)	0.2160(2)	0.5258(8)	−0.1686(2)
C(6)	0.1531(2)	0.4422(8)	−0.1374(2)
C(7)	0.0879(2)	0.1681(8)	−0.0394(2)
O(1)	0.0866(1)	0.0089(6)	0.0180(1)
O(2)	0.0276(1)	0.2586(6)	−0.0754(1)
H(1)	−0.009(2)	0.135(10)	−0.053(2)
H(2)	0.224(2)	0.065(8)	0.011(2)
H(3)	0.330(2)	0.238(8)	−0.044(2)
H(4)	0.390(2)	0.484(9)	−0.139(2)
H(5)	0.346(2)	0.678(10)	−0.199(2)
H(6)	0.215(2)	0.666(9)	−0.217(2)
H(7)	0.106(1)	0.490(7)	−0.163(1)
	Molecule B		
N′	0.3304(1)	−0.5445(7)	0.3401(1)
C(1′)	0.4286(1)	−0.2821(7)	0.1519(2)
C(2′)	0.4624(2)	−0.2073(8)	0.2194(2)
C(3′)	0.4309(8)	−0.2967(8)	0.2817(2)
C(4′)	0.3642(8)	−0.4648(8)	0.2783(2)
C(5′)	0.3300(8)	−0.5452(8)	0.2111(2)
C(6′)	0.3622(8)	−0.4530(8)	0.1489(2)
C(7′)	0.4615(8)	−0.1727(8)	0.0862(2)
O(1′)	0.5213(6)	−0.0236(6)	0.0883(1)
O(2′)	0.4248(6)	−0.2343(6)	0.0262(1)
H(1′)	0.450(2)	−0.111(11)	−0.013(2)
H(2′)	0.508(2)	−0.067(8)	0.220(2)
H(3′)	0.457(2)	−0.250(8)	0.330(2)
H(4′)	0.359(2)	−0.517(9)	0.383(2)
H(5′)	0.301(2)	−0.711(11)	0.341(2)
H(6′)	0.282(1)	−0.671(7)	0.206(1)
H(7′)	0.337(1)	−0.506(7)	0.100(2)

The resulting structure is shown in Figure XVAIII,74. Its two different molecules have the bond dimensions of Figure XVAIII,75. The benzene

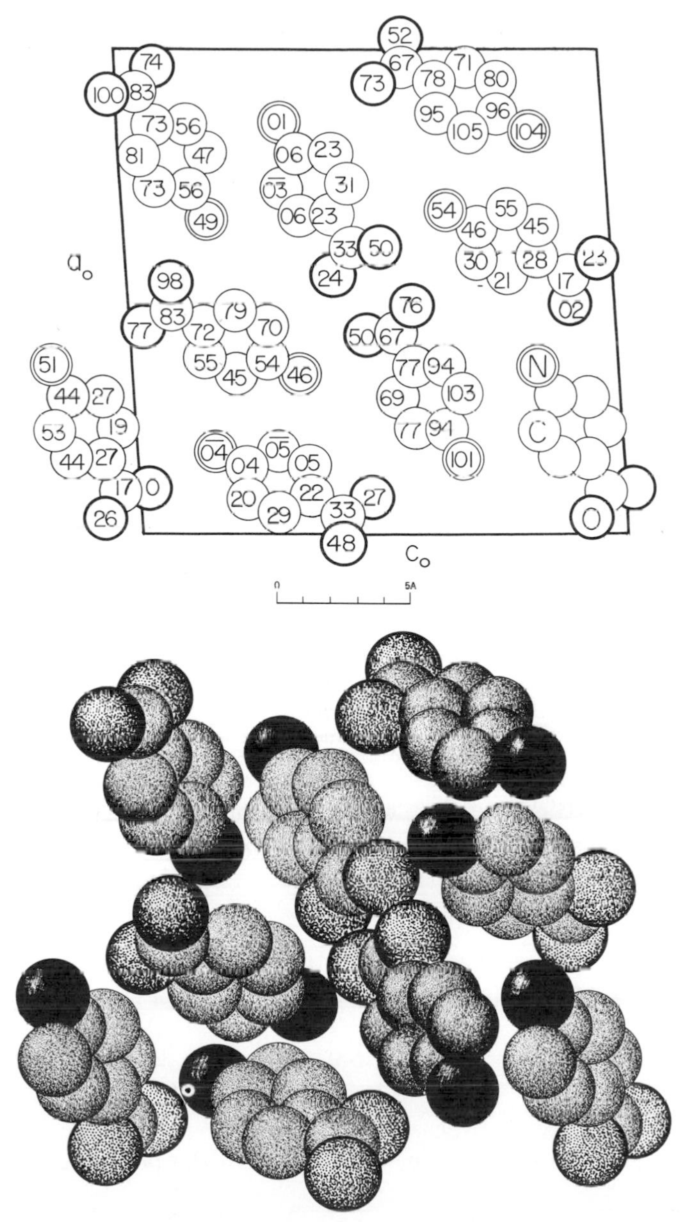

Fig. XVAIII,74a (top). The monoclinic structure of *p*-aminobenzoic acid projected along its b_0 axis.

Fig. XVAIII,74b (bottom). A packing drawing of the monoclinic structure of p-aminobenzoic acid viewed along its b_0 axis. The amino groups are black; the oxygen atoms are heavily dotted and outlined. Atoms of carbon are lightly dot shaded.

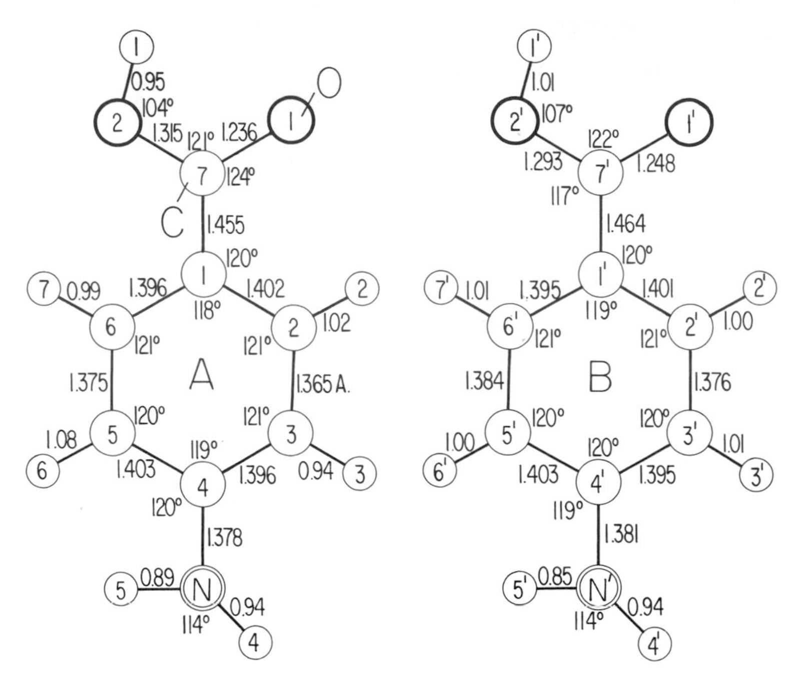

Fig. XVAIII,75. Bond dimensions in the crystallographically different molecules (A and B) in *p*-aminobenzoic acid.

ring of each is strictly planar, with the attached nitrogen and carboxyl carbon atoms displaced from the plane by 0.06–0.07 A. Carboxyl oxygens are as much as 0.14 A. from the plane.

XV,aIII,49. The monoclinic crystals of *methyl-m-bromocinnamate*, $BrC_6H_4(CH)_2COOCH_3$, have a tetramolecular unit of the dimensions:

$$a_0 = 7.830 \text{ A.}; \quad b_0 = 5.976 \text{ A.}; \quad c_0 = 21.208 \text{ A.}$$
$$\beta = 99°31'$$

The space group is C_{2h}^5 ($P2_1/a$) with atoms in the positions:

$$(4e) \quad \pm(xyz;\ x+{}^1/_2,{}^1/_2-y,z)$$

The parameters are those of Table XVAIII,43.

TABLE XVAIII,43

Parameters of the Atoms in Methyl *m*-Bromocinnamate

Atom	x	y	z
Br	0.1969	0.1897	0.0496
O(1)	0.5888	−0.0924	0.3823
O(2)	0.5684	0.2615	0.4225
C(1)	0.6866	−0.1596	0.4421
C(2)	0.5370	0.1332	0.3793
C(3)	0.4502	0.1831	0.3121
C(4)	0.3905	0.3795	0.2982
C(5)	0.3005	0.4459	0.2328
C(6)	0.2968	0.3147	0.1795
C(7)	0.2024	0.3787	0.1196
C(8)	0.1186	0.5900	0.1131
C(9)	0.1249	0.7379	0.1692
C(10)	0.2119	0.6578	0.2268
H(1)	0.429	0.074	0.279
H(2)	0.406	0.504	0.336
H(3)	0.207	0.746	0.264
H(4)	0.054	0.865	0.164
H(5)	0.056	0.632	0.074
H(6)	0.356	0.184	0.184

In this structure (Fig. XVAIII,76) molecules have the bond dimensions of Figure XVAIII,77. They can be described in terms of three planes: one containing the benzene ring, another the ethylenic group C(2)–C(5), and the third the group $C(3)CO_2CH_3$. The only significant departure from these planes involves the methyl C(1) atom which is 0.10 A. away from the last. Between molecules there is a short C(1)–O(2) = 3.21 A.; other intermolecular distances range upwards from 3.53 A.

XV,aIII,50. *Methyl-p-bromocinnamate,* $BrC_6H_4(CH)_2COOCH_3$, forms monoclinic crystals which have a tetramolecular unit of the dimensions:

$$a_0 = 8.485 \text{ A.}; \quad b_0 = 20.703 \text{ A.}; \quad c_0 = 5.764 \text{ A.}$$
$$\beta = 92°12'$$

The space group is C_{2h}^5 ($P2_1/n$) with all atoms in the positions:

$$(4e) \quad \pm(xyz;\ x+{}^1/_2,{}^1/_2-y,z+{}^1/_2)$$

The determined parameters are listed in Table XVAIII,44.

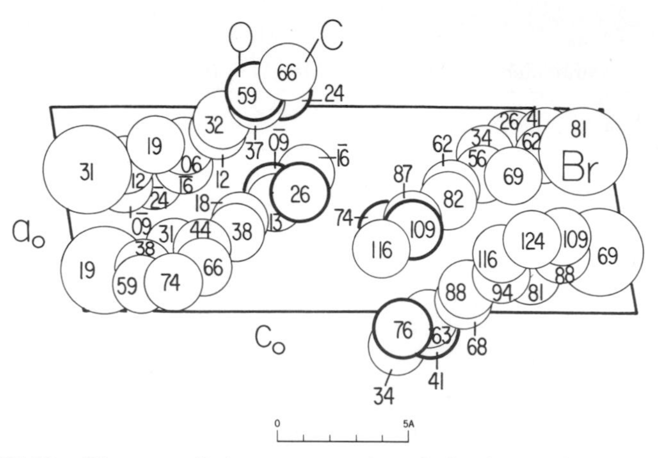

Fig. XVAIII,76. The monoclinic structure of methyl-*m*-bromocinnamate projected along its b_0 axis.

TABLE XVAIII,44

Parameters of the Atoms in Methyl *p*-Bromocinnamate

Atom	x	y	z
Br	−0.0913	0.2353	0.0186
O(1)	0.2526	0.5592	0.0105
O(2)	0.4218	0.5561	0.7647
C(1)	0.3559	0.6077	0.1542
C(2)	0.3031	0.5381	0.8468
C(3)	0.1961	0.4874	0.7326
C(4)	0.2359	0.4582	0.5498
C(5)	0.1517	0.4068	0.4304
C(6)	0.2084	0.3820	0.2146
C(7)	0.1409	0.3315	0.0918
C(8)	0.0052	0.3053	0.1835
C(9)	−0.0576	0.3278	0.3806
C(10)	0.0143	0.3775	0.5007

The structure is shown in Figure XVAIII,78. Its molecule has bond dimensions (Fig. XVAIII,79) similar to those of the *meta* compound (**XV, aIII,49**). All the atoms in the three planes of this molecule are, however, strictly coplanar. Between molecules the shortest separation is an O(2)–C(6) = 3.39 A.

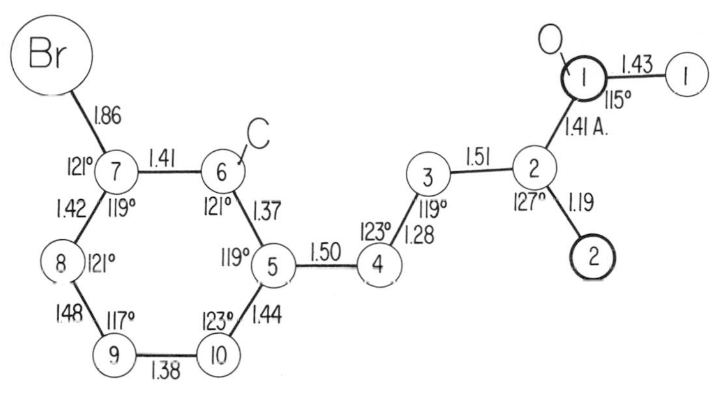

Fig. XVAIII,77. Bond dimensions in the molecule of methyl-*m*-bromocinnamate.

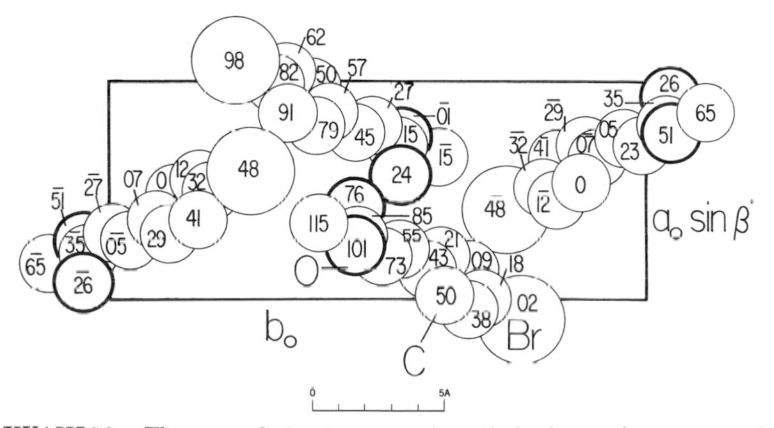

Fig. XVAIII,78. The monoclinic structure of methyl-*p*-bromocinnamate projected along its c_0 axis.

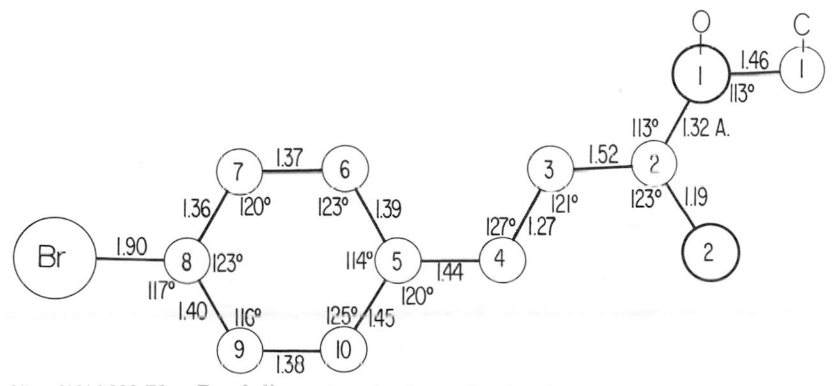

Fig. XVAIII,79. Bond dimensions in the molecule of methyl-*p*-bromocinnamate.

XV,aIII,51. Crystals of *o-phthalic acid*, $C_6H_4(COOH)_2$, are monoclinic with a tetramolecular unit of the dimensions:

$$a_0 = 5.03(2) \text{ A.}; \quad b_0 = 14.30(2) \text{ A.}; \quad c_0 = 9.59(2) \text{ A.}$$
$$\beta = 93°11(26)'$$

Atoms are in the positions

$$(8f) \quad \pm(xyz;\ x,\bar{y},z+{}^1/_2;\ x+{}^1/_2,y+{}^1/_2,z;\ x+{}^1/_2,{}^1/_2-y,z+{}^1/_2)$$

of the space group C_{2h}^6 ($C2/c$). Their parameters are listed in Table XVAIII,45.

TABLE XVAIII,45

Parameters of the Atoms in Phthalic Acid

Atom	x	y	z
C(1)	0.1693	0.1492	0.6252
C(2)	0.0923	0.0627	0.7000
C(3)	0.1912	0.0220	0.1522
C(4)	0.0977	0.1065	0.2032
O(1)	0.0147	0.2128	0.6022
O(2)	0.0945	0.3535	0.4238

The structure, shown in Figure XVAIII,80, is built of molecules having the bond dimensions of Figure XVAIII,81. Their benzene rings and the attached carboxyl carbon atoms are coplanar but the carboxyl oxygens are rotated out of this plane. The molecules are tied into chains by hydrogen bonds of length 2.68 A. involving the O(2) atoms.

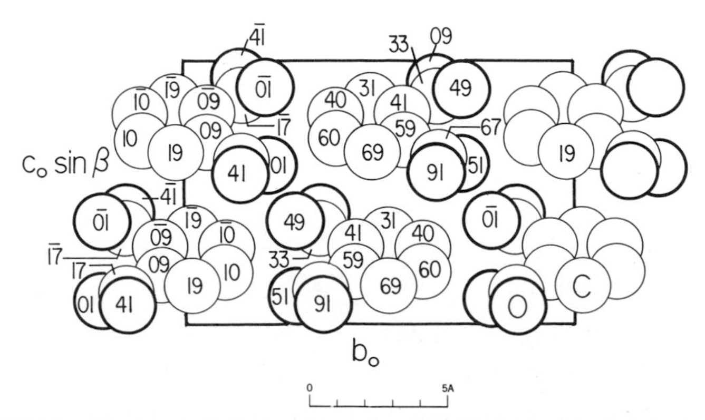

Fig. XVAIII,80. The monoclinic structure of *o*-phthalic acid projected along its a_0 axis.

XV,aIII,52. *Potassium acid phthalate*, $KHC_6H_4(COO)_2$, forms orthorhombic crystals which have a tetramolecular cell of the edge lengths:

$$a_0 = 6.466 \text{ A.}; \quad b_0 = 9.609 \text{ A.}; \quad c_0 = 13.857 \text{ A.}$$

The space group has been chosen as C_{2v}^5 ($P2_1ab$) with all atoms in the positions:

$$(4a) \quad xyz;\ x+{}^1/_2,\bar{y},\bar{z};\ x+{}^1/_2,{}^1/_2-y,z;\ x,y+{}^1/_2,\bar{z}$$

The determined parameters are stated in Table XVAIII,46.

TABLE XVAIII,46

Parameters of the Atoms in $KHC_6H_4(COO)_2$

Atom	x	y	z
C(I)	0.0056	−0.1778	0.2176
O(I)	0.0787	−0.2983	0.2626
O(I′)	0.0175	−0.1611	0.1273
C(II)	−0.2384	0.0689	0.1568
O(II)	−0.1586	0.1452	0.0931
O(II′)	−0.4038	0.0001	0.1440
C(1)	−0.0252	−0.0592	0.2894
C(2)	0.0626	−0.0637	0.3847
C(3)	0.0382	0.0473	0.4505
C(4)	−0.0735	0.1628	0.4206
C(5)	0.1580	0.1693	0.3251
C(6)	−0.1359	0.0580	0.2584
H(0)	0.038	−0.380	0.235
H(2)	0.140	−0.146	0.400
H(3)	0.096	0.048	0.521
H(4)	−0.091	0.237	0.464
H(5)	0.231	0.252	0.304
K	0.2500	0.0990	0.0388

The structure is shown in Figure XVAIII,82. Its anions have the bond dimensions of Figure XVAIII,83, the carboxyl group labeled (1) being considered non-ionized. The benzene ring is planar to within 0.01 A. but carboxyl carbons are 0.029 A. [for C(I)] below and 0.036 A. [for C(II)] above this plane. The carboxyl groups make angles of 31.7° (for I) and 77.4° (for II) with the benzene plane. Coordination about the potassium cations is sixfold, with three K–O(II) distances of 2.629, 2.774, and 2.986 A., two K–O(II′) of 2.787 and 2.803 A. and one K–O(I′) of 2.863 A. There is a short hydrogen bond of 2.546 A. between O(I) and O(II′).

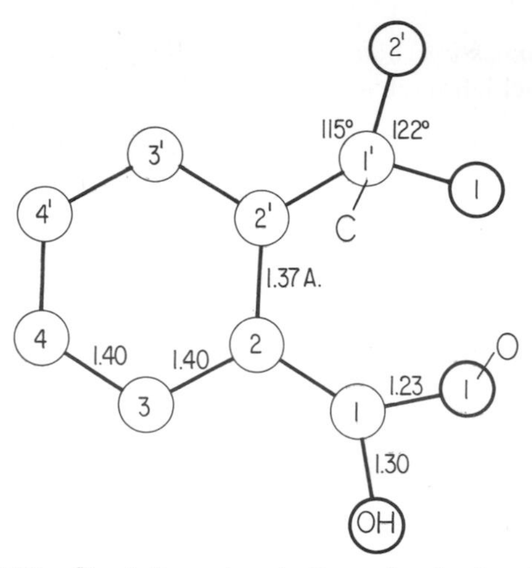

Fig. XVAIII,81. Bond dimensions in the molecule of *o*-phthalic acid.

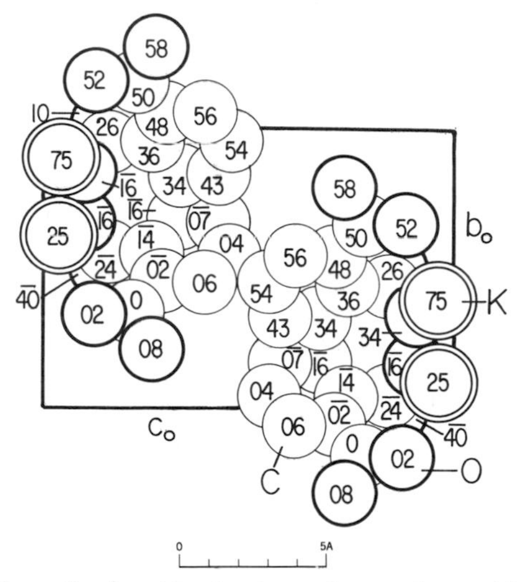

Fig. XVAIII,82. The orthorhombic structure of potassium acid phthalate projected along its a_0 axis.

In an earlier study this compound was erroneously assigned the different space group V^3 ($P2_12_12$). It was there shown that other alkali salts have similar units; undoubtedly they are isostructural. In line with the more recent investigation the correct space group presumably is C_{2v}^5 ($P2_1ab$). The units of these other compounds have the edge lengths:

$RbHC_6H_4(COO)_2$: $a_0 = 6.55$ A.; $b_0 = 10.02$ A.; $c_0 = 12.99$ A.
$CsHC_6H_4(COO)_2$: $a_0 = 6.58$ A.; $b_0 = 10.81$ A.; $c_0 = 12.84$ A.
$TlHC_6H_4(COO)_2$: $a_0 = 6.63$ A.; $b_0 = 10.54$ A.; $c_0 = 12.95$ A.

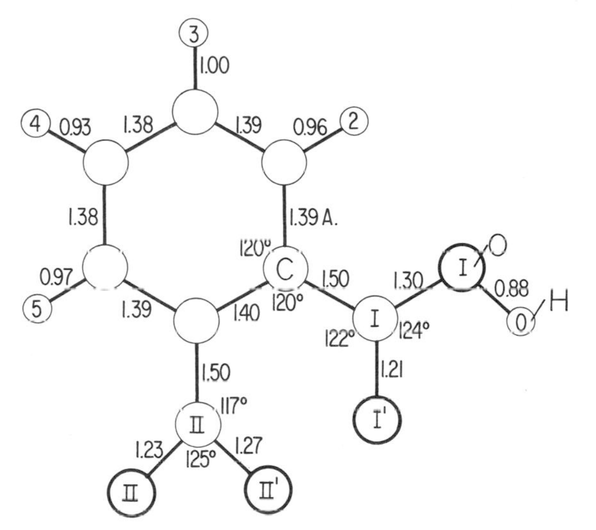

Fig. XVAIII,83. Bond dimensions in the anion of potassium acid phthalate.

XV,aIII,53. *Terephthalic acid,* p-$C_6H_4(COOH)_2$, occurs in two triclinic forms, both of which have been subjected to x-ray analysis. For modification I, a unimolecular cell of the following dimensions was chosen:

$$a_0 = 7.730 \text{ A.}; \quad b_0 = 6.443 \text{ A.}; \quad c_0 = 3.749 \text{ A.}$$
$$\alpha = 92.75°; \quad \beta = 109.15°; \quad \gamma = 95.95°$$

All atoms are in the positions $(2i)$ $\pm(xyz)$ of C_i^1 $(P\bar{1})$. Determined parameters are listed in Table XVAIII,47.

TABLE XVAIII,47

Parameters of the Atoms in Terephthalic Acid (Form I)

Atom	x	y	z
C(1)	0.1451	0.1160	0.0063
C(2)	−0.0361	0.1895	−0.1615
C(3)	0.1801	−0.0442	0.1687
C(4)	0.3012	0.3017	0.0078
O(1)	0.4633	0.2614	0.1856
O(2)	0.2681	0.4638	−0.1660
H(1)	0.575	0.375	0.177
H(2)	−0.062	0.336	−0.284
H(3)	0.320	−0.076	0.299

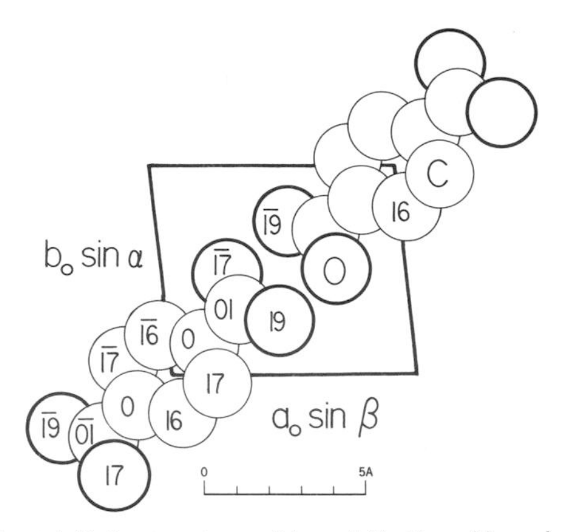

Fig. XVAIII,84. The triclinic structure of terephthalic acid projected along its c_0 axis.

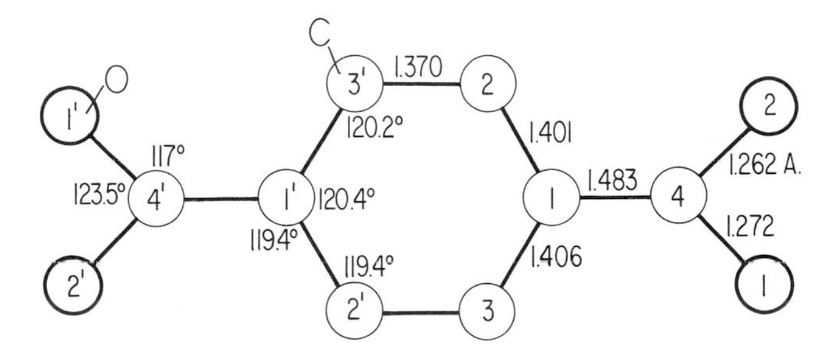

Fig. XVAIII,85. Bond dimensions in the molecule of terephthalic acid.

The molecules in its structure (Fig. XVAIII,84) have a center of symmetry and the bond dimensions of Figure XVAIII,85. The central benzene ring is planar but the carboxyl groups are inclined by 5.25° to the plane. In the crystal the molecules are tied together into chains by hydrogen bonds between their carboxyl oxygens, with O–H . . . O = 2.608 A.

XV,aIII,54. The second form (II) of *terephthalic acid*, p-$C_6H_4(COOH)_2$, has been assigned a unimolecular triclinic cell of the dimensions:

$$a_0 = 9.54 \text{ A.}; \quad b_0 = 5.34 \text{ A.}; \quad c_0 = 5.02 \text{ A.}$$
$$\alpha = 86°57'; \quad \beta = 134°39'; \quad \gamma = 94°48'$$

Atoms, in the general positions of C_i^1 ($P\bar{1}$), $(2i)$ $\pm(xyz)$, have the approximate parameters of Table XVAIII,48. As in form I, the molecules are linked into chains by hydrogen bonds.

TABLE XVAIII,48

Parameters of the Atoms in Terephthalic Acid (Form II)

Atom	x	y	z
C(1)	0.147	−0.002	0.000
C(2)	0.175	0.217	−0.184
C(3)	0.031	−0.224	0.186
C(4)	0.301	−0.007	0.003
O(1)	0.461	−0.196	0.181
O(2)	0.275	0.209	−0.158

XV,aIII,55. *Diethyl terephthalate*, $C_6H_4(COOC_2H_5)_2$, is monoclinic with a bimolecular unit of the dimensions:

$$a_0 = 9.12 \text{ A.}; \quad b_0 = 15.39 \text{ A.}; \quad c_0 = 4.21 \text{ A.}$$
$$\beta = 93°24'$$

All atoms are in the positions

$$(4e) \quad \pm(xyz;\ x+1/2,1/2-y,z+1/2)$$

of the space group C_{2h}^5 in the orientation $P2_1/n$. Parameters are those of Table XVAIII,49.

TABLE XVAIII,49

Parameters of the Atoms in Diethyl Terephthalate

Atom	x	y	z
CH(1)	0.153	−0.008	0.990
CH(2)	−0.070	0.067	0.849
C(3)	0.083	0.062	0.841
C(4)	0.172	0.125	0.678
CH_2(5)	0.408	0.175	0.511
CH_3(6)	0.562	0.133	0.517
O(1)	0.315	0.110	0.672
O(2)	0.112	0.192	0.539

The structure that results is shown in Figure XVAIII,86. The molecule is planar except for the terminal CH_3 radical of the ethyl group; the CH_2(5)–CH_3(6) bond makes an angle of 9° with the molecular plane. Bond lengths found in this early structure are shown in Figure XVAIII,87.

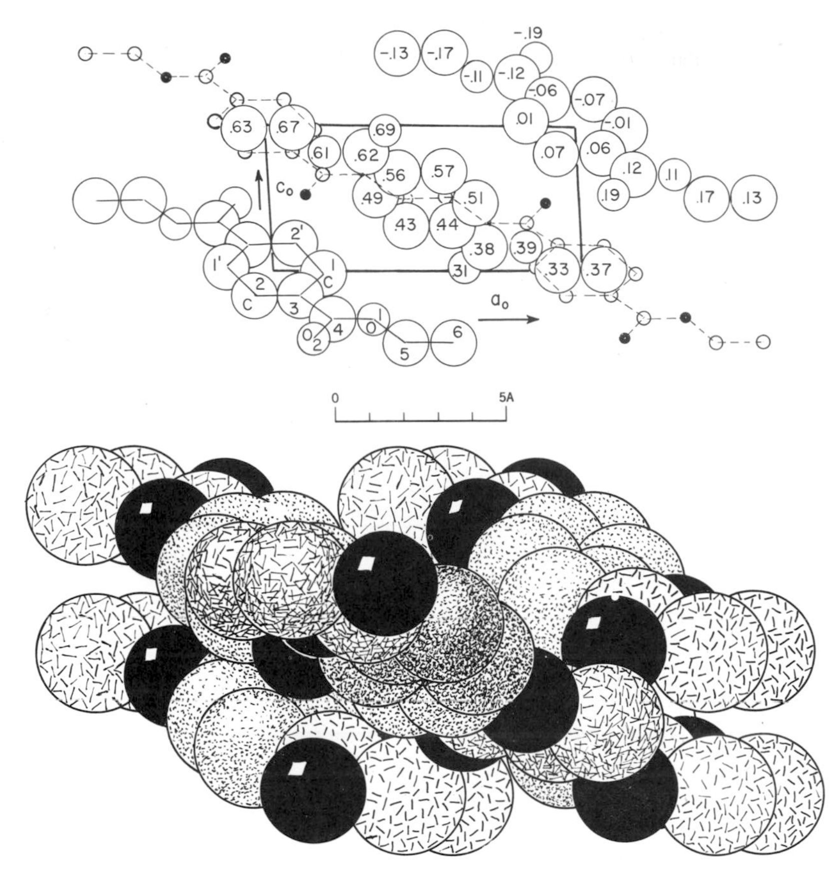

Fig. XVAIII,86a (top). The monoclinic structure of diethyl terephthalate projected along its b_0 axis. Left hand axes.

Fig. XVAIII,86b (bottom). A packing drawing of the monoclinic structure of diethyl terephthalate seen along its b_0 axis. The oxygen atoms are black. Benzene carbon atoms are dotted; the others are line shaded. Left hand axes.

XV,aIII,56. A complex formed by interaction between a molecule of *p-benzoquinone* and one of *p-chlorophenol*, $p\text{-}C_6H_4O_2 \cdot p\text{-}ClC_6H_4OH$, is triclinic with a bimolecular cell of the dimensions:

$$a_0 = 6.80(3) \text{ A.}; \quad b_0 = 7.90(3) \text{ A.}; \quad c_0 = 12.25(3) \text{ A.}$$
$$\alpha = 106.9(5)°; \quad \beta = 106.5(5)°; \quad \gamma = 58.0(5)°$$

Atoms are in the positions $(2i)$ $\pm(xyz)$ of C_i^1 $(P\bar{1})$.

There is a corresponding *bromophenol* complex, $p\text{-}C_6H_4O_2 \cdot p\text{-}BrC_6H_4OH$, which is isostructural and has the cell dimensions:

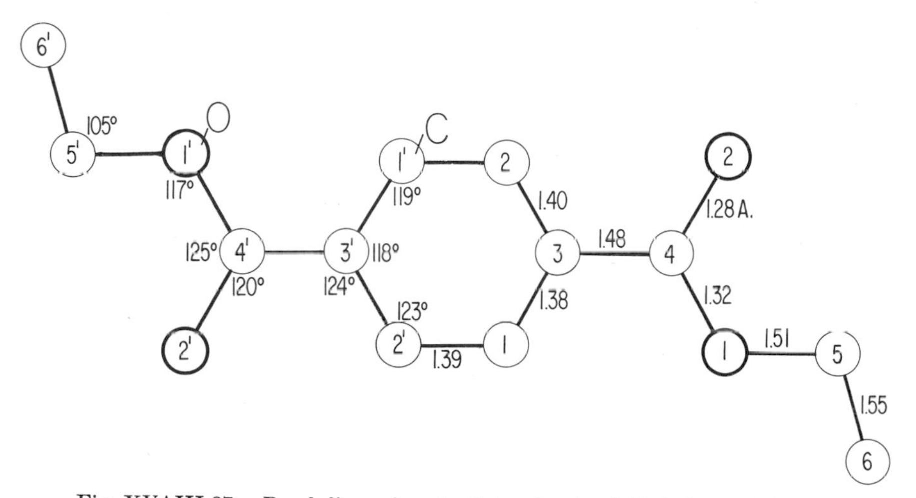

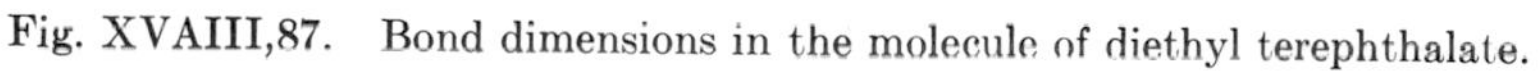

Fig. XVAIII,87. Bond dimensions in the molecule of diethyl terephthalate.

$$a_0 = 6.84(3)\text{ A.};\quad b_0 = 7.86(3)\text{ A.};\quad c_0 = 12.49(3)\text{ A.}$$
$$\alpha = 105.8(5)^\circ;\quad \beta = 104.4(5)^\circ;\quad \gamma = 58.7(5)^\circ$$

The parameters of the chlorine compound, and in square brackets, the less accurate bromine parameters, are given in Table XVAIII,50.

TABLE XVAIII,50

Parameters of the Atoms in $C_6H_4O_2 \cdot ClC_6H_4OH$

Atom	x	y	z
Cl[Br]	0.8515(6) [0.868]	0.3359(6) [0.319]	0.0918(2) [0.080]
O(1)	0.3803(19) [0.365]	0.3023(17) [0.312]	0.4310(9) [0.410]
O(2)	0.8141(21) [0.811]	0.8693(18) [0.870]	0.1183(10) [0.123]
O(3)	0.3711(18) [0.376]	0.8133(16) [0.831]	0.3816(8) [0.406]
C(1)	0.7006(23) [0.714]	0.3349(19) [0.338]	0.1924(9) [0.185]
C(2)	0.4826(24) [0.472]	0.3779(20) [0.379]	0.1601(12) [0.156]
C(3)	0.3712(25) [0.360]	0.3691(21) [0.372]	0.2353(11) [0.229]
C(4)	0.4952(25) [0.447]	0.3196(20) [0.317]	0.3441(11) [0.339]
C(5)	0.7195(27) [0.716]	0.2832(21) [0.271]	0.3752(11) [0.367]
C(6)	0.8328(27) [0.828]	0.2940(22) [0.279]	0.3013(12) [0.295]
C(7)	0.7123(26) [0.714]	0.8598(21) [0.786]	0.1768(12) [0.185]
C(8)	0.4703(25) [0.472]	0.8977(20) [0.905]	0.1395(12) [0.156]
C(9)	0.3628(25) [0.360]	0.8736(20) [0.900]	0.2055(11) [0.229]
C(10)	0.4783(26) [0.477]	0.8230(22) [0.840]	0.3206(11) [0.338]
C(11)	0.7148(25) [0.716]	0.7869(23) [0.797]	0.3572(13) [0.367]
C(12)	0.8328(29) [0.828]	0.8066(23) [0.804]	0.2949(14) [0.295]

Note: Values for the bromine analogue in square brackets.

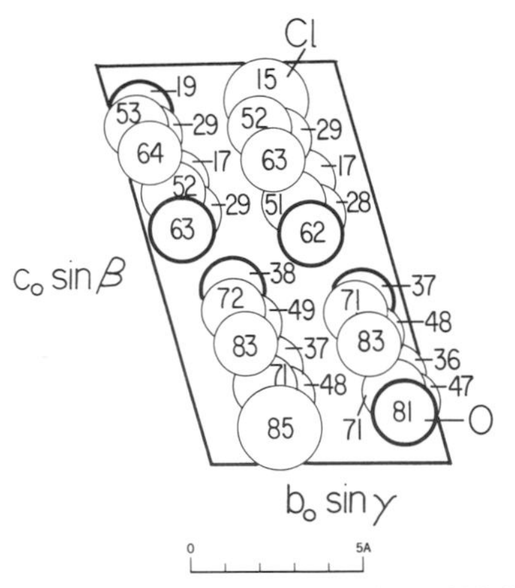

Fig. XVAIII,88. The triclinic structure of p-$C_6H_4O_2 \cdot p$-ClC_6H_4OH projected along its a_0 axis.

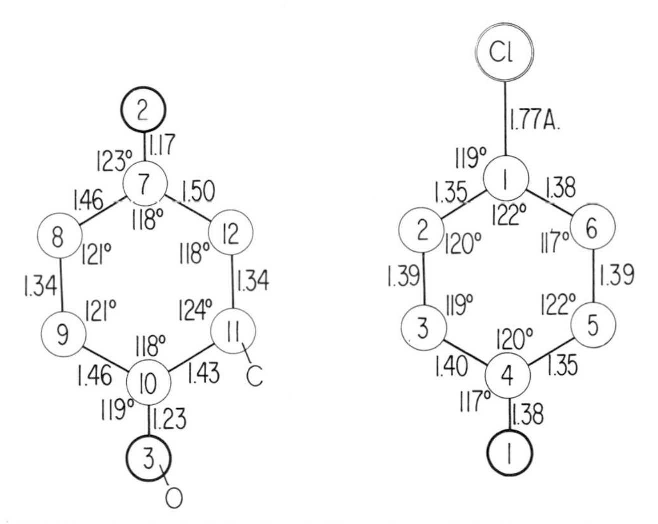

Fig. XVAIII,89. On the left the bond dimensions of the benzoquinone molecule as it occurs in the complex p-$C_6H_4O_2 \cdot p$-ClC_6H_4OH. On the right the bond dimensions of the chlorophenol molecule in this complex.

The resulting structure is shown in Figure XVAIII,88. Bond dimensions in the chloro complex are those of Figure XVAIII,89. As Figure XVAIII,88 indicates, quinone oxygens are roughly centered over benzene rings viewed normal to c_0, suggesting strong interaction between them; in this direction the shortest intermolecular distances are C–C = 3.31 A. and C–O = 3.24 A. Along c_0 the molecules are held together by hydrogen bonds O(1)–H . . . O(3) = 2.70 A.

XV,aIII,57. The complex formed by one molecule of *p-benzoquinone* and two of *p-chlorophenol*, $p\text{-}C_6H_4O_2 \cdot 2p\text{-}ClC_6H_4OH$, is monoclinic with a bimolecular unit of the dimensions:

$$a_0 = 11.70(3) \text{ A.}; \quad b_0 = 6.10(3) \text{ A.}; \quad c_0 = 11.83(3) \text{ A.}$$
$$\beta = 97.0(5)°$$

The corresponding isostructural *bromophenol* compound has the cell:

$$a_0 = 11.91(3) \text{ A.}; \quad b_0 = 6.16(3) \text{ A.}; \quad c_0 = 12.03(3) \text{ A.}$$
$$\beta = 96.0(5)°$$

The space group is C_{2h}^5 ($P2_1/c$) with atoms in the positions:

$$(4e) \quad \pm(xyz;\ x,{}^1/_2-y,z+{}^1/_2)$$

The parameters are those of Table XVAIII,51, the less accurate values for the bromine compound being in square brackets.

TABLE XVAIII,51

Parameters of the Atoms in $C_6H_4O_2 \cdot 2ClC_6H_4OH$

Atom	x	y	z
C(1)	0.1320(8) [0.140]	0.3450(20) [0.374]	0.6148(10) [0.614]
C(2)	0.1913(10) [0.203]	0.5011(22) [0.502]	0.6875(10) [0.689]
C(3)	0.2504(10) [0.267]	0.6742(24) [0.671]	0.6401(10) [0.640]
C(4)	0.2450(9) [0.265]	0.6870(20) [0.752]	0.5238(10) [0.516]
C(5)	0.1864(9) [0.198]	0.5326(20) [0.623]	0.4477(9) [0.448]
C(6)	0.1293(9) [0.134]	0.3582(23) [0.462]	0.5000(9) [0.500]
C(7)	0.4521(9) [0.449]	0.3185(19) [0.304]	0.5557(10) [0.562]
C(8)	0.4411(9) [0.447]	0.3317(21) [0.344]	0.4316(9) [0.429]
C(9)	0.4864(9) [0.500]	0.4987(19) [0.498]	0.3761(10) [0.379]
O(1)	0.3009(7) [0.317]	0.8489(15) [0.893]	0.4693(7) [0.474]
O(2)	0.4141(7) [0.407]	0.1640(14) [0.138]	0.6052(7) [0.612]
Cl[Br]	0.0557(3) [0.058]	0.1394(6) [0.135]	0.6753(3) [0.678]

Note: Values for the bromine analogue in square brackets.

The structure is shown in Figure XVAIII,90. Its molecules, the benzoquinone having a center of symmetry, exhibit the bond dimensions of Figure XVAIII,91. Between O(1) and O(2) of different molecules there are hydrogen bonds of length 2.74 A. Other intermolecular separations range upwards from an O–C = 3.11 A. and a C–C = 3.25 A.

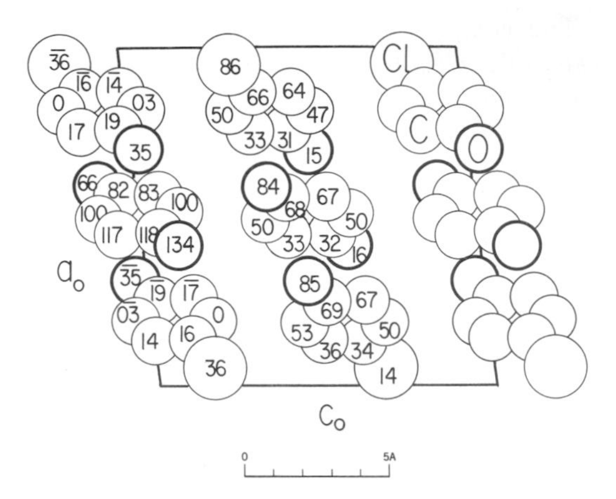

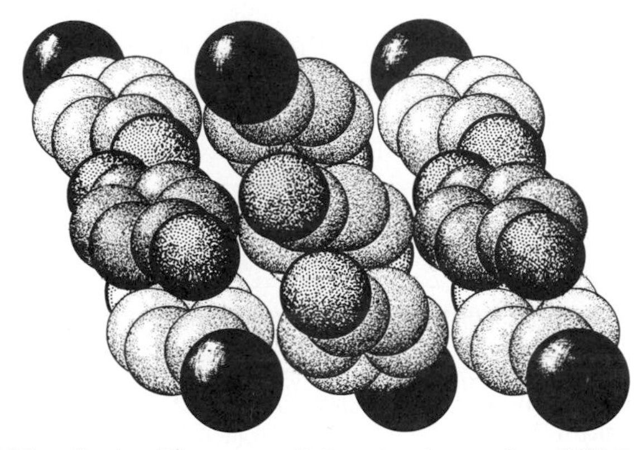

Fig. XVAIII,90a (top). The monoclinic structure of p-$C_6H_4O_2 \cdot 2p$-ClC_6H_4OH projected along its b_0 axis.

Fig. XVAIII,90b (bottom). A packing drawing of the monoclinic structure of p-$C_6H_4O_2 \cdot 2p$-ClC_6H_4OH viewed along its b_0 axis. The chlorine atoms are large and black. The atoms of oxygen are more coarsely dot shaded than those of carbon.

XV,aIII,58. Crystals of *p-cresol*, $CH_3C_6H_4OH$, are monoclinic with a large unit containing eight molecules. Its dimensions are:

$$a_0 = 5.72(2) \text{ A.}; \quad b_0 = 11.74(3) \text{ A.}; \quad c_0 = 18.68(4) \text{ A.}$$
$$\beta = 98°49(30)'$$

Atoms are in the positions

$$(4e) \quad \pm(xyz;\ \ x,{}^1/_2-y,z+{}^1/_2)$$

of $C_{2h}^5(P2_1/c)$. Their parameters are listed in Table XVAIII,52.

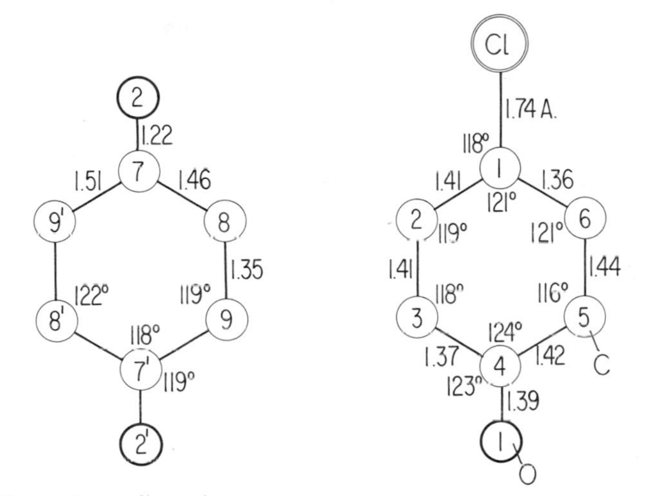

Fig. XVAIII,91. Bond dimensions in the benzoquinone (on the left) and chlorophenol (on the right) molecules in p-$C_6H_4O_2 \cdot 2p$-ClC_6H_4OH.

TABLE XVAIII,52

Parameters of the Atoms in p-Cresol

Atom	x	y	z
C(1)	0.167(3)	0.291(1)	0.036(1)
C(2)	0.949(3)	0.271(1)	0.001(1)
C(3)	0.831(3)	0.174(2)	0.017(1)
C(4)	0.928(3)	0.099(1)	0.074(1)
C(5)	0.154(3)	0.126(1)	0.110(1)
C(6)	0.277(3)	0.224(2)	0.095(1)
C(7)	0.799(3)	0.992(2)	0.090(1)
O(1)	0.286(2)	0.392(1)	0.019(1)
C(8)	0.502(3)	0.089(1)	0.355(1)
C(9)	0.425(3)	0.206(1)	0.351(1)
C(10)	0.544(3)	0.281(1)	0.308(1)
C(11)	0.732(3)	0.247(1)	0.274(1)
C(12)	0.811(3)	0.129(2)	0.283(1)
C(13)	0.680(3)	0.052(1)	0.324(1)
C(14)	0.866(3)	0.323(2)	0.229(1)
O(2)	0.384(2)	0.019(1)	0.397(1)

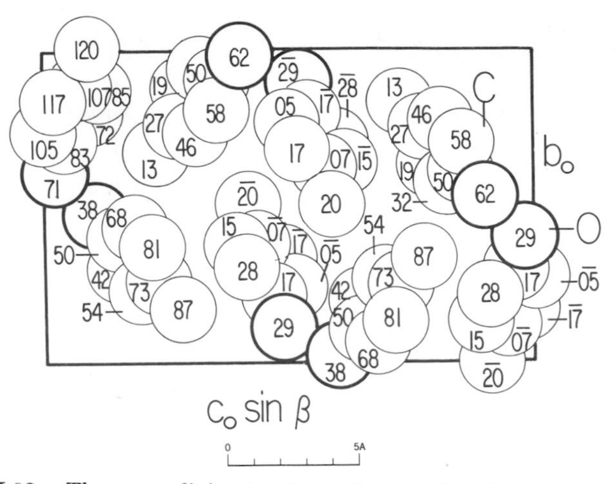

Fig. XVAIII,92. The monoclinic structure of *p*-cresol projected along its a_0 axis.

The structure (Fig. XVAIII,92) contains two crystallographically different molecules. Both are planar and they have bond dimensions which are the same to within the limit of accuracy of the determination. Their average C–O distance is 1.41 A. In the crystal they are united by hydrogen bonds of lengths 2.65 and 2.71 A.

XV,aIII,59. The orthorhombic crystals of *p-aminophenol*, $NH_2C_6H_4OH$, have a tetramolecular unit of the edge lengths:

$$a_0 = 12.90 \text{ A.}; \quad b_0 = 8.19 \text{ A.}; \quad c_0 = 5.25 \text{ A.}$$

The space group is C_{2v}^9 ($P2_1nb$) with atoms in the positions:

$$(4a) \quad xyz;\ x+\tfrac{1}{2},\bar{y},\bar{z};\ x+\tfrac{1}{2},\tfrac{1}{2}-y,z+\tfrac{1}{2};\ x,y+\tfrac{1}{2},\tfrac{1}{2}-z$$

Parameters are those of Table XVAIII,53.

TABLE XVAIII,53

Parameters of the Atoms in *p*-Aminophenol

Atom	x	y	z
C(1)	0.350	0.353	0.012
CH(2)	0.308	0.258	−0.184
CH(3)	0.204	0.227	−0.186
C(4)	0.140	0.295	−0.008
CH(5)	0.177	0.389	0.184
CH(6)	0.283	0.420	0.191
NH_2	0.456	0.381	0.021
OH	0.029	0.255	−0.017

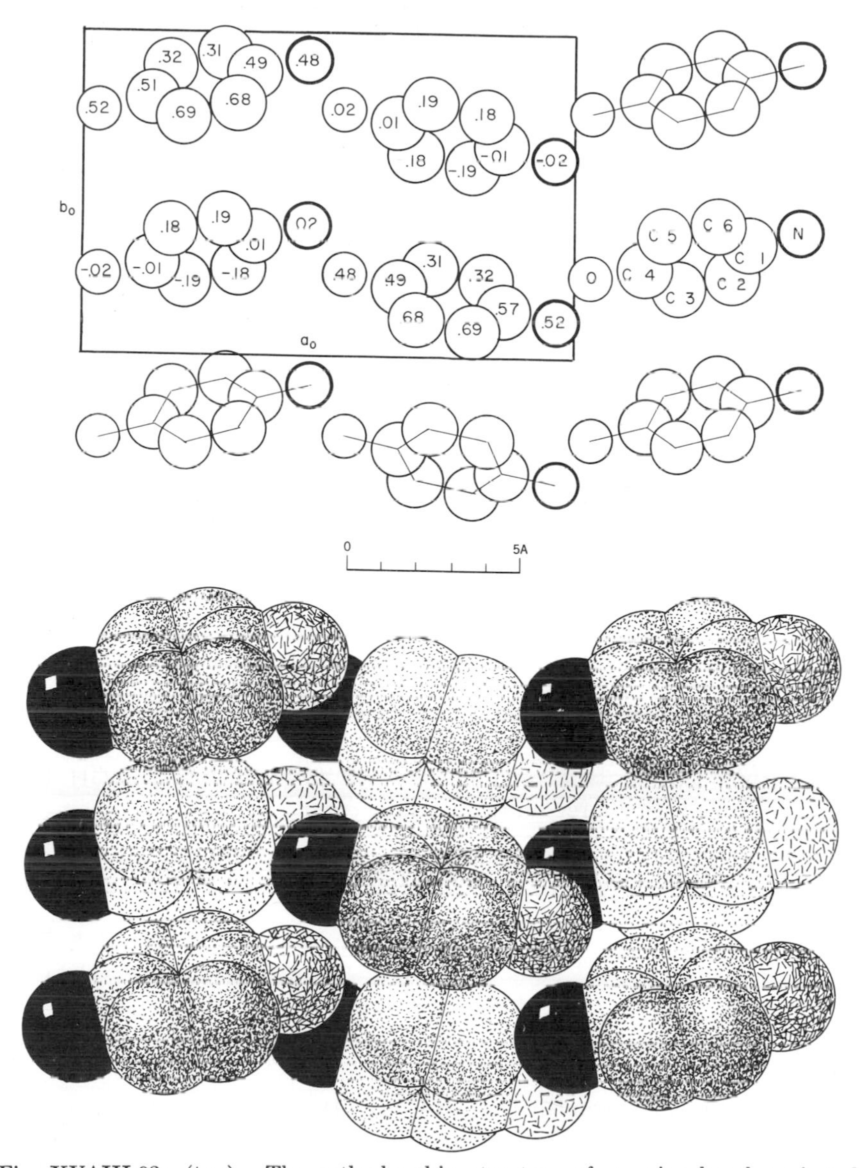

Fig. XVAIII,93a (top). The orthorhombic structure of *p*-aminophenol projected along its c_0 axis.

Fig. XVAIII,93b (bottom). A packing drawing of the orthorhombic *p*-aminophenol arrangement viewed along its c_0 axis. The oxygen atoms are black; the nitrogen atoms are line shaded. The carbon atoms are dotted.

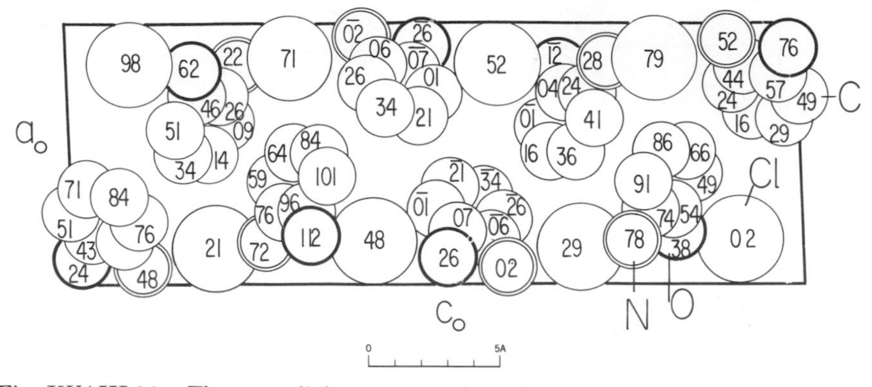

Fig. XVAIII,94. The monoclinic structure of *o*-aminophenol hydrochloride projected along its b_0 axis.

The resulting structure is shown in Figure XVAIII,93. In it the long axis of the molecules runs roughly parallel to the a_0 axis but the planar molecules are tilted at considerable angles to one another and to the a_0c_0 face of the crystal. The distribution of the molecules is polar, as would be expected, with the NH_2 and OH groups brought close together. Within the molecule the C–C separations lie between 1.36 and 1.40 A.; C–N = 1.39 A. and C–O = 1.47 A. Each nitrogen atom has three oxygen neighbors in adjacent molecules, presumably bound by hydrogen bonds of lengths 3.13 and 3.18 A.

XV,aIII,60. Crystals of *o-aminophenol hydrochloride*, $NH_2C_6H_4OH \cdot HCl$, are monoclinic with a large unit containing eight molecules. Its dimensions are:

$$a_0 = 10.280(6) \text{ A.}; \quad b_0 = 4.938(2) \text{ A.}; \quad c_0 = 28.010(6) \text{ A.}$$
$$\beta = 92.12(8)^\circ$$

Atoms are in the positions

$$(4e) \quad \pm(xyz;\ x,{}^1/_2-y,z+{}^1/_2)$$

of the space group C_{2h}^5 ($P2_1/c$). The parameters are listed in Table XVAIII,54.

The resulting arrangement (Fig. XVAIII,94) is built up of two crystallographically different molecules having the bond dimensions of Figure XVAIII,95. They are essentially planar but in one molecule, O(1) is 0.06 A. and in the other molecule, N(2) is 0.04 A. outside the plane of the rest. The structure is held together by numerous hydrogen bonds, with N–H–Cl = 3.134–3.237 A. and O–H–Cl = 3.005 or 3.072 A. Each chloride anion participates in four of these bonds.

TABLE XVAIII,54

Parameters of the Atoms in o-$NH_2C_6H_4OH \cdot HCl$

Atom	x	y	z
Cl(1)	0.1579(2)	0.0184(4)	−0.0882(1)
Cl(2)	0.1271(2)	0.2112(4)	0.1935(1)
O(1)	0.1778(5)	0.3800(12)	−0.1739(2)
O(2)	0.0945(5)	0.2374(11)	0.0109(2)
N(1)	0.1600(6)	0.7840(13)	−0.2371(2)
N(2)	0.0374(6)	0.4803(13)	0.0936(2)
C(1)	0.2830(6)	0.7435(15)	−0.2092(2)
C(2)	0.3861(7)	0.9062(18)	−0.2164(3)
C(3)	0.5017(8)	0.8633(20)	−0.1902(3)
C(4)	0.5065(9)	0.6576(20)	−0.1564(3)
C(5)	0.4004(8)	0.4874(17)	−0.1484(3)
C(6)	0.2867(7)	0.5353(15)	−0.1763(2)
C(7)	0.1558(6)	0.5625(15)	0.0693(2)
C(8)	0.1807(6)	0.4328(15)	0.0270(2)
C(9)	0.2895(8)	0.5094(16)	0.0019(3)
C(10)	0.3713(9)	0.7130(19)	0.0210(3)
C(11)	0.3435(9)	0.8378(20)	0.0632(3)
C(12)	0.2375(8)	0.7639(18)	0.0882(3)
H(1)	0.109	0.798	−0.218
H(2)	0.152	0.914	−0.265
H(3)	0.128	0.656	−0.262
H(4)	0.137	0.237	−0.145
H(5)	−0.019	0.600	0.090
H(6)	0.055	0.420	0.129
H(7)	0.012	0.317	0.066
H(8)	0.125	0.165	−0.023
H(9)	0.589	0.627	−0.136
H(10)	0.579	0.981	−0.195
H(11)	0.381	0.055	−0.240
H(12)	0.405	0.338	−0.124
H(13)	0.403	0.984	0.076
H(14)	0.450	0.768	0.003
H(15)	0.309	0.417	−0.029
H(16)	0.219	0.855	0.119

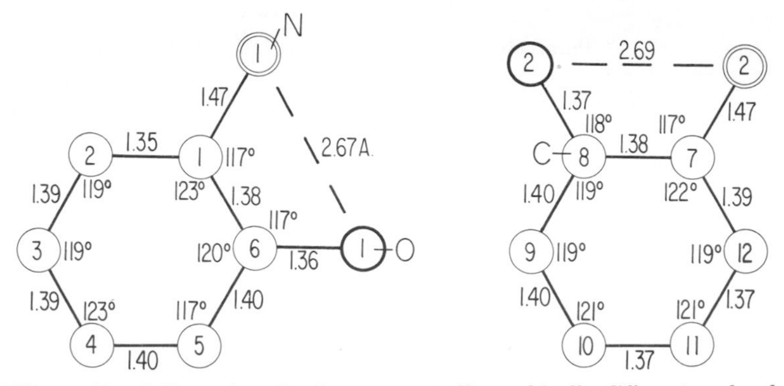

Fig. XVAIII,95. Bond dimensions in the two crystallographically different molecules of *o*-aminophenol in crystals of its hydrochloride.

XV,aIII,61. The crystals from a solution of *p*-aminophenol in acetone are of the condensation product *4-isopropylidene-aminophenol*, $C_6H_4(OH)$-$N{=}C(CH_3)_2$. Its orthorhombic unit contains four of these molecules and has the edges:

$$a_0 = 5.74 \text{ A.};\quad b_0 = 12.01 \text{ A.};\quad c_0 = 12.16 \text{ A.}$$

Atoms are in general positions of V^3 ($P22_12_1$):

$$(4c)\quad xyz;\ x\bar{y}\bar{z};\ \bar{x},y+{}^1/_2,{}^1/_2-z;\ \bar{x},{}^1/_2-y,z+{}^1/_2$$

with the parameters of Table XVAIII,55.

TABLE XVAIII,55

Parameters of the Atoms in 4-Isopropylidene-Aminophenol

Atom	x	y	z
C(1)	0.785	0.923	0.214
C(2)	0.993	0.929	0.272
C(3)	0.024	0.019	0.343
C(4)	0.856	0.104	0.343
C(5)	0.646	0.098	0.284
C(6)	0.617	0.006	0.214
C(7)	0.760	0.231	0.487
C(8)	0.819	0.340	0.543
C(9)	0.557	0.171	0.530
N	0.902	0.202	0.405
OH	0.738	0.830	0.151

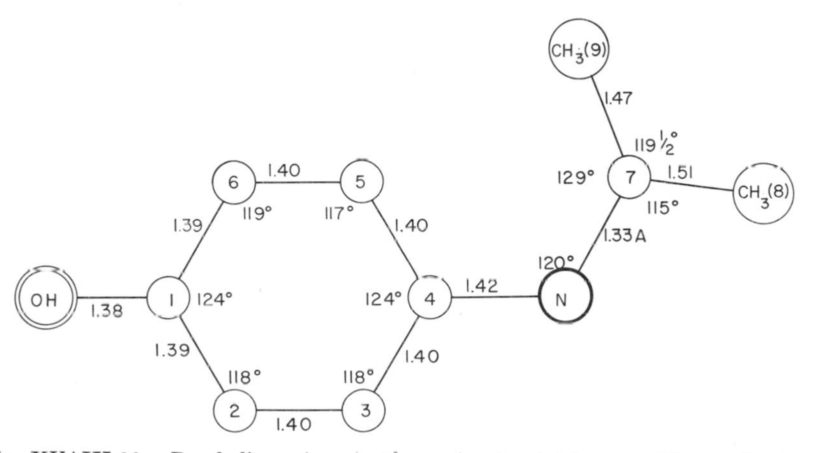

Fig. XVAIII,96. Bond dimensions in the molecule of 4-isopropylideneaminophenol.

The principal bond dimensions in the resulting molecule are indicated in Figure XVAIII,96. The benzene ring with its hydroxyl group and the nitrogen atom are planar. As can be seen from Figure XVAIII,97, which shows the molecular packing in the crystal, the rest of the molecule lies outside this plane. These molecules are bound into chains along b_0 by hydrogen bonds between nitrogen and hydroxyl (N–H–O = 2.66 A.).

XV,aIII,62. Crystals of *m-hydroxybenzamide*, $OHC_6H_4CO \cdot NH_2$, are monoclinic with a tetramolecular unit of the dimensions:

$$a_0 = 11.59(3) \text{ A.}; \quad b_0 = 5.03(3) \text{ A.}; \quad c_0 = 15.53(3) \text{ A.}$$
$$\beta = 136°6(30)'$$

Atoms are in the positions

$$(4e) \quad \pm(xyz;\ x,{}^1/_2-y,z+{}^1/_2)$$

of the space group C_{2h}^5 ($P2_1/c$). The parameters are those of Table XVAIII,56.

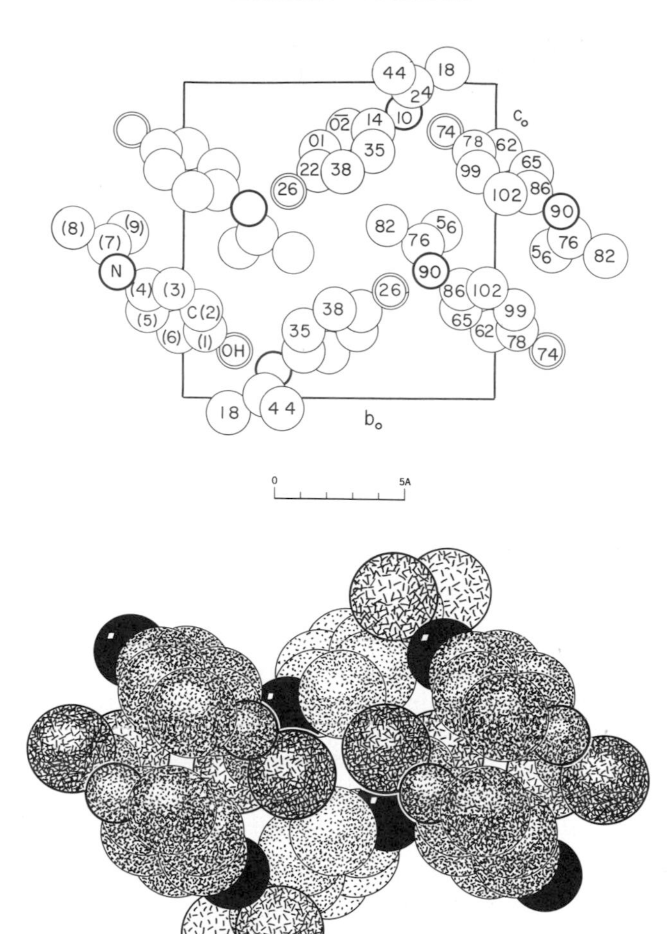

Fig. XVAIII,97a (top). The orthorhombic structure of 4-isopropylideneaminophenol projected along its a_0 axis. Left hand axes.

Fig. XVAIII,97b (bottom). A packing drawing of the orthorhombic structure of 4-isopropylideneaminophenol seen along its a_0 axis. The hydroxyl oxygens are black; the nitrogen atoms are small, dotted and heavily outlined. Of the large carbon atoms those belonging to the benzene ring are dotted, the rest are line shaded.

The arrangement of the molecules, which have the bond dimensions of Figure XVAIII,99, is that shown in Figure XVAIII,98. The benzene ring and its attached C(7) and O(2) atoms are planar while the plane defined

TABLE XVAIII,56

Parameters of the Atoms in $OHC_6H_4CO \cdot NH_2$

Atom	x	y	z
C(1)	0.3662	0.2182	0.3524
C(2)	0.1971	0.2519	0.2572
C(3)	0.1292	0.4526	0.1706
C(4)	0.2324	0.6251	0.1813
C(5)	0.4037	0.5984	0.2782
C(6)	0.4700	0.3985	0.3630
C(7)	0.6541	0.3551	0.4710
N	0.7492	0.5606	0.5089
O(1)	0.7059	0.1305	0.5196
O(2)	0.0940	0.0890	0.2466
H(1)	0.417	0.047	0.425
H(2)	0.133	−0.094	0.298
H(3)	0.000	0.442	0.089
H(4)	0.182	0.801	0.110
H(5)	0.475	0.745	0.269
H(N)	0.703	0.777	0.483
H(N′)	0.877	0.564	0.594

by C(7), O(1), and N is turned about C(6)–C(7) by 24.4°. Three types of hydrogen bond hold the molecules together; they are O–H–O(amide) = 2.76 A., N–H–O(amide) = 2.94 A., and N–H–O(hydroxy) = 2.98 A.

XV,aIII,63. The α-modification of *p-nitrophenol*, $NO_2C_6H_4OH$, has been studied at 90°K. Its crystals are monoclinic with a tetramolecular unit of the dimensions:

$$a_0 = 11.66 \text{ A.}; \quad b_0 = 8.78 \text{ A.}; \quad c_0 = 6.098 \text{ A.}$$
$$\beta = 107°32'$$

This form can be obtained at room temperature but it passes on irradiation to the β-modification described in **XV,aIII,64.** Studied at low temperature the space group is C_{2h}^5 ($P2_1/n$) with atoms in the positions:

$$(4e) \quad \pm(xyz;\ x+{}^1/_2,{}^1/_2-y,z+{}^1/_2)$$

The parameters are listed in Table XVAIII,57.

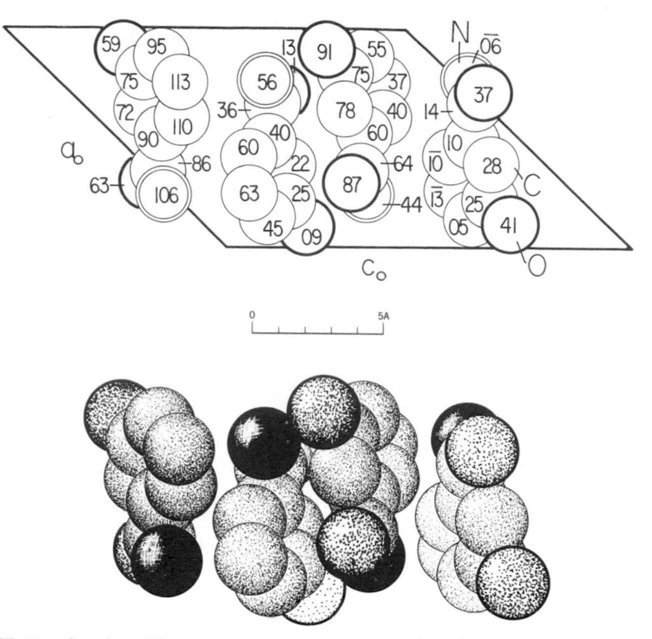

Fig. XVAIII,98a (top). The monoclinic structure of *m*-hydroxybenzamide projected along its b_0 axis.

Fig. XVAIII,98b (bottom). A packing drawing of the monoclinic *m*-hydroxybenzamide arrangement viewed along its b_0 axis. The nitrogen atoms are black. The oxygen atoms are heavily outlined and coarsely dotted; the carbon atoms, of similar size, are lightly dotted and outlined.

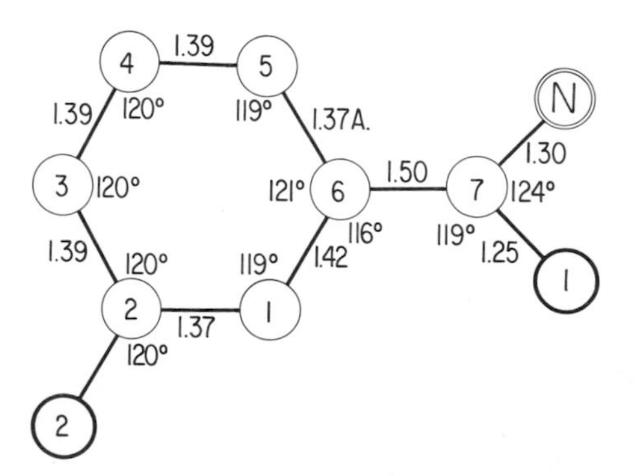

Fig. XVAIII,99. Bond dimensions in the molecule of *m*-hydroxybenzamide.

TABLE XVAIII,57

Parameters of the Atoms in α-p-$NO_2C_6H_4OH$

Atom	x	y	z
O(1)	−0.0586(3)	0.2931(4)	0.8930(6)
O(2)	−0.0864(3)	0.1267(4)	0.6215(6)
O(3)	0.3188(3)	0.5238(4)	0.4370(6)
N	−0.0350(3)	0.2409(5)	0.7240(7)
C(1)	0.0554(3)	0.3144(5)	0.6431(8)
C(2)	0.1138(3)	0.4413(5)	0.7612(7)
C(3)	0.2013(4)	0.5104(5)	0.6860(8)
C(4)	0.2298(4)	0.4526(5)	0.4971(8)
C(5)	0.1689(3)	0.3267(5)	0.3777(7)
C(6)	0.0800(3)	0.2580(5)	0.4503(9)
H(O)	0.335	0.489	0.361
H(C,2)	0.093	0.485	0.882
H(C,3)	0.241	0.609	0.777
H(C,5)	0.187	0.292	0.249
H(C,6)	0.043	0.160	0.367

The structure, illustrated in Figure XVAIII,100, has molecules with the bond dimensions of Figure XVAIII,101. They are nearly planar, the

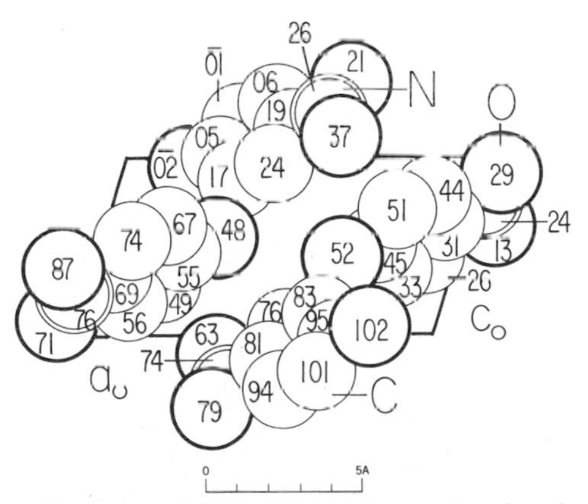

Fig. XVAIII,100. The monoclinic structure of α-p-nitrophenol projected along its b_0 axis.

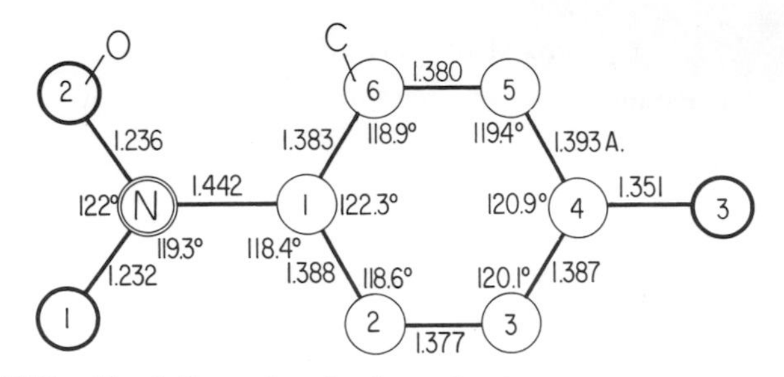

Fig. XVAIII,101. Bond dimensions in the molecule of p-nitrophenol in crystals of its α modification.

angle between the plane of the benzene ring and that of the NO_2 group being only 1°32′. Atoms of the nitro group as well as the phenolic oxygen therefore depart by no more than a few hundredths of an angstrom unit from the benzene plane. Chains of hydrogen bonds between the nitro and hydroxy oxygens (O–H . . . O = 2.82 A.) link the molecules to one another.

XV,aIII,64. Crystals of the β-form of *p-nitrophenol*, $NO_2C_6H_4OH$, are also monoclinic but with a unit of the dimensions:

$$a_0 = 15.403(3) \text{ A.}; \quad b_0 = 11.117(2) \text{ A.}; \quad c_0 = 3.785(1) \text{ A.}$$
$$\beta = 107°4(2)' \quad \text{(room temperature)}$$
$$a_0 = 15.21(2) \text{ A.}; \quad b_0 = 11.04(2) \text{ A.}; \quad c_0 = 3.622(6) \text{ A.}$$
$$\beta = 106°49(10)' \quad \text{(ca. 90°K.)}$$

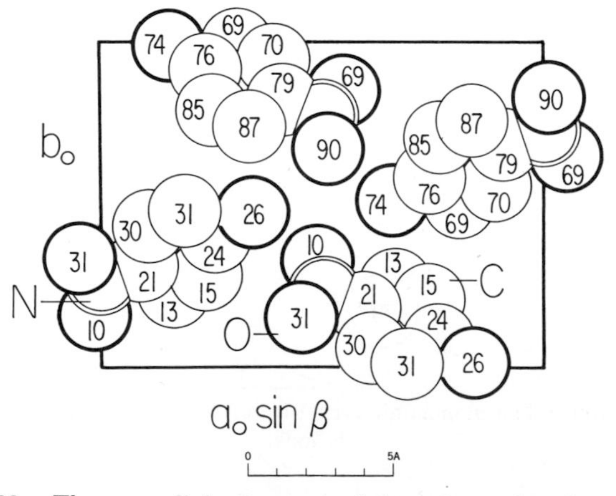

Fig. XVAIII,102. The monoclinic structure of the β form of p-nitrophenol projected along its c_0 axis.

All atoms are in the positions

$$(4e) \quad \pm(xyz;\ x+{}^1/_2,{}^1/_2-y,z)$$

of C_{2h}^5 ($P2_1/a$). A recent refinement has led to the parameters listed in Table XVAIII,58.

TABLE XVAIII,58

Parameters of the Atoms in β-p-$NO_2C_6H_4OH$

Atom	x	y	z
O(1)	−0.0469(1)	0.3229(2)	0.3131(5)
O(2)	−0.0117(1)	0.1575(1)	0.0983(5)
O(3)	0.3389(1)	0.4605(1)	0.2571(4)
N	0.0063(1)	0.2609(2)	0.2067(4)
C(1)	0.0933(1)	0.3109(1)	0.2114(4)
C(2)	0.1096(1)	0.4314(2)	0.3008(4)
C(3)	0.1926(1)	0.4792(1)	0.3132(5)
C(4)	0.2590(1)	0.4069(2)	0.2393(4)
C(5)	0.2417(1)	0.2863(2)	0.1508(5)
C(6)	0.1582(1)	0.2378(2)	0.1331(5)
H(O)	0.377	0.414	0.205
H(C,2)	0.063	0.478	0.340
H(C,3)	0.206	0.563	0.376
H(C,5)	0.286	0.237	0.109
H(C,6)	0.141	0.159	0.076

In the crystal, molecules having the bond dimensions of Figure XVAIII,103 are arranged as shown in Figure XVAIII,102. The atoms

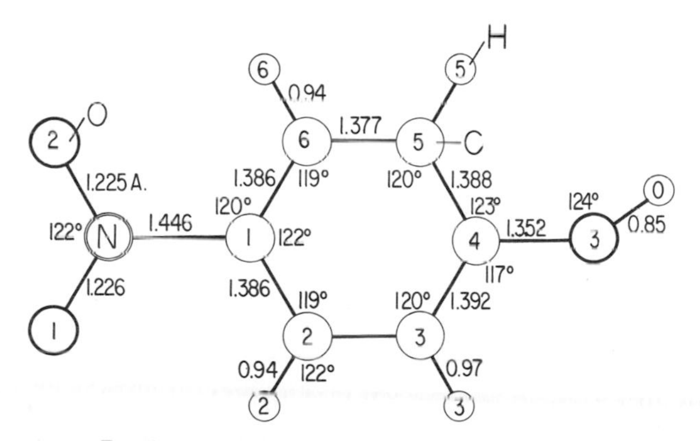

Fig. XVAIII,103. Bond dimensions in the molecule of p-nitrophenol in crystals of its β form.

of a molecule are coplanar except for the nitro group which departs somewhat from the plane of the rest. The nitrogen atom of this group is 0.04 A. from the plane and its oxygens are on either side of the plane by −0.07 A. [O(2)] and 0.18 A. [O(1)]; its plane is thus twisted by 7°9′ from the benzene plane. Between molecules there is one hydrogen bond O(2)–H–O(3) for which H–O(3) = 1.89 A. and O(2)–O(3) = 2.86 A. All other intermolecular separations are 3.31 A. or more.

XV,aIII,65. Crystals of *potassium o-nitrophenol hemihydrate*, $K(NO_2$-$C_6H_4O) \cdot {}^1/_2H_2O$, are monoclinic with a unit that contains eight molecules and has the dimensions:

$$a_0 = 24.73 \text{ A.}; \quad b_0 = 5.21 \text{ A.}; \quad c_0 = 11.96 \text{ A.}; \quad \beta = 105.15°$$

The water oxygens are in the positions

$$(4e) \quad \pm(0\,u\,{}^1/_4;\ {}^1/_2,u+{}^1/_2,{}^1/_4) \qquad \text{with } u = -0.478$$

of the space group C_{2h}^6 ($C2/c$). All other atoms are in the general positions:

$$(8f) \quad \pm(xyz;\ x,\bar{y},z+{}^1/_2;\ x+{}^1/_2,y+{}^1/_2,z;\ x+{}^1/_2,{}^1/_2-y,z+{}^1/_2)$$

with the parameters of Table XVAIII,59.

TABLE XVAIII,59

Parameters of Atoms in $K(NO_2C_6H_4O) \cdot {}^1/_2H_2O$

Atom	x	y	z
K	0.0415	−0.183	−0.077
O(1)	0.0542	−0.088	0.162
O(2)	0.0607	0.321	0.031
O(3)	0.1378	0.523	0.061
N	0.1096	0.367	0.090
C(1)	0.1015	−0.008	0.218
C(2)	0.1316	0.215	0.188
C(3)	0.1861	0.282	0.254
C(4)	0.2107	0.130	0.352
C(5)	0.1842	−0.060	0.385
C(6)	0.1313	−0.133	0.323

In the resulting structure (Fig. XVAIII,104) the anion is planar with the bond dimensions of Figure XVAIII,105. Each potassium cation has around it seven oxygen atoms; two of these are phenolic oxygens, (K–O = 2.71 and

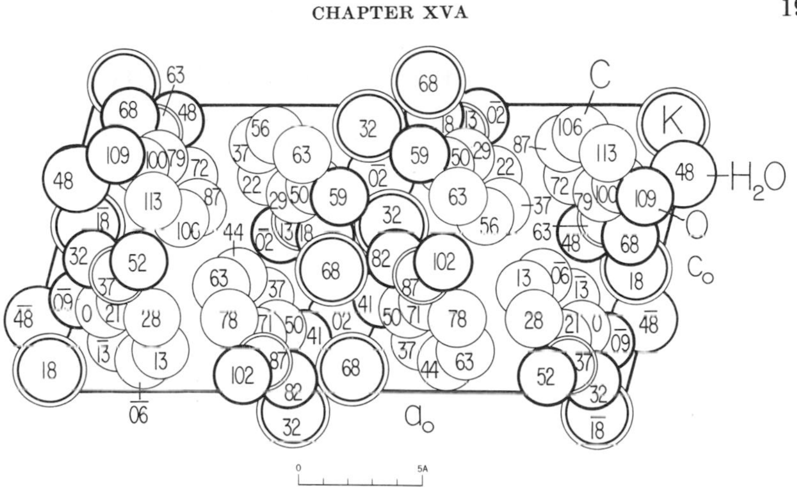

Fig. XVAIII,104. The monoclinic structure of $K(NO_2C_6H_4O)\cdot{}^1/_2H_2O$ projected along its b_0 axis.

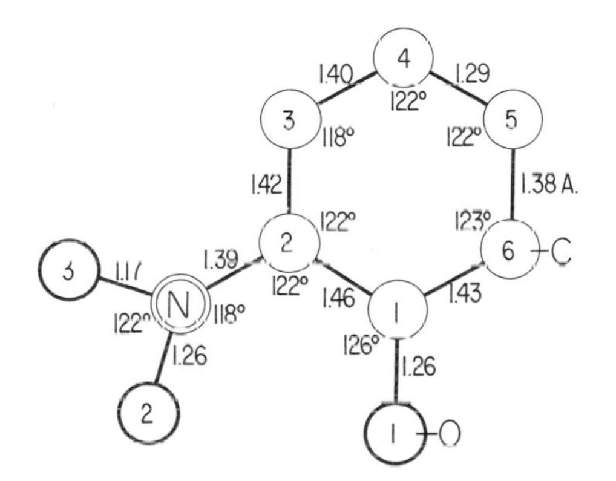

Fig. XVAIII,105. The bond dimensions in the anion in crystals of $K(NO_2C_6H_4O)\cdot{}^1/_2H_2O$.

2.82 A.), four are nitro oxygens, (K–O = 2.82–2.94 A.) and one is a water oxygen (K–O = 2.69 A.). The anions are linked together in pairs by hydrogen bonds to an intervening water molecule (O–H–O = 2.76 A.).

XV,aIII,66. *Catechol* (pyrocatechol), *o-dihydroxybenzene*, $o\text{-}C_6H_4(OH)_2$, is monoclinic with a tetramolecular cell of the dimensions:

$$a_0 = 10.941 \text{ A.};\quad b_0 = 5.509 \text{ A.};\quad c_0 = 10.069 \text{ A.}$$
$$\beta = 119°0'$$

The space group is C_{2h}^5 ($P2_1/a$) with all atoms in the positions:

$$(4e)\quad \pm(xyz;\ x+{}^1/_2,{}^1/_2-y,z)$$

The determined parameters are those listed in Table XVAIII,60.

TABLE XVAIII,60

Parameters of the Atoms in Catechol

Atom	x	y	z
C(1)	0.1105	0.3682	0.1761
C(2)	−0.0139	0.3556	0.1811
C(3)	−0.0360	0.5045	0.2778
C(4)	0.0662	0.6711	0.3683
C(5)	0.1893	0.6837	0.3624
C(6)	0.2125	0.5301	0.2662
O(1)	0.1251	0.2137	0.0767
O(2)	−0.1180	0.2009	0.0888
H(1)	0.2242	0.2468	0.0813
H(2)	−0.1208	0.0410	0.0249
H(3)	−0.1325	0.4948	0.2823
H(4)	0.0488	0.7892	0.4435
H(5)	0.2687	0.8116	0.4329
H(6)	0.3090	0.5401	0.2618

Molecules in the resulting structure (Fig. XVAIII,106) have the bond dimensions of Figure XVAIII,107. They are essentially planar, the two oxygen atoms being less than 0.1 A. from the plane of the benzene ring. In the crystal, pairs of these molecules, related to one another by a center of symmetry, are tied together by two hydrogen bonds of lengths 2.795 and 2.806 A.

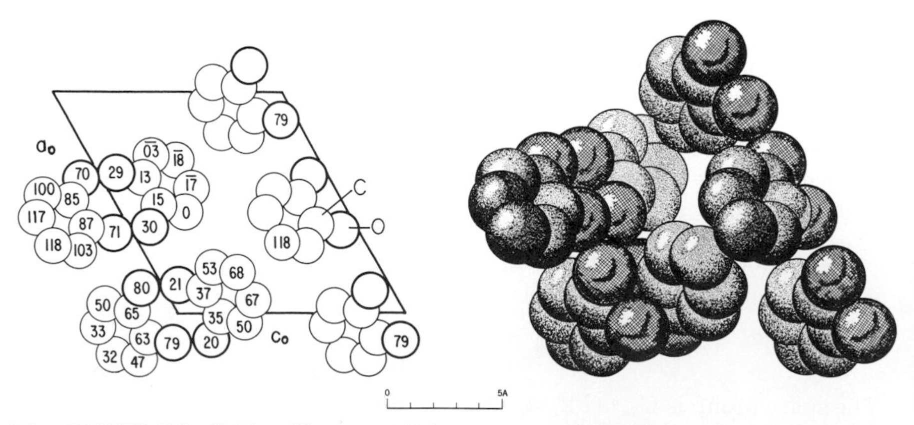

Fig. XVAIII,106a (left). The monoclinic structure of catechol projected along its b_0 axis.

Fig. XVAIII,106b (right). A packing drawing of the monoclinic structure of catechol seen along its b_0 axis. The carbon atoms are dotted; the oxygens are square shaded.

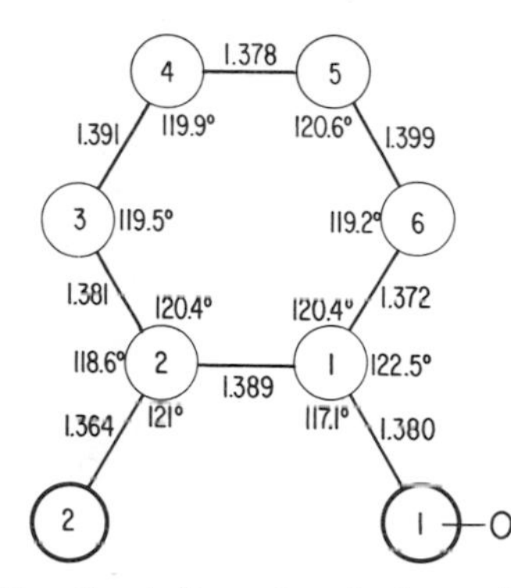

Fig. XVAIII,107. Bond dimensions in the molecule of catechol.

XV,aIII,67. The crystal structure of the α-form of *resorcinol*, m-C_6H_4-$(OH)_2$, has been investigated with both x-rays and neutrons. Its orthorhombic tetramolecular cell has the edge lengths:

$$a_0 = 10.53 \text{ A.}; \quad b_0 = 9.53 \text{ A.}; \quad c_0 = 5.66 \text{ A.}$$

Atoms are in the positions

$$(4a) \quad xyz;\ \bar{x},\bar{y},z+\tfrac{1}{2};\ \tfrac{1}{2}-x,y+\tfrac{1}{2},z+\tfrac{1}{2};\ x+\tfrac{1}{2},\tfrac{1}{2}-y,z$$

of the space group C_{2v}^9 ($Pna2_1$). The earlier x-ray parameters and, in parentheses, the incomplete neutron values are listed in Table XVAIII,61.

TABLE XVAIII,61

Parameters of the Atoms in α-Resorcinol[a]

Atom	x	y	z
C(1)	0.0172 (0.0167)	0.2058 (0.2142)	0.0642
C(2)	0.0661 (0.0658)	0.1014 (0.1092)	0.2097
C(3)	0.1761 (0.1756)	0.0308 (0.0350)	0.1456
C(4)	0.2375 (0.2350)	0.0644 (0.0658)	−0.0642
C(5)	0.1889 (0.1867)	0.1692 (0.1692)	−0.2097
C(6)	0.0789 (0.0792)	0.2397 (0.2392)	−0.1456
O(7)	0.0336 (0.0358)	0.3411 (0.3425)	−0.2914
O(8)	0.3456 (0.3417)	−0.0022 (0.0000)	−0.1322
H(1)	— (−0.0717)	— (0.2642)	—
H(2)	— (0.0233)	— (0.0892)	—
H(3)	— (0.2133)	— (−0.0461)	—
H(5)	— (0.2389)	— (0.1925)	—
H(7)	— (−0.0394)	— (0.3925)	—
H(8)	— (0.1242)	— (0.4292)	—
H(8′)	— (0.1242)	— (0.4292)	—

[a] Neutron values in parentheses.

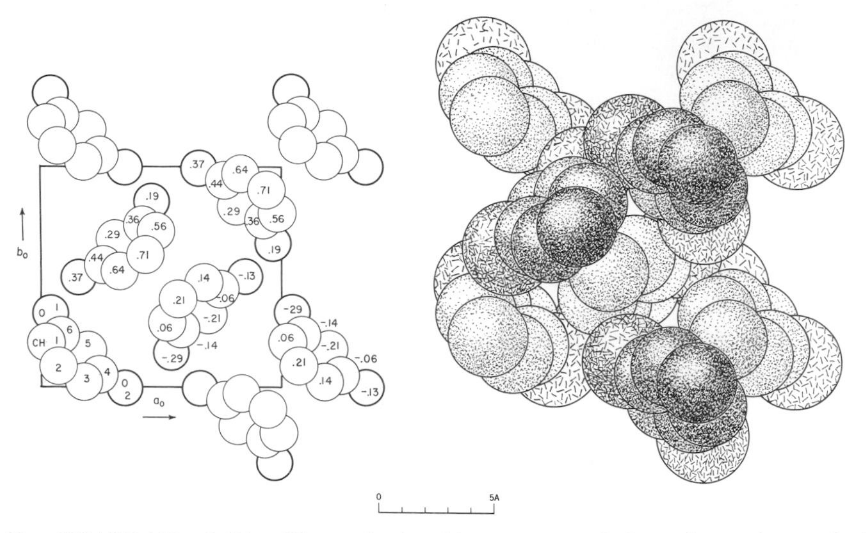

Fig. XVAIII,108a (left). The orthorhombic structure of the α form of resorcinol projected along its c_0 axis.

Fig. XVAIII,108b (right). A packing drawing of the orthorhombic structure of α-resorcinol viewed along its c_0 axis. The oxygen atoms are line shaded; the atoms of carbon are dotted.

The structure is shown in Figure XVAIII,108. Its molecules are planar with interatomic distances agreeing well with more recent and accurate values. The hydrogen bonds that tie the molecules to one another provide a shortest separation of O–H . . . O of 2.75 A.

XV,aIII,68. The β-form of *resorcinol*, m-$C_6H_4(OH)_2$, like the more common α-modification, is orthorhombic with a tetramolecular cell of the edge lengths:

$$a_0 = 7.91 \text{ A.}; \quad b_0 = 12.57 \text{ A.}; \quad c_0 = 5.50 \text{ A.}$$

The space group is the same C_{2v}^9 ($Pna2_1$) for both forms; atoms therefore are in the positions:

$$(4a) \quad xyz;\ \bar{x},\bar{y},z+1/2;\ 1/2-x,y+1/2,z+1/2;\ x+1/2,1/2-y,z$$

Parameters found for this modification are listed in Table XVAIII,62.

TABLE XVAIII,62

Parameters of the Atoms in β-Resorcinol

Atom	x	y	z
CH(1)	0.202	0.177	0.013
CH(2)	0.127	0.106	0.176
CH(3)	0.162	−0.002	0.163
C(4)	0.272	−0.040	−0.013
CH(5)	0.347	0.028	−0.176
C(6)	0.312	0.138	−0.163
OH(1)	0.382	0.210	−0.319
OH(2)	0.320	−0.143	−0.026

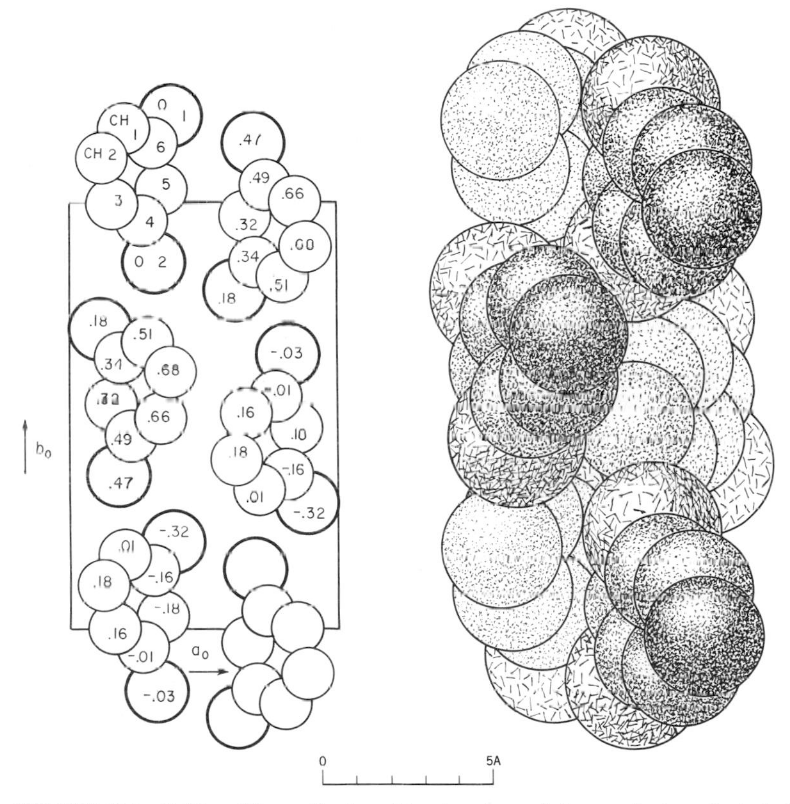

Fig. XVAIII,109a (left). The orthorhombic structure of β-resorcinol projected along its c_0 axis.

Fig. XVAIII,109b (right). A packing drawing of the orthorhombic structure of the β modification of resorcinol viewed along its c_0 axis. The oxygen atoms are line shaded; the atoms of carbon are dotted.

The structure of this high temperature form, stable above 74°C., is shown in Figure XVAIII,109. Atomic separations are normal and between molecules the shortest distances are O–H . . . O = 2.70 and 2.75 A.

XV,aIII,69. The monoclinic γ-form of *hydroquinone*, *p*-$C_6H_4(OH)_2$, has a tetramolecular unit of the dimensions:

$$a_0 = 8.07 \text{ A.};\quad b_0 = 5.20 \text{ A.};\quad c_0 = 13.20 \text{ A.};\quad \beta = 107°$$

The space group is C_{2h}^5 ($P2_1/c$) with all atoms in the positions:

$$(4e)\quad \pm(xyz;\ x,{}^1/_2-y,z+{}^1/_2)$$

Determined parameters are those in Table XVAIII,63.

TABLE XVAIII,63

Parameters of the Atoms in γ-Hydroquinone

Atom	x	y	z
	Molecule A		
O	0.0262	−0.0009	0.2140
C(1)	0.0119	−0.0024	0.1063
C(2)	0.0905	0.1891	0.0666
C(3)	−0.0788	−0.1959	0.0413
H(O)	−0.0078	−0.1893	0.2497
H(C2)	0.1598	0.3601	0.1132
H(C3)	−0.1367	−0.3529	0.0713
	Molecule B		
O′	0.4779	0.5178	0.2053
C(1′)	0.4914	0.5053	0.1025
C(2′)	0.5836	0.3148	0.0724
C(3′)	0.4076	0.6926	0.0320
H(O′)	0.5248	0.3548	0.2552
H(C2′)	0.6471	0.1526	0.1215
H(C3′)	0.3382	0.8478	0.0574

In this structure (Fig. XVAIII,110) each of the two crystallographically different molecules has a center of symmetry. Their bond dimensions are given in Figure XVAIII,111. Both molecules, including the hydrogen atoms attached to the ring, have been found planar to within the limit of error of the determination. In the crystal the angle between these molecular planes is 111.8°.

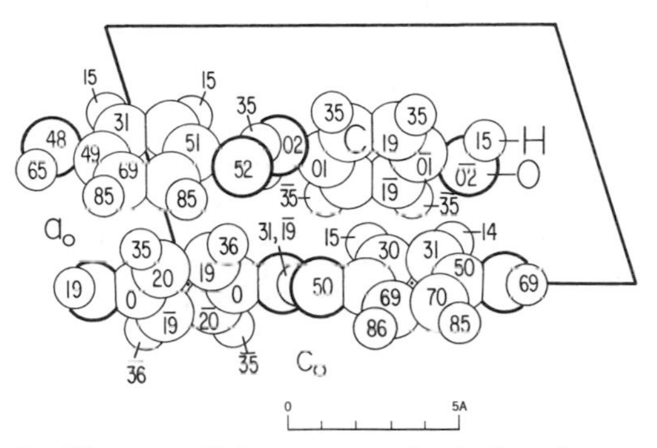

Fig. XVAIII,110. The monoclinic structure of γ-hydroquinone projected along its b_0 axis. Positions for the hydrogen atoms are included in this drawing.

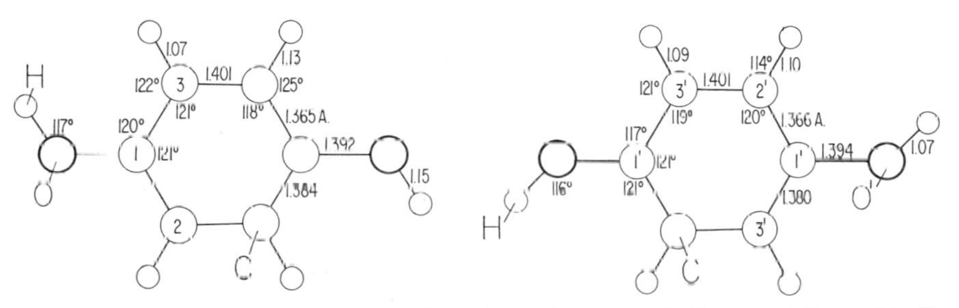

Fig. XVAIII,111a (left). Bond dimensions in one of the crystallographically different molecules of γ-hydroquinone.

Fig. XVAIII,111b (right). Bond dimensions in the second of the two crystallographically different molecules of γ-hydroquinone.

XV,aIII,70. The ordinary, monoclinic, form of *quinhydrone*, p-$C_6H_4O_2 \cdot p$-$C_6H_4(OH)_2$, has a tetramolecular unit of the dimensions:

$$a_0 = 7.674(1)\text{ A.};\quad b_0 = 6.001(1)\text{ A.};\quad c_0 = 21.788(2)\text{ A.}$$
$$\beta = 90°38(3)'$$

Its space group C_{2h}^5 has, with reference to the foregoing cell, the unusual orientation $B2_1/d$. In such a cell atoms are in the positions:

$$(8e)\quad \pm(xyz;\ x+{}^1/_2,y,z+{}^1/_2;\ x+{}^1/_4,{}^1/_2-y,z+{}^3/_4;\ x+{}^3/_4,{}^1/_2-y,z+{}^1/_4)$$

Parameters are listed in Table XVAIII,64.

TABLE XVAIII,64

Parameters of the Atoms in Monoclinic Quinhydrone

Atom	x	y	z
C(1)	0.1039	0.1640	0.0216
C(2)	0.0518	0.0077	0.0610
C(3)	0.0549	0.1695	−0.0412
C(4)	0.6021	0.1854	0.0195
C(5)	0.5523	0.0121	0.0621
C(6)	0.5439	0.1632	−0.0453
O(1)	0.2128	0.3355	0.0440
O(2)	0.6849	0.3496	0.0374

Bond dimensions for the separate quinone and hydroquinone molecules in the structure (Fig. XVAIII,112) are those of Figure XVAIII,113. The

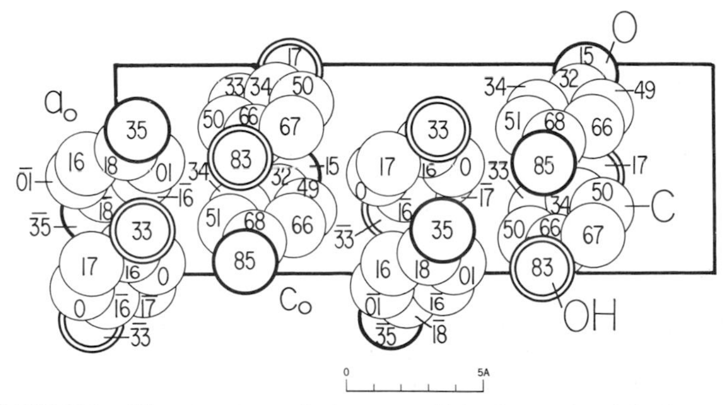

Fig. XVAIII,112. The structure of the monoclinic form of quinhydrone projected along its b_0 axis.

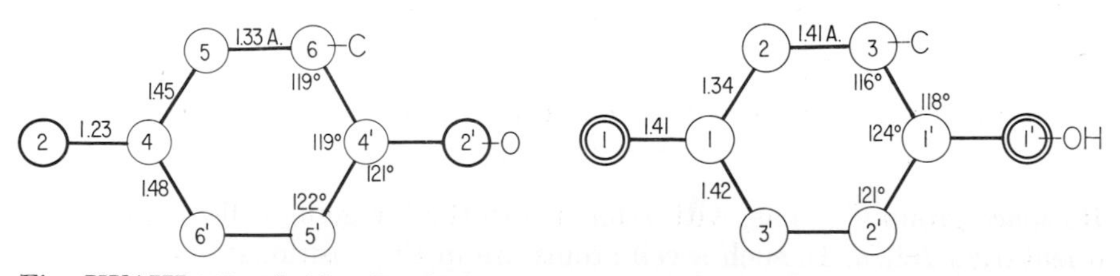

Fig. XVAIII,113a (left). Bond dimensions in the benzoquinone part of the quinhydrone molecule in its monoclinic crystals.

Fig. XVAIII,113b (right). Bond dimensions in the hydroquinone part of the quinhydrone molecule in its monoclinic form.

rings of the hydroquinone as well as the quinone molecules are appreciably distorted from a true hexagonal shape though both are essentially planar. In the crystal they are tied together by hydrogen bonds of length 2.71 A. to form chains in which the kinds of molecule alternate.

XV,aIII,71. *Quinhydrone*, p-$C_6H_4O_2 \cdot p$-$C_6H_4(OH)_2$, also possesses a rare triclinic modification for which the unimolecular cell has the dimensions:

$$a_0 = 7.652(22) \text{ A.}; \quad b_0 = 5.956(13) \text{ A.}; \quad c_0 = 6.770(20) \text{ A.}$$
$$\alpha = 107°37(7)'; \quad \beta = 121°56(3)'; \quad \gamma = 90°17(9)'$$

All atoms are in the general positions of C_i^1 $(P\bar{1})$: $(2i)$ $\pm(xyz)$ with the parameters of Table XVAIII,65.

TABLE XVAIII,65

Parameters of the Atoms in Triclinic Quinhydrone

Atom	x	y	z
	Hydroquinone		
O(1)	0.1302(6)	0.2796(6)	−0.1756(7)
C(1)	0.0659(8)	0.1419(8)	−0.0846(9)
C(2)	0.1239(8)	0.2278(8)	0.1623(9)
C(3)	0.0560(8)	0.0818(8)	0.2441(9)
H(1)	0.214(7)	0.419(8)	−0.040(8)
H(2)	0.213(7)	0.405(8)	0.270(8)
H(3)	0.080(7)	0.138(8)	0.412(8)
	Quinone		
O(2)	0.6115(6)	0.2834(6)	−0.1665(7)
C(4)	0.5610(8)	0.1490(8)	−0.0897(9)
C(5)	0.6279(8)	0.2327(8)	0.1751(9)
C(6)	0.5666(8)	0.0897(8)	0.2551(9)
H(4)	0.718(7)	0.410(8)	0.278(8)
H(5)	0.588(7)	0.144(8)	0.412(8)

The molecular distribution is that of Figure XVAIII,114. The molecules themselves have the bond dimensions of Figure XVAIII,115. As this figure suggests, the hydroxyl hydrogen atom of the hydroquinone part of the molecule has been placed off the line joining its oxygen and the oxygen of the quinone part. The shortest intermolecular distances are C–O = 3.175–3.501 A. and C–C = 3.204–3.425 A.

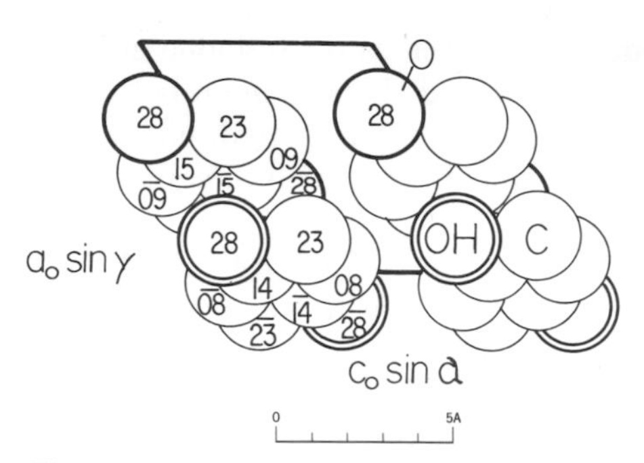

Fig. XVAIII,114. The structure of quinhydrone in its triclinic modification projected along its b_0 axis.

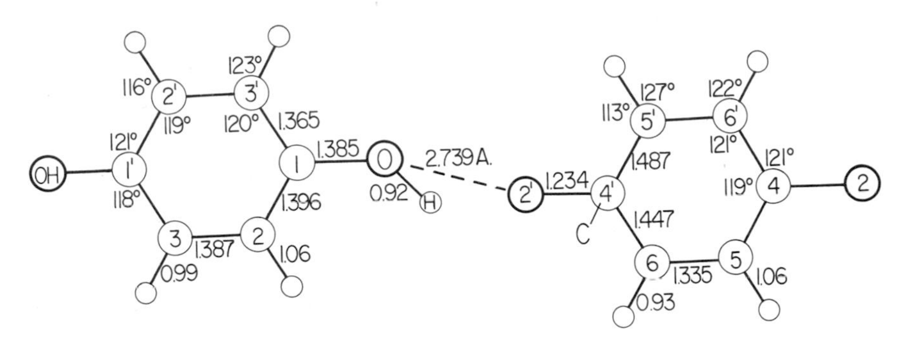

Fig. XVAIII,115. Bond dimensions in the molecule of quinhydrone in its triclinic modification.

XV,aIII,72. The monoclinic crystals of the *hydroquinone-acetone* complex $C_6H_4(OH)_2 \cdot (CH_3)_2CO$ have a tetramolecular unit of the dimensions:

$$a_0 = 18.64(4) \text{ A.}; \quad b_0 = 7.14(2) \text{ A.}; \quad c_0 = 7.41(2) \text{ A.}$$
$$\beta = 111°0(60)'$$

The space group is C_{2h}^6 ($C2/c$) with atoms in the positions:

$$\text{C(5)}: (4e) \quad \pm(0\,u\,{}^1/_4;\ {}^1/_2,u+{}^1/_2,{}^1/_4)$$
$$\text{with } u = 0.702$$
$$\text{O(1)}: (4e) \quad \text{with } u = 0.521$$

The other atoms are in:

$$(8f) \quad \pm(xyz;\ x,\bar{y},z+{}^1/_2;\ x+{}^1/_2,y+{}^1/_2,z;\ x+{}^1/_2,{}^1/_2-y,z+{}^1/_2)$$

with the parameters:

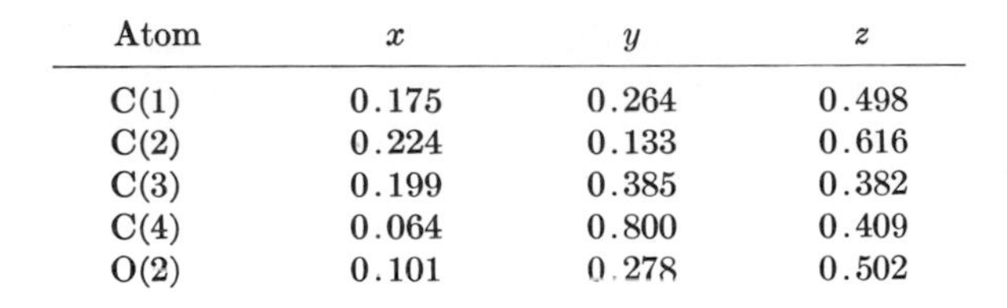

Atom	x	y	z
C(1)	0.175	0.264	0.498
C(2)	0.224	0.133	0.616
C(3)	0.199	0.385	0.382
C(4)	0.064	0.800	0.409
O(2)	0.101	0.278	0.502

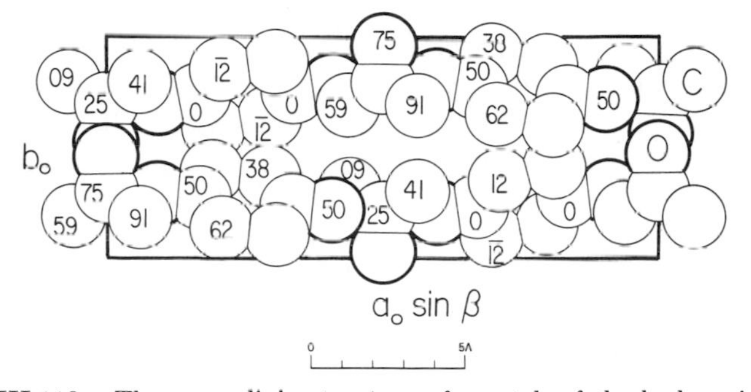

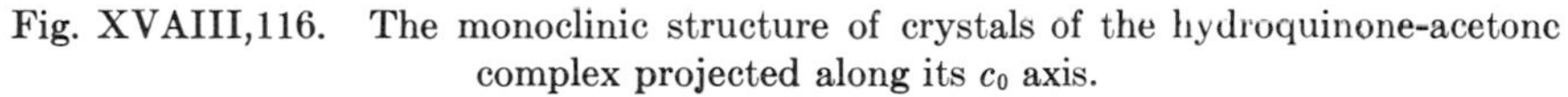
Fig. XVAIII,116. The monoclinic structure of crystals of the hydroquinone-acetone complex projected along its c_0 axis.

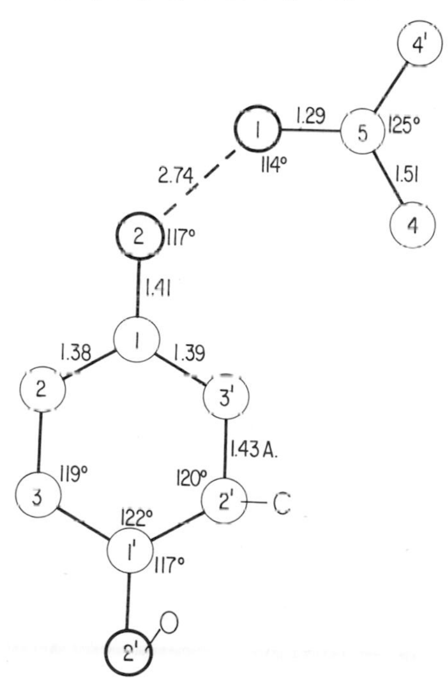

Fig. XVAIII,117. Bond dimensions in the hydroquinone and acetone components of the complex they form.

The structure, as shown in Figure XVAIII,116, consists of chains of hydrogen bond-linked quinol and acetone molecules running parallel to the c_0 axis. These molecules have the bond dimensions of Figure XVAIII,117. Between them there is a short C(5)–O(2) = 3.07 A.; the next shortest atomic approach is C(4)–O(2) = 3.37 A.

XV,aIII,73. Many years ago it was found that *p-dihydroxy benzene* (β-quinol), $C_6H_4(OH)_2$, forms a broad series of *clathrate* compounds. In its crystal structure there are large completely enclosed holes (not channels, as in the zeolites, for instance) and many different small alien molecules can be trapped in these holes. The number of trapped molecules, and therefore the limiting composition of the clathrate crystal, obviously depends on the number of holes: for β-quinol there is one hole for every three quinol molecules and hence the limiting composition is $3C_6H_4(OH)_2 \cdot R$. Examples of such quinol-clathrates, illustrative of the wide range of molecules that can be entrapped, are given in Table XVAIII,66.

TABLE XVAIII,66

Cell Dimensions of Hexagonal Clathrate Quinol Compounds

Compound	a_0, A.	c_0, A.
$3C_6H_4(OH)_2 \cdot 0.88SO_2$	16.32	5.82
$3C_6H_4(OH)_2 \cdot 0.74CO_2$	16.20	5.83
$3C_6H_4(OH)_2 \cdot 0.64H_2S$	16.61	5.50
$3C_6H_4(OH)_2 \cdot 0.85HCl$	16.58	5.47
$3C_6H_4(OH)_2 \cdot 0.36HBr$	16.60	5.49
$3C_6H_4(OH)_2 \cdot 0.62C_2H_2$	16.66	5.47
$3C_6H_4(OH)_2 \cdot 0.99CH_3CN$	15.98	6.25
$3C_6H_4(OH)_2 \cdot 0.97CH_3OH$	16.59	5.56
$3C_6H_4(OH)_2 \cdot 0.82HCOOH$	16.45	5.66
$3C_6H_4(OH)_2 \cdot 0.88Xe$ [a]	—	—

[a] Cell dimensions do not seem to have been published for this xenon compound, nor for the corresponding argon compound $3C_6H_4(OH)_2 \cdot 0.8A$.

A detailed analysis was made years ago of the *β-quinol sulfur dioxide clathrate.* Its symmetry is hexagonal (rhombohedral) and there are nine quinol molecules in a hexagonal unit that has the edge lengths:

$$a_0' = 16.32 \text{ A.}, \qquad c_0' = 5.82 \text{ A.}$$

The space group is C_{3i}^2 ($R\bar{3}$) with all atoms in the positions:

$$(18f) \quad \pm(xyz;\ \bar{y},x-y,z;\ y-x,\bar{x},z);\quad \text{rh}$$

The following parameters were assigned:

Atom	x	y	z
C(1)	0.263	0.135	0.665
C(2)	0.334	0.113	0.645
C(3)	0.262	0.188	0.855
O	0.195	0.105	0.500

Molecules in the resulting structure (Fig. XVAIII,118) are planar, with C–C = 1.39 A. and C–O = 1.36 A. According to this arrangement, each point of the rhombohedral lattice has about it six equidistant hydroxyls from six different molecules. Hydrogen bonds of length 2.75 A. between these hydroxyls bind the molecules together into a three-dimensional network that leaves holes centered in these lattice points, each big enough to be filled by a molecule of SO_2.

XV,aIII,74. The monoclinic crystals of L-*tyrosine hydrochloride*, $OHC_6H_4CH_2CH(NH_2)COOH \cdot HCl$, have a bimolecular unit of the dimensions:

$$a_0 = 11.07 \text{ A.}; \quad b_0 = 9.03 \text{ A.}; \quad c_0 = 5.09 \text{ A.}; \quad \beta = 91°48'$$

The space group is C_2^2 ($P2_1$) with atoms in the positions:

$$(2a) \quad xyz; \bar{x},y+{}^1/_2,\bar{z}$$

Parameters are listed in Table XVAIII,67.

TABLE XVAIII,67

Parameters of the Atoms in Tyrosine Hydrochloride

Atom	x	y	z
C(1)	0.0990	0.035	0.362
C(2)	0.0815	0.127	0.157
C(3)	0.9730	0.128	0.019
C(4)	0.8785	0.033	0.088
C(5)	0.8970	−0.059	0.304
C(6)	0.0080	−0.060	0.445
C(7)	0.7620	0.035	−0.078
C(8)	0.6530	0.087	0.075
C(9)	0.6690	0.247	0.155
O(1)	0.2020	0.034	0.520
O(2)	0.0390	0.353	0.003
O(3)	0.7230	0.273	0.373
N	0.5400	0.074	−0.088
Cl	0.4130	0.250	0.405

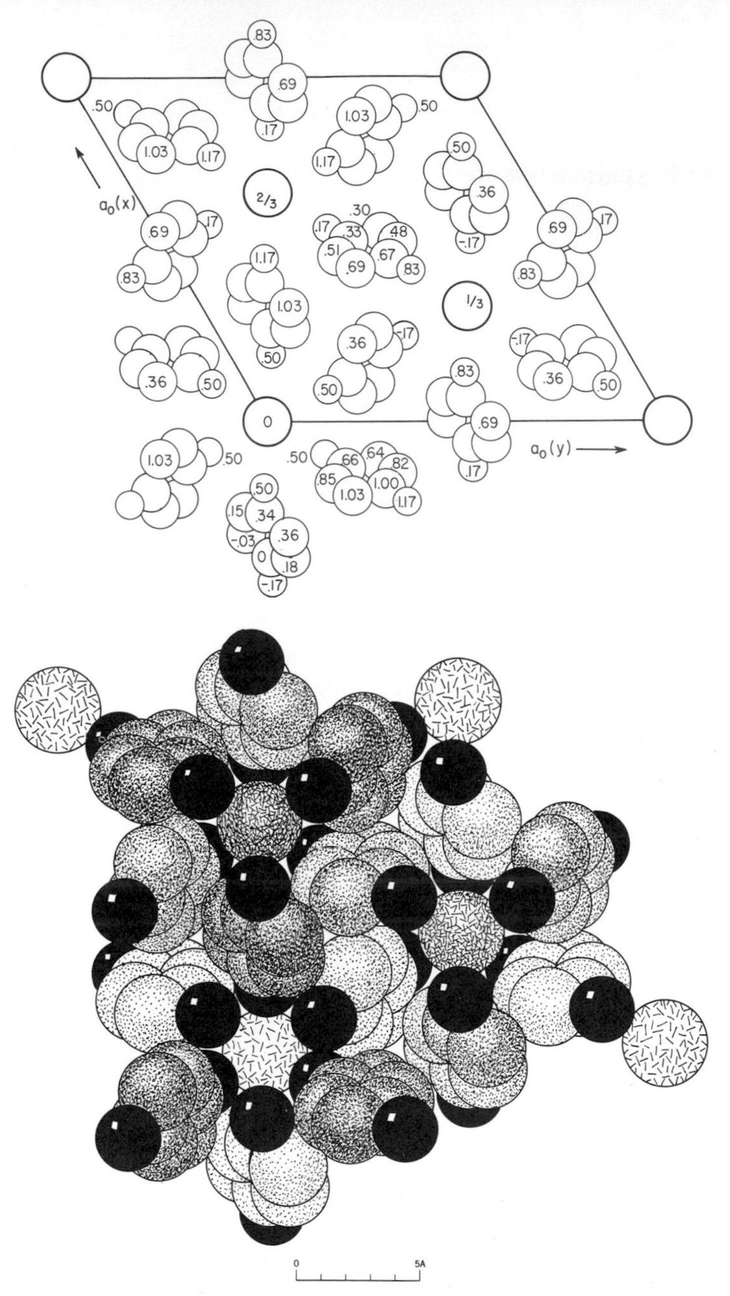

Fig. XVAIII,118a (top). The hexagonal structure of the quinol-sulfur dioxide clathrate projected along its c_0 axis. The smallest circles are the hydroxyl radicals; the large circles are the centers of the sulfur dioxide (or other enclosed) molecules. Left hand axes.

Fig. XVAIII,118b (bottom). A packing drawing of the hexagonal quinol-SO_2 clathrate structure viewed along its c_0 axis. The enclosed SO_2 molecules are large and line shaded; the hydroxyl groups are black. Atoms of carbon are dotted.

The structure is shown in Figure XVAIII,119. As with other amino acids the molecules are held together by a system of hydrogen bonds, the shortest of which in this case are N–H–O = 2.79 A. and O–H–O = 2.50 A.

The corresponding hydrobromide is isostructural. For it:

$$a_0 = 11.38 \text{ A.}; \quad b_0 = 9.12 \text{ A.}; \quad c_0 = 5.175 \text{ A.}; \quad \beta = 91°12'$$

XV,aIII,75. Crystals of *glycyl-L-tyrosine hydrochloride monohydrate*, $HOC_6H_4CH_2CH(COOH)NH \cdot C(O)CH_2NH_2 \cdot HCl \cdot H_2O$, and the corresponding hydrobromide are monoclinic. Their bimolecular cells have the dimensions:

For the chloride:

$$a_0 = 14.05 \text{ A.}; \quad b_0 = 9.35 \text{ A.}; \quad c_0 = 5.10 \text{ A.}; \quad \beta = 91°20'$$

For the bromide:

$$a_0 = 14.25 \text{ A.}; \quad b_0 = 9.55 \text{ A.}; \quad c_0 = 5.08 \text{ A.}; \quad \beta = 91°4'$$

All atoms are in the general positions of C_2^2 ($P2_1$):

$$(2a) \quad xyz; \; \bar{x}, y+{}^1/_2, \bar{z}$$

They have been assigned the parameters of Table XVAIII,68.

TABLE XVAIII,68
Parameters of the Atoms in Glycyl-L-tyrosine Hydrochloride Monohydrate

Atom	x	y	z
C(1)	0.912	0.517	0.378
C(2)	0.921	0.610	0.178
C(3)	0.008	0.609	0.028
C(4)	0.082	0.516	0.107
C(5)	0.070	0.424	0.312
C(6)	0.985	0.424	0.460
C(7)	0.173	0.515	0.946
C(8)	0.262	0.560	0.107
C(9)	0.255	0.718	0.168
C(10)	0.430	0.565	0.043
C(11)	0.510	0.508	0.874
O(1)	0.829	0.516	0.517
O(2)	0.278	0.811	0.016
O(3)	0.220	0.748	0.386
O(4)	0.452	0.633	0.224
O(5)	0.427	0.833	0.644
N(1)	0.344	0.529	0.941
N(2)	0.605	0.568	0.955
Cl	0.320	0.250	0.540

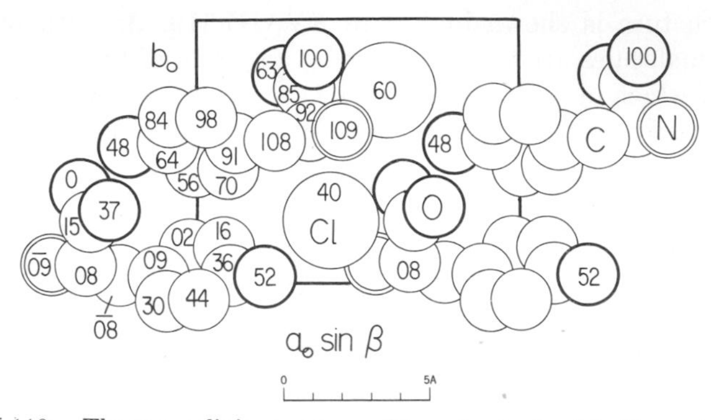

Fig. XVAIII,119. The monoclinic structure of tyrosine hydrochloride projected along its c_0 axis.

The molecular cation thus established has the bond dimensions of Figure XVAIII,120. Its chain atoms from N(2) to C(7) are approximately in a plane that makes a considerable angle with the benzene ring. In the structure (Fig. XVAIII,121) these complex cations lie parallel to the a_0 axis and form sheets normal to b_0. The crystal is held together by a complicated combination of ionic attractions and hydrogen bonds.

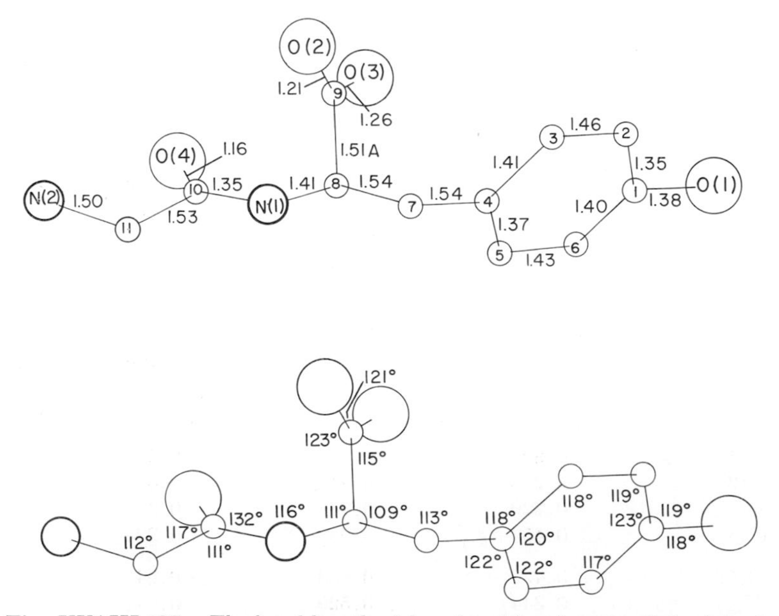

Fig. XVAIII,120. The bond lengths (above) and bond angles (below) in the molecule of glycyl-L-tyrosine hydrochloride in its monohydrate.

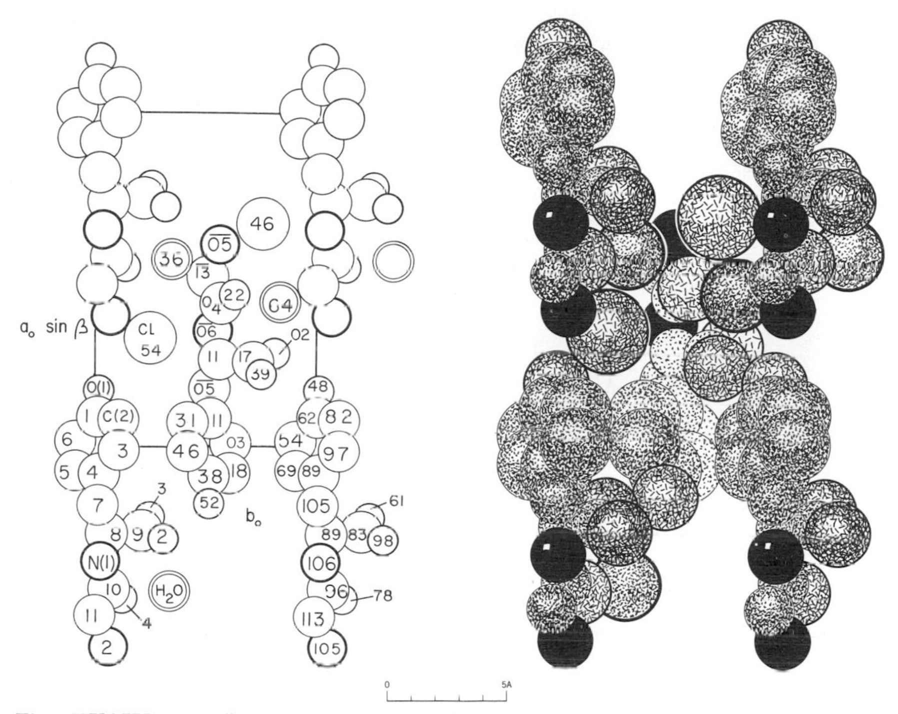

Fig. XVAIII,121a (left). The monoclinic structure of glycyl-L-tyrosine hydrochloride monohydrate projected along its c_0 axis. Left hand axes.

Fig. XVAIII,121b (right). A packing drawing of the monoclinic structure of glycyl-L- tyrosine hydrochloride monohydrate seen along its c_0 axis. The nitrogen atoms are black. Chloride ions are the larger; oxygen atoms are the smaller of the line shaded circles. The water molecules are heavily outlined and dotted. The carbon atoms are also dotted but lightly outlined; the aliphatic carbons are smaller than the others.

XV,aIII,76. Crystals of *potassium hydrogen bis(p-hydroxybenzoate) monohydrate*, $KH(OHC_6H_4COO)_2 \cdot H_2O$, have a bimolecular monoclinic structure. The unit has the dimensions:

$$a_0 = 16.40 \text{ A.}; \quad b_0 = 3.82 \text{ A.}; \quad c_0 = 11.30 \text{ A.}; \quad \beta = 92°30'$$

For the isomorphous rubidium salt:

$$a_0 = 16.47 \text{ A.}; \quad b_0 = 3.91 \text{ A.}; \quad c_0 = 11.50 \text{ A.}; \quad \beta = 93°30'$$

The determined space group is C_{2h}^4 ($P2/c$). Atomic positions have been found for the potassium compound; those for the rubidium salt must be essentially the same. As can be seen from Table XVAIII,69, the potassium ions are in special positions

$$(2e) \quad \pm(0\ u\ {}^1/_4)$$

and the water oxygen atoms in

$$(2f) \quad \pm(^1/_2\, u\, ^1/_4)$$

All other atoms are in general positions

$$(4g) \quad \pm(xyz;\ x,\bar{y},z+^1/_2)$$

the parameters being those of the table.

TABLE XVAIII,69

Parameters of the Atoms in Monoclinic Crystals of Potassium Hydrogen Bis(*p*-Hydroxybenzoate) Monohydrate

Atom	Position	x	y	z
K	(2*e*)	0	−0.333	$^1/_4$
H_2O	(2*f*)	$^1/_2$	0.017	$^1/_4$
O(1)	(4*g*)	0.0675	0.125	−0.0395
O(2)	(4*g*)	0.0994	0.133	0.1509
OH(3)	(4*g*)	0.4432	0.350	−0.0728
C(1)	(4*g*)	0.1191	0.150	0.0500
C(2)	(4*g*)	0.2056	0.208	0.0149
CH(3)	(4*g*)	0.2309	0.133	−0.0982
CH(4)	(4*g*)	0.3080	0.167	−0.1246
C(5)	(4*g*)	0.3661	0.308	−0.0412
CH(6)	(4*g*)	0.3389	0.383	0.0693
CH(7)	(4*g*)	0.2617	0.350	0.0939

The structure that results is shown in Figure XVAIII,122. The benzene ring of its molecular anion is planar but there is an inequality in the C–C bond lengths (Fig. XVAIII,123) thought indicative of some quinone-like character. The short distance (2.61 A.) between the O(1) atoms of adjacent ions was presumed due to hydrogen bonding involving the ionic hydrogens midway between them. Thus the structure is layer-like with the organic anions separated on the carboxyl side by the potassium and hydrogen ions and on the hydroxyl side by water molecules. There are short H_2O–O(3) separations (2.58 A.) which presumably correspond to hydrogen bonds. Between adjacent anions the shortest C–C separation is 3.72 A.; the shortest C–H_2O distance is 3.55 A. Each potassium ion has about it six oxygen atoms at the corners of a considerably distorted octahedron; the K–O distances lie between 2.7 and 3.1 A.

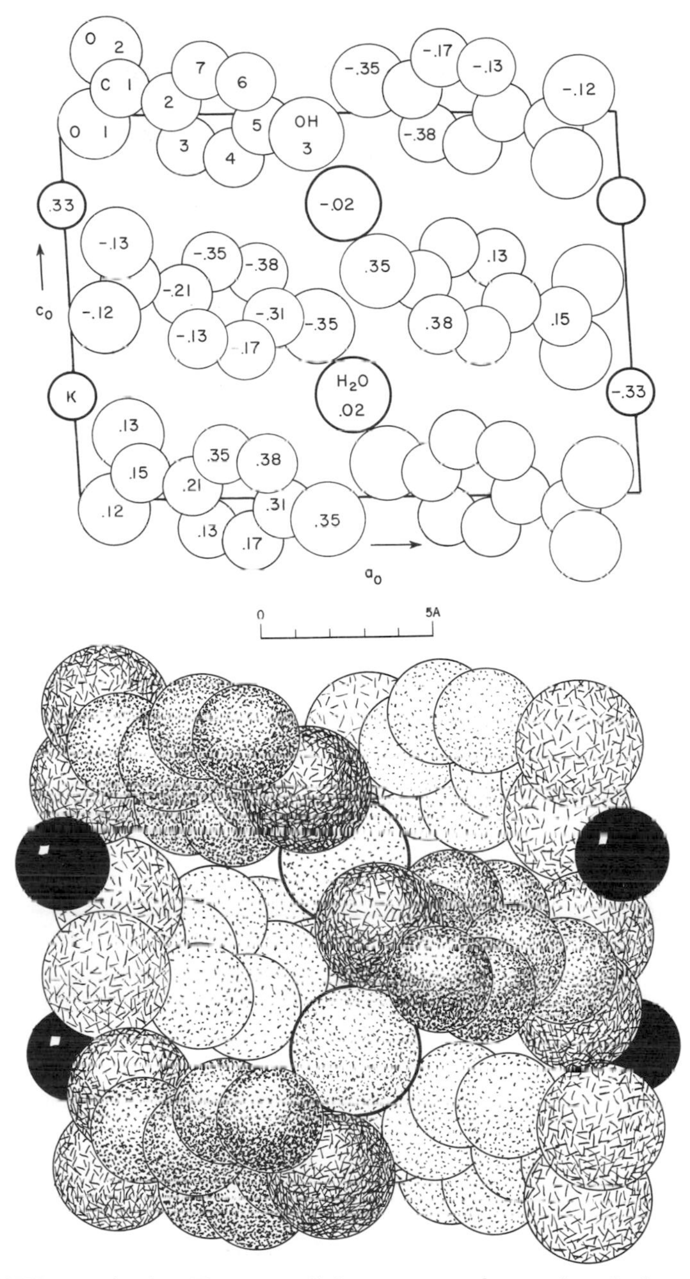

Fig. XVAIII,122a (top). The monoclinic structure of potassium hydrogen bis(*p*-hydroxybenzoate) monohydrate projected along its b_0 axis. Left hand axes.

Fig. XVAIII,122b (bottom). A packing drawing of the monoclinic potassium hydrogen bis(*p*-hydroxybenzoate) arrangement viewed along its b_0 axis. The potassium atoms are black; the oxygen atoms are line shaded. Of the dotted circles the carbon atoms are the smaller; the water molecules are larger and heavily outlined.

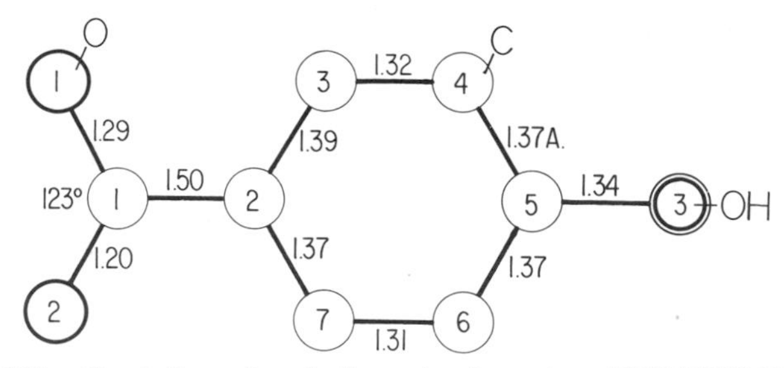

Fig. XVAIII,123. Bond dimensions in the molecular anion of $KH(OHC_6H_4COO)_2 \cdot H_2O$

XV,aIII,77. Crystals of *salicylic acid*, *o*-$C_6H_4(OH)COOH$, are monoclinic with a tetramolecular unit of the dimensions:

$$a_0 = 11.52(1) \text{ A.}; \quad b_0 = 11.21(1) \text{ A.}; \quad c_0 = 4.92(1) \text{ A.}$$
$$\beta = 90°50(2)'$$

All atoms are in the positions

$$(4e) \quad \pm(xyz; x+{}^1/_2, {}^1/_2-y, z)$$

of the space group C_{2h}^5 ($P2_1/a$). The parameters, as determined in the more recent study, are those of Table XVAIII,70.

TABLE XVAIII,70
Parameters of the Atoms in Salicylic Acid

Atom	x	y	z
O(1)	−0.0505	0.1373	0.8380
O(2)	0.0988	0.0168	0.7703
O(3)	0.2509	0.0710	0.3992
C(1)	0.0836	0.1926	0.5060
C(2)	0.1847	0.1698	0.3587
C(3)	0.2204	0.2506	0.1647
C(4)	0.1572	0.3526	0.1137
C(5)	0.0573	0.3761	0.2574
C(6)	0.0214	0.2969	0.4501
C(7)	0.0450	0.1088	0.7126
H(1)	−0.069	0.078	0.983
H(2)	0.210	0.021	0.545
H(3)	0.294	0.238	0.067
H(4)	0.180	0.412	−0.034
H(5)	0.008	0.436	0.214
H(6)	−0.051	0.309	0.549

For oxygen, $\sigma(x) = \sigma(y) = 0.0002$, $\sigma(z) = 0.0004$–0.0005
For carbon, $\sigma(x) = 0.0003$–0.0004, $\sigma(y) = 0.0003$, $\sigma(z) = 0.0005$–0.0007
For hydrogen, $\sigma(x) = \sigma(y) = 0.003$–$0.004$, $\sigma(z) = 0.007$–0.008

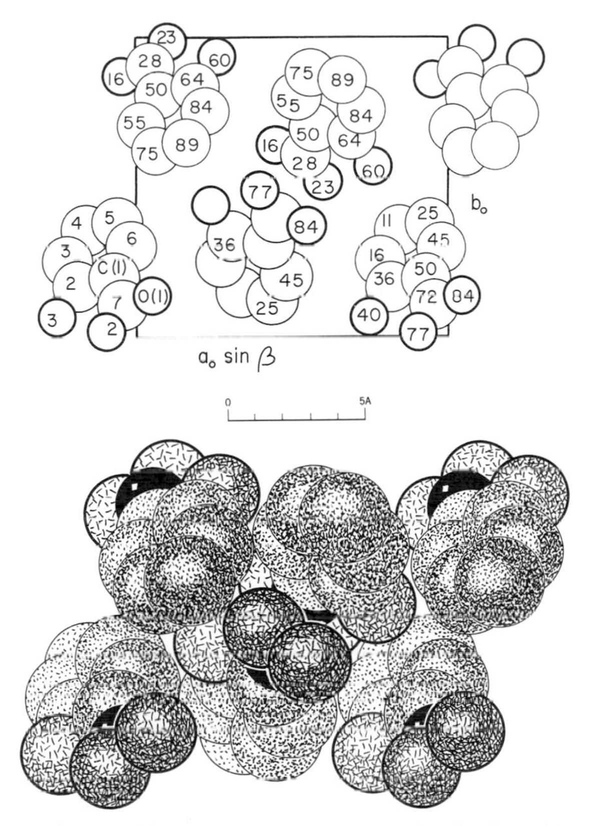

Fig. XVAIII,124a (top). The monoclinic structure of salicylic acid projected along its c_0 axis. Left hand axes.

Fig. XVAIII,124b (bottom). A packing drawing of the monoclinic salicylic acid structure seen along its c_0 axis. The carboxyl carbon atom is black; the others are dot shaded. Oxygen atoms are line shaded and more heavily outlined.

The structure is shown in Figure XVAIII,124. The molecule, which according to the latest determination has the bond dimensions of Figure XVAIII,125, is planar with only the O(2) atom departing by as much as 0.03 A. from the best plane through it; atoms of hydrogen are, however, as much as 0.12 A. away from the plane.

XV,aIII,78. *Alkali acid salicylates* crystallizing with one molecule of water have monoclinic tetramolecular cells of the dimensions:
For $KH[C_6H_4(OH)COO]_2 \cdot H_2O$:

$$a_0 = 16.90 \text{ A.}; \quad b_0 = 3.88 \text{ A.}; \quad c_0 = 22.40 \text{ A.}; \quad \beta = 96°0'$$

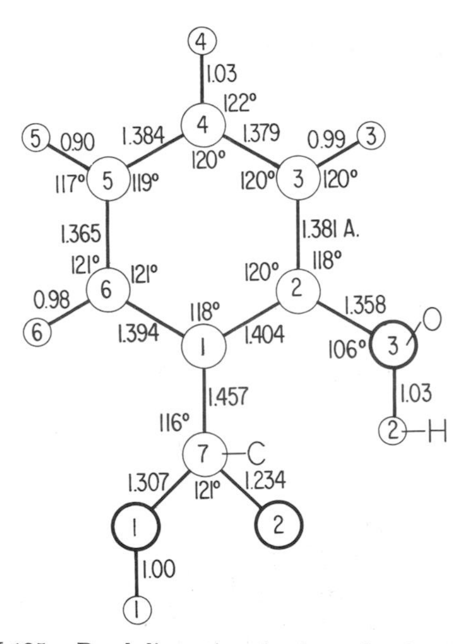

Fig. XVAIII,125. Bond dimensions in the molecule of salicylic acid.

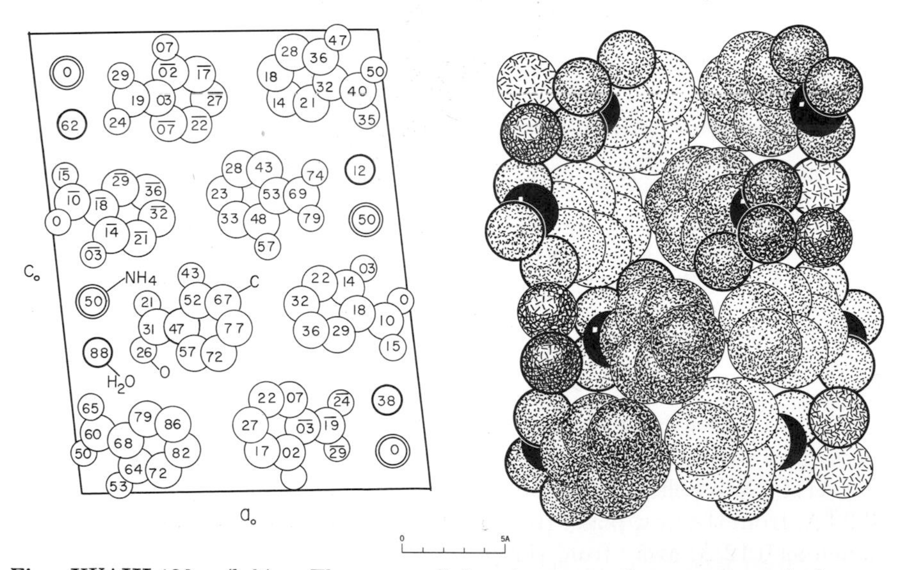

Fig. XVAIII,126a (left). The monoclinic structure of ammonium hydrogen salicylate monohydrate projected along its b_0 axis. Left hand axes.

Fig. XVAIII, 126b (right). A packing drawing of the monoclinic $NH_4H[OHC_6H_4CO\text{-}O]_2 \cdot H_2O$ arrangement seen along its b_0 axis. The ammonium ions are line shaded and lightly outlined; the water molecules are the more heavily outlined line shaded circles. Oxygen atoms are heavily outlined and dotted. The carboxyl carbons are black; the others are dotted and lightly outlined.

For $NH_4H[C_6H_4(OH)COO]_2 \cdot H_2O$:

$$a_0 = 17.28 \text{ A.}; \quad b_0 = 3.89 \text{ A.}; \quad c_0 = 22.33 \text{ A.}; \quad \beta = 98°48'$$

For $RbH[C_6H_4(OH)COO]_2 \cdot H_2O$:

$$a_0 = 17.13 \text{ A.}; \quad b_0 = 3.99 \text{ A.}; \quad c_0 = 22.74 \text{ A.}; \quad \beta = 98°30'$$

All atoms are in general positions of C_{2h}^5 ($P2_1/c$):

$$(4e) \quad \pm(xyz;\ x,{}^1/_2-y,z+{}^1/_2)$$

with the parameters, for the ammonium salt, listed in Table XVAIII,71.

TABLE XVAIII,71

Parameters of the Atoms in $NH_4H(C_6H_4OHCOO)_2 \cdot H_2O$

Atom	x	y	z
NH_4	0.0992	0.000	−0.0863
H_2O	0.0906	−0.380	−0.1994
O(1)	0.2237	0.243	−0.1949
O(2)	0.2491	0.2925	−0.0951
O(3)	0.3894	0.069	−0.0306
O(4)	0.0587	0.650	0.1878
O(5)	0.0208	0.500	0.0892
O(6)	0.1163	0.530	0.0187
C(1)	0.2740	0.189	−0.1439
C(2)	0.3475	0.033	−0.1445
C(3)	0.3678	−0.067	−0.2007
C(4)	0.4398	−0.222	−0.2023
C(5)	0.4893	−0.272	−0.1490
C(6)	0.4702	−0.174	−0.0937
C(7)	0.3976	−0.018	−0.0915
C(8)	0.0718	0.604	0.1327
C(9)	0.1505	0.676	0.1177
C(10)	0.1753	0.643	0.0621
C(11)	0.2497	0.716	0.0505
C(12)	0.3073	0.825	0.0939
C(13)	0.2862	0.865	0.1513
C(14)	0.2115	0.794	0.1624

The resulting molecular arrangement is illustrated in Figure XVAIII,126. Each ammonium ion has five oxygen and one water neighbors with NH_4–H_2O = 2.91 A. and NH_4–O = 2.84–2.95 A. Water molecules are hydrogen-bonded to O(1) and O(4) with O–H–O = 2.76 and 2.65 A., respectively.

The salicylate ions have substantially the same dimensions as in the acid (**XV,aIII,77**).

XV,aIII,79. *Copper salicylate tetrahydrate,* $Cu(OHC_6H_4CO_2)_2 \cdot 4H_2O$, forms monoclinic crystals whose bimolecular cell has the dimensions:

$$a_0 = 3.728(8) \text{ A.}; \quad b_0 = 17.70(3) \text{ A.}; \quad c_0 = 12.27(2) \text{ A.}$$
$$\beta = 93°16'$$

The space group is C_{2h}^5 ($P2_1/c$). The copper atoms are in the special positions:

$$(2a) \quad 000; 0\ {}^1/_2\ {}^1/_2$$

The other atoms are in the general positions:

$$(4e) \quad \pm(xyz;\ x,{}^1/_2-y,z+{}^1/_2)$$

with the parameters of Table XVAIII,72.

TABLE XVAIII,72

Parameters of the Atoms in $Cu(OHC_6H_4CO_2)_2 \cdot 4H_2O$

Atom	x	y	z
O(1)	0.036	0.033	0.142
O(2)	0.757	0.145	0.107
OH(3)	0.768	0.249	0.251
$H_2O(4)$	0.324	0.423	0.457
$H_2O(5)$	0.557	0.398	0.268
C(1)	0.005	0.123	0.284
C(2)	0.927	0.197	0.319
C(3)	0.992	0.225	0.425
C(4)	0.113	0.328	0.003
C(5)	0.214	0.100	0.478
C(6)	0.126	0.076	0.370
C(7)	0.915	0.103	0.172

The salicylate anion in this structure (Fig. XVAIII,127) has the bond dimensions of Figure XVAIII,128. It is planar except for O(2) which is 0.26 A. outside the plane of the rest. The coordination of the copper atom is fourfold and planar, with an additional two O(4) atoms at a distance of 2.89 A. These O(4) are hydrogen-bonded to O(2) atoms (O–H–O = 2.67 A.).

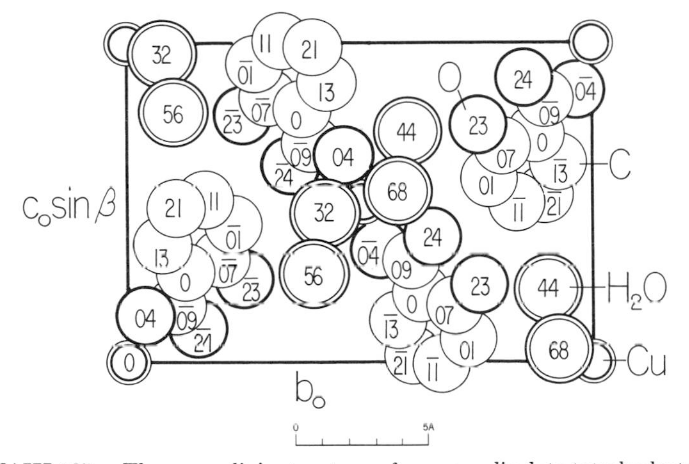

Fig. XVAIII,127. The monoclinic structure of copper salicylate tetrahydrate projected along its a_0 axis.

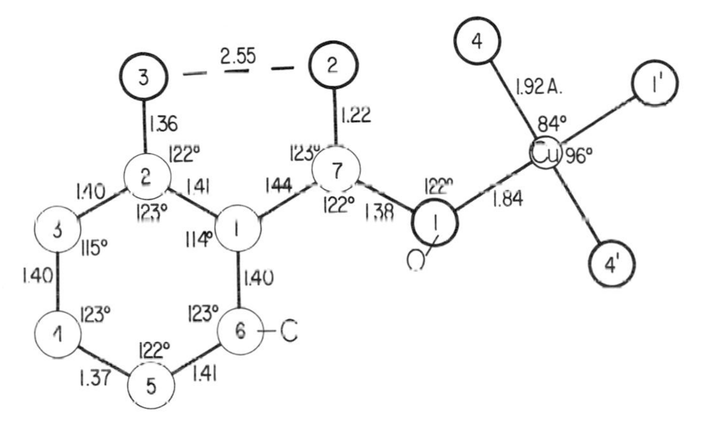

Fig. XVAIII,128. Bond dimensions in the molecular anion in copper salicylate tetrahydrate.

The water molecules O(5) ,not associated with copper, fill holes between the salicylate groups and are tied by hydrogen bonds to the O(3) [O–H–O = 2.76 A.] and O(4) [O–H–O = 2.56 A.] atoms.

XV,aIII,80. Crystals of *zinc salicylate dihydrate*, $Zn(OHC_6H_4COO)_2 \cdot 2H_2O$, are monoclinic with a bimolecular unit of the dimensions:

$$a_0 = 15.43(2) \text{ A.}; \quad b_0 = 5.35(2) \text{ A.}; \quad c_0 = 9.18(2) \text{ A.}$$
$$\beta = 93°48'$$

The zinc atoms are in the positions:

$$(2a)\quad 0u0;\ {}^1/_2,u+{}^1/_2,0 \qquad \text{with } u = 0 \text{ (arbitrary)}$$

of C_2^3 ($C2$). The other atoms are in

$$(4c)\quad xyz;\ \bar{x}y\bar{z};\ x+{}^1/_2,y+{}^1/_2,z;\ {}^1/_2-x,y+{}^1/_2,\bar{z}$$

with the parameters of Table XVAIII,73.

TABLE XVAIII,73

Parameters of Atoms in $Zn(OHC_6H_4COO)_2 \cdot 2H_2O$

Atom	x	y	z
C(1)	0.432	0.813	0.176
C(2)	0.377	0.997	0.245
C(3)	0.287	0.019	0.208
C(4)	0.236	0.198	0.272
C(5)	0.273	0.348	0.375
C(6)	0.361	0.320	0.409
C(7)	0.412	0.150	0.349
O(1)	0.514	0.822	0.202
O(2)	0.399	0.680	0.086
O(3)	0.250	0.848	0.109
O(H_2O)	0.075	0.730	0.111

The structure that results (Fig. XVAIII,129) is composed of Zn^{2+} and $(OHC_6H_4CO_2)^-$ ions and water molecules, the last two providing hydrogen bonds that help to tie the structure together in the a_0 and b_0 directions. Coordination about Zn^{2+} is tetrahedral with Zn–2O(2) = 2.03 A. and Zn–2(H_2O) = 2.06 A.; the next shortest approach to zinc is Zn–O(1) = 2.52 A. The salicylate anion is planar except for its O(1) and O(3) atoms which are 0.17 and 0.12 A. (in opposite directions) outside this plane. Bond dimensions are those of Figure XVAIII,130. There are short O–O separations, some presumably due to hydrogen bonds, of H_2O–O(1) = 2.55 A., H_2O–O(3) = 2.77 A., and O(3)–O(3) = 3.06 A. The short intramolecular O(2)–O(3) = 2.51 A. may represent internal hydrogen bonding.

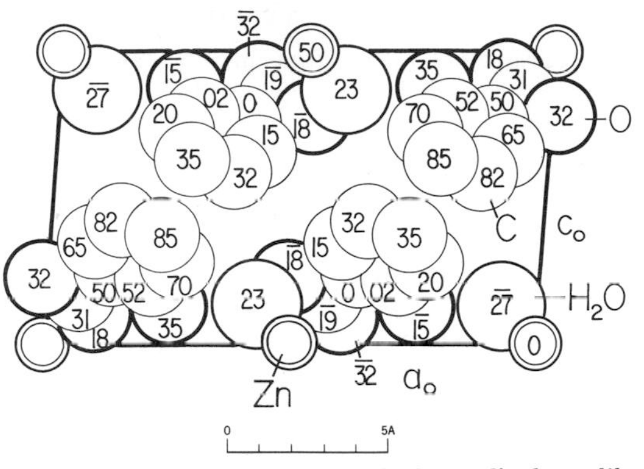

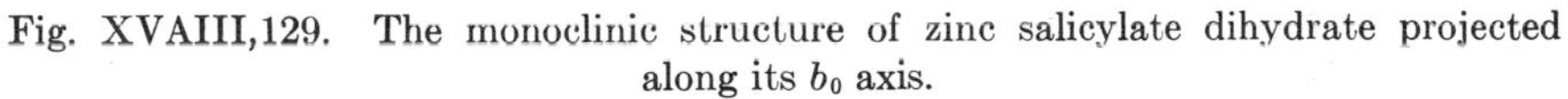

Fig. XVAIII,129. The monoclinic structure of zinc salicylate dihydrate projected along its b_0 axis.

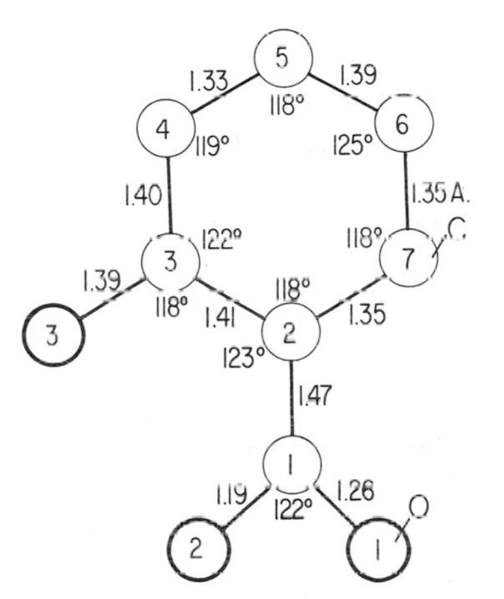

Fig. XVAIII,130. Bond dimensions in the molecular anion of zinc salicylate dihydrate.

XV,aIII,81. *Salicylamide*, $C_6H_4(OH)CONH_2$, forms monoclinic crystals. Their unit cells, containing eight molecules, have the dimensions:

$$a_0 = 12.92(4) \text{ A.}; \quad b_0 = 4.98(2) \text{ A.}; \quad c_0 = 21.04(5) \text{ A.}$$
$$\beta = 91.8(4)°$$

The space group is C_{2h}^6 in the orientation $I2/a$. Atoms accordingly are in the positions:

$$(8f) \quad \pm(xyz;\ x+{}^1/_2,\bar{y},z); \quad \text{B.C.}$$

with the parameters of Table XVAIII,74.

TABLE XVAIII,74

Parameters of the Atoms in Salicylamide

Atom	x	y	z
C(1)	0.0799	0.177	0.3720
C(2)	0.1861	0.230	0.3601
C(3)	0.2600	0.067	0.3934
C(4)	0.2284	−0.124	0.4360
C(5)	0.1254	−0.174	0.4468
C(6)	0.0524	−0.020	0.4150
C(7)	0.2120	0.452	0.3134
N(8)	0.3099	0.480	0.3005
O(9)	0.1424	0.580	0.2870
O(10)	0.0050	0.326	0.3418

The structure, shown in Figure XVAIII,131, is composed of molecules which are planar except for a nitrogen atom that is 0.08 A. outside the plane of the benzene ring. They have the bond dimensions of Figure XVAIII,132. Between them are hydrogen bonds of lengths N–H–O(9) = 2.94 A. and N–H–O(10) = 3.05 A.

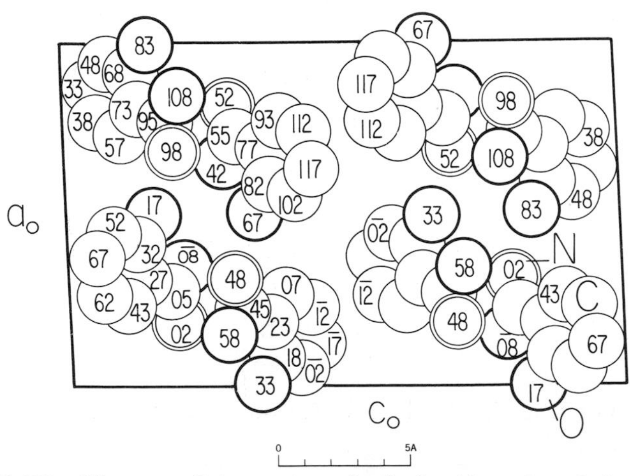

Fig. XVAIII,131. The monoclinic structure of salicylamide projected along its b_0 axis.

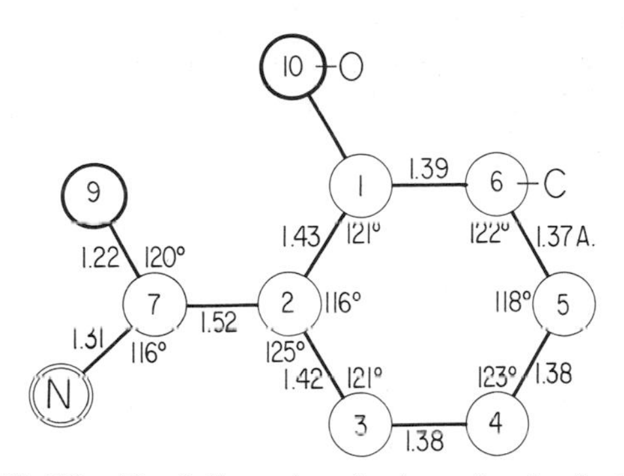

Fig. XVAIII,132. Bond dimensions in the molecule of salicylamide.

XV,aIII,82. A structure has been briefly described for *salicylic aldehyde*, OHC_6H_4CHO. Its monoclinic unit contains four molecules and has the dimensions:

$$a_0 = 6.33 \text{ A.}; \quad b_0 = 13.90 \text{ A.}; \quad c_0 = 7.08 \text{ A.}; \quad \beta = 102°56'$$

Atoms are in general positions

$$(4e) \quad \pm(xyz;\ x,{}^1/_2-y,z+{}^1/_2)$$

of C_{2h}^5 ($P2_1/c$). Their parameters are listed in Table XVAIII,75.

TABLE XVAIII,75

Parameters of the Atoms in Salicylic Aldehyde

Atom	x	y	z
C(1)	0.900	0.233	0.475
C(2)	0.953	0.330	0.488
C(3)	0.783	0.397	0.446
C(4)	0.566	0.369	0.391
C(5)	0.521	0.272	0.381
C(6)	0.683	0.200	0.421
C(7)	0.633	0.104	0.408
OH	0.080	0.173	0.520
O	0.770	0.038	0.442

The molecules in this structure are flat, with C–C within the ring between 1.37 and 1.42 A. The separation of the ring carbon from the aldehyde carbon (C7) is 1.37 A. The C–O distances are 1.25 A. in the aldehyde group and 1.39 A. for C(1)–OH.

XV,aIII,83. One of the two forms of *bis(salicylaldehydato)copper(II)*, $[C_6H_4(O)CHO]_2Cu$, forms monoclinic crystals whose bimolecular unit has the dimensions:

$$a_0 = 8.72(2) \text{ A.}; \quad b_0 = 6.19(2) \text{ A.}; \quad c_0 = 11.26(3) \text{ A.}$$
$$\beta = 104.8°$$

The space group is C_{2h}^5 in the axial orientation $P2_1/n$. The atoms of copper are in the positions $(2a)$ 000; $\frac{1}{2}\,\frac{1}{2}\,\frac{1}{2}$. All the other atoms are in the general positions

$$(4e) \quad \pm(xyz;\ x+\tfrac{1}{2},\tfrac{1}{2}-y,z+\tfrac{1}{2})$$

with the determined parameters of Table XVAIII,76.

TABLE XVAIII,76

Parameters of the Atoms in Bis(Salicylaldehydato)copper [Form I]

Atom	x	y	z
O(1)	0.3393	0.7035	0.4819
O(2)	0.6280	0.6184	0.6572
C(1)	0.5950	0.7960	0.6995
C(2)	0.4630	0.9284	0.6586
C(3)	0.3390	0.8787	0.5488
C(4)	0.2150	0.0346	0.5146
C(5)	0.2135	0.2064	0.5825
C(6)	0.3380	0.2580	0.6887
C(7)	0.4573	0.1140	0.7270

The structure is shown in Figure XVAIII,133. Its molecules are approximately planar with a maximum departure from the plane of 0.13 A. [for O(1)]. The bond dimensions are those of Figure XVAIII,134. There are several intermolecular C–C and C–O separations in the range between 3.41 and 3.50 A. As Figure XVAIII,133 suggests, the coordination of the copper atoms is planar and square.

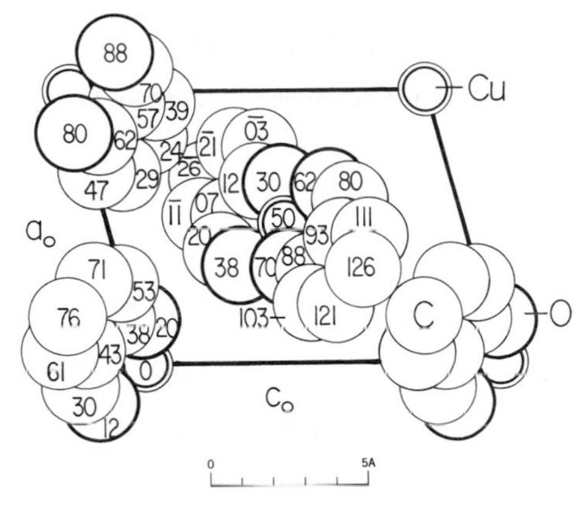

Fig. XVAIII,133. The monoclinic structure of the first form of bis(salicylaldehydato) copper projected along its b_0 axis.

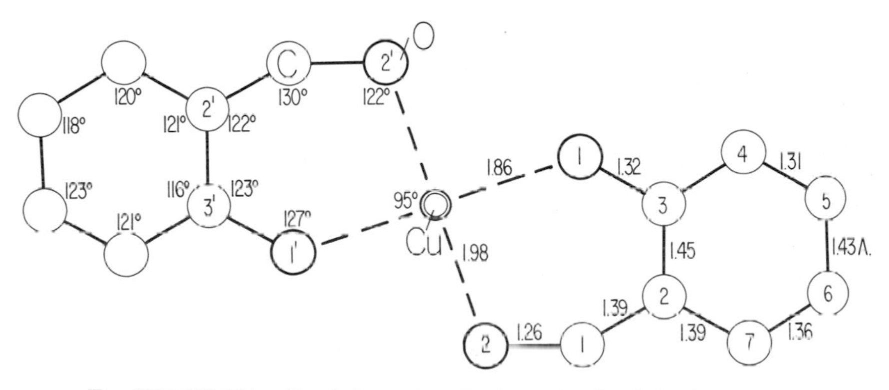

Fig. XVAIII,134. Bond dimensions in the molecule of the first form of bis(salicylaldehydato)copper.

XV,aIII,84. A second modification of *bis(salicylaldehydato)copper(II)*, $[C_6H_4(O)CHO]_2Cu$, is also monoclinic with a bimolecular unit of the dimensions:

$$a_0 = 11.75(3) \text{ A.}; \quad b_0 = 4.00(2) \text{ A.}; \quad c_0 = 12.42(3) \text{ A.}$$
$$\beta = 90°18'$$

The space group is C_{2h}^5 ($P2_1/c$). Atoms of copper are in the positions:

$$(2a) \quad 000; 0\ {}^1/_2\ {}^1/_2$$

All the other atoms are in:

$$(4e) \quad \pm(xyz; x,{}^1/_2-y,z+{}^1/_2)$$

with the parameters of Table XVAIII,77.

TABLE XVAIII,77

Parameters of Atoms in the Second Form of $Cu(C_7H_5O_2)_2$

Atom	x	y	z
O(1)	0.1234	0.3024	−0.0191
O(2)	0.0247	−0.0505	0.1553
C(1)	0.1078	0.0829	0.2046
C(2)	0.1967	0.2722	0.1620
C(3)	0.2020	0.3673	0.0547
C(4)	0.2972	0.5681	0.0209
C(5)	0.3831	0.6474	0.0915
C(6)	0.3767	0.5668	0.2010
C(7)	0.2843	0.3695	0.2356

The resulting structure, shown in Figure XVAIII,135, contains molecules having the bond dimensions of Figure XVAIII,136. The atoms of a salicylaldehydato group depart from its best plane by as much as 0.1 A. Copper(II) has its usual square coordination but the planes of the two ligands associated with it are 0.37 A. apart. Between separate molecules there are, however, Cu–O distances of 3.15 A. and these can be interpreted as giving the copper a distorted octahedral coordination.

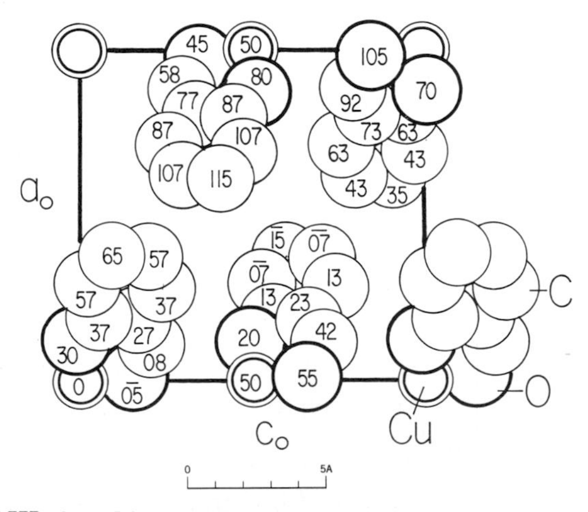

Fig. XVAIII,135. The monoclinic structure of the second modification of bis(salicylaldehydato)copper projected along its b_0 axis.

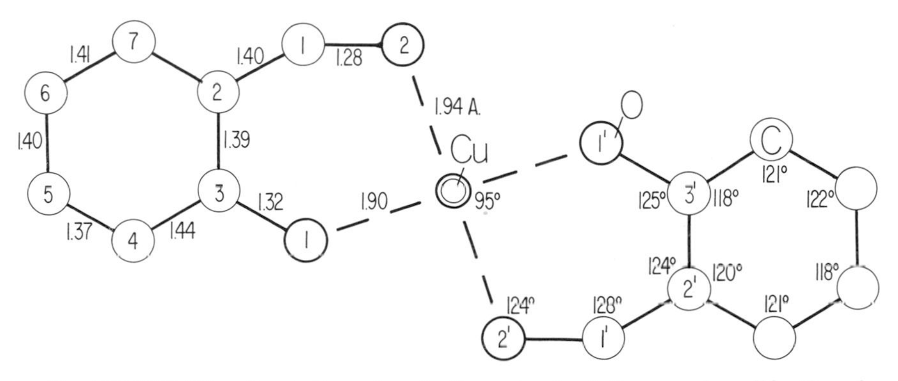

Fig. XVAIII,136. Bond dimensions in the molecule of the second modification of bis(salicylaldehydato)copper.

XV,aIII,85. *Bis(salicylaldehydato)nickel dihydrate*, $[C_6H_4(O)CHO]_2Ni \cdot 2H_2O$, forms monoclinic, pseudo-orthorhombic, crystals whose bimolecular unit has the dimensions:

$$a_0 = 12.93 \text{ A.}; \quad b_0 = 7.32 \text{ A.}; \quad c_0 = 7.41 \text{ A.}; \quad \beta = 90°15'$$

The space group is C_{2h}^3 ($A2/m$) with atoms in the positions:

Ni: $(2a)$ $000; 0\,{}^1/_2\,{}^1/_2$

$O(H_2O)$: $(4g)$ $\pm(0u0; 0,u+{}^1/_2,{}^1/_2)$ with $u = 0.2788$

Other atoms are in the positions:

$$(4i) \quad \pm(u0v;\ u,{}^1/_2,v+{}^1/_2)$$

with the parameters of Table XVAIII,78. Parameters for the hydrogen atoms are stated in the original.

TABLE XVAIII,78

Parameters of Atoms in Bis(salicylaldehydato)nickel Dihydrate

Atom	u	v
O(2)	0.1267	0.1603
O(3)	0.0870	−0.2262
C(1)	0.2163	0.1076
C(2)	0.2495	−0.0784
C(3)	0.3604	−0.0964
C(4)	0.4048	−0.2671
C(5)	0.3416	−0.4161
C(6)	0.2343	−0.4035
C(7)	0.1878	−0.2330

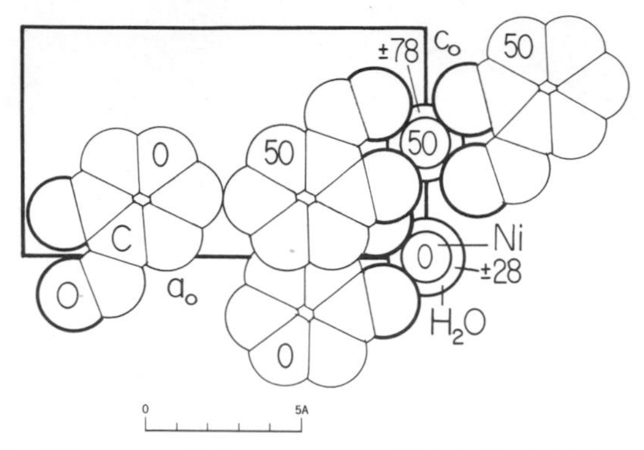

Fig. XVAIII,137. The monoclinic structure of bis(salicylaldehydato)nickel dihydrate projected along its b_0 axis.

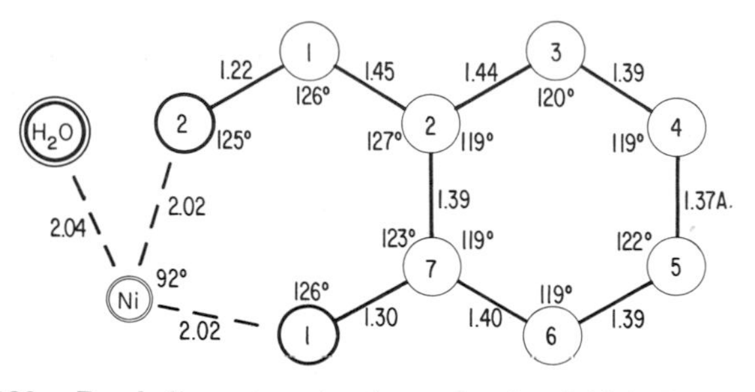

Fig. XVAIII,138. Bond dimensions in the molecule of bis(salicylaldehydato)nickel dihydrate.

The resulting structure, shown in Figure XVAIII,137, has planar molecules with the bond dimensions of Figure XVAIII,138. Coordination around nickel is octahedral and very nearly regular.

XV,aIII,86. *Diethyl (salicylaldehydato) thallium(III)*, $Tl[(C_2H_5)_2OC_6H_4CHO]$, is triclinic. It has been assigned a large, non-primitive tetramolecular cell of the dimensions:

$$a_0 = 8.00(2) \text{ A.}; \quad b_0 = 20.71(1) \text{ A.}; \quad c_0 = 7.71(2) \text{ A.}$$
$$\alpha = 100.7(2)°; \quad \beta = 101.6(2)°; \quad \gamma = 88.4(3)°$$

The space group is C_i^1 ($C\bar{1}$) and in the side-centered arrangement used here atoms are in the general positions:

$$(4i) \quad \pm(xyz;\ x+{}^1/_2,y+{}^1/_2,z)$$

The determined parameters are listed in Table XVAIII,79.

TABLE XVAIII,79

Parameters of the Atoms in $Tl(C_2H_5)_2(C_7H_5O_2)$

Atom	x	y	z
Tl	−0.2619(2)	0.5160(1)	0.0004(2)
O(1)	−0.442(3)	0.423(1)	0.008(3)
O(2)	−0.069(3)	0.414(1)	0.037(3)
C(1)	−0.082(5)	0.365(2)	0.106(5)
C(2)	−0.413(4)	0.371(2)	0.086(4)
C(3)	−0.563(4)	0.339(2)	0.119(5)
C(4)	−0.531(5)	0.285(2)	0.219(6)
C(5)	−0.376(5)	0.259(2)	0.273(5)
C(6)	−0.237(5)	0.291(2)	0.238(5)
C(7)	−0.258(5)	0.345(2)	0.144(5)
C(8)	−0.205(6)	0.624(2)	0.351(6)
C(9)	−0.192(5)	0.548(2)	0.298(5)
C(10)	−0.318(5)	0.498(2)	−0.286(5)
C(11)	−0.311(6)	0.426(2)	−0.379(6)

The molecules in the structure thus defined (Fig. XVAIII,139) have the bond dimensions of Figure XVAIII,140. The thallium atom is chemically bound to the two ethyl radicals and chelated to the pair of salicylaldehydato oxygens of one molecule; but it also comes close to these oxygens belonging to the next molecule along the a_0 axis. In this way the structure is

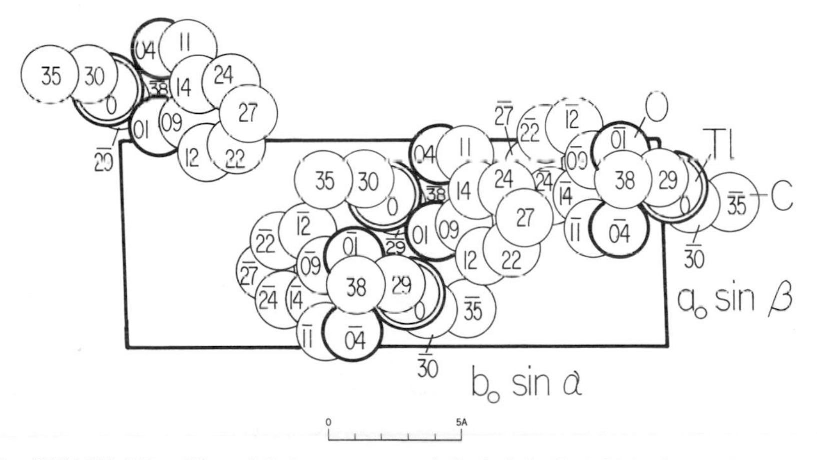

Fig. XVAIII,139. The triclinic structure of diethyl (salicylaldehydato) thallium projected along its c_0 axis.

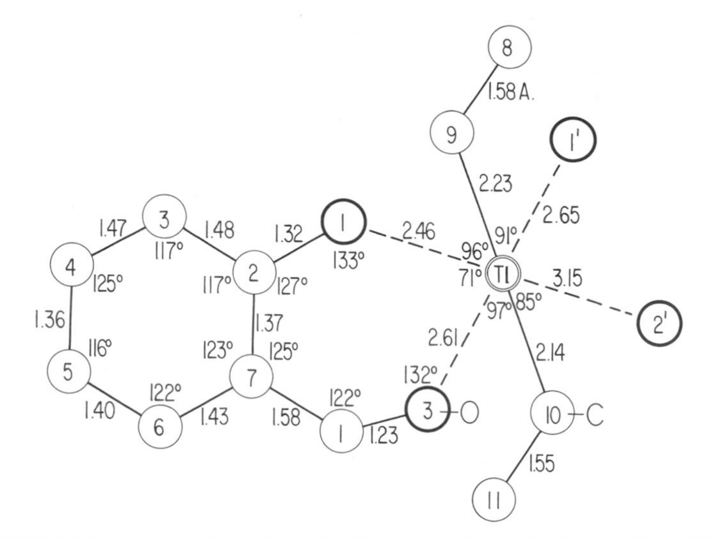

Fig. XVAIII,140. Bond dimensions in the molecule of diethyl (salicylaldehydato) thallium.

built up of chains of molecules. The salicylaldehydato group is planar but the thallium atom lies 0.85 A. from this plane.

XV,aIII,87. Crystals of *acetylsalicylic acid* (aspirin), $(HOOC)C_6H_4$-$OC(O)CH_3$, are monoclinic with a tetramolecular unit of the dimensions:

$$a_0 = 11.446(13) \text{ A.}; \quad b_0 = 6.596(6) \text{ A.}; \quad c_0 = 11.388(9) \text{ A.}$$
$$\beta = 95°33(2)'$$

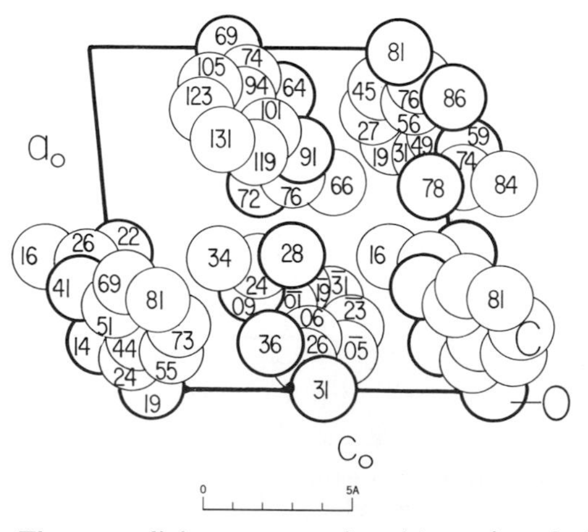

Fig. XVAIII,141. The monoclinic structure of aspirin projected along its b_0 axis.

Atoms are in the positions:

$$(4e) \quad \pm(xyz;\ x,{}^1/_2-y,z+{}^1/_2)$$

of the space group C_{2h}^5 ($P2_1/c$). Parameters, presumably applying to these axes, are stated in Table XVAIII,80.

TABLE XVAIII,80

Parameters of the Atoms in Aspirin

Atom	x	y	z
O(1)	0.0106	0.1876	0.0966
O(2)	0.1202	0.1387	−0.0508
O(3)	0.2852	0.4120	−0.0884
O(4)	0.4030	0.2200	0.0343
C(1)	0.1530	0.4366	0.0674
C(2)	0.2462	0.5118	0.0092
C(3)	0.2992	0.6947	0.0419
C(4)	0.2607	0.8129	0.1347
C(5)	0.1697	0.7317	0.1939
C(6)	0.1167	0.5506	0.1602
C(7)	0.0808	0.2417	0.0372
C(8)	0.3651	0.2611	−0.0631
C(9)	0.3968	0.1618	−0.1738
H(1)	0.071	−0.001	−0.081
H(3)	0.361	0.738	0.001
H(4)	0.300	0.940	0.163
H(5)	0.144	0.806	0.257
H(6)	0.056	0.507	0.199
H(7)	0.310	0.121	−0.236
H(8)	0.431	0.037	−0.170
H(9)	0.430	0.230	−0.216

The resulting structure, shown in Figure XVAIII,141, is built up of molecules having the bond dimensions of Figure XVAIII,142. Their atoms lie in two planes, one for the salicylic acid portion, the other for the acetyl group. Between the two planes the angle is 84°45′. In the crystal the molecules are united in pairs by hydrogen bonds of length 2.645 A. between carboxyl oxygens.

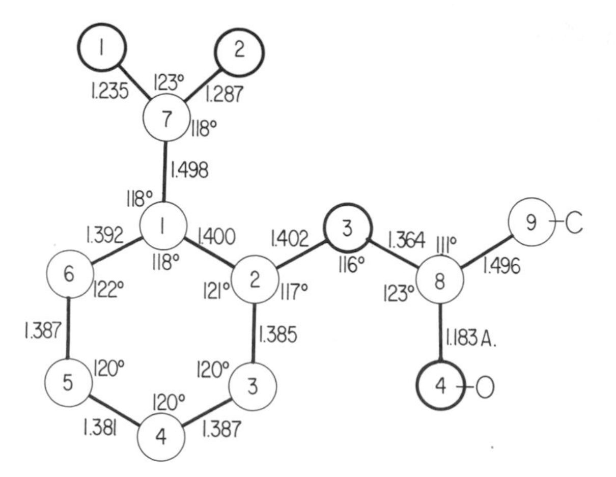

Fig. XVAIII,142. Bond dimensions in the molecule of aspirin.

XV,aIII,88. *Potassium hydrogen bisacetylsalicylate* (diaspirinate), KH-$[O_2CC_6H_4OC(O)CH_3]_2$, is monoclinic with a tetramolecular cell of the dimensions:

$$a_0 = 26.63(2) \text{ A.}; \quad b_0 = 7.207(7) \text{ A.}; \quad c_0 = 10.68(2) \text{ A.}$$
$$\beta = 116.24(13)°$$

The *rubidium* salt is isostructural; its unit has the dimensions:

$$a_0 = 26.34 \text{ A.}; \quad b_0 = 7.38 \text{ A.}; \quad c_0 = 10.72 \text{ A.}; \quad \beta = 116°$$

Both compounds have their atoms in the following positions of the space group C_{2h}^6 ($C2/c$):

K[Rb]: (4*e*) $\pm(0\ u\ ^1/_4;\ ^1/_2,u+^1/_2,^1/_4)$
with u(K) = 0.24358 and u(Rb) = 0.2420(3)
H(1): (4*a*) $000;\ 0\ 0\ ^1/_2;\ ^1/_2\ ^1/_2\ 0;\ ^1/_2\ ^1/_2\ ^1/_2$

All other atoms are in the general positions:

$$(8f) \quad \pm(xyz;\ x,\bar{y},z+{}^1/_2;\ x+{}^1/_2,y+{}^1/_2,z;\ x+{}^1/_2,{}^1/_2-y,z+{}^1/_2)$$

with the parameters of Table XVAIII,81.

TABLE XVAIII,81

Parameters of Atoms in $KH(C_9H_7O_4)_2$ and (in parentheses) of $RbH(C_9H_7O_4)_2$

Atom	x	y	z
O(1)	0.03889 (0.0390)	0.89483 (0.8994)	0.01822 (0.0186)
O(2)	0.07298 (0.0765)	0.98846 (0.9949)	0.23789 (0.2375)
O(3)	0.15179 (0.1557)	0.72105 (0.7280)	0.40609 (0.4055)
O(4)	0.07910 (0.0820)	0.52902 (0.5429)	0.29326 (0.2953)
C(1)	0.16533 (0.1675)	0.72124 (0.7273)	0.29429 (0.2930)
C(2)	0.21527 (0.2168)	0.63683 (0.6391)	0.31435 (0.3090)
C(3)	0.23119 (0.2324)	0.63673 (0.6353)	0.20847 (0.2013)
C(4)	0.19766 (0.1972)	0.72147 (0.7207)	0.08175 (0.0749)
C(5)	0.14790 (0.1484)	0.80589 (0.8085)	0.06195 (0.0612)
C(6)	0.13106 (0.1321)	0.80747(0.8081)	0.16827 (0.1664)
C(7)	0.07755 (0.0802)	0.90443 (0.9033)	0.14414 (0.1456)
C(8)	0.10544 (0.1097)	0.62607 (0.6370)	0.39190 (0.3924)
C(9)	0.09291 (0.0992)	0.65853 (0.6626)	0.51171 (0.5182)
H(2)	0.2364 (0.233)	0.5690 (0.577)	0.3952 (0.377)
H(3)	0.2666 (0.275)	0.5708 (0.594)	0.2151 (0.218)
H(4)	0.2068 (0.201)	0.7261 (0.755)	0.0146 (0.024)
H(5)	0.1237 (0.128)	0.8690 (0.875)	−0.0297 (−0.012)
H(7)	0.0680 —	0.5714 —	0.5150 —
H(8)	0.1052 —	0.7559 —	0.5559 —
H(9)	0.1244 —	0.5810 —	0.6058 —

The resulting structure is shown in Figure XVAIII,143. Bond dimensions for the more precisely determined potassium salt are those of Figure XVAIII,144. The atoms O(3) and C(7) depart by only about 0.03 A. from the plane of the benzene ring. The atoms O(3), C(8), O(4), and C(9) also are coplanar and their plane makes an angle of 69° with that of the ring. A third plane through the carboxyl group, C(7)–O(1)–O(2), is turned through 34.5° with respect to the ring. The H(1) atom is considered to provide a "symmetrical" bond between O(1) and O(1′) atoms; like all such bonds it is especially short, with O(1)–H(1)–O(1′) = 2.455 A. The potassium cations have seven oxygen neighbors with K–O = 2.720–2.832 A.

The basic structure of the rubidium salt is similar, with Rb–7O = 2.84–2.90 A.

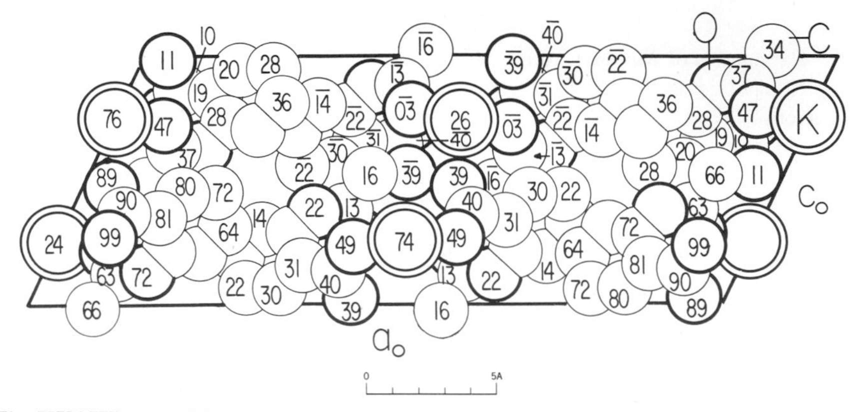

Fig. XVAIII,143. The monoclinic structure of potassium hydrogen bis(acetylsalicylate) projected along its b_0 axis.

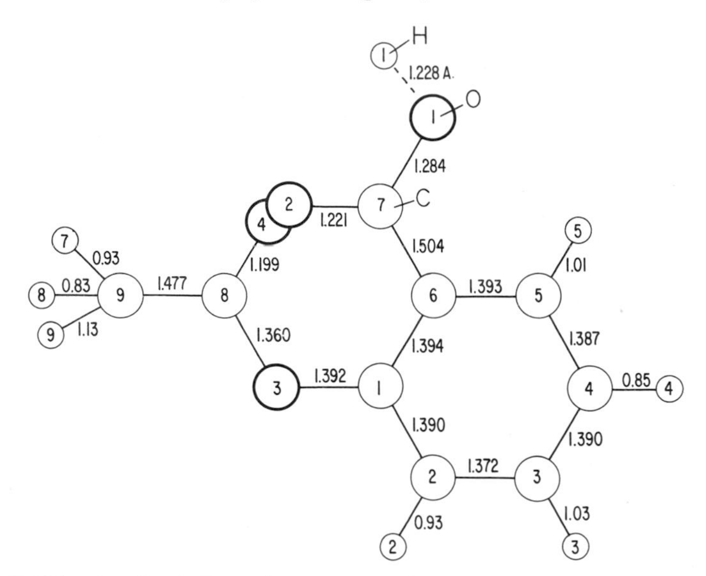

Fig. XVAIII,144. Bond dimensions in the molecular anion in potassium hydrogen bis(acetylsalicylate).

XV,aIII,89. The monoclinic crystals of *bis(salicylaldiminato)copper(II)*, $[C_6H_4(O)CHNH]_2Cu$, possess a bimolecular unit of the dimensions:

$$a_0 = 12.99(2) \text{ A.}; \quad b_0 = 5.85(1) \text{ A.}; \quad c_0 = 8.08(2) \text{ A.}$$
$$\beta = 94°30(6)'$$

The copper atoms are in

$$(2a) \quad 000; 0\ {}^1/_2\ {}^1/_2$$

of C_{2h}^5 ($P2_1/c$). All other atoms are in the positions:

$$(4e) \quad \pm(xyz;\ x,{}^1/_2-y,z+{}^1/_2)$$

with the parameters of Table XVAIII,82.

TABLE XVAIII,82

Parameters of the Atoms in Bis(salicylaldiminato)copper(II) and, in Parentheses, the Nickel Compound

Atom	x	y	z
O	0.0990 (0.0963)	0.1609 (0.1563)	0.1479 (0.1391)
N	0.0906 (0.0866)	−0.2544 (−0.2452)	−0.0165 (−0.0245)
C(1)	0.1841 (0.1807)	−0.2816 (−0.2811)	0.0468 (0.0378)
C(2)	0.2379 (0.2370)	−0.1064 (−0.1125)	0.1462 (0.1400)
C(3)	0.3422 (0.3413)	−0.1619 (−0.1585)	0.2017 (0.1984)
C(4)	0.4017 (0.3999)	−0.0108 (−0.0100)	0.3014 (0.3025)
C(5)	0.3580 (0.3540)	0.1992 (0.1979)	0.3466 (0.3431)
C(6)	0.2556 (0.2533)	0.2524 (0.2500)	0.2948 (0.2870)
C(7)	0.1934 (0.1924)	0.0944 (0.0950)	0.1957 (0.1860)

Bond dimensions of the molecules in this structure (Fig. XVAIII,145) are given in Figure XVAIII,146. Each salicylaldiminato group is essentially planar but their two planes are stepped with respect to one another, the separation being 0.29 A. Thus the coordination about copper is approximately, but not exactly, square.

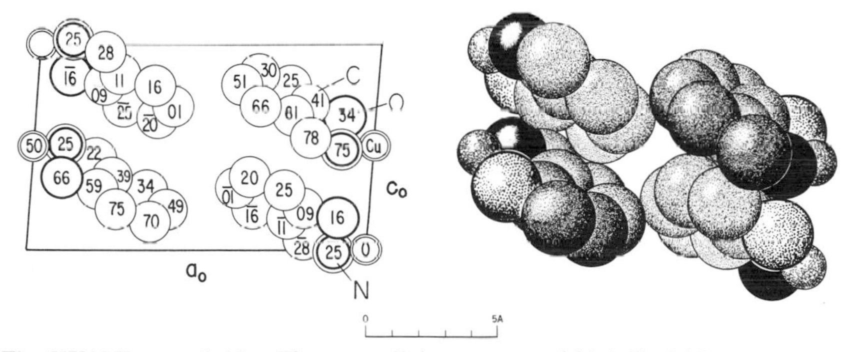

Fig. XVAIII,145a (left). The monoclinic structure of bis(salicylaldiminato) copper projected along its b_0 axis.

Fig. XVAIII,145b (right). A packing drawing of the monoclinic structure of bis (salicylaldiminato)copper viewed along its b_0 axis. The small circles are copper; the nitrogen atoms are black. The atoms of oxygen are coarsely dot shaded; those of carbon are finely dot shaded.

The isostructural *nickel* compound $[C_6H_4(O)CHNH]_2Ni$ was investigated several years before the copper. Its unit has the dimensions:

$$a_0 = 12.96 \text{ A.}; \quad b_0 = 5.83 \text{ A.}; \quad c_0 = 8.11 \text{ A.}; \quad \beta = 95°35'$$

The parameters found for it are given in parentheses in the table; those for hydrogen were stated in the original article.

XV,aIII,90. Crystals of *bis(salicylaldoximato)copper(II)*, $(OC_6H_4\text{-}CHNOH)_2Cu$, are monoclinic with a bimolecular unit of the dimensions:

$$a_0 = 13.98(3) \text{ A.}; \quad b_0 = 6.08(1) \text{ A.}; \quad c_0 = 8.00(1) \text{ A.}$$
$$\beta = 97°35'$$

The copper atoms are in the positions

$$(2a) \quad 000;\ 0\ {}^1/_2\ {}^1/_2$$

of the space group C_{2h}^5 ($P2_1/c$). All other atoms are in the general positions:

$$(4e) \quad \pm(xyz;\ x,{}^1/_2-y,z+{}^1/_2)$$

Their parameters are stated in Table XVAIII,83.

TABLE XVAIII,83

Parameters of Atoms in $(OC_6H_4CHNOH)_2Cu$

Atom	x	y	z
O(1)	0.0060(6)	0.9118(19)	0.6759(11)
O(2)	0.1187(6)	0.3472(19)	0.4884(11)
N	0.0678(8)	0.7440(23)	0.6199(13)
C(1)	0.2100(10)	0.4139(29)	0.5429(17)
C(2)	0.2877(10)	0.2842(29)	0.5123(17)
C(3)	0.3806(10)	0.3518(29)	0.5680(17)
C(4)	0.4007(10)	0.5451(29)	0.6501(17)
C(5)	0.3261(10)	0.6758(29)	0.6811(17)
C(6)	0.2276(10)	0.6164(29)	0.6234(17)
C(7)	0.1560(10)	0.7732(29)	0.6632(17)
H(1)	0.275	0.129	0.450
H(2)	0.438	0.250	0.543
H(3)	0.475	0.594	0.694
H(4)	0.341	0.828	0.746
H(5)	0.179	0.922	0.729
H(6)	0.042	0.189	0.387

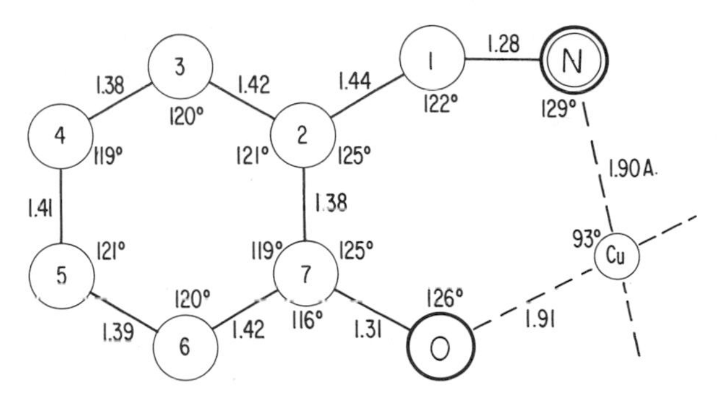

Fig. XVAIII,146. Bond dimensions in the molecule of bis(salicylaldiminato)copper.

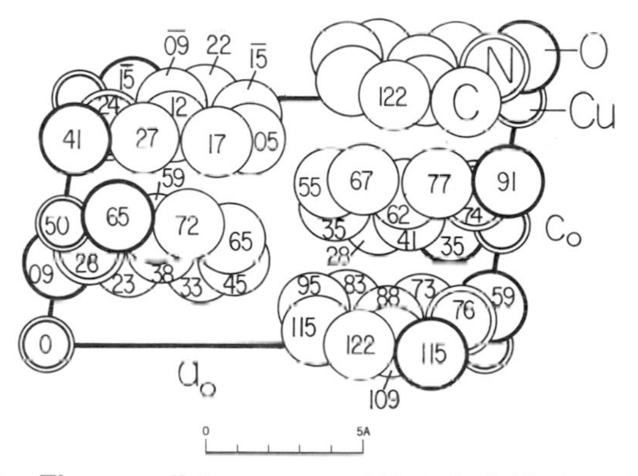

Fig. XVAIII,147. The monoclinic structure of bis(salicylaldoximato)copper projected along its b_0 axis.

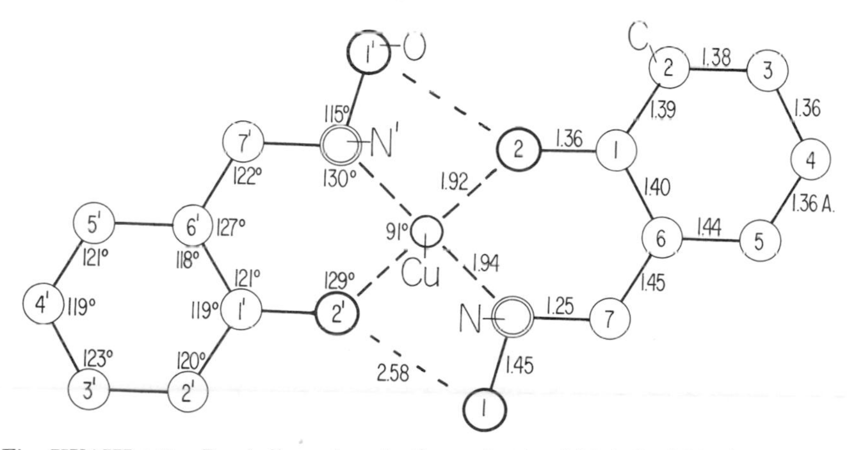

Fig. XVAIII,148. Bond dimensions in the molecule of bis(salicylaldoximato)copper.

The structure, shown in Figure XVAIII,147, is composed of molecules having the bond dimensions of Figure XVAIII,148. The Cu(II) atom has its usual approximately square coordination provided by nitrogen and O(2) atoms. There are also the two more distant O(1) neighbors 2.66 A. away; the line through them departs by 15° from the normal to the plane of the other bonds. The total molecule is almost but not exactly planar, the planes of the two benzene rings being 0.13 A. apart.

XV,aIII,91. Crystals of *bis(salicylaldoximato)nickel*, $(OC_6H_4CHNOH)_2$-Ni, have a structure somewhat different from that of the copper compound (**XV,aIII,90**). They, too, are monoclinic but their bimolecular cell has the dimensions:

$$a_0 = 13.83 \text{ A.}; \quad b_0 = 4.88 \text{ A.}; \quad c_0 = 10.20 \text{ A.}; \quad \beta = 110°26'$$

The space group is C_{2h}^5 with the axial orientation $P2_1/n$. The nickel atoms are in:

$$(2a) \quad 000; \ {}^1/_2\ {}^1/_2\ {}^1/_2$$

The other atoms are in:

$$(4e) \quad \pm(xyz;\ x+{}^1/_2,{}^1/_2-y,z+{}^1/_2)$$

According to a recent reworking of the original data the parameters are those of Table XVAIII,84.

TABLE XVAIII,84

Parameters of the Atoms in $(OC_6H_4CHNOH)_2Ni$

Atom	x	y	z
O(2)	−0.0932(4)	0.2698(10)	0.0002(5)
C(3)	−0.0926(6)	0.4386(16)	0.1024(8)
C(4)	−0.1762(6)	0.6165(17)	0.0815(8)
C(5)	−0.1785(7)	0.7969(16)	0.1839(10)
C(6)	−0.0980(6)	0.8112(14)	0.3106(8)
C(7)	−0.0148(6)	0.6394(17)	0.3335(8)
C(8)	−0.0105(6)	0.4512(19)	0.2332(11)
C(9)	0.0791(6)	0.2870(16)	0.2615(7)
N(10)	0.0909(4)	0.0887(14)	0.1770(5)
O(11)	0.1818(5)	−0.0485(15)	0.2345(7)

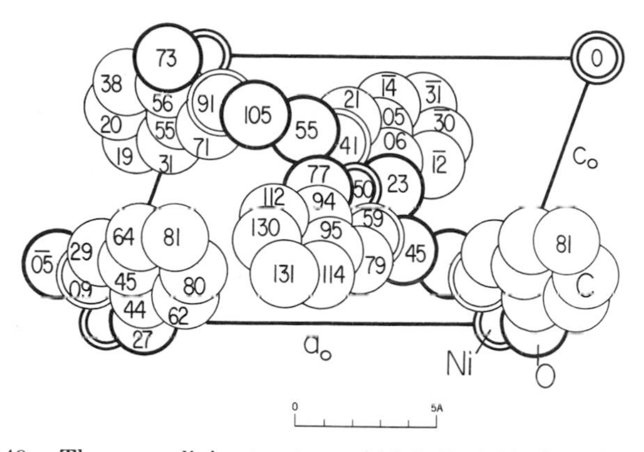

Fig. XVAIII,149. The monoclinic structure of bis(salicylaldoximato)nickel projected along its b_0 axis.

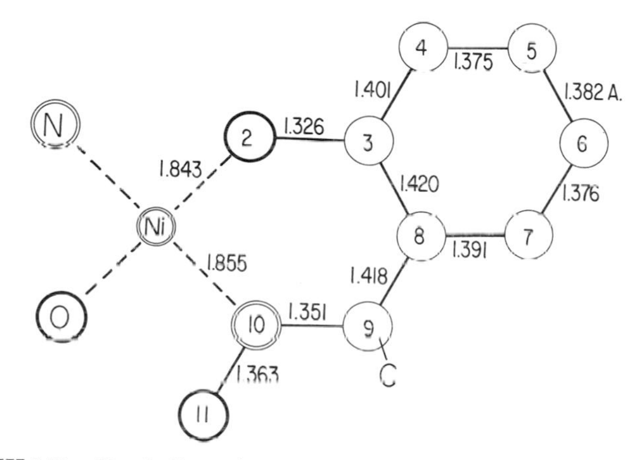

Fig. XVAIII,150. Bond dimensions in the molecule of bis(salicylaldoximato)nickel.

The structure (Fig. XVAIII,149) is composed of molecules having the bond dimensions of Figure XVAIII,150. They are planar except that the nitrogen and oxygen atoms lie 0.06 and 0.08 A. and the nickel atoms 0.12 A. from the plane of the rest.

XV,aIII,92. The monoclinic form of *bis(N-methylsalicylaldiminato)-nickel*, $(OC_6H_4CHNCH_3)_2Ni$, has a bimolecular unit of the dimensions:

$$a_0 = 11.94(5) \text{ A.}; \quad b_0 = 7.00(3) \text{ A.}; \quad c_0 = 8.30(4) \text{ A.}$$
$$\beta = 92°30(20)'$$

The nickel atoms are in

$$(2a) \quad 000;\ 0\ ^1/_2\ ^1/_2$$

and all other atoms in the general positions

$$(4e) \quad \pm(xyz;\ x,^1/_2-y,z+^1/_2)$$

of C_{2h}^5 ($P2_1/c$). Parameters are given in Table XVAIII,85.

TABLE XVAIII,85

Parameters of Atoms in Monoclinic Bis(*N*-methylsalicylaldiminato)nickel

Atom	x	y	z
C(1)	0.198	0.040	0.175
C(2)	0.250	0.102	0.322
C(3)	0.365	0.080	0.350
C(4)	0.428	−0.010	0.238
C(5)	0.375	−0.080	0.096
C(6)	0.260	−0.053	0.066
C(7)	0.220	−0.086	−0.100
C(8)	0.095	−0.112	−0.326
N	0.116	−0.055	−0.145
O	0.093	0.071[a]	0.164

[a] The −0.071 given in the original leads to an intermolecular C–O of little more than 2.0 A. Perhaps the +0.071 used here was intended.

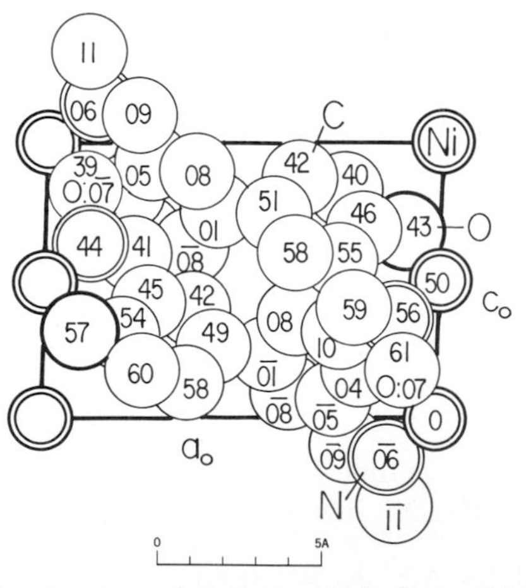

Fig. XVAIII,151. The structure of the monoclinic form of bis(*N*-methylsalicylaldiminato)nickel projected along its b_0 axis.

The structure is shown in Figure XVAIII,151. Its molecules are planar (if y for oxygen has the changed sign of the table) except for the terminal methyl carbon whose bond with nitrogen makes an angle of 12° with the plane. Coordination about the nickel is square, with Ni–O = 1.80 A. and Ni–N = 1.90 A.; the angle N–Ni–O = 96°.

XV,aIII,93. The α form of *bis(N-methylsalicylaldiminato)copper(II)*, $(OC_6H_4CHNCH_3)_2Cu$, has orthorhombic symmetry. Its tetramolecular cell has the edge lengths:

$$a_0 = 24.71(2) \text{ A.}; \quad b_0 = 9.25(1) \text{ A.}; \quad c_0 = 6.66(2) \text{ A.}$$

The copper atoms are in the positions

$$(4c) \quad 000; 0\,0\,{}^1/_2; \quad \text{B.C.}$$

of the space group V_h^{26} (*Ibam*). All the heavier and nearly all the hydrogen atoms have been placed in the special positions

$$(8j) \quad \pm(uv0; u\,\bar{v}\,{}^1/_2); \quad \text{B.C.}$$

with the parameters stated in square brackets in Table XVAIII,86. Values for hydrogen are given in the original article.

TABLE XVAIII,86

Parameters of Atoms in $(C_8H_8NO)_2Ni$ and $(C_8H_8NO)_2Cu$, in Square Brackets

Atom	u	v
N	0.0635(4) [0.0655]	−0.1222(13) [−0.1250]
O	0.0406(3) [0.0435]	0.1694(9) [0.1695]
C(1)	0.0948(5) [0.0965]	0.1892(20) [0.1870]
C(2)	0.1146(6) [0.1170]	0.3287(16) [0.3285]
C(3)	0.1708(6) [0.1725]	0.3539(18) [0.3515]
C(4)	0.2063(6) [0.2085]	0.2456(17) [0.2340]
C(5)	0.1889(6) [0.1885]	0.1036(17) [0.0955]
C(6)	0.1312(6) [0.1330]	0.0705(16) [0.0690]
C(7)	0.1145(5) [0.1155]	−0.0778(14) [−0.0795]
C(8)	0.0563(5) [0.0590]	−0.2836(14) [−0.2885]

The structure as shown in Figure XVAIII,152 differs from most of the compounds of this type in requiring absolutely planar molecules.

A more precise determination has been made of the isostructural *nickel* compound. For it:

$$a_0 = 24.46(1) \text{ A.}; \quad b_0 = 9.22(3) \text{ A.}; \quad c_0 = 6.58 \text{ A.}$$

The parameters are those of Table XVAIII,86. The bond dimensions of its molecules, close to those prevailing in the copper compound, are shown in Figure XVAIII,153.

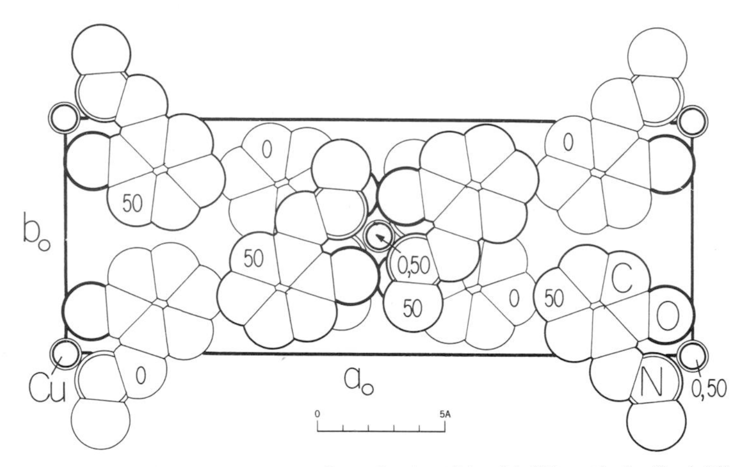

Fig. XVAIII,152. The structure of orthorhombic bis(*N*-methylsalicylaldiminato) copper projected along its c_0 axis.

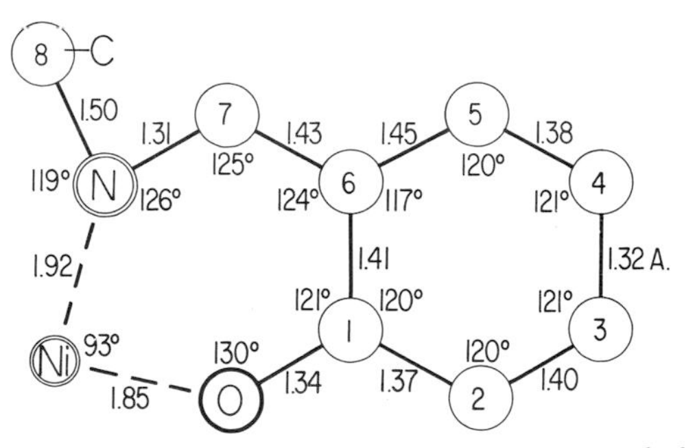

Fig. XVAIII,153. Bond dimensions in the molecule of bis(*N*-methylsalicylaldiminato) nickel as it occurs in its **α**, orthorhombic, modification.

XV,aIII,94. The triclinic crystals of *bis(N-methylsalicylaldiminato)zinc*, $(OC_6H_4CHNCH_3)_2Zn$, possess a bimolecular unit of the dimensions:

$$a_0 = 9.48 \text{ A.};\quad b_0 = 10.53 \text{ A.};\quad c_0 = 8.45 \text{ A.}$$
$$\alpha = 99°45';\quad \beta = 92°58';\quad \gamma = 117°58'$$

Atoms are in the positions $(2i)$ $\pm(xyz)$ of the space group C_i^1 $(P\bar{1})$. Parameters are those of Table XVAIII,87; values for hydrogen are to be found in the original article.

TABLE XVAIII,87

Parameters of the Atoms in $(C_8H_8NO)_2Zn$

Atom	x	y	z
Zn	0.0531(1)	0.1726(1)	0.0505(1)
O(1)	0.2178(6)	0.3598(5)	0.0080(7)
O(2)	−0.0979(5)	−0.0332(5)	0.1011(7)
N(1)	−0.1296(7)	0.1979(6)	−0.0544(9)
N(2)	0.1307(6)	0.2392(5)	0.2979(8)
C(1)	0.0459(8)	0.4113(7)	−0.1612(10)
C(2)	0.1983(8)	0.4338(7)	−0.0952(10)
C(3)	0.3384(8)	0.5471(8)	−0.1370(11)
C(4)	0.3267(10)	0.6332(9)	−0.2403(13)
C(5)	0.1743(11)	0.6093(10)	−0.3053(13)
C(6)	0.0379(10)	0.5019(9)	−0.2651(12)
C(7)	−0.1046(8)	0.3014(8)	−0.1342(11)
C(8)	−0.2980(10)	0.0943(9)	−0.0438(13)
C(9)	−0.1158(8)	0.0526(7)	0.3695(10)
C(10)	−0.1803(7)	−0.0514(7)	0.2272(10)
C(11)	−0.3326(9)	−0.1750(8)	0.2123(11)
C(12)	−0.4213(10)	−0.1934(9)	0.3394(13)
C(13)	−0.3600(11)	−0.0891(10)	0.4817(14)
C(14)	−0.2057(10)	0.0309(9)	0.4998(13)
C(15)	0.0429(8)	0.1813(7)	0.4003(10)
C(16)	0.2865(9)	0.3715(8)	0.3603(12)

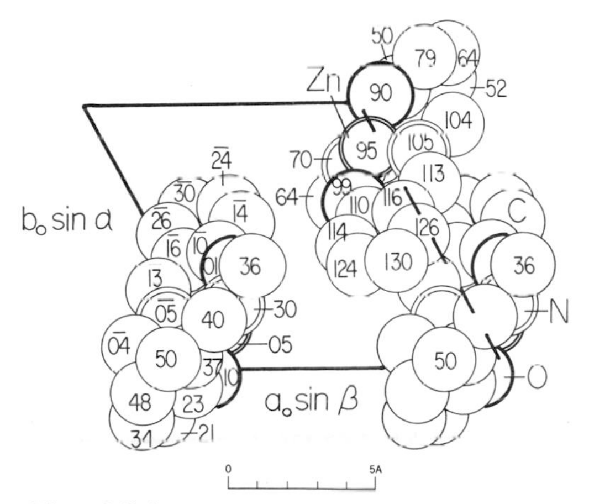

Fig. XVAIII,154. The triclinic structure of bis(*N*-methylsalicylaldiminato)zinc projected along its c_0 axis.

In this structure (Fig. XVAIII,154) the zinc atom has an unusual fivefold coordination. Bond dimensions for the dimeric molecular association about a zinc atom are given in Figure XVAIII,155; the fifth atom about zinc is an O(2) from a neighboring molecule. As a result, pairs of zinc atoms are associated by holding two of these O(2) atoms in common (Fig. XVAIII, 156). The two salicylaldimine portions are planar except that O(1) is 0.11 A. out of the plane of the benzene ring to which it is attached, and for the other molecular group the C(15) is 0.10 A. and the N(2) atom 0.15 A. out of the plane of their ring.

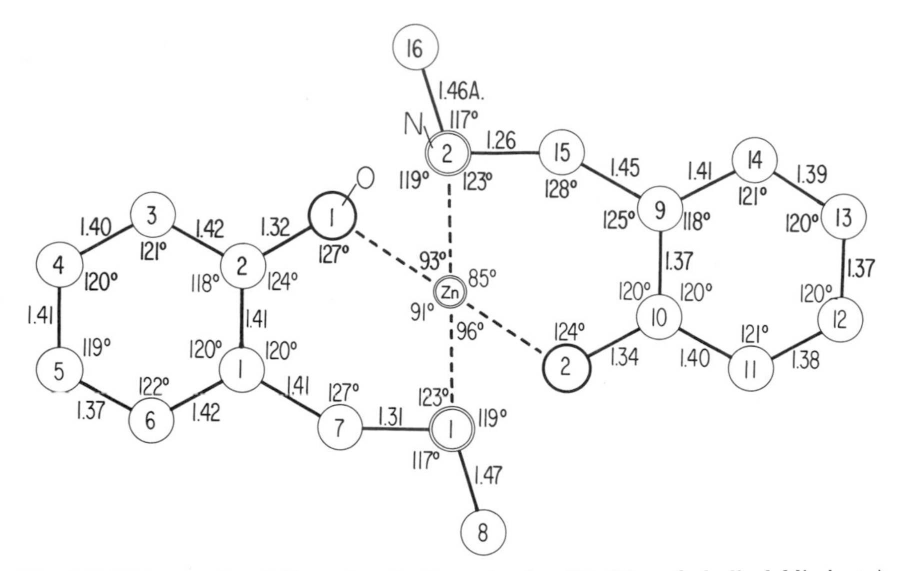

Fig. XVAIII,155. Bond dimensions in the molecule of bis(*N*-methylsalicylaldiminato) zinc.

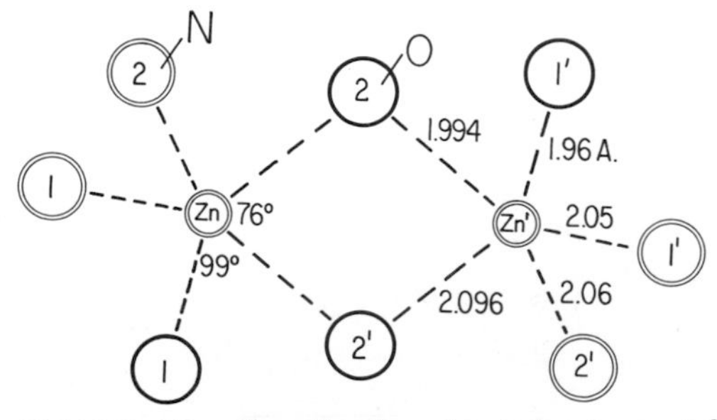

Fig. XVAIII,156. Coordination about zinc in crystals of bis(*N*-methylsalicylaldiminato)zinc.

The manganese and cobalt compounds are isostructural, with the cell dimensions:

For $(OC_6H_4CHNCH_3)_2Mn$:

$$a_0 = 9.35 \text{ A.}; \quad b_0 = 10.76 \text{ A.}; \quad c_0 = 8.45 \text{ A.}$$
$$\alpha = 100°28'; \quad \beta = 92°29'; \quad \gamma = 117°51'$$

For $(OC_6H_4CHNCH_3)_2Co$:

$$a_0 = 9.40 \text{ A.}; \quad b_0 = 10.72 \text{ A.}; \quad c_0 = 8.30 \text{ A.}$$
$$\alpha = 96°23'; \quad \beta = 95°16'; \quad \gamma = 118°10'$$

XV,aIII,95. *Bis(N-ethylsalicylaldiminato)copper(II)*, $(OC_6H_4CHNC_2H_5)_2Cu$, is monoclinic with a tetramolecular cell of the dimensions:

$$a_0 = 9.80(2) \text{ A.}; \quad b_0 = 8.42(2) \text{ A.}; \quad c_0 = 21.12(2) \text{ A.}$$
$$\beta = 101.3(3)°$$

All atoms are in the positions

$$(4e) \quad \pm(xyz;\ x,{}^1/_2-y,z+{}^1/_2)$$

of C_{2h}^5 ($P2_1/c$). The parameters are those of Table XVAIII,88.

TABLE XVAIII,88

Parameters of the Atoms in Bis(*N*-Ethylsalicylaldiminato)copper(II)

Atom	x	y	z
Cu	0.2178(1) [0.2179]	0.1139(2) [0.1144]	0.3362(1) [0.3363]
O	0.2867(7) [0.2868]	0.0444(9) [0.0449]	0.4214(3) [0.4217]
O′	0.2071(7) [0.2087]	0.1041(9) [0.1026]	0.2447(3) [0.2454]
N	0.0373(9) [0.0374]	0.0158(11) [0.0142]	0.3298(4) [0.3310]
N′	0.3529(8) [0.3523]	0.2906(11) [0.2893]	0.3473(4) [0.3464]
C(1)	0.0025(11) [0.0002]	−0.0701(14) [−0.0700]	0.3761(5) [0.3756]
C(2)	0.0903(10) [0.0899]	−0.1097(13) [−0.1107]	0.4360(5) [0.4374]
C(3)	0.0355(12) [0.0330]	−0.2140(16) [−0.2096]	0.4774(5) [0.4769]
C(4)	0.1097(13) [0.1059]	−0.2609(16) [−0.2576]	0.5382(6) [0.5369]
C(5)	0.2464(12) [0.2478]	−0.2049(16) [−0.1989]	0.5564(5) [0.5552]
C(6)	0.3036(12) [0.3039]	−0.1016(14) [−0.1022]	0.5170(5) [0.5177]
C(7)	0.2266(10) [0.2303]	−0.0566(14) [−0.0529]	0.4553(5) [0.4557]
C(8)	−0.0740(12) [−0.0734]	0.0356(16) [0.0405]	0.2698(5) [0.2697]
C(9)	−0.1473(17) [−0.1483]	0.1903(20) [0.1943]	0.2717(7) [0.2739]
C(1′)	0.4183(10) [0.4163]	0.3393(14) [0.3389]	0.3024(5) [0.3030]

(continued)

TABLE XVAIII,88 (*continued*)

Atom	x	y	z
C(2′)	0.3962(10) [0.3959]	0.2807(14) [0.2801]	0.2381(4) [0.2378]
C(3′)	0.4832(12) [0.4829]	0.3437(15) [0.3459]	0.1981(5) [0.1977]
C(4′)	0.4668(12) [0.4665]	0.3020(16) [0.3002]	0.1333(5) [0.1331]
C(5′)	0.3619(12) [0.3629]	0.1978(16) [0.1964]	0.1069(5) [0.1075]
C(6′)	0.2753(12) [0.2755]	0.1250(15) [0.1278]	0.1447(5) [0.1459]
C(7′)	0.2917(10) [0.2889]	0.1701(14) [0.1724]	0.2127(5) [0.2118]
C(8′)	0.3843(12) [0.3813]	0.3709(14) [0.3737]	0.4110(5) [0.4090]
C(9′)	0.2589(16) [0.2567]	0.4717(20) [0.4686]	0.4185(7) [0.4194]

The structure, as shown in Figure XVAIII,157, has molecules with the bond dimensions of Figure XVAIII,158. The benzene rings and their attached carbon, nitrogen, and oxygen atoms are planar to within a few hundredths of an angstrom unit but the terminal carbon atoms of the ethyl groups, C(9) and C(9′), are more than one angstrom unit outside this plane. The planes of the two salicylidene residues about copper are, however, twisted with respect to one another. As a result, the plane Cu–O–N makes an angle of 35.6° with that of Cu–O′–N′. The fourfold coordination about copper is in consequence something between square and tetrahedral: the angle between the planes would be zero if it were square and 90° if it were tetrahedral.

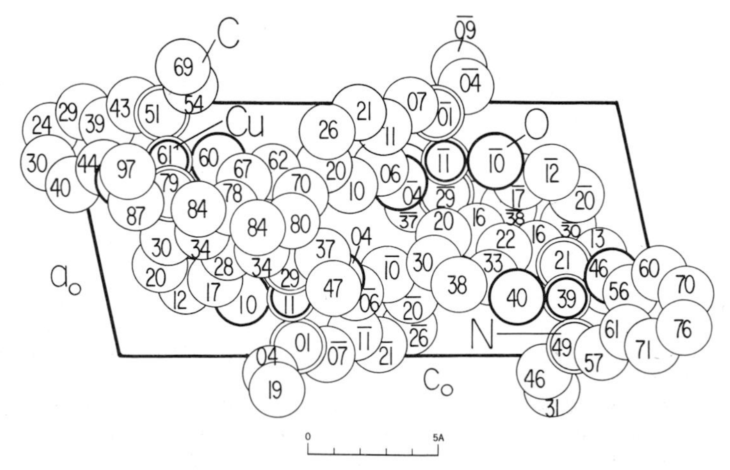

Fig. XVAIII,157. The monoclinic structure of bis(N-ethylsalicylaldiminato)copper projected along its b_0 axis.

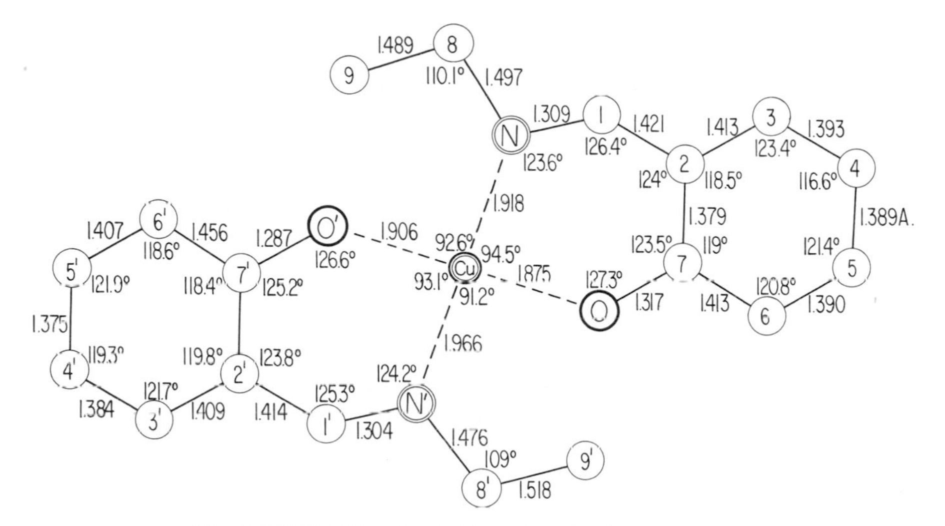

Fig. XVAIII,158. Bond dimensions in the molecule of bis(N-ethylsalicylaldiminato)copper.

Another determination of structure (1967: P,B&G) has just been published. According to it:

$$a_0 = 9.824(12) \text{ A.}; \quad b_0 = 8.432(12) \text{ A.}; \quad c_0 = 20.989(26) \text{ A.}$$
$$\beta = 101°0(10)'$$

Choosing the a_0 and c_0 axes and the origin to agree with those employed in the preceding study, the parameters in square brackets of the table were obtained. They agree well with those found in the other investigation.

XV,aIII,96. The monoclinic crystals of *bis(N-ethylsalicylaldiminato)-palladium(II)*, $(OC_6H_4CHNC_2H_5)_2Pd$, have a bimolecular unit of the dimensions:

$$a_0 = 8.43(3) \text{ A.}; \quad b_0 = 5.60(2) \text{ A.}; \quad c_0 = 17.97(6) \text{ A.}$$
$$\beta = 94°42(20)'$$

The space group is C_{2h}^5 ($P2_1/c$) with atoms in the positions:

$$\text{Pd: } (2a) \quad 000;\ 0\ {}^1/_2\ {}^1/_2$$

Other atoms are in:

$$(4e) \quad \pm(xyz;\ x,{}^1/_2-y,z+{}^1/_2)$$

The parameters are listed in Table XVAIII,89.

TABLE XVAIII,89

Parameters of Atoms in $(OC_6H_4CHNC_2H_5)_2Pd$

Atom	x	y	z
O	−0.1461	−0.0620	0.0759
N	0.0941	0.2480	0.0552
C(1)	−0.1811	0.0800	0.1319
C(2)	−0.2941	−0.0100	0.1819
C(3)	−0.3347	0.1200	0.2381
C(4)	−0.2501	0.3333	0.2600
C(5)	−0.1313	0.4180	0.2149
C(6)	−0.0938	0.2950	0.1520
C(7)	0.0355	0.3760	0.1080
C(8)	0.2349	0.4150	0.0299
C(9)	0.3831	0.2900	0.0567

The structure (Fig. XVAIII,159) is composed of molecules having the bond dimensions of Figure XVAIII,160. In this crystal, as in the butyl (**XV,aIII,101**) and certain other analogues, the chelating part of the molecule is not planar.

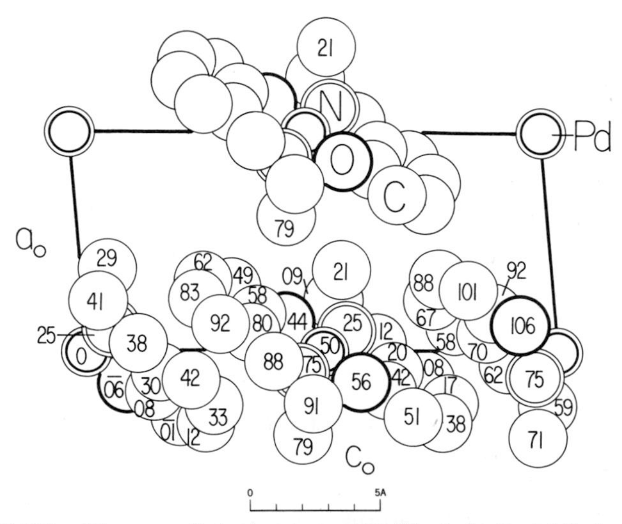

Fig. XVAIII,159. The monoclinic structure of bis(N-ethylsalicylaldiminato)palladium projected along its b_0 axis.

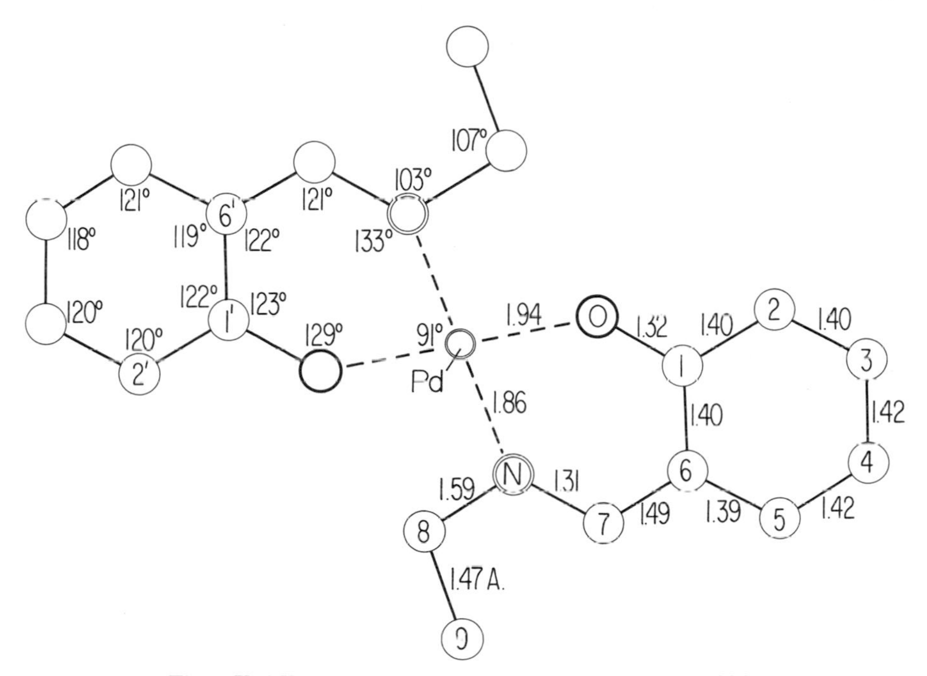

Fig. XVAIII,160. Bond dimensions in the molecule of bis (*N*-ethylsalicylaldiminato)palladium.

XV,aIII,97. Crystals of *bis(N-ethylsalicylaldiminato)nickel(II)*, $(OC_6H_4CHNC_2H_5)_2Ni$, though monoclinic, have been given a structure different from that of either the copper (**XV,aIII,95**) or palladium (**XV,aIII,96**) compounds. Two molecules are contained in a unit of the dimensions:

$$a_0 = 14.45(2) \text{ A.}; \quad b_0 = 7.15(1) \text{ A.}; \quad c_0 = 8.09(1) \text{ A.}$$
$$\beta = 92°$$

The space group is C_{2h}^5 as for the other two substances but with the orientation $P2_1/a$. Atoms thus are in the positions:

$$\text{Ni: } (2a) \quad 000;\ {}^1/_2\ {}^1/_2\ 0$$

All other atoms are in:

$$(4e) \quad \pm(xyz;\ x+{}^1/_2,{}^1/_2-y,z)$$

with the parameters of Table XVAIII,90. Calculated hydrogen parameters are stated in the original.

TABLE XVAIII,90

Parameters of Atoms in $(OC_6H_4CHNC_2H_5)_2Ni$

Atom	x	y	z
O	0.057	0.052	−0.195
N	0.074	0.188	0.112
C(1)	0.139	0.111	−0.220
C(2)	0.179	0.079	−0.370
C(3)	0.268	0.139	−0.402
C(4)	0.319	0.235	−0.280
C(5)	0.282	0.273	−0.124
C(6)	0.191	0.205	−0.097
C(7)	0.150	0.253	0.059
C(8)	0.044	0.268	0.272
C(9)	−0.038	0.406	0.246

The structure is illustrated in Figure XVAIII,161. Except for atoms of the ethyl radicals the chelate groups, having the bond dimensions of Figure XVAIII,162, are each planar to within about 0.1 A. They are staggered with respect to one another, the plane of the nickel and its coordinated nitrogen and oxygen atoms making an angle of 23° with each.

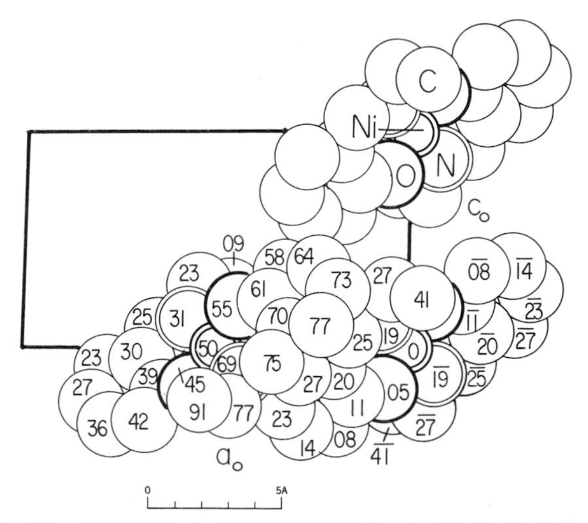

Fig. XVAIII,161. The monoclinic structure of bis(*N*-ethylsalicylaldiminato)nickel projected along its b_0 axis.

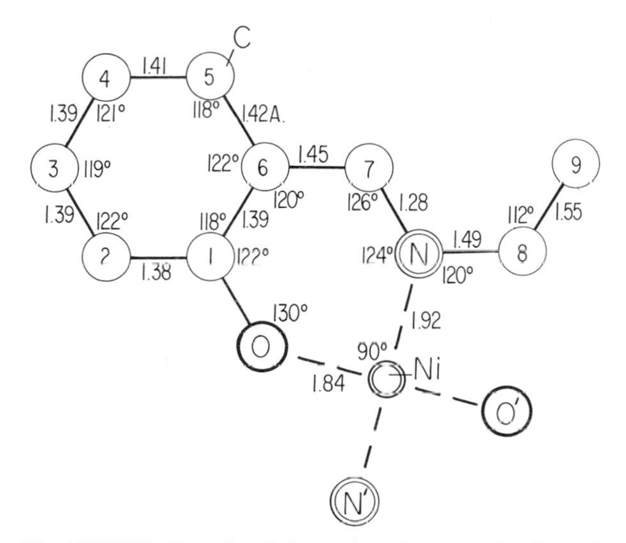

Fig. XVAIII,162. Bond dimensions in the molecule of bis (*N*-ethylsalicylaldiminato)nickel.

Coordination of the nickel atom is square, with two nitrogen and two oxygen atoms at opposite corners of the square.

XV,aIII,98. Crystals of *bis(N-2-hydroxyethylsalicylaldiminato)copper-(II)*, $(OC_6H_4CHNC_2H_4OH)_2Cu$, have monoclinic symmetry. Their tetramolecular unit has the dimensions:

$$a_0 = 18.66(4)\text{ A.};\quad b_0 = 4.71(1)\text{ A.};\quad c_0 = 19.99(4)\text{ A.}$$
$$\beta = 97°48(6)'$$

The space group is C_{2h}^5 ($P2_1/c$) with atoms in the positions:

$$\text{Cu(1)}: (2a)\quad 000;\ 0\ {}^1/_2\ {}^1/_2$$
$$\text{Cu(2)}: (2d)\quad {}^1/_2\ 0\ {}^1/_2;\ {}^1/_2\ {}^1/_2\ 0$$

All other atoms are in:

$$(4e)\quad \pm(xyz;\ x,{}^1/_2-y,z+{}^1/_2)$$

Parameters are stated in Table XVAIII,91.

TABLE XVAIII,91

Parameters of Atoms in Bis(*N*-2-Hydroxyethylsalicylaldiminato)copper(II)

Atom	*x*	*y*	*z*
		Part A	
O(1)	−0.0049	−0.1437	0.0863
N	−0.0885	0.2259	0.0066
C(1)	−0.1301	0.2196	0.0532
C(2)	−0.1187	0.2596	0.1094
C(3)	−0.0588	−0.1357	−0.1261
C(4)	−0.0491	−0.3195	0.1841
C(5)	−0.1058	−0.3223	−0.2250
C(6)	−0.1675	−0.1508	0.2121
C(7)	−0.1723	0.0281	0.1572
C(8)	−0.1117	0.4511	−0.0472
C(9)	−0.1597	0.2882	−0.1104
O(2)	−0.2320	0.2293	−0.0930
		Part B	
O(1′)	0.5732	0.5245	0.0761
N′	0.4513	0.2052	0.0483
C(1′)	0.4766	0.1061	0.1086
C(2′)	0.4577	0.8014	−0.1517
C(3′)	0.4130	0.5886	−0.1358
C(4′)	0.3478	0.5001	−0.1791
C(5′)	0.3299	0.6501	−0.2436
C(6′)	0.3750	0.8516	−0.2626
C(7′)	0.4397	0.9318	−0.2173
C(8′)	0.3854	0.0479	0.0743
C(9′)	0.3154	0.2460	0.0110
O(2′)	0.2970	0.2767	0.0801

In the structure there is a square and planar coordination of two oxygen and two nitrogen atoms about copper. The two crystallographically different chelating groups have the general bond dimensions of Figure XVAIII,163. The B group is planar but A is distorted so that the planes of the two benzene rings lie 0.62 A. apart.

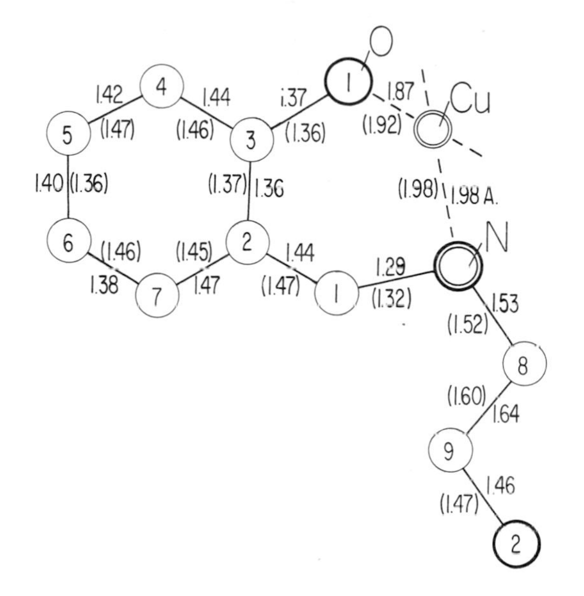

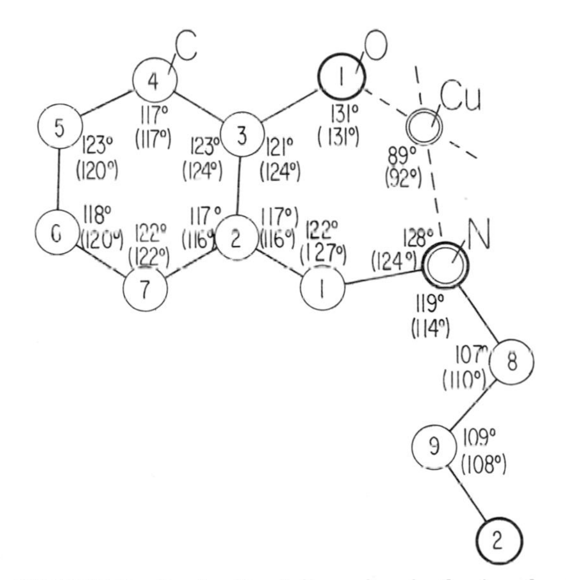

Fig. XVAIII,163a (top). Bond dimensions in the A molecule of bis(N-2-hydroxyethylsalicylaldiminato)copper.

Fig. XVAIII,163b (bottom). Bond dimensions in the B molecule of bis(N-2-hydroxyethylsalicylaldiminato)copper.

XV,aIII,99. The orthorhombic crystals of *bis(N-isopropylsalicylaldiminato)nickel,* $(OC_6H_4CHNC_3H_7)_2Ni$, have a unit containing eight molecules. Its edge lengths are:

$$a_0 = 13.219(6) \text{ A.}; \quad b_0 = 19.697(8) \text{ A.}; \quad c_0 = 15.140(18) \text{ A.}$$

All atoms are in the positions

$$(8c) \quad \pm(xyz; \ ^1/_2-x,y+^1/_2,z; \ x,^1/_2-y,z+^1/_2; \ ^1/_2-x,\bar{y},z+^1/_2)$$

of the space group V_h^{15} (*Pbca*). The parameters are those of Table XVAIII,92.

TABLE XVAIII,92

Parameters of the Atoms in $[OC_6H_4CHNC_3H_7]_2Ni$

Atom	x	y	z
C(1)	−0.0505(5)	0.3012(3)	0.2202(7)
C(2)	−0.0491(5)	0.3672(3)	0.2555(7)
C(3)	−0.1335(5)	0.3956(4)	0.2970(7)
C(4)	−0.2220(5)	0.3582(4)	0.3052(7)
C(5)	−0.2273(5)	0.2932(3)	0.2717(7)
C(6)	−0.1433(4)	0.2619(3)	0.2289(6)
C(7)	−0.1541(5)	0.1940(3)	0.1987(6)
C(8)	−0.1182(5)	0.0889(3)	0.1337(8)
C(9)	−0.1072(7)	0.0773(3)	0.0347(9)
C(10)	−0.0541(6)	0.0371(4)	0.1819(7)
C(11)	0.2106(4)	0.1019(3)	0.0859(8)
C(12)	0.2785(5)	0.0485(3)	0.1142(7)
C(13)	0.3441(5)	0.0159(4)	0.0545(9)
C(14)	0.3435(5)	0.0336(4)	−0.0348(9)
C(15)	0.2801(6)	0.0831(3)	−0.0676(7)
C(16)	0.2125(4)	0.1168(3)	−0.0059(8)
C(17)	0.1510(5)	0.1720(3)	−0.0451(7)
C(18)	0.0396(5)	0.2638(3)	−0.0567(7)
C(19)	0.0949(7)	0.3302(4)	−0.0327(8)
C(20)	−0.0757(6)	0.2701(4)	−0.0392(7)
N(1)	−0.0884(4)	0.1587(2)	0.1545(5)
N(2)	0.0844(4)	0.2077(2)	−0.0019(5)
O(1)	0.0300(3)	0.2771(2)	0.1822(4)
O(2)	0.1525(3)	0.1303(2)	0.1439(4)
Ni	0.0476(1)	0.1939(1)	0.1216(1)

(*continued*)

TABLE XVAIII,92 (*continued*)

Atom	x	y	z
H(1)	0.017	0.396	0.250
H(2)	−0.130	0.445	0.322
H(3)	−0.284	0.379	0.337
H(4)	−0.294	0.266	0.280
H(5)	−0.222	0.170	0.212
H(6)	−0.193	0.082	0.153
H(7)	−0.152	0.113	0.002
H(8)	−0.131	0.029	0.018
H(9)	−0.032	0.084	0.016
H(10)	−0.061	0.044	0.249
H(11)	0.021	0.043	0.163
H(12)	−0.078	−0.012	0.165
H(13)	0.281	0.035	0.181
H(14)	0.392	−0.023	0.074
H(15)	0.393	0.010	−0.077
H(16)	0.281	0.096	−0.135
H(17)	0.162	0.183	−0.111
H(18)	0.051	0.253	−0.122
H(19)	0.172	0.326	−0.043
H(20)	0.083	0.341	0.035
H(21)	0.065	0.370	−0.070
H(22)	−0.112	0.225	−0.055
H(23)	−0.104	0.310	0.077
H(24)	−0.087	0.281	0.028

A portion of the structure is shown in Figure XVAIII,164. The bond dimensions of its molecular group are those of Figure XVAIII,165. Except for the isopropyl radicals, the two organic parts of the molecule are planar; the planes of these radicals, however, are almost normal to the salicylaldimine planes. In this compound the coordination of the nickel atom is distortedly tetrahedral with the O(1)–Ni–N(1) and O(2)–Ni–N(2) planes making 81.5° with one another. Between molecules the interatomic distances range upwards from a C(5)–O(1) = 3.30 A.

The *copper* compound $(OC_6H_4CHNC_3H_7)_2Cu$ is isostructural. Its cell has the dimensions:

$$a_0 = 12.87(4)\ \text{A.};\quad b_0 = 20.68(3)\ \text{A.};\quad c_0 = 14.58(3)\ \text{A.}$$

Atomic parameters are listed in Table XVAIII,93; values for hydrogen are given in the original. Though the structure is in essence the same as

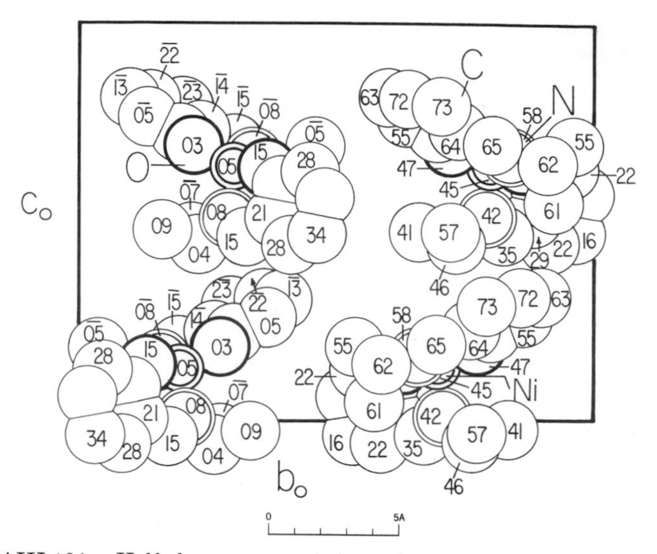

Fig. XVAIII,164. Half the contents of the orthorhombic unit cell of bis(N-isopropylsalicylaldiminato)nickel projected along the a_0 axis. The other molecules are obtained from these by inversion through the origin.

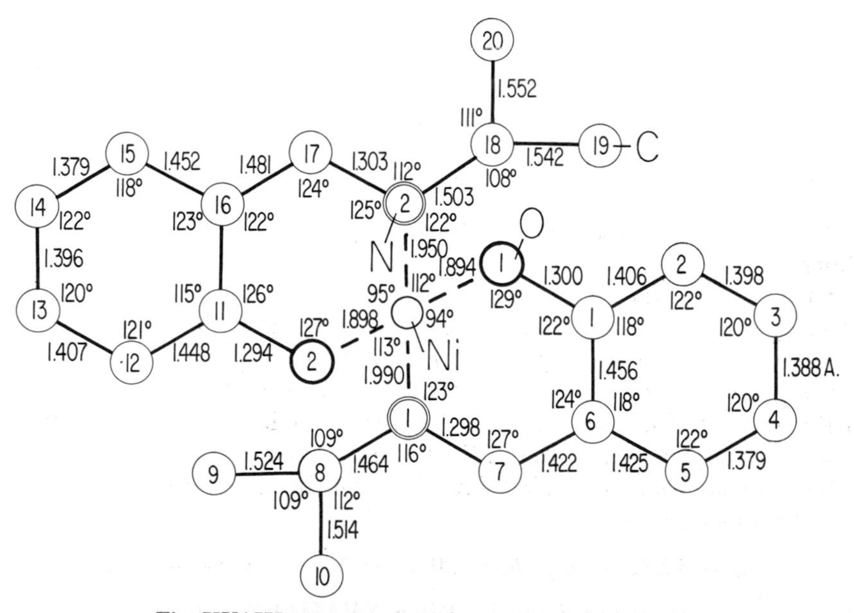

Fig. XVAIII,165. Bond dimensions in the molecule of bis(N-isopropylsalicylaldiminato)nickel.

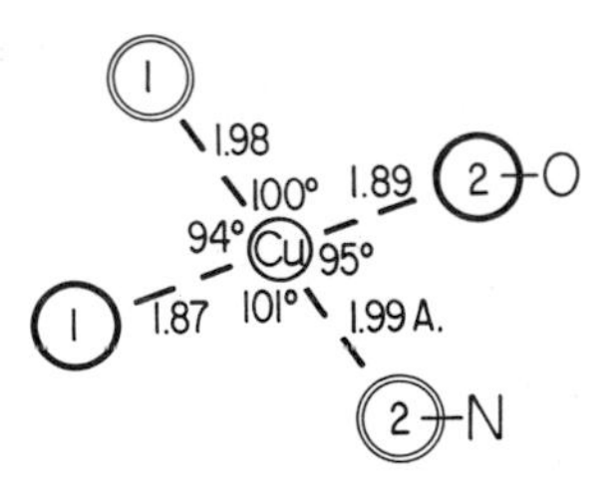

Fig. XVAIII,166. Bond dimensions involving the atoms coordinated with copper in bis(*N*-isopropylsalicylaldiminato)copper.

that of the nickel compound, there are minor differences in the coordination about the metal atom (Fig. XVAIII,166) and one salicylaldimine group shows an appreciable departure from planarity.

TABLE XVAIII,93

Parameters of the Atoms in $[OC_6H_4CHNC_3H_7]_2Cu$

Atom	x	y	z
Cu	0.0613(2)	0.1938(1)	0.1253(1)
O(1)	0.0563(8)	0.2762(3)	0.1778(4)
O(2)	0.1493(8)	0.1199(3)	0.1370(4)
N(1)	−0.0645(9)	0.1622(4)	0.1845(5)
N(2)	0.1127(10)	0.2159(4)	−0.0005(5)
C(1)	−0.0192(11)	0.3010(4)	0.2264(6)
C(2)	−0.0055(11)	0.3659(4)	0.2593(6)
C(3)	−0.0880(14)	0.3945(6)	0.3108(8)
C(4)	−0.1821(14)	0.3624(6)	0.3286(8)
C(5)	−0.1934(12)	0.2988(4)	0.2977(7)
C(6)	−0.1112(11)	0.2671(4)	0.2478(6)
C(7)	−0.1312(12)	0.2009(4)	0.2274(6)
C(8)	−0.1036(14)	0.0934(5)	0.1792(8)
C(9)	−0.0881(15)	0.0696(6)	0.0783(9)
C(10)	−0.0307(13)	0.0540(5)	0.2461(9)
C(11)	0.2083(13)	0.0946(5)	0.0710(8)
C(12)	0.2613(12)	0.0364(5)	0.0931(7)
C(13)	0.3278(13)	0.0076(5)	0.0288(8)
C(14)	0.3396(15)	0.0340(7)	−0.0579(10)
C(15)	0.2892(14)	0.0900(5)	−0.0814(8)
C(16)	0.2206(12)	0.1221(5)	−0.0177(7)
C(17)	0.1712(12)	0.1797(5)	−0.0467(7)
C(18)	0.0706(13)	0.2731(5)	−0.0496(8)
C(19)	0.1387(11)	0.3297(4)	−0.0318(6)
C(20)	−0.0452(15)	0.2827(6)	−0.0294(9)

XV,aIII,100. Crystals of *bis(N-γ-dimethylaminopropylsalicylaldiminato)-nickel*, $[OC_6H_4CHN(CH_2)_3N(CH_3)_2]_2Ni$, are monoclinic with a tetramolecular cell of the dimensions:

$$a_0 = 10.226(8)\text{ A.};\quad b_0 = 15.254(12)\text{ A.};\quad c_0 = 15.333(8)\text{ A.}$$
$$\beta = 107°12(4)'$$

All atoms are in the positions:

$$(4e)\quad \pm(xyz;\ x,{}^1/_2-y,z+{}^1/_2)$$

of the space group C_{2h}^5 ($P2_1/c$). The parameters are listed in Table XVAIII,94.

TABLE XVAIII,94
Parameters of the Atoms in $[OC_6H_4CHN(CH_2)_3N(CH_3)_2]_2Ni$

Atom	x	y	z
Ni	0.2560(1)	0.1220(1)	0.2446(1)
O(1)	0.3973(4)	0.0613(2)	0.3466(3)
O(2)	0.1171(4)	0.1839(3)	0.1442(3)
N(1)	0.4088(5)	0.2030(3)	0.2320(3)
N(2)	0.1025(6)	0.0404(3)	0.2580(3)
N(3)	0.2290(6)	0.2329(4)	0.3412(4)
N(4)	0.2810(5)	0.0128(3)	0.1487(3)
C(1)	0.6035(8)	0.1211(5)	0.3257(4)
C(2)	0.5290(8)	0.0600(4)	0.3644(4)
C(3)	0.6091(8)	−0.0021(5)	0.4250(5)
C(4)	0.7471(10)	−0.0061(6)	0.4460(6)
C(5)	0.8191(8)	0.0541(6)	0.4079(6)
C(6)	0.7483(8)	0.1154(6)	0.3508(5)
C(7)	0.5375(9)	0.1924(4)	0.2684(5)
C(8)	−0.0895(7)	0.1208(4)	0.1611(4)
C(9)	−0.0152(7)	0.1837(4)	0.1248(4)
C(10)	−0.0952(7)	0.2483(5)	0.0654(4)
C(11)	−0.2333(8)	0.2498(6)	0.0431(5)
C(12)	−0.3088(7)	0.1865(6)	0.0762(5)
C(13)	−0.2333(8)	0.1236(5)	0.1334(5)
C(14)	−0.0254(7)	0.0504(3)	0.2191(4)
C(15)	0.3651(8)	0.2890(4)	0.1913(5)
C(16)	0.3336(8)	0.3479(4)	0.2610(6)
C(17)	0.2116(9)	0.3176(4)	0.2938(5)
C(18)	0.3455(8)	0.2346(5)	0.4254(5)
C(19)	0.1059(9)	0.2172(5)	0.3685(6)
C(20)	0.1481(7)	−0.0445(4)	0.3002(5)
C(21)	0.1817(7)	−0.1038(4)	0.2319(5)
C(22)	0.3000(8)	−0.0730(4)	0.1964(5)
C(23)	0.1625(7)	0.0084(5)	0.0659(5)
C(24)	0.4029(8)	0.0281(5)	0.1194(5)

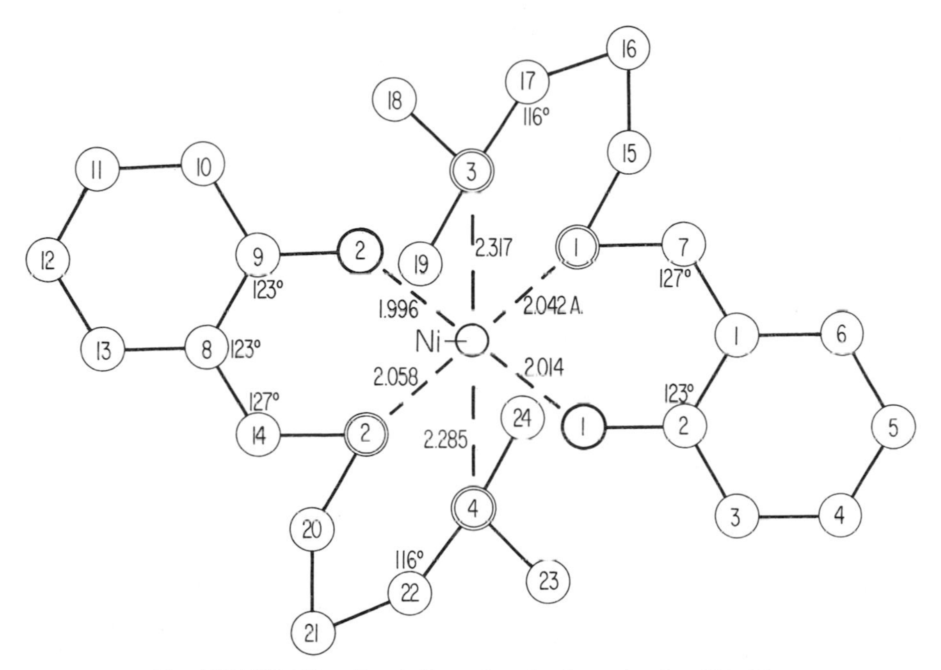

Fig. XVAIII,167. Bond dimensions in the molecule of bis(*N*-γ-dimethylaminopropylsalicylaldiminato)nickel.

The molecules in this structure have the bond dimensions of Figure XVAIII,167. Coordination about the central nickel atom is in this instance octahedral with four nearest nitrogen and two oxygen atoms. The angles of their bonds to nickel, however, depart considerably from a right angle, ranging between 81.5 and 98.6°. Though it is not a structural requirement, the molecules have an apparent center of symmetry.

XV,aIII,101. The monoclinic crystals of *bis(N-butylsalicylaldiminato)-palladium(II)*, $(OC_6H_4CHNC_4H_9)_2Pd$, have a bimolecular unit of the dimensions:

$$a_0 = 10.99(4) \text{ A.}; \quad b_0 = 7.25(3) \text{ A.}; \quad c_0 = 14.72(5) \text{ A.}$$
$$\beta = 120°0(30)'$$

The space group is C_{2h}^5 ($P2_1/c$) with atoms in the positions:

$$\text{Pd: } (2a) \quad 000; 0\,{}^1/_2\,{}^1/_2$$

The other atoms are in:

$$(4e) \quad \pm(xyz;\ x,{}^1/_2-y,z+{}^1/_2)$$

Their parameters are stated in Table XVAIII,95.

TABLE XVAIII,95

Parameters of Atoms in $(OC_6H_4CHNC_4H_9)_2Pd$

Atom	x	y	z
O	0.1808	0.5153	0.4980
N	−0.1077	0.2999	0.3944
C(1)	0.1828	0.4026	0.4069
C(2)	0.3090	0.4306	0.4080
C(3)	0.3253	0.3500	0.3303
C(4)	0.2246	0.2341	0.2585
C(5)	0.1008	0.2016	0.2600
C(6)	0.0800	0.2838	0.3360
C(7)	−0.0457	0.2414	0.3350
C(8)	−0.2228	0.1863	0.3722
C(9)	−0.2000	0.1166	0.4667
C(10)	−0.3365	0.0741	0.4584
C(11)	−0.4451	−0.0051	0.3589

The structure is shown in Figure XVAIII,168. Its molecules, possessing a center of symmetry, have the bond dimensions of Figure XVAIII,169. In this crystal the atoms involved in chelation are reported as not coplanar.

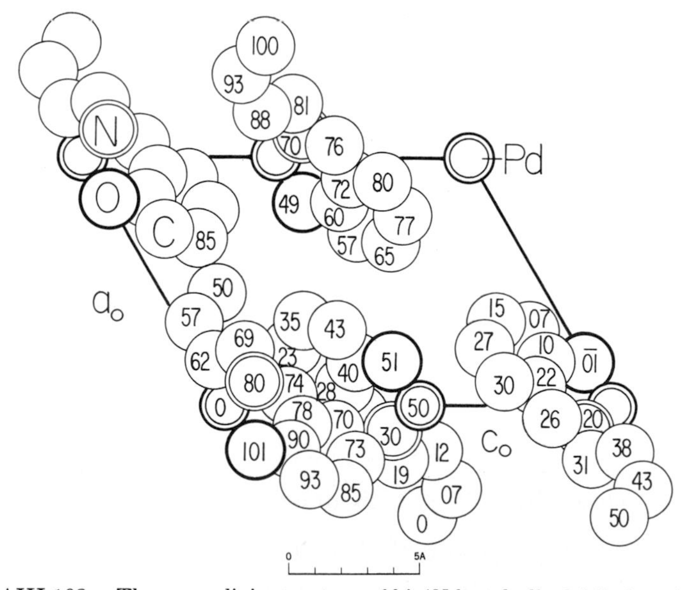

Fig. XVAIII,168. The monoclinic structure of bis(*N*-butylsalicylaldiminato)palladium projected along its b_0 axis.

XV,aIII,102. Crystals of *bis(N-t-butylsalicylaldiminato)copper(II)*, $(OC_6H_4CHNC_4H_9)_2Cu$, are orthorhombic with a tetramolecular unit of the edge lengths:

$$a_0 = 9.109(7) \text{ A.}; \quad b_0 = 11.166(5) \text{ A.}; \quad c_0 = 21.349(7) \text{ A.}$$

Atoms are in the positions

$$(4a) \quad xyz;\ {}^1/_2-x,\bar{y},z+{}^1/_2;\ x+{}^1/_2,{}^1/_2-y,\bar{z};\ \bar{x},y+{}^1/_2,{}^1/_2-z$$

of the space group V^4 ($P2_12_12_1$). The parameters are listed in Table XVAIII,96.

TABLE XVAIII,96

Parameters of the Atoms in Bis(*N*-*t*-butylsalicylaldiminato)copper(II)

Atom	x	y	z
Cu	0.1867(1)	0.2052(1)	0.1538(1)
C(1)	0.3367(10)	0.2661(8)	0.2675(4)
C(2)	0.3741(13)	0.2402(10)	0.3309(5)
C(3)	0.4681(14)	0.3136(11)	0.3657(6)
C(4)	0.5320(15)	0.4118(12)	0.3371(6)
C(5)	0.5020(13)	0.4386(10)	0.2743(5)
C(6)	0.4056(10)	0.3667(8)	0.2392(4)
C(7)	0.3826(11)	0.4015(8)	0.1740(4)
C(8)	0.3003(11)	0.4037(8)	0.0679(4)
C(9)	0.1450(15)	0.3984(12)	0.0461(6)
C(10)	0.4041(14)	0.3237(11)	0.0273(6)
C(11)	0.3548(14)	0.5345(11)	0.0657(6)
C(12)	0.1568(10)	0.0068(8)	0.0717(4)
C(13)	0.2256(12)	0.0687(9)	0.0267(5)
C(14)	0.1703(13)	−0.1850(10)	0.0176(5)
C(15)	0.0553(14)	−0.2266(10)	0.0512(5)
C(16)	−0.0151(12)	−0.1550(9)	0.0949(5)
C(17)	0.0359(10)	−0.0344(8)	0.1052(4)
C(18)	−0.0504(10)	0.0373(8)	0.1480(4)
C(19)	−0.1214(12)	0.2077(11)	0.2072(5)
C(20)	−0.1367(18)	0.1510(14)	0.2717(7)
C(21)	−0.0655(15)	0.3359(11)	0.2167(6)
C(22)	−0.2716(19)	0.2112(17)	0.1729(8)
N(1)	0.3067(9)	0.3509(6)	0.1329(3)
N(2)	−0.0119(8)	0.1431(6)	0.1676(3)
O(1)	0.2441(8)	0.1962(6)	0.2389(3)
O(2)	0.2141(7)	0.1141(6)	0.0794(3)

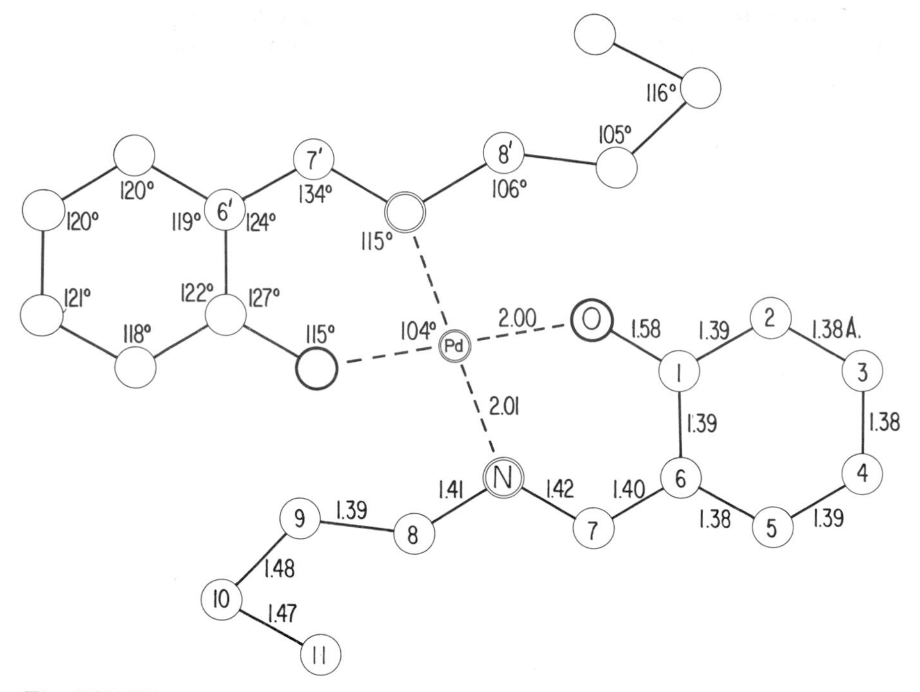

Fig. XVAIII,169. Bond dimensions in the molecule of bis(*N*-butylsalicylaldiminato) palladium.

This structure is made up of molecules having the bond dimensions of Figure XVAIII,170. The two chelated groups are not parallel; there is instead an angle of 61.9° between the planes of the two benzene rings. In one group the salicylaldiminato portion is planar, in the other it is bent at the line through N(2) and O(2). There is a considerable distortion from the usual square coordination about the copper atom, with O(1)–Cu–O(2) = 137.4° and N(2)–Cu–N(1) = 145.3°.

XV,aIII,103. The monoclinic crystals of *N-salicylideneglycinatoaquo-copper(II) hemihydrate*, $(CuC_9H_7NO_3 \cdot H_2O) \cdot {}^1/_2H_2O$, possess a unit cell that contains eight molecules and has the dimensions:

$$a_0 = 17.16(3) \text{ A.}; \quad b_0 = 6.84(2) \text{ A.}; \quad c_0 = 17.57(1) \text{ A.}$$
$$\beta = 111.29(5)°$$

The oxygen atom of the water of crystallization is in the positions

$$(4e) \quad \pm(0\,u\,{}^1/_4;\ {}^1/_2,u+{}^1/_2,{}^1/_4) \qquad \text{with } u = 0.5964$$

of C_{2h}^6 ($C2/c$). All the other atoms are in:

$$(8f) \quad \pm(xyz;\ x,\bar{y},z+{}^1/_2;\ x+{}^1/_2,y+{}^1/_2,z;\ x+{}^1/_2,{}^1/_2-y,z+{}^1/_2)$$

Parameters, including those selected for hydrogen, are listed in Table XVAIII,97.

TABLE XVAIII,97

Parameters of Atoms in $(CuC_9H_7NO_3 \cdot H_2O) \cdot {}^1/_2H_2O$

Atom	x	y	z
Cu	0.12172	0.09512	0.20439
O(1)	0.0068	0.1730	0.1500
O(2)	0.2287	0.9576	0.2457
O(3)	0.3421	0.8799	0.2162
O(H_2O)	0.0943	0.9736	0.2962
N	0.1530	0.1430	0.1099
C(1)	0.2743	0.9673	0.2018
C(2)	0.2400	0.0987	0.1262
C(3)	0.1025	0.1887	0.0382
C(4)	0.0141	0.2352	0.0168
C(5)	0.4704	0.2073	0.4346
C(6)	0.3863	0.1570	0.4081
C(7)	0.3446	0.1582	0.4620
C(8)	0.1133	0.2840	0.4563
C(9)	0.0282	0.2287	0.4280
H(1)	0.048	0.060	0.319
H(2)	0.089	0.857	0.299
H(3)	0.047	0.499	0.251
H(4)	0.285	0.204	0.140
H(5)	0.246	0.012	0.077
H(6)	0.130	0.211	−0.007
H(7)	−0.001	0.296	−0.103
H(8)	−0.133	0.384	−0.139
H(9)	−0.205	0.417	−0.045
H(10)	−0.141	0.263	0.078

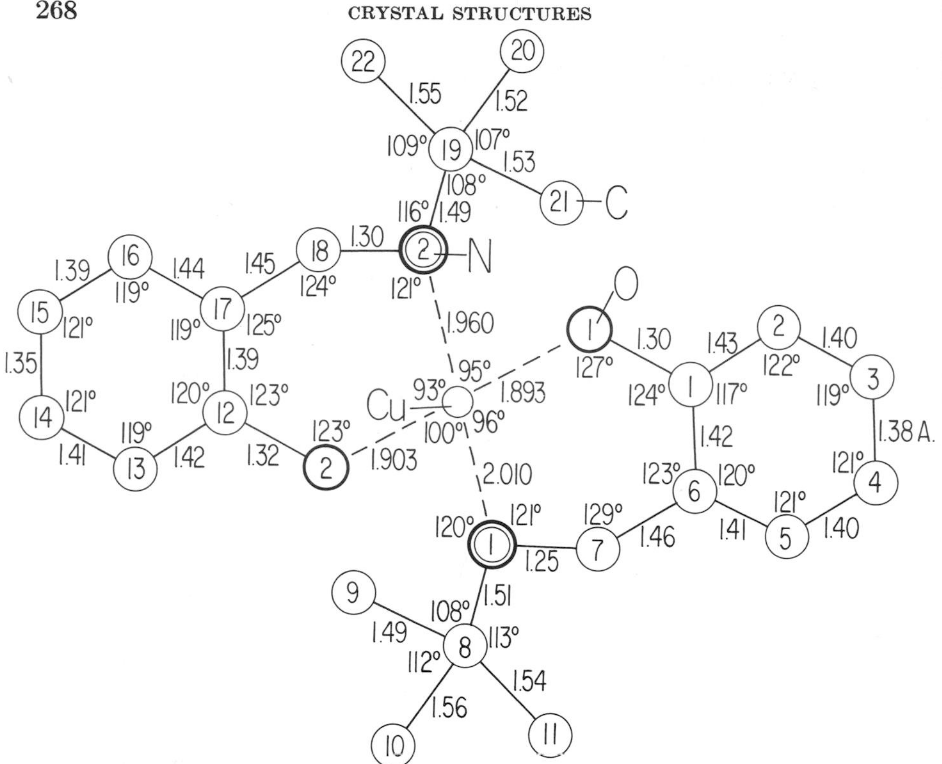

Fig. XVAIII,170. Bond dimensions in the molecule of bis(*N*-*t*-butylsalicylaldiminato) copper.

Molecules having the bond dimensions of Figure XVAIII,172 are arranged as in Figure XVAIII,171. The benzene ring is planar and so is the salicylaldiminato–copper group (except for nitrogen which is 0.06 A. apart). Considering only the closest nitrogen and oxygen atoms the coordination of the copper is approximately square though it is about 0.2 A. outside the plane through these atoms. There is, however, a fifth oxygen, O(3), belonging to an adjacent molecule, distant 2.334 A. from the copper; taking it into consideration the total coordination can be described as fivefold and that of a tetragonal pyramid.

XV,aIII,104. The triclinic crystals of *acetylacetonemono(o-hydroxyanil)-copper(II)*, $[OC_6H_4NC(CH_3)CHC(CH_3)O]Cu$, have a large tetramolecular unit of the dimensions:

$$a_0 = 8.91(1) \text{ A.}; \quad b_0 = 10.48(1) \text{ A.}; \quad c_0 = 11.73(1) \text{ A.}$$
$$\alpha = 101.6(3)°; \quad \beta = 110.6(3)°; \quad \gamma = 92.5(3)°$$

Atoms are in the positions $(2i)$ $\pm(xyz)$ of C_i^1 $(P\bar{1})$. The parameters are given in Table XVAIII,98.

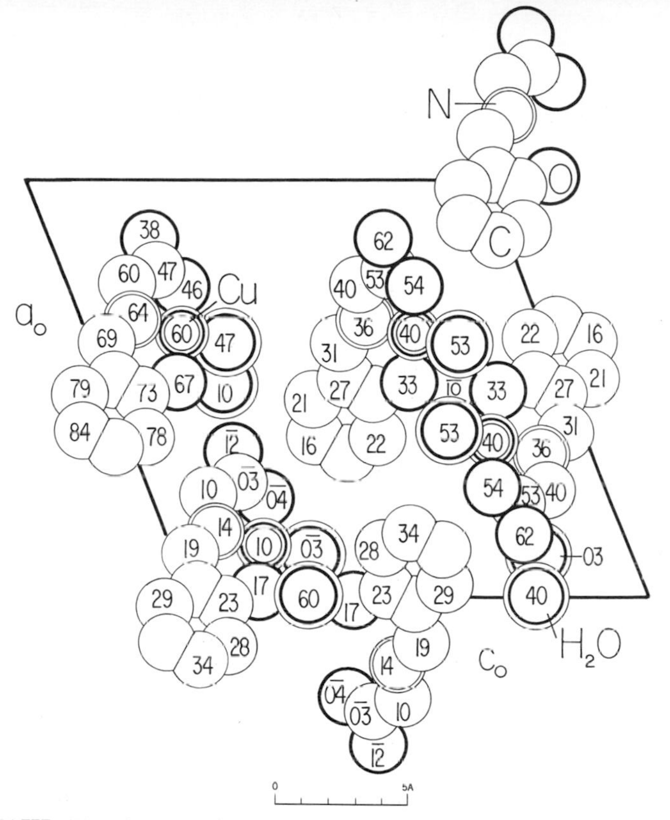

Fig. XVAIII,171. A portion of the monoclinic structure of *N*-salicylideneglycinatoaquocopper hemihydrate projected along its b_0 axis. Five of the eight molecules in the unit are shown.

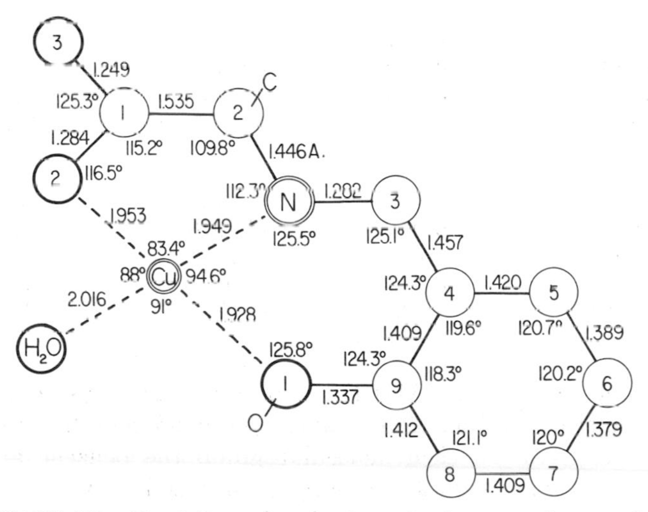

Fig. XVAIII,172. Bond dimensions in the molecule present in crystals of *N*-salicylideneglycinatoaquocopper hemihydrate.

TABLE XVAIII,98

Parameters of the Atoms in Acetylacetone-mono(*o*-hydroxyanil) Copper

Atom	x	y	z
Cu(1)	0.3371	0.1119	0.1673
O(1)	0.4095	0.1617	0.3429
O(2)	0.2601	0.0510	−0.0096
N(1)	0.4895	0.2353	0.1418
C(1)	0.5422	0.3010	0.5440
C(2)	0.5075	0.2699	0.4062
C(3)	0.5863	0.3574	0.3512
C(4)	0.5736	0.3392	0.2326
C(5)	0.6562	0.4518	0.2098
C(6)	0.4772	0.2046	0.0139
C(7)	0.5806	0.2573	−0.0360
C(8)	0.5508	0.2106	−0.1599
C(9)	0.4307	0.1102	−0.2378
C(10)	0.3368	0.0598	−0.1832
C(11)	0.3529	0.0967	−0.0624
Cu(2)	0.1145	−0.1081	−0.0331
O(1′)	0.0878	−0.1953	−0.1987
O(2′)	0.1368	−0.0118	0.1342
N(1′)	−0.0037	−0.2508	0.0051
C(1′)	−0.0536	−0.3294	−0.3979
C(2′)	−0.0231	−0.2936	−0.2594
C(3′)	−0.1086	−0.3706	−0.2036
C(4′)	−0.1019	−0.3465	−0.0884
C(5′)	−0.2102	−0.4405	−0.0612
C(6′)	0.0247	−0.2214	0.1345
C(7′)	0.0004	−0.3055	0.2053
C(8′)	0.0216	−0.2653	0.3252
C(9′)	0.0940	−0.1347	0.3904
C(10′)	0.1297	−0.0515	0.3259
C(11′)	0.1063	−0.0930	0.2021

The structure (Fig. XVAIII,173) contains two crystallographically different molecular groups. Their averaged bond lengths are those of Figure XVAIII,174. In each molecular group the benzene and acetyl-acetone–imine components are roughly planar and they make angles of

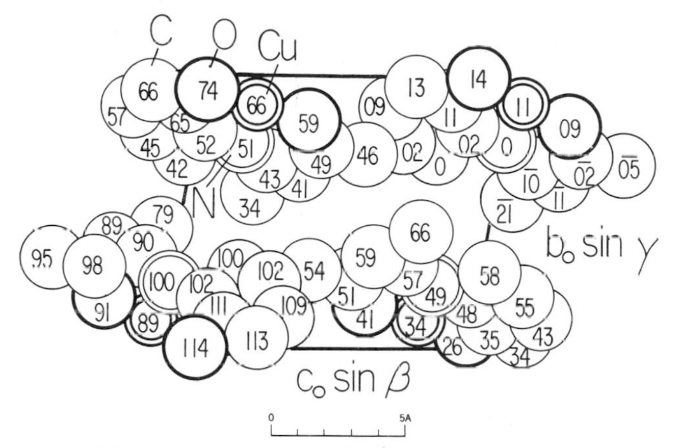

Fig. XVAIII,173. The triclinic structure of acetylacetone-mono(*o*-hydroxyanil)copper projected along its a_0 axis.

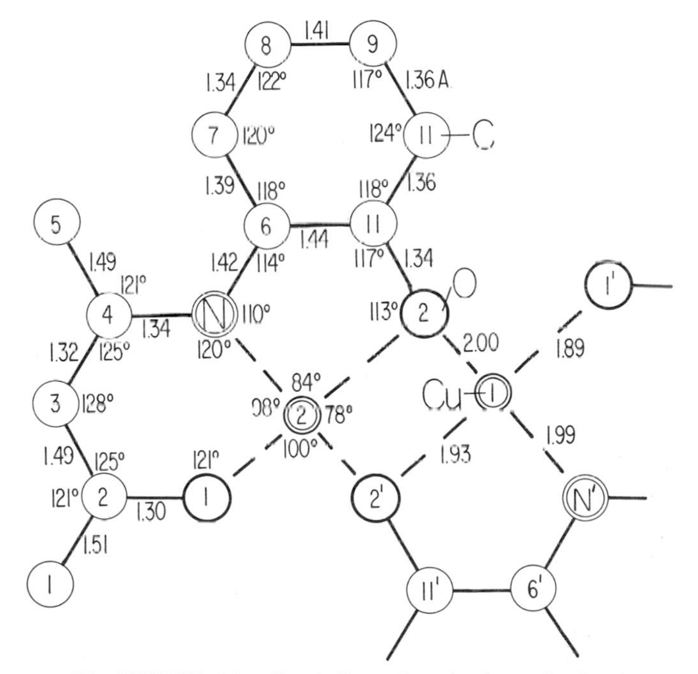

Fig. XVAIII,174. Bond dimensions in the molecule of acetylacetone-mono(*o*-hydroxyanil)copper.

about 15° with one another. Coordination about the two copper atoms per cell is not the same. About Cu(1) it is essentially square, with two more distant carbons octahedrally situated. About Cu(2) there is a distorted square of oxygen neighbors and a fifth oxygen at a distance of 2.64 A.; together they form a square pyramid about the copper.

XV,aIII,105. According to a preliminary note, crystals of *p-chlorophenylmethyl sulfone*, $p\text{-ClC}_6\text{H}_4(\text{CH}_3)\text{SO}_2$, are triclinic with a bimolecular unit of the dimensions:

$$a_0 = 7.04(2)\text{ A.};\quad b_0 = 11.04(2)\text{ A.};\quad c_0 = 5.45(5)\text{ A.}$$
$$\alpha = 95°;\quad \beta = 91°;\quad \gamma = 95.5°$$

Atoms are in the positions $(2i)\ \pm(xyz)$ / of C_i^1 ($P\bar{1}$). Parameters are stated to have the values of Table XVAIII,99.

TABLE XVAIII,99

Parameters of the Atoms in $p\text{-ClC}_6\text{H}_4(\text{CH}_3)\text{SO}_2$

Atom	x	y	z
Cl	0.819	0.472	0.792
S	0.237	0.110	0.172
O(1)	0.310	0.091	−0.087
O(2)	0.043	0.147	0.131
C(1)	0.260	−0.027	0.324
C(2)	0.392	0.220	0.255
C(3)	0.596	0.218	0.341
C(4)	0.715	0.291	0.549
C(5)	0.642	0.370	0.684
C(6)	0.458	0.371	0.604
C(7)	0.332	0.305	0.417

XV,aIII,106. *Orthanilic acid*, $\text{NH}_3\text{C}_6\text{H}_4\text{SO}_3$, is monoclinic with a tetramolecular unit of the dimensions:

$$a_0 = 7.935(4)\text{ A.};\quad b_0 = 6.570(6)\text{ A.};\quad c_0 = 14.225(5)\text{ A.}$$
$$\beta = 105°54(6)'$$

The space group is C_{2h}^5 ($P2_1/c$) with atoms in the positions:

$$(4e)\quad \pm(xyz;\ x,{}^1/_2-y,z+{}^1/_2)$$

The parameters, as determined from combined x-ray and neutron diffraction data, are listed in Table XVAIII,100.

TABLE XVAIII,100

Parameters of the Atoms in $NH_3C_6H_4SO_3$

Atom	x	y	z
S	0.33196(9)	0.73142(13)	0.10660(5)
O(1)	0.26816(45)	0.77709(90)	0.00446(21)
O(2)	0.41077(50)	0.53527(71)	0.12541(35)
O(3)	0.43982(51)	0.88753(78)	0.16302(30)
N	0.33546(36)	0.71997(50)	0.32333(19)
C(1)	0.14425(38)	0.72277(50)	0.15185(21)
C(2)	0.16129(39)	0.71739(50)	0.25220(21)
C(3)	0.01460(43)	0.71176(55)	0.28655(25)
C(4)	−0.15149(45)	0.71026(61)	0.22104(29)
C(5)	−0.16923(44)	0.71417(70)	0.12208(29)
C(6)	−0.02364(46)	0.72136(66)	0.08655(25)
H(1)	0.0345	0.7132	0.3657
H(2)	−0.2594	0.7000	0.2517
H(3)	−0.3012	0.7001	0.0649
H(4)	−0.0411	0.7155	0.0063
H(5)	0.3350	0.7208	0.3922
H(6)	0.4026	0.8463	0.3142
H(7)	0.3944	0.5870	0.3170

The resulting structure is shown in Figure XVAIII,175. In its molecules the benzene ring and attached sulfur, nitrogen and hydrogen atoms are

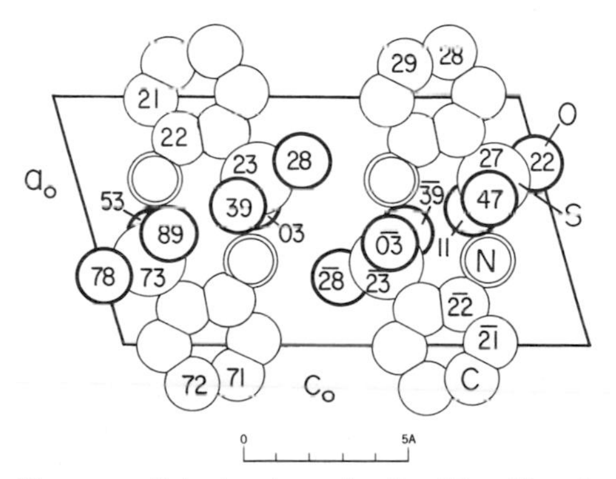

Fig. XVAIII,175. The monoclinic structure of orthanilic acid projected along its b_0 axis.

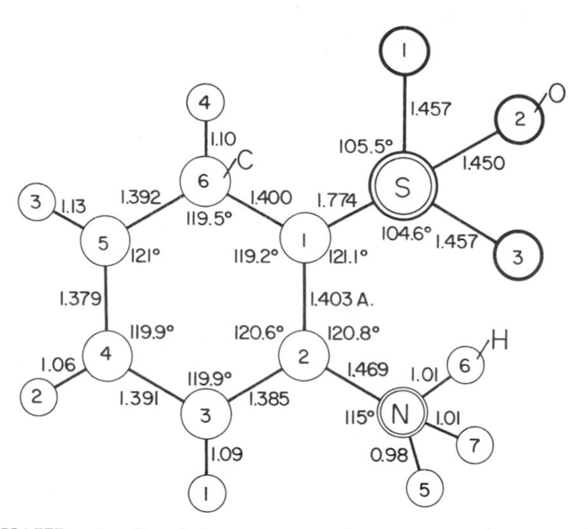

Fig. XVAIII,176. Bond dimensions in the molecule of orthanilic acid.

coplanar; bond dimensions are those of Figure XVAIII,176. Close contacts are between the nitrogen atoms of one molecule and the oxygen atoms of others through hydrogen bonds of lengths between 2.753 and 2.852 A. There are three such bonds for each nitrogen corresponding to the cationic character of its part of the molecule.

XV,aIII,107. Crystals of *metanilic acid* (aniline *m*-sulfonic acid), $NH_3C_6H_4SO_3$, are orthorhombic with a tetramolecular cell of the edge lengths:

$$a_0 = 8.500(1) \text{ A.}; \quad b_0 = 11.944(1) \text{ A.}; \quad c_0 = 6.756(5) \text{ A.}$$

All atoms are in the positions

(4*c*) $\pm(u\ v\ {}^1/_4;\ u+{}^1/_2,{}^1/_2-v,{}^1/_4)$

(8*d*) $\pm(xyz;\ x+{}^1/_2,{}^1/_2-y,{}^1/_2-z;\ x,y,{}^1/_2-z;\ x+{}^1/_2,{}^1/_2-y,z)$

of the space group V_h^{16} (*Pnam*). Parameters are listed in Table XVAIII,101, those for hydrogen having been assumed.

TABLE XVAIII,101

Positions and Parameters of the Atoms in Metanilic Acid

Atom	Position	x	y	z
S	(4c)	0.3114	0.6226	$^1/_4$
O(1)	(4c)	0.2400	0.7320	$^1/_4$
O(2)	(8d)	0.4000	0.5990	0.4270
N	(4c)	−0.2880	0.5300	$^1/_4$
C(1)	(4c)	0.1460	0.5280	$^1/_4$
C(2)	(4c)	0.1790	0.4120	$^1/_4$
C(3)	(4c)	0.0560	0.3380	$^1/_4$
C(4)	(4c)	−0.0970	0.3760	$^1/_4$
C(5)	(4c)	−0.1270	0.4920	$^1/_4$
C(6)	(4c)	−0.0060	0.5670	$^1/_4$
H(1)	(4c)	0.296	0.383	$^1/_4$
H(2)	(4c)	0.082	0.252	$^1/_4$
H(3)	(4c)	−0.189	0.316	$^1/_4$
H(4)	(4c)	−0.029	0.652	$^1/_4$
H(5)	(8d)	−0.341	0.500	0.3650
H(6)	(4c)	−0.289	0.614	$^1/_4$

The structure (Fig. XVAIII,177) is composed of molecules which are exactly planar except for oxygen and hydrogen. Bond dimensions are

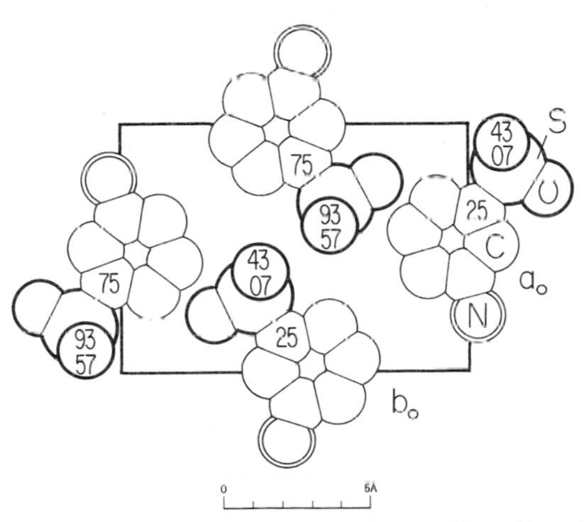

Fig. XVAIII,177. The orthorhombic structure of metanilic acid projected along its c_0 axis.

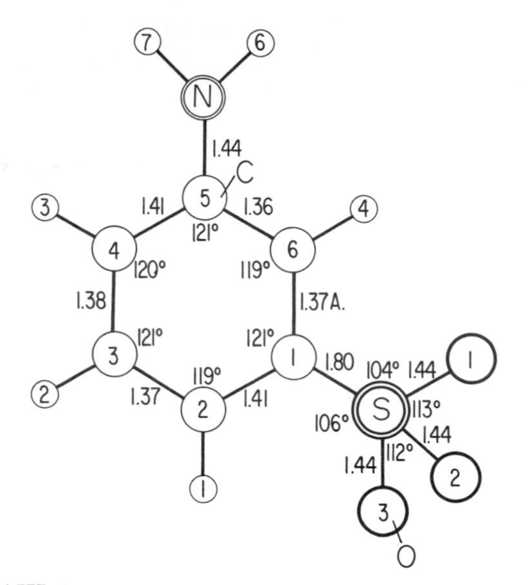

Fig. XVAIII,178. Bond dimensions in the molecule of metanilic acid.

shown in Figure XVAIII,178. The distribution around the sulfur atom is that of a flattened tetrahedron. Between molecules there are short N–O separations of 2.84 and 2.85 A. attributable to hydrogen bonding. They indicate that this molecule, like that of orthanilic acid (**XV,aIII,106**), has the zwitter-ion configuration which makes the NH_3 end of the molecule positive and the SO_3 end negative.

XV,aIII,108. The monoclinic crystals of *p-sulfanilic acid monohydrate*, $NH_3C_6H_4SO_3 \cdot H_2O$, possess a tetramolecular unit of the dimensions:

$$a_0 = 6.473(1) \text{ A.}; \quad b_0 = 18.308(2) \text{ A.}; \quad c_0 = 6.812(1) \text{ A.}$$
$$\beta = 93°38(6)'$$

All atoms are in the positions

$$(4e) \quad \pm(xyz; x,{}^1/_2-y,z+{}^1/_2)$$

of C_{2h}^5 ($P2_1/c$). Determined parameters are listed in Table XVAIII,102; values assigned the hydrogen atoms are to be found in the original article.

TABLE XVAIII,102

Parameters of the Atoms in Sulfanilic Acid Monohydrate

Atom	x	y	z
S	0.2358	0.1402	0.2610
C(1)	0.3119	0.0474	0.2501
C(2)	0.1633	−0.0060	0.2351
C(3)	0.2160	−0.0804	0.2204
C(4)	0.4215	−0.0967	0.2351
C(5)	0.5828	−0.0443	0.2528
C(6)	0.5215	0.0289	0.2633
N	0.4892	−0.1740	0.2234
O(1)	0.3947	0.1776	0.3736
O(2)	0.0445	0.1415	0.3626
O(3)	0.2028	0.1628	0.0588
O(H_2O)	0.1842	−0.2510	0.4867

The structure (Fig. XVAIII,179) contains molecules having the bond dimensions of Figure XVAIII,180. In them the benzene ring and attached

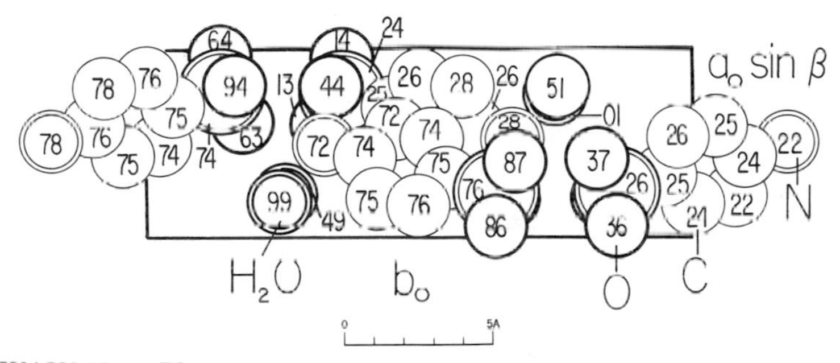

Fig. XVAIII,179. The monoclinic structure of p-sulfanilic acid monohydrate projected along its c_0 axis.

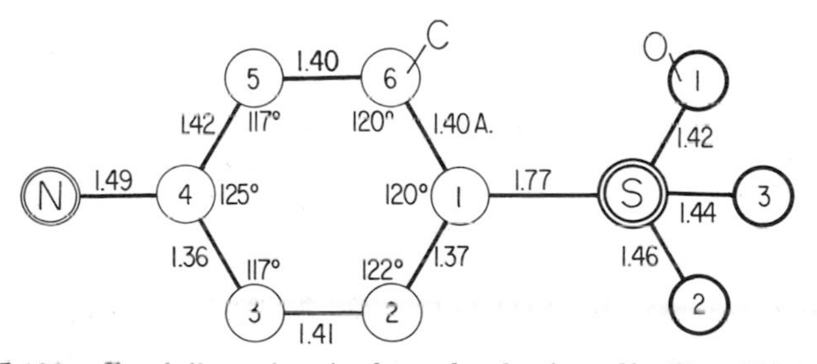

Fig. XVAIII,180. Bond dimensions in the molecule of p-sulfanilic acid in crystals of its monohydrate.

sulfur and nitrogen atoms are coplanar to less than 0.02 A.; the averaged angles C–S–O = 106° and O–S–O = 113°. In the crystal the benzene rings lie one above another in the c_0 direction at a distance of 3.41 A. The molecules in successive layers and in the same layers along the b_0 direction are held together by hydrogen bonds of the types N–H–O = 2.80–2.91 A. and O–H–O = 2.73–2.96 A.

XV,aIII,109. The α-form of *sulfanilamide*, p-$NH_2C_6H_4SO_2NH_2$, is orthorhombic with a large unit that contains eight molecules and has the edge lengths:

$$a_0 = 5.65(5)\ \text{A.};\quad b_0 = 18.509(5)\ \text{A.};\quad c_0 = 14.794(4)\ \text{A.}$$

The space group is V_h^{15} ($Pbca$) with atoms in the positions:

$$(8c)\quad \pm(xyz;\ {}^1/_2-x,y+{}^1/_2,z;\ x,{}^1/_2-y,z+{}^1/_2;\ x+{}^1/_2,y,{}^1/_2-z)$$

The parameters are those of Table XVAIII,103.

TABLE XVAIII,103

Parameters of the Atoms in α-Sulfanilamide

Atom	x	y	z
S	0.2484	0.0602	0.1054
O(1)	0.2518	0.0449	0.0117
O(2)	0.4580	0.0444	0.1603
N(1)	0.0505	0.3741	0.1327
N(2)	0.0355	0.0129	0.1470
C(1)	0.0972	0.3000	0.1250
C(2)	0.2974	0.2719	0.1657
C(3)	0.3400	0.1979	0.1631
C(4)	0.1819	0.1515	0.1178
C(5)	−0.0236	0.1801	0.0788
C(6)	−0.0737	0.2530	0.0838
H(2)	0.424	0.308	0.199
H(3)	0.502	0.181	0.194
H(5)	−0.150	0.145	0.046
H(6)	−0.234	0.274	0.053

The structure (Fig. XVAIII,181) is made up of molecules having the bond dimensions of Figure XVAIII,182. The benzene ring and attached amino nitrogen are coplanar and the sulfur atom departs by only 0.15 A.

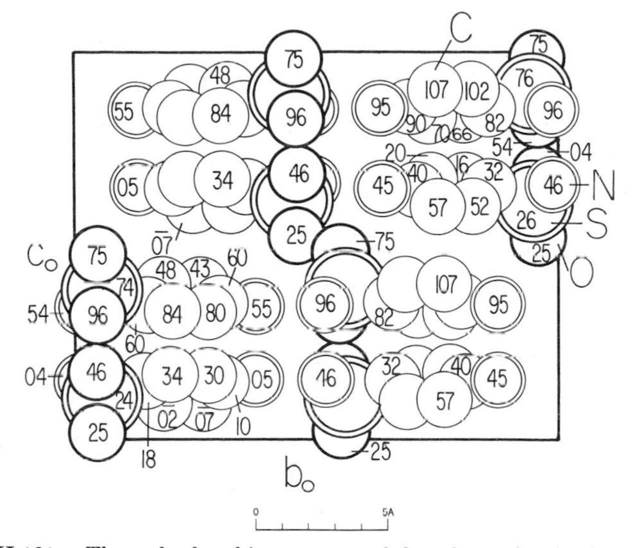

Fig. XVAIII,181. The orthorhombic structure of the α form of sulfanilamide projected along its a_0 axis.

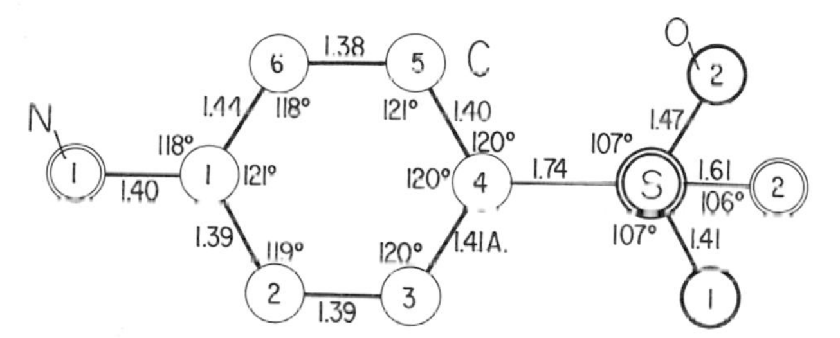

Fig. XVAIII,182. Bond dimensions in the molecule of sulfanilamide in crystals of its α form.

from this plane. Around sulfur the distribution is tetrahedral. The molecules are in layers parallel to the c_0 axis, being held together by four N–H . . . O hydrogen bonds (N–O = 2.94 or 3.12 A.). Between layers there are other probable hydrogen bonds, with N–H . . . O = 3.06 A.

XV,aIII,110. Two recent studies have been made of the structure of *β-sulfanilamide*, *p*-$NH_2C_6H_4SO_2NH_2$. It is monoclinic with a tetramolecular cell of the dimensions:

$$a_0 = 8.975(3) \text{ A.}; \quad b_0 = 9.005(3) \text{ A.}; \quad c_0 = 10.039(4) \text{ A.}$$
$$\beta = 111°26(3)'$$

Atoms are in the positions

$$(4e) \quad \pm(xyz;\ x,{}^1/_2-y,z+{}^1/_2)$$

of the space group C_{2h}^5 ($P2_1/c$). The two determinations have provided parameters that agree well with one another. In the later study both neutrons and x-rays were used and this has furnished accurate parameters for the atoms of hydrogen. These, together with the x-ray parameters obtained in the same investigation, are listed in Table XVAIII,104.

TABLE XVAIII,104

Parameters of the Atoms in β-Sulfanilamide

Atom	x	y	z
S	0.08243(4)	0.85326(4)	0.28753(4)
O(1)	0.00713(17)	0.86047(16)	0.13418(15)
O(2)	0.11695(17)	0.98978(14)	0.36595(17)
N(1)	−0.03952(20)	0.76159(19)	0.34060(19)
N(2)	0.69212(19)	0.53272(22)	0.41005(19)
C(1)	0.26272(19)	0.75661(19)	0.32723(17)
C(2)	0.26607(21)	0.62849(19)	0.25025(18)
C(3)	0.40872(22)	0.55368(21)	0.28011(18)
C(4)	0.55062(20)	0.60579(22)	0.38458(18)
C(5)	0.54441(21)	0.73402(22)	0.46142(19)
C(6)	0.40144(21)	0.80837(20)	0.43327(19)
Hydrogen parameters (x-ray results in parentheses)			
H(1)	0.1545 (0.1722)	0.5894 (0.5856)	0.1665 (0.1878)
H(2)	0.4160 (0.4108)	0.4550 (0.4615)	0.2188 (0.2290)
H(3)	0.6520 (0.6502)	0.7722 (0.7707)	0.5343 (0.5274)
H(4)	0.4004 (0.4011)	0.9067 (0.9095)	0.4934 (0.4822)
H(5)	−0.0076 (−0.0151)	0.7648 (0.7578)	0.4467 (0.4342)
H(6)	−0.0774 (−0.0628)	0.6671 (0.6661)	0.2871 (0.3197)
H(7)	0.6930 (0.6924)	0.4274 (0.4108)	0.3692 (0.3974)
H(8)	0.7837 (0.7816)	0.5585 (0.5547)	0.4943 (0.4800)

The resulting structure is shown in Figure XVAIII,183. In its molecules, having the bond dimensions of Figure XVAIII,184, the benzene ring is planar, with the nitrogen and sulfur atoms 0.04 A. outside this plane. Except for H(7) which is 0.47 A. from the plane, the atoms of hydrogen do not depart by more than 0.10 A. from it.

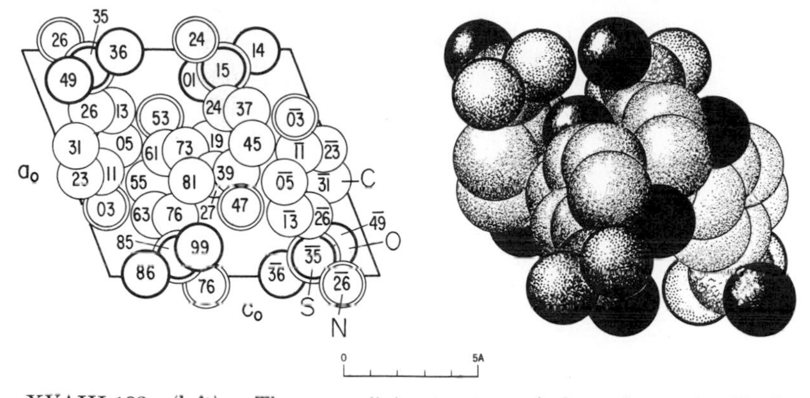

Fig. XVAIII,183a (left). The monoclinic structure of the β form of sulfanilamide projected along its b_0 axis.

Fig. XVAIII,183b (right). A packing drawing of the monoclinic structure of β-sulfanilamide viewed along its b_0 axis. The nitrogen atoms are black; the oxygens are heavily outlined and dotted. The atoms of carbon are lightly dot shaded; the sulfurs are larger than the others and lightly curl shaded.

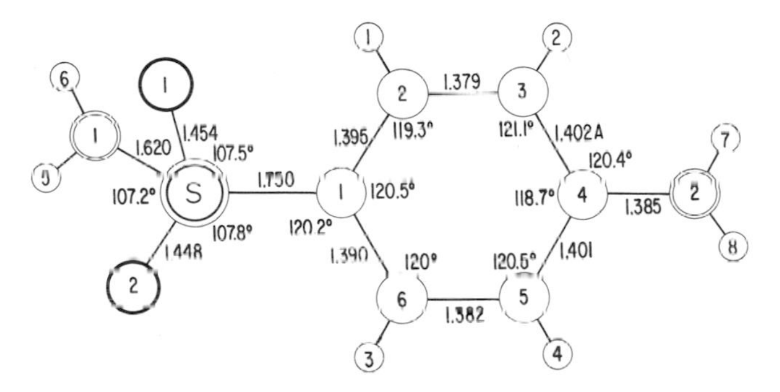

Fig. XVAIII,184. Bond dimensions in the molecule in crystals of β-sulfanilamide.

XV,aIII,111. The third, high temperature, γ, modification of *sulfanilamide*, p-$NH_2C_6H_4SO_2NH_2$, is monoclinic. Its tetramolecular unit has the dimensions:

$$a_0 = 7.95(1)\text{ A.};\quad b_0 = 12.945(5)\text{ A.};\quad c_0 = 7.79(1)\text{A.}$$
$$\beta = 106°30(10)'$$

All atoms are in the positions

$$(4e)\quad \pm(xyz;\ x,{}^1/_2-y,z+{}^1/_2)$$

of C_{2h}^5 ($P2_1/c$). The parameters are those of Table XVAIII,105.

TABLE XVAIII,105

Parameters of the Atoms in γ-Sulfanilamide

Atom	x	y	z
S	0.3916	0.1619	0.9627
O(1)	0.3968	0.2732	0.9876
O(2)	0.4354	0.0983	0.1207
N(1)	0.5340	0.1342	0.8483
N(2)	0.6856	0.0476	0.5110
C(1)	0.1824	0.1296	0.8311
C(2)	0.0762	0.2019	0.7261
C(3)	0.9135	0.1754	0.6166
C(4)	0.8494	0.0765	0.6182
C(5)	0.9545	0.0034	0.7259
C(6)	0.1183	0.0304	0.8367
H(1)	0.123	0.274	0.716
H(2)	0.840	0.228	0.538
H(3)	0.908	0.930	0.730
H(4)	0.192	0.978	0.917

The resulting structure (Fig. XVAIII,185) contains molecules like those in the other modifications. Their benzene rings and attached N(2) and sulfur atoms are coplanar, the distribution around sulfur being tetrahedral. Bond dimensions in this case are those of Figure XVAIII,186. The molecules are held together by hydrogen bonds: N(1)–O(1) = 2.96 A., N(1)–

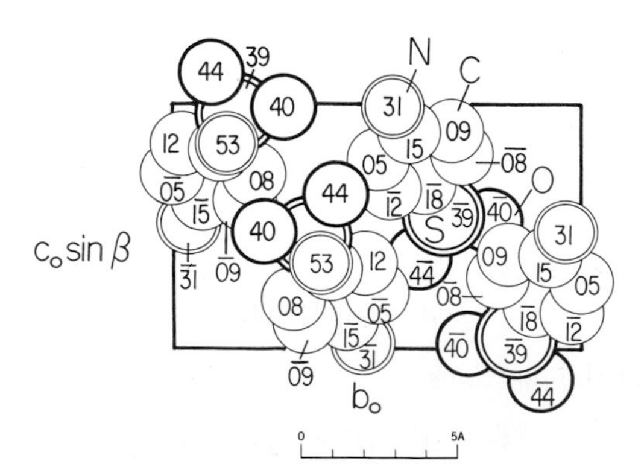

Fig. XVAIII,185. The monoclinic structure of the γ-modification of sulfanilamide projected along its a_0 axis.

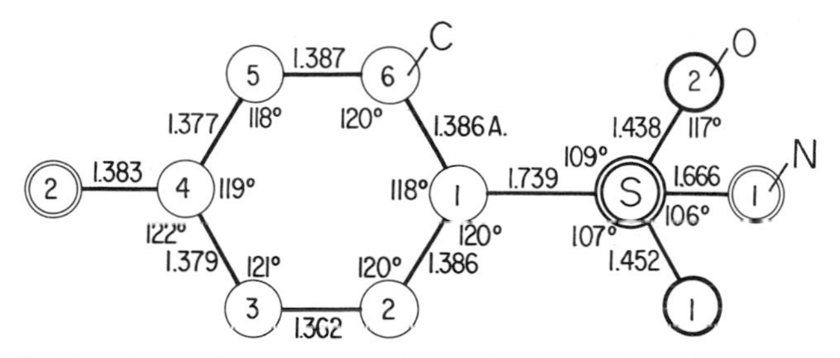

Fig. XVAIII,186. Bond dimensions in the molecule of the γ-form of sulfanilamide.

O(2) = 3.03 A., N(2)–O(1) = 3.24 A., N(2)–O(2) = 3.19 A.; there is also a short N(2)–N(2) = 3.16 A.

XV,aIII,112. The orthorhombic crystals of *sulfanilamide monohydrate*, $p\text{-}NH_2C_6H_4SO_2NH_2 \cdot H_2O$, have a tetramolecular unit of the edge lengths:

$$a_0 = 6.02(1) \text{ A.}; \quad b_0 = 7.36(1) \text{ A.}; \quad c_0 = 19.26(1) \text{ A.}$$

The space group is V^4 ($P2_12_12_1$) with atoms in the positions:

$$(4a) \quad xyz;\ {}^1/_2-x,\bar{y},z+{}^1/_2;\ x+{}^1/_2,{}^1/_2-y,\bar{z};\ \bar{x},y+{}^1/_2,{}^1/_2-z$$

The parameters as stated in a preliminary note are those of Table XVAIII,106. They do not lead to a possible atomic arrangement but if the x parameter for C(5) is 0.58 rather than 0.42, a reasonable molecule results.

TABLE XVAIII,106

Parameters of the Atoms in Sulfanilamide Monohydrate

Atom	x	y	z
S	0.940	−0.005	0.140
O(1)	0.840	−0.165	0.170
O(2)	0.180	−0.025	0.140
H_2O	0.440	0.105	0.250
N(1)	0.855	0.170	0.185
N(2)	0.655	0.005	−0.155
C(1)	0.855	0.005	0.055
C(2)	0.995	0.075	0.005
C(3)	0.930	0.075	0.930
C(4)	0.720	0.005	0.915
C(5)	0.420	−0.065	−0.035
C(6)	0.630	−0.065	0.035

XV,aIII,113. The monoclinic crystals of *zinc p-toluene sulfonate hexahydrate*, $Zn(CH_3C_6H_4SO_3)_2 \cdot 6H_2O$, have a bimolecular unit of the dimensions:

$$a_0 = 25.24(1) \text{ A.}; \quad b_0 = 6.295(2) \text{ A.}; \quad c_0 = 6.98(4) \text{ A.}$$
$$\beta = 91°18(15)'$$

The atoms of zinc are in

$$(2a) \quad 000; \; {}^1/_2 \, {}^1/_2 \, {}^1/_2$$

of C_{2h}^5 in the orientation $P2_1/n$. The other atoms are in

$$(4e) \quad \pm(xyz; \; x+{}^1/_2, {}^1/_2-y, z+{}^1/_2)$$

with the parameters of Table XVAIII,107.

TABLE XVAIII,107

Parameters of Atoms in $Zn(CH_3C_6H_4SO_3)_2 \cdot 6H_2O$

Atom	x	y	z
S	0.0950	0.3917	0.5152
O(1)	0.0950	0.1587	0.5152
O(2)	0.0773	0.4801	0.6880
O(3)	0.0709	0.4767	0.3458
O(4,H_2O)	0.0502	0.8986	0.2342
O(5,H_2O)	0.0378	0.2880	0.0117
O(6,H_2O)	0.0551	0.8999	0.8024
C(1)	0.1626	0.4510	0.5175
C(2)	0.2027	0.3125	0.5557
C(3)	0.2425	0.8600	0.9480
C(4)	0.2299	0.0527	0.0064
C(5)	0.2327	0.7256	0.4370
C(6)	0.1775	0.6480	0.4609
C(7)	0.1708	0.1384	0.0233

The structure is shown in Figure XVAIII,187. In its anions sulfur is at the center of a tetrahedron of three oxygens and the C(1) atom, at distances of 1.40–1.47 and 1.74 A. The coordination of zinc is octahedral rather than the more usual tetrahedral, with Zn–O = 2.05–2.14 A., all of these being water oxygens. Hydrogen bonds of lengths 2.73–2.82 A. link the water molecules with each of the sulfonate oxygens.

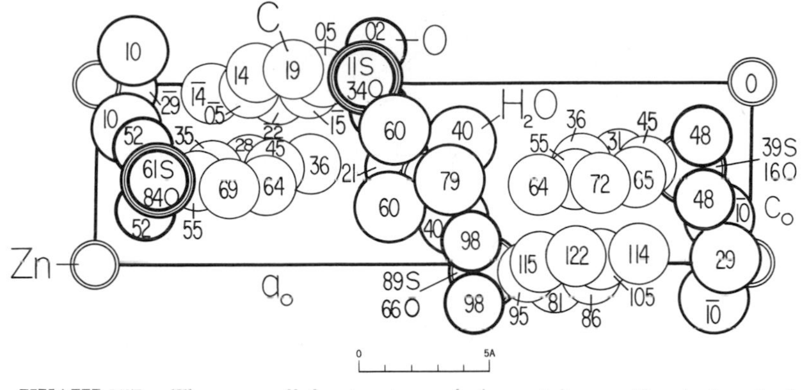

Fig. XVAIII,187. The monoclinic structure of zinc *p*-toluenesulfonate hexahydrate projected along its b_0 axis.

The isostructural magnesium salt has the cell dimensions:

$$a_0 = 25.2(1) \text{ A.}; \quad b_0 = 6.26(4) \text{ A.}; \quad c_0 = 6.95(4) \text{ A.}$$
$$\beta = 91°54(25)'$$

XV,aIII,114. Crystals of *p-chlorobenzene seleninic acid*, $ClC_6H_4SeO_2H$, are monoclinic with a tetramolecular unit of the dimensions:

$$a_0 = 12.14(3) \text{ A.}; \quad b_0 = 5.14(2) \text{ A.}; \quad c_0 = 12.55(3) \text{ A.}$$
$$\beta = 111°5(15)'$$

All atoms are in the positions

$$(4e) \quad \pm(xyz;\ x,{}^1/_2-y,z+{}^1/_2)$$

of the space group C_{2h}^5 ($P2_1/c$). The parameters are those of Table XVAIII,108.

TABLE XVAIII,108
Parameters of the Atoms in $ClC_6H_4SeO_2H$

Atom	x	y	z
Se	0.0889(4)	0.9969(29)	0.3803(4)
Cl	0.4520(12)	0.9014(65)	0.6375(11)
O(1)	0.0826(22)	0.041 —	0.2449(22)
O(2)	0.9552(22)	0.144 —	0.3799(23)
C(1)	0.1886(26)	0.270 —	0.4470(31)
C(2)	0.2622(35)	0.406 —	0.4050(37)
C(3)	0.3445(42)	0.507	0.4554(37)
C(4)	0.3523(41)	0.673 —	0.5643(42)
C(5)	0.2920(51)	0.548 —	0.6194(30)
C(6)	0.2144(45)	0.345 —	0.5706(32)

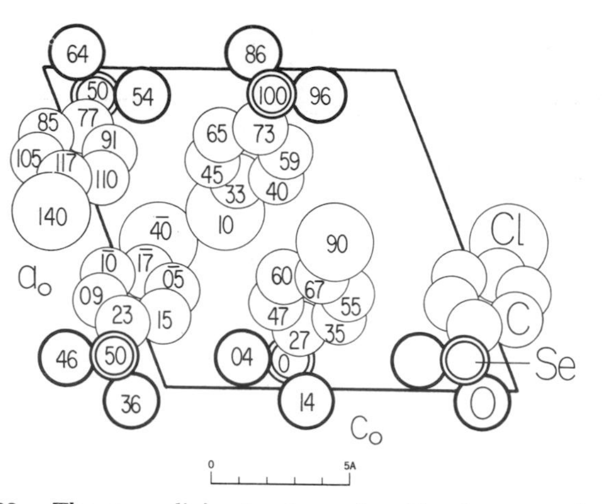

Fig. XVAIII,188. The monoclinic structure of *p*-chlorobenzene seleninic acid projected along its b_0 axis.

The structure (Fig. XVAIII,188) is built up of molecules having bond dimensions that are similar to, but less accurately determined than, those found for benzene seleninic acid (**XV,aII,49**). In the molecule, C–Cl = 1.69 A.

Though the unit cells of this compound and of benzene seleninic acid resemble one another, the crystals are not isostructural. The following two acids, however, have units very close to that of the chlorine-substituted acid and the same space group; they well may have the atomic arrangement described here.

p-Toluene seleninic acid, $CH_3C_6H_4SeO_2H$:

$$a_0 = 12.32 \text{ A.}; \quad b_0 = 5.14 \text{ A.}; \quad c_0 = 12.52 \text{ A.}; \quad \beta = 111°40'$$

p-Nitrobenzene seleninic acid, $NO_2C_6H_4SeO_2H$:

$$a_0 = 12.34 \text{ A.}; \quad b_0 = 5.14 \text{ A.}; \quad c_0 = 12.66 \text{ A.}; \quad \beta = 113°50'$$

XV,aIII,115. *Ammonium acid o-carboxybenzene sulfonate,* $NH_4(C_6H_4\text{-}COOHSO_3)$, forms orthorhombic crystals. Their unit, containing eight molecules, has the cell edges:

$$a_0 = 7.045(3) \text{ A.}; \quad b_0 = 10.535(3) \text{ A.}; \quad c_0 = 25.515(5) \text{ A.}$$

The space group is V_h^{15} (*Pcab*) with atoms in the positions:

$$(8c) \quad \pm(xyz;\ x,y+{}^1/_2,{}^1/_2-z;\ x+{}^1/_2,{}^1/_2-y,z;\ {}^1/_2-x,y,z+{}^1/_2)$$

The parameters are listed in Table XVAIII,109.

TABLE XVAIII,109

Parameters of the Atoms in $NH_4(C_6H_4COOHSO_3)$

Atom	x	y	z
NH_4	0.69062	0.12177	0.02801
S	0.16759	0.06241	0.08477
O(1)	0.28435	0.10823	0.04228
O(2)	0.01317	0.14690	0.09636
O(3)	0.10928	−0.06810	0.07580
C(7)	0.48170	−0.14283	0.11513
O(7,1)	0.46178	−0.25560	0.13705
O(7,2)	0.53409	−0.12690	0.07091
C(1)	0.43878	−0.03741	0.15245
C(2)	0.54199	−0.03485	0.19846
C(3)	0.52247	0.06565	0.23348
C(4)	0.39889	0.16245	0.22242
C(5)	0.29310	0.16036	0.17635
C(6)	0.31252	0.06038	0.14168
H(1)	0.485	−0.319	0.109
H(2)	0.635	−0.105	0.205
H(3)	0.584	0.072	0.265
H(4)	0.336	0.217	0.249
H(5)	0.218	0.230	0.164

The structure, shown in Figure XVAIII,189, is ionic. Its complex anion, in which the SO_3 is considered to be the ionized radical and the carboxyl group neutral, has the bond dimensions of Figure XVAIII,190. The benzene ring is planar with the sulfur atom 0.055 A. on one side and the carboxyl carbon 0.125 A. on the other side of the plane. The plane of the carboxyl group is twisted through 50.7° with respect to the ring. Coordination about sulfur is tetrahedral. In the crystal these anions are tied together by hydrogen bonds of length 2.64 A. between the carboxyl oxygens. The ammonium cations, for which definite hydrogen positions could not be found, have an eightfold coordination, with NH_4–O = 2.889–3.247 A.

Ammonium acid phthalate, $NH_4HC_6H_4(COO)_2$, is isostructural. Its unit has the edges:

$$a_0 = 6.40 \text{ A.}; \quad b_0 = 10.23 \text{ A.}; \quad c_0 = 26.14 \text{ A.}$$

The parameters are listed in Table XVAIII,110.

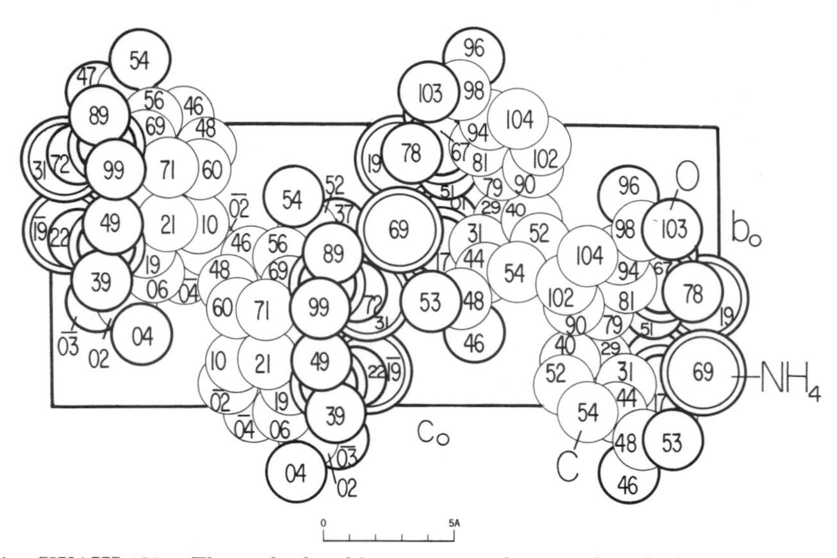

Fig. XVAIII,189. The orthorhombic structure of ammonium hydrogen *o*-carboxybenzenesulfonate projected along its a_0 axis.

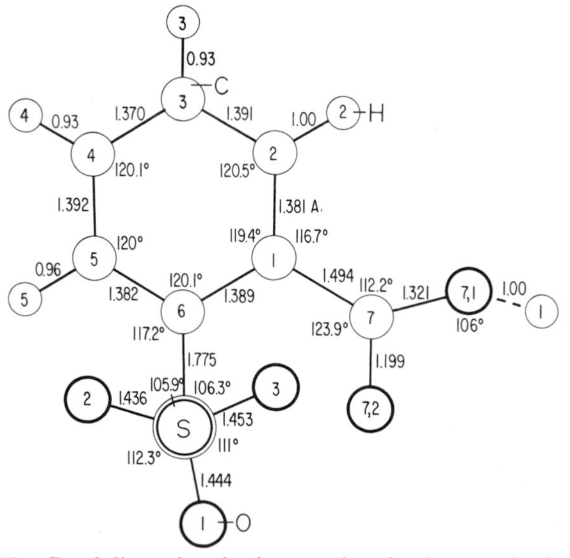

Fig. XVAIII,190. Bond dimensions in the organic anion in crystals of ammonium acid *o*-carboxybenzenesulfonate.

TABLE XVAIII,110

Parameters of the Atoms in $NH_4HC_6H_4(COO)_2$

Atom	x	y	z
NH_4	0.219	−0.120	0.0175
C(1)	0.389	−0.050	0.2250
C(2)	0.380	0.061	0.1935
C(3)	0.486	0.062	0.1475
C(4)	0.586	−0.050	0.1298
C(5)	0.593	−0.158	0.1620
C(6)	0.505	−0.152	0.2090
C(7)	0.479	0.192	0.1170
C(8)	0.724	−0.050	0.0821
O(7,1)	0.443	0.290	0.1385
O(7,2)	0.506	0.177	0.0703
O(8,1)	0.899	0.005	0.0820
O(8,2)	0.652	−0.120	0.0475

The anion, as determined in this earlier study, has the bond dimensions of Figure XVAIII,191. Except for the oxygen atoms it is planar with the C(8)–O(8,1)–O(8,2) carboxyl group turned 65° and the C(7)–O(7,1)–O(7,2) turned 21° with respect to the ring. The latter carboxyl is apparently

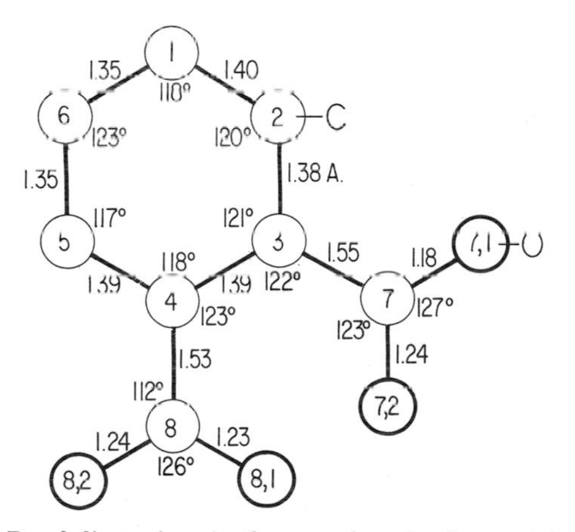

Fig. XVAIII,191. Bond dimensions in the organic anion in crystals of ammonium acid phthalate.

un-ionized. There are only six oxygen atoms (rather than the eight of the sulfonate) about each ammonium cation, at distances between 2.81 and 3.10 A.

The monomethyl ammonium acid phthalate, $(CH_3NH_3)HC_6H_4(COO)_2$, apparently has the same structure. For it the cell edges are:

$$a_0 = 7.22 \text{ A.}; \quad b_0 = 11.01 \text{ A.}: \quad c_0 = 24.46 \text{ A.}$$

Atomic parameters have not been established.

The sodium salt $NaH(C_6H_4COO)_2$ has an eight-molecule cell of the same shape. Its cell edges are:

$$a_0 = 6.76 \text{ A.}; \quad b_0 = 9.31 \text{ A.}; \quad c_0 = 26.42 \text{ A.}$$

A different space group, C_{2v}^{17} ($B2ab$), was, however, assigned this substance.

XV,aIII,116. The monoclinic crystals of *m-aminobenzenearsonic acid*, $NH_2C_6H_4AsO(OH)_2$, have an eight-molecule unit of the dimensions:

$$a_0 = 13.35 \text{ A.}; \quad b_0 = 15.60 \text{ A.}; \quad c_0 = 7.46 \text{ A.}; \quad \beta = 92°42'$$

The space group is C_{2h}^5 in the unusual non-primitive, base-centered orientation $B2_1/a$. Atoms therefore are in the positions:

$$(8e') \quad \pm(xyz;\ x+\tfrac{1}{2},\tfrac{1}{2}-y,z;\ x+\tfrac{1}{2},y,z+\tfrac{1}{2};\ x,\tfrac{1}{2}-y,z+\tfrac{1}{2})$$

The parameters are stated in Table XVAIII,111.

TABLE XVAIII,111

Parameters of the Atoms in *m*-Aminobenzenearsonic Acid

Atom	x	y	z
As	0.0874	0.1625	0.1351
O(1)	0.172	0.238	0.215
O(2)	−0.040	0.196	0.136
O(3)	0.112	0.138	−0.092
C(1)	0.111	0.059	0.268
C(2)	0.109	−0.022	0.184
C(3)	0.130	−0.092	0.300
C(4)	0.137	−0.089	0.488
C(5)	0.141	−0.008	0.563
C(6)	0.124	0.061	0.457
N	0.143	−0.179	0.221

In this crystal the benzene ring and its attached nitrogen atom are planar and the arsenic atom is tetrahedrally surrounded by three oxygens and a carbon atom. The distances As–O average 1.76 A., and As–C = 1.91 A. Molecules are held together by a system of hydrogen bonds of lengths 2.56, 2.64, 2.66, and 2.95 A.; the first of these is between oxygen atoms, the other three between nitrogen and oxygen.

XV,aIII,117. Crystals of *p-aminobenzenearsonic acid*, $NH_2C_6H_4AsO(OH)_2$, are monoclinic with a bimolecular unit of the dimensions:

$$a_0 = 7.35 \text{ A.}; \quad b_0 = 6.35 \text{ A.}; \quad c_0 = 8.81 \text{ A.}; \quad \beta = 101°$$

The space group, chosen as C_2^2 ($P2_1$), places the atoms in the positions:

$$(2a) \quad xyz; \bar{x},y+{}^1/_2,\bar{z}$$

The parameters are listed in Table XVAIII,112.

TABLE XVAIII,112

Parameters of the Atoms in *p*-Aminobenzenearsonic Acid

Atom	x	y	z
As	0.1632	0.000	0.1831
O(1)	0.338	0.063	0.093
O(2)	−0.041	0.078	0.086
O(3)	0.156	−0.293	0.185
C(1)	0.213	0.093	0.398
C(2)	0.299	0.295	0.418
C(3)	0.331	0.386	0.572
C(4)	0.298	0.254	0.698
C(5)	0.234	0.047	0.675
C(6)	0.184	−0.028	0.523
N	0.327	0.318	0.863

In this structure (Fig. XVAIII,192), as in the *meta* compound of the preceding paragraph, arsenic atoms are tetrahedrally surrounded by three oxygens and one carbon, with As–C = 1.95 A. and an average As–O = 1.73 A. The benzene ring and its attached nitrogen atom are planar. Molecules are bound together by an O–H–O = 2.52 A. and three N–H–O = 2.58, 2.84, and 2.87 A.

XV,aIII,118. A brief description has recently been published of a structure for *cobalt(II)-bis(o-phenylene bis dimethylarsine) diperchlorate,* $[C_6H_4\{As(CH_3)_2\}_2]_2Co(ClO_4)_2$. Its crystals are monoclinic with a bimolecular unit of the dimensions:

$$a_0 = 17.07(10) \text{ A.}; \quad b_0 = 18.66(10) \text{ A.}; \quad c_0 = 9.49(5) \text{ A.}$$
$$\beta = 97.5(5)°$$

The cobalt atoms are in the positions

$$(2a) \quad 000; 0\ {}^1/_2\ {}^1/_2$$

of the space group C_{2h}^5 ($P2_1/c$). All other atoms are in the positions:

$$(4e) \quad \pm(xyz;\ x,{}^1/_2-y,z+{}^1/_2)$$

Parameters are given in Table XVAIII,113.

TABLE XVAIII,113

Parameters of the Atoms in $[C_6H_4\{As(CH_3)_2\}_2]_2Co(ClO_4)_2$

Atom	x	y	z
As(1)	0.1078	0.0102	0.0785
As(2)	−0.0437	0.1960	0.0584
Cl	−0.1337	−0.2620	0.1220
$CH_3(1)$	0.2189	−0.0195	0.0406
$CH_3(2)$	0.1058	−0.1310	0.1654
$CH_3(3)$	−0.1208	0.1504	0.1427
$CH_3(4)$	−0.0749	0.3628	−0.0078
C(1)	0.1063	0.1984	0.1313
C(2)	0.0342	0.2795	0.1149
C(3)	0.0428	0.4097	0.1489
C(4)	0.0959	0.4658	0.1951
C(5)	0.1656	0.3576	0.2076
C(6)	0.1758	0.2392	0.1755
O(1)	−0.1244	−0.2465	0.0499
O(2)	−0.1513	−0.4142	0.1227
O(3)	−0.1882	−0.2192	0.1697
O(4)	−0.0650	−0.2811	0.1599

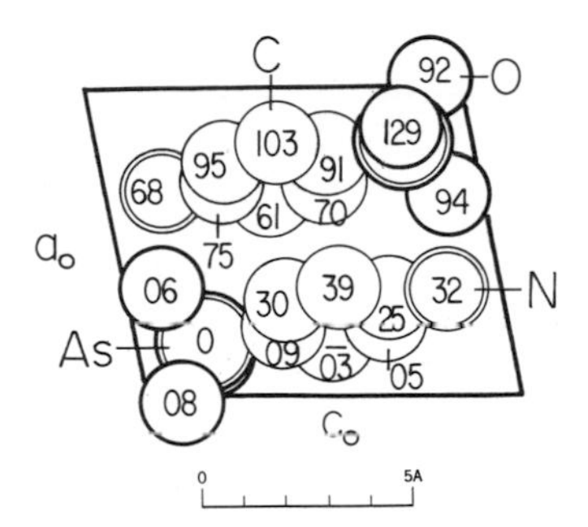

Fig. XVAIII,192. The monoclinic structure of *p*-aminobenzenearsonic acid projected along its b_0 axis.

XV,aIII,119. Crystals of *dichloro-di(o-phenylene bis dimethylarsine) platinum(II)*, $[C_6H_4\{As(CH_3)_2\}_2]_2PtCl_2$, are orthorhombic with a tetramolecular cell of the edge lengths:

$$a_0 = 9.66 \text{ A.}; \quad b_0 = 16.60 \text{ A.}; \quad c_0 = 18.02 \text{ A.}$$

The platinum atoms are in the positions:

$$(4a) \quad 000;\ 0\ 0\ {}^1/_2;\ {}^1/_2\ {}^1/_2\ 0;\ {}^1/_2\ {}^1/_2\ {}^1/_2$$

of the space group V_h^{14} (*Pcan*). All other atoms are in:

$$(8d) \quad \pm(xyz;\ x+{}^1/_2,y+{}^1/_2,{}^1/_2-z;\ x+{}^1/_2,{}^1/_2-y,z;\ \bar{x},y,z+{}^1/_2)$$

The parameters are those of Table XVAIII,114.

TABLE XVAIII,114

Parameters of Atoms in $PtCl_2[C_6H_4\{As(CH_3)_2\}_2]_2$

Atom	x	y	z
As(2)	0.5179	0.3886	0.0824
As(3)	0.2044	0.0450	0.4400
Cl(4)	0.2252	0.3626	0.3749
C(5)	0.279	0.040	0.116
C(6)	0.401	0.037	0.151
C(7) [a]	0.195	0.098	0.198
C(8)	0.242	0.111	0.126
C(9)	0.242	0.169	0.175
C(10)	0.368	0.165	0.211
C(11)	0.006	0.218	0.043
C(12)	0.155	0.129	0.362
C(13)	0.380	0.392	0.169
C(14)	0.347	0.090	0.497

[a] There is an error here, apparently in the x value for C(7). It leads to an impossibly long C(6)–C(7) bond.

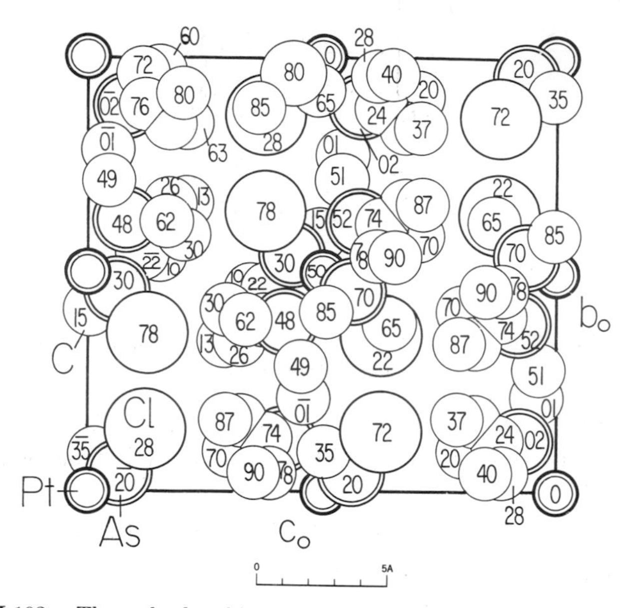

Fig. XVAIII,193. The orthorhombic structure of dichloro-di(*o*-phenylene bis dimethylarsine) platinum projected along its a_0 axis.

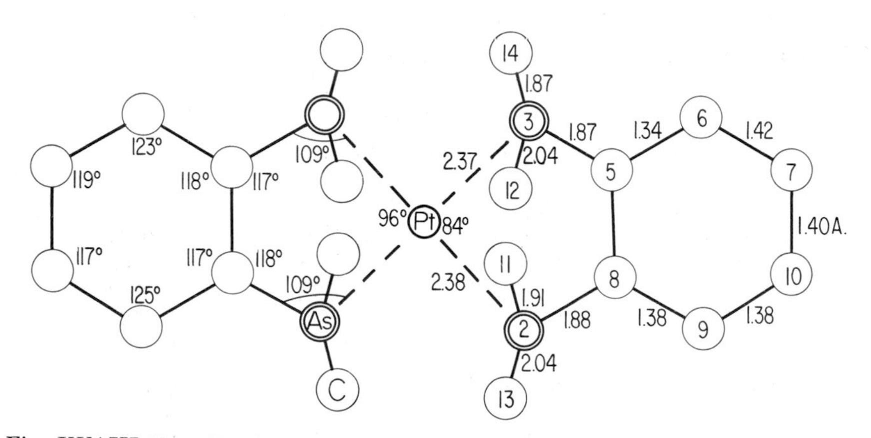

Fig. XVAIII,194. Bond dimensions in the molecule of dichloro-di(*o*-phenylene bis dimethylarsine) platinum.

The resulting structure (Fig. XVAIII,193) has molecules with the bond dimensions of Figure XVAIII,194. In them the arsenic atoms are -0.11 A. [for As(2)] and -0.19 A. [for As(3)] out of the plane of the benzene ring. The platinum atom has its expected square coordination supplied by four atoms of arsenic; the two chloride anions are much more distant but in a line that is roughly normal to the Pt–As plane. Between molecules the shortest C–C distance is about 3.5 A.

XV,aIII,120. *Diiodo-di(o-phenylene bis dimethylarsine) nickel,* $[C_6H_4\{As(CH_3)_2\}_2]_2NiI_2$, forms monoclinic crystals. Their bimolecular unit has the dimensions:

$$a_0 = 9.49 \text{ A.}; \quad b_0 = 9.25 \text{ A.}; \quad c_0 = 16.94 \text{ A.}; \quad \beta = 114.0°$$

The nickel atoms are in

$$(2a) \quad 000; 0\ {}^1/_2\ {}^1/_2$$

and all others in

$$(4e) \quad \pm(xyz; x,{}^1/_2-y,z+{}^1/_2)$$

of the space group C_{2h}^5 ($P2_1/c$). The determined parameters are those of Table XVAIII,115.

TABLE XVAIII,115

Parameters of Atoms in $NiI_2[C_6H_4\{As(CH_3)_2\}_2]_2$

Atom	*x*	*y*	*z*
As(1)	0.2054	0.1534	0.0579
As(2)	0.0671	−0.0908	0.1371
I	0.1849	−0.2462	−0.0527
C(1)	0.2886	0.0186	0.3035
C(2)	0.2321	0.0293	0.2178
C(3)	0.3911	0.1014	0.0456
C(4)	0.3970	0.1092	0.3591
C(5)	0.2875	0.1425	0.1834
C(6)	0.4521	0.2170	0.3232
C(7)	0.4035	0.2377	0.2372
C(8)	0.1861	0.3645	0.0277
C(9)	−0.0847	−0.0914	0.1857
C(10)	0.1491	−0.2766	0.1644

The structure, illustrated in Figure XVAIII,195, is composed of centrosymmetric molecules which have the bond dimensions of Figure XVAIII, 196. Their benzene ring is planar but the As(1) is 0.021 A. to one side and the As(2) 0.094 A. to the other side of the ring. The plane through them

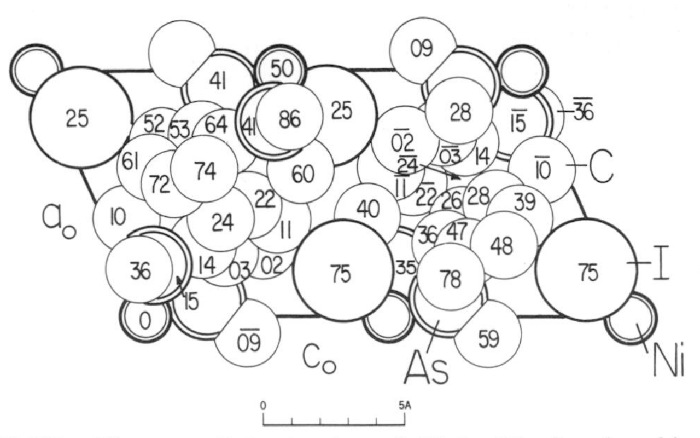

Fig. XVAIII,195. The monoclinic structure of diiodo-di(*o*-phenylene bis dimethylarsine) nickel projected along its b_0 axis.

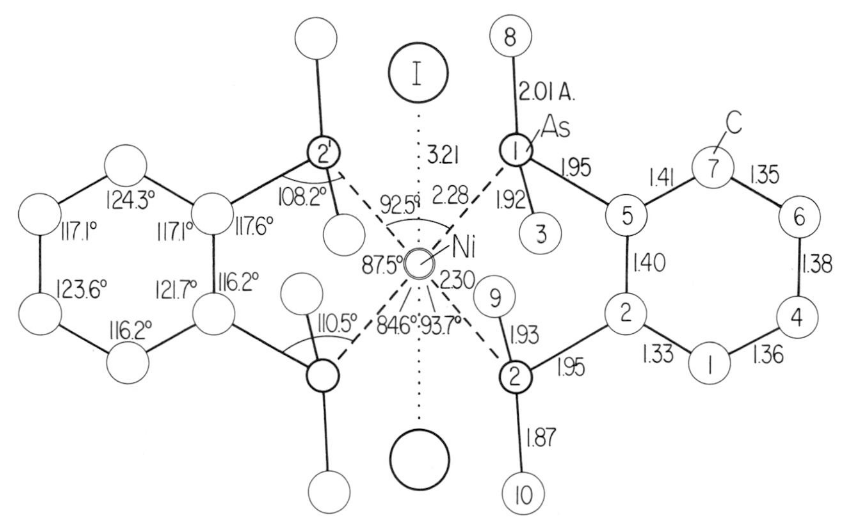

Fig. XVAIII,196. Bond dimensions in the molecule of diiodo-di(*o*-phenylene bis dimethylarsine) nickel in its crystals.

and the nickel atom thus makes 10.4° with that of the ring. About the arsenic atoms the distribution of bonds is roughly tetrahedral. The line through the iodide anions is nearly normal to the Ni–As plane, thereby giving nickel an elongated octahedral coordination.

The corresponding *platinum* and *palladium* compounds are isostructural. Their units have the dimensions:

For $[C_6H_4\{As(CH_3)_2\}_2]_2PdI_2$:

$$a_0 = 9.64 \text{ A.}; \quad b_0 = 9.24 \text{ A.}; \quad c_0 = 17.23 \text{ A.}; \quad \beta = 114°$$

For $[C_6H_4\{As(CH_3)_2\}_2]_2PtI_2$:

$$a_0 = 9.68 \text{ A.}; \quad b_0 = 9.27 \text{ A.}; \quad c_0 = 17.27 \text{ A.}; \quad \beta = 114°30'$$

The atomic parameters assigned them, when given the same designations as for the nickel compound, are listed in Table XVAIII,116. The carbon parameters show only fair agreement between themselves and with the perhaps better values for the nickel compound.

TABLE XVAIII,116

Parameters of the Atoms in $PdI_2[C_6H_4\{As(CH_3)_2\}_2]_2$ and (in parentheses) in the Platinum Compound

Atom	x	y	z
As(1)	0.2106 (0.2083)	0.1610 (0.1624)	0.0613 (0.0590)
As(2)	0.0710 (0.0715)	−0.0870 (−0.0836)	0.1426 (0.1419)
I	0.1971 (0.2050)	0.7416 (0.7384)	−0.0536 (−0.0549)
C(1)	0.315 (0.227)	0.035 (0.025)	0.302 (0.290)
C(2)	0.237 (0.200)	0.043 (0.029)	0.212 (0.203)
C(3)	0.353 (0.359)	0.196 (0.141)	0.022 (0.021)
C(4)	0.448 (0.342)	0.102 (0.109)	0.348 (0.347)
C(5)	0.297 (0.267)	0.137 (0.136)	0.181 (0.183)
C(6)	0.486 (0.438)	0.210 (0.194)	0.306 (0.328)
C(7)	0.440 (0.398)	0.233 (0.217)	0.220 (0.234)
C(8)	0.208 (0.149)	0.344 (0.347)	0.028 (0.031)
C(9)	0.910 (0.935)	0.931 (0.930)	0.190 (0.198)
C(10)	0.151 (0.188)	0.725 (0.756)	0.138 (0.171)

XV,aIII,121. A structure, which may be approximate only, has been described for *mono(o-phenylene bis dimethylarsine) iron tricarbonyl*, $[C_6H_4\{As(CH_3)_2\}_2]\cdot Fe(CO)_3$. Its crystals are orthorhombic with a tetramolecular unit of the edge lengths:

$$a_0 = 16.73(3) \text{ A.}; \quad b_0 = 10.37(3) \text{ A.}; \quad c_0 = 9.71(3) \text{ A.}$$

The chosen space group V_h^{16} (*Pnma*) places a majority of the atoms in the special positions:

$$(4c) \quad \pm(u\,{}^1/_4\,v;\ u+{}^1/_2,{}^1/_4,{}^1/_2-v)$$

The other atoms are in the general positions:

$$(8d) \quad \pm(xyz;\ {}^1/_2-x,y+{}^1/_2,z+{}^1/_2;\ x,{}^1/_2-y,z;\ x+{}^1/_2,y,{}^1/_2-z)$$

Positions and parameters are listed in Table XVAIII,117.

TABLE XVAIII,117

Positions and Parameters of the Atoms in $[C_6H_4\{As(CH_3)_2\}_2]\cdot Fe(CO)_3$

Atom	Position	x	y	z
C(1)	(4*c*)	0.103	1/4	0.011
C(2)	(4*c*)	−0.076	1/4	0.395
C(3)	(4*c*)	−0.152	1/4	0.416
C(4)	(4*c*)	−0.012	1/4	0.493
C(5)	(4*c*)	−0.173	1/4	0.554
C(6)	(4*c*)	−0.034	1/4	0.647
C(7)	(4*c*)	−0.118	1/4	0.666
C(8)	(8*d*)	0.140	0.392	0.530
C(9)	(8*d*)	0.090	0.599	0.887
C(10)	(8*d*)	0.156	0.403	0.192
O(1)	(4*c*)	0.101	1/4	0.887
O(2)	(8*d*)	0.197	0.491	0.192
Fe	(4*c*)	0.1039	1/4	0.1848
As(1)	(4*c*)	0.0970	1/4	0.4277
As(2)	(4*c*)	0.0339	1/4	0.8037

The structure is made up of molecules having the bond dimensions of Figure XVAIII,197. Distribution about the iron atom is that of a trigonal bipyramid, the distortion from regularity being small.

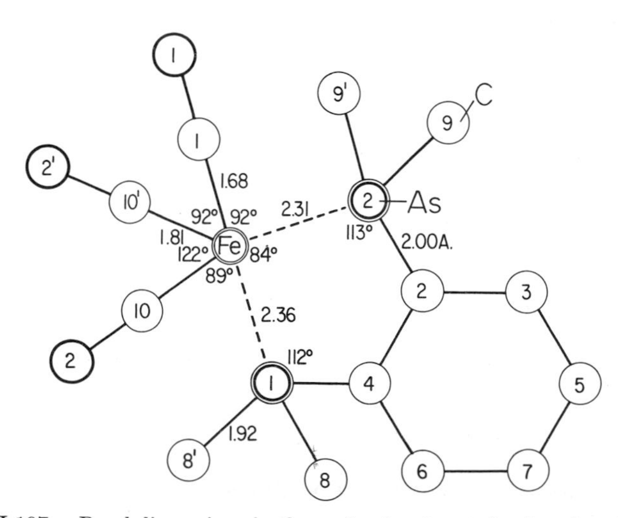

Fig. XVAIII,197. Bond dimensions in the molecule of mono(*o*-phenylene bis dimethylarsine) iron tricarbonyl in the crystals it forms.

IV. Compounds containing a Tri-substituted Benzene Ring

XV,aIV,1. Crystals of *1,3,5-tribromobenzene*, $C_6H_3Br_3$, and of *1,3,5-trichlorobenzene*, $C_6H_3Cl_3$, at −183°C. as well as at room temperature are orthorhombic with a tetramolecular unit. Their dimensions are as follows:

$C_6H_3Br_3$: a_0 = 14.23 A.; b_0 = 13.55 A.; c_0 = 4.08 A. (20°C.)
$C_6H_3Cl_3$: a_0 = 13.93 A.; b_0 = 13.19 A.; c_0 = 3.91 A. (20°C.)
$C_6H_3Cl_3$: a_0 = 13.77 A.; b_0 = 13.07 A.; c_0 = 3.82 A. (−183°C.)

Atoms are in the positions

$$(4a) \quad xyz;\ x+1/2,1/2-y,\bar{z};\ \bar{x},y+1/2,1/2-z;\ 1/2-x,\bar{y},z+1/2$$

of the space group V^4 ($P2_12_12_1$). The parameters are listed in Tables XVAIV,1 and XVAIV,2.

TABLE XVAIV,1

Parameters of the Atoms in $C_6H_3Br_3$

Atom	x	y	z
Br(1)	0.3494	0.2261	0.597
Br(3)	0.3591	0.6403	0.686
Br(5)	0.6568	0.4363	0.074
C(1)	0.4096	0.3449	0.513
C(2)	0.3685	0.4335	0.614
C(3)	0.4138	0.5227	0.553
C(4)	0.5005	0.5237	0.390
C(5)	0.5416	0.4351	0.290
C(6)	0.4963	0.3457	0.351
H(2)	0.301	0.433	0.734
H(4)	0.536	0.593	0.340
H(6)	0.528	0.277	0.271

The structure, shown in Figure XVAIV,1, is built of molecules that are planar, with a mean C–Br = 1.857 A. and C–Cl = 1.711 A. (20°C.). The mean C–C = 1.397 A. for the bromo and 1.387 A. for the chloro compound.

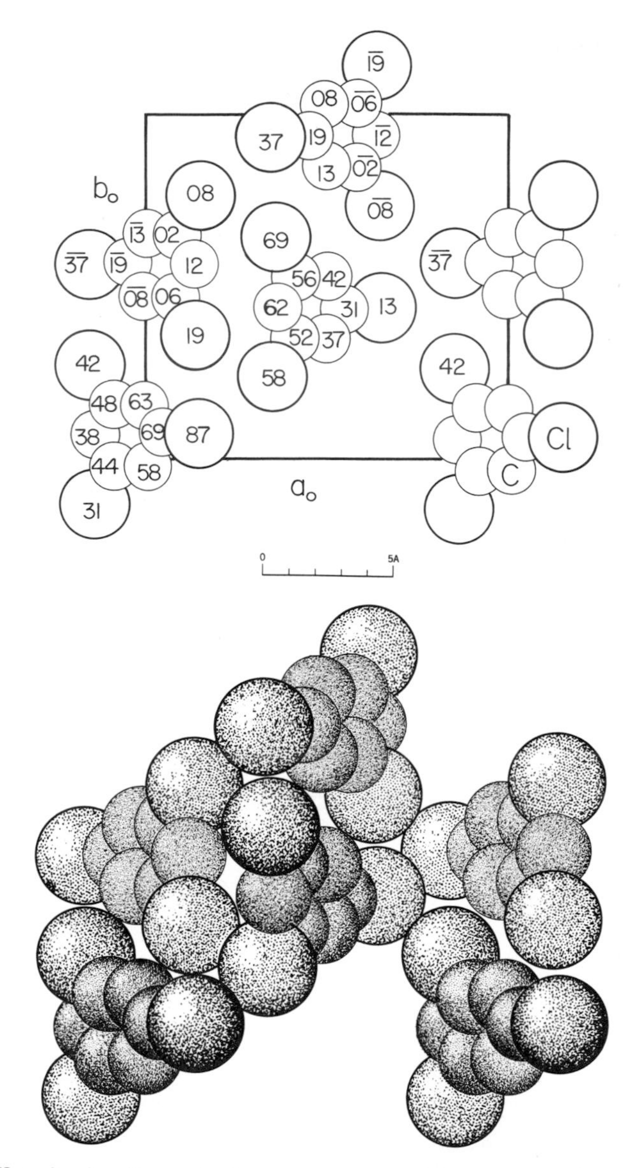

Fig. XVAIV,1a (top). The orthorhombic structure of 1,3,5-trichlorobenzene projected along its c_0 axis.

Fig. XVAIV,1b (bottom). A packing drawing of the orthorhombic 1,3,5-trichlorobenzene arrangement viewed along its c_0 axis. The larger circles are the chlorine atoms.

TABLE XVAIV,2

Parameters of the Atoms in $C_6H_3Cl_3$ (−183°C. values in parentheses)

Atom	x	y	z
Cl(1)	0.3490 (0.3469)	0.2336 (0.2316)	0.584
Cl(3)	0.3626 (0.3608)	0.6359 (0.6419)	0.687
Cl(5)	0.6588 (0.6600)	0.4339 (0.4364)	0.129
C(1)	0.4086 (0.4071)	0.3452 (0.3446)	0.521
C(2)	0.3653 (0.3640)	0.4338 (0.4364)	0.622
C(3)	0.4163 (0.4130)	0.5248 (0.5289)	0.564
C(4)	0.5058 (0.5051)	0.5245 (0.5286)	0.416
C(5)	0.5474 (0.5486)	0.4347 (0.4364)	0.312
C(6)	0.4992 (0.4987)	0.3441 (0.3442)	0.368
H(2)	0.295 (0.292)	0.435 (0.436)	0.737
H(4)	0.543 (0.543)	0.596 (0.600)	0.374
H(6)	0.532 (0.532)	0.273 (0.272)	0.290

XV,aIV,2. According to a preliminary note, crystals of *2-methyl-3-bromophenol*, $C_6H_3Br(CH_3)OH$, are orthorhombic. Their tetramolecular unit has the edge lengths:

$$a_0 = 12.13(4) \text{ A.}; \quad b_0 = 12.45(4) \text{ A.}; \quad c_0 = 4.68(5) \text{ A.}$$

The space group, chosen as C_{2v}^9 ($Pna2_1$), puts atoms in the positions:

$$(4a) \quad xyz; \bar{x},\bar{y},z+1/2; 1/2-x,y+1/2,z+1/2; x+1/2,1/2-y,z$$

The parameters have been stated to be those of Table XVAIV,3.

TABLE XVAIV,3

Parameters of the Atoms in $C_6H_3Br(CH_3)OH$

Atom	x	y	z
Br	0.223	0.086	0.00
O	0.047	0.456	0.28
C(1)	0.035	0.322	0.28
C(2)	0.120	0.285	0.13
C(3)	0.132	0.163	0.17
C(4)	0.041	0.124	0.43
C(5)	0.044	0.294	0.47
C(6)	0.037	0.186	0.56
C(7)	0.200	0.340	0.03

XV,aIV,3. The orthorhombic crystals of *2,3-dimethylphenol*, $(CH_3)_2C_6$-H_3OH, have a tetramolecular unit of the dimensions:

$$a_0 = 4.81(2) \text{ A.}; \quad b_0 = 5.92(3) \text{ A.}; \quad c_0 = 24.65(4) \text{ A.}$$

The space group is V^4 $(P2_12_12_1)$ with atoms in the positions:

$$(4a) \quad xyz;\ {}^1/_2-x,\bar{y},z+{}^1/_2;\ x+{}^1/_2,{}^1/_2-y,\bar{z};\ \bar{x},y+{}^1/_2,{}^1/_2-z$$

Two determinations have been published; the parameters most recently found are those of Table XVAIV,4.

TABLE XVAIV,4

Parameters of the Atoms in $(CH_3)_2C_6H_3OH$

Atom	x	y	z
C(1)	0.989(4)	0.916(5)	0.077(1)
C(2)	0.811(4)	0.766(5)	0.100(1)
C(3)	0.772(4)	0.781(5)	0.158(1)
C(4)	0.906(4)	0.944(5)	0.185(1)
C(5)	0.087(5)	0.102(6)	0.161(1)
C(6)	0.136(4)	0.090(5)	0.104(1)
C(7)	0.579(5)	0.616(5)	0.185(1)
C(8)	0.666(5)	0.569(6)	0.065(1)
O	0.023(3)	0.900(3)	0.019(5)

The structure is shown in Figure XVAIV,2. Its molecules, having the bond dimensions of Figure XVAIV,3, are practically planar, the maximum departures from the best plane through their atoms being 0.03 A. They are

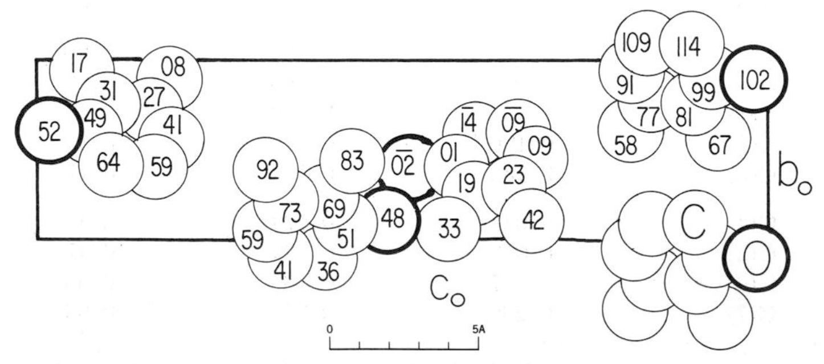

Fig. XVAIV,2. The orthorhombic structure of 2,3-dimethylphenol projected along its a_0 axis.

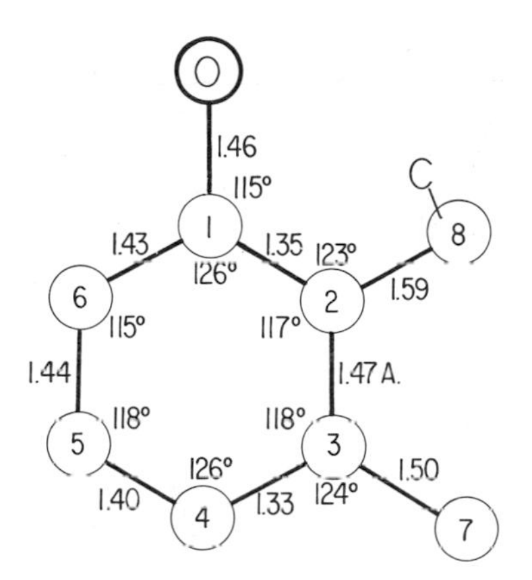

Fig. XVAIV,3. Bond dimensions in the molecule of 2,3-dimethylphenol.

tied together into strings by hydrogen bonds of length 2.84 A. between oxygen atoms.

XV,aIV,4. *2,5-Dimethylphenol*, $(CH_3)_2C_6H_3OH$, forms monoclinic crystals whose bimolecular unit has the dimensions:

$$a_0 = 5.94(2) \text{ A.}; \quad b_0 = 4.91(2) \text{ A.}; \quad c_0 = 12.48(3) \text{ A.}$$
$$\beta = 109°42(30)'$$

The space group is C_2^2 ($P2_1$) with atoms in the positions:

$$(2a) \quad xyz; \bar{x},y+{}^1/_2,\bar{z}$$

The chosen parameters are listed in Table XVAIV,5.

TABLE XVAIV,5

Parameters of the Atoms in $(CH_3)_2C_6H_3OH$

Atom	x	y	z
C(1)	0.469(2)	0.093(4)	0.154(1)
C(2)	0.303(2)	0.030(5)	0.198(1)
C(3)	0.313(3)	0.060(5)	0.305(1)
C(4)	0.481(3)	0.247(5)	0.364(1)
C(5)	0.652(2)	0.356(4)	0.322(1)
C(6)	0.640(2)	0.271(5)	0.214(1)
C(7)	0.113(3)	0.227(5)	0.130(1)
C(8)	0.838(3)	0.562(5)	0.387(1)
O	0.454(2)	0.000—	0.043(1)

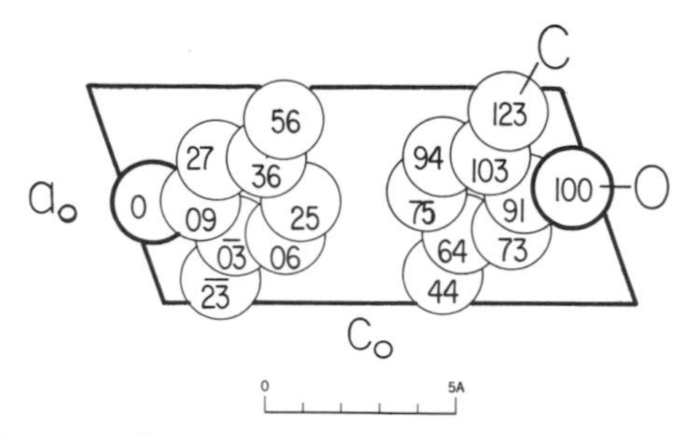

Fig. XVAIV,4. The monoclinic structure of 2,5-dimethylphenol projected along its b_0 axis.

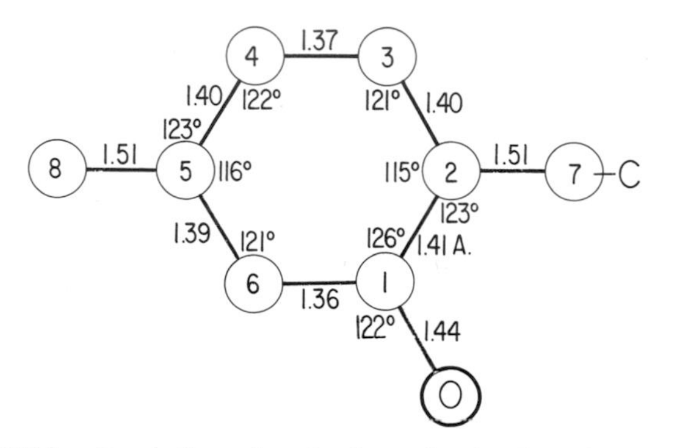

Fig. XVAIV,5. Bond dimensions in the molecule of 2,5-dimethylphenol.

The resulting structure (Fig. XVAIV,4) is built of molecules which are planar to within less than 0.02 A. and have the bond dimensions of Figure XVAIV,5. Between molecules there are hydrogen bonds of length O–H–O = 2.81 A.

XV,aIV,5. The orthorhombic crystals of anhydrous *phloroglucinol*, $C_6H_3(OH)_3$, have a tetramolecular cell of the dimensions:

$$a_0 = 4.83(1) \text{ A.}; \quad b_0 = 9.37(2) \text{ A.}; \quad c_0 = 12.56(3) \text{ A.}$$

All atoms are in the positions

$$(4a) \quad xyz;\ ^1/_2-x,\bar{y},z+{}^1/_2;\ x+{}^1/_2,{}^1/_2-y,\bar{z};\ \bar{x},y+{}^1/_2,{}^1/_2-z$$

of the space group V^4 ($P2_12_12_1$). The determined parameters are those of Table XVAIV,6.

TABLE XVAIV,6

Parameters of the Atoms in $C_6H_3(OH)_3$

Atom	x	y	z
C(1)	0.4082	0.4929	0.4815
C(2)	0.5106	0.3693	0.4379
C(3)	0.6993	0.3786	0.3556
C(4)	0.7758	0.5097	0.3133
C(5)	0.6660	0.6318	0.3578
C(6)	0.4792	0.6255	0.4421
O(7)	0.2251	0.4810	0.5659
O(8)	0.8150	0.2582	0.3121
O(9)	0.7424	0.7647	0.3230
H(10)	0.472	0.2763	0.464
H(11)	0.931	0.506	0.266
H(12)	0.399	0.711	0.487
H(13)	0.263	0.558	0.616
H(14)	0.785	0.155	0.375
H(15)	0.850	0.762	0.269

In this structure the molecules have the distribution shown in Figure XVAIV,6; their bond dimensions are those of Figure XVAIV,7. They are themselves held together by hydrogen bonds which range between 2.73 and 2.76 A.

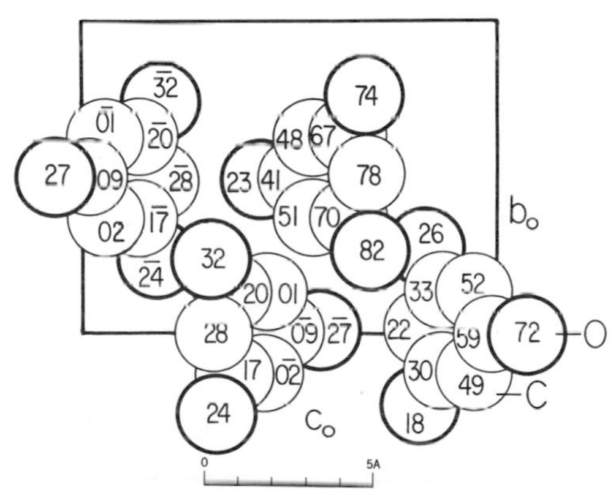

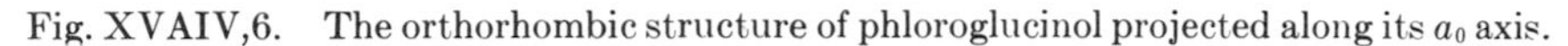
Fig. XVAIV,6. The orthorhombic structure of phloroglucinol projected along its a_0 axis.

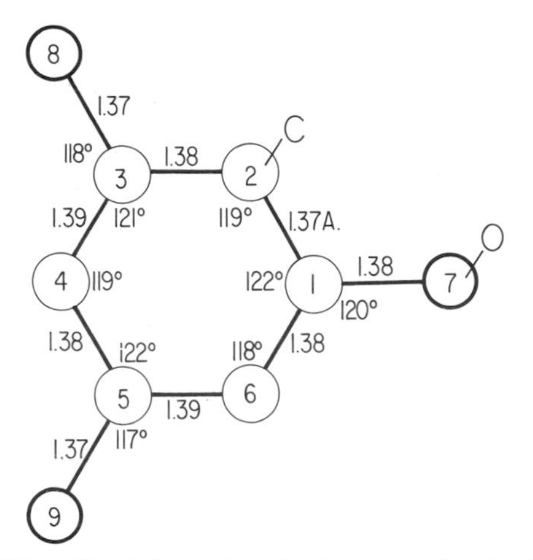

Fig. XVAIV,7. Bond dimensions in the molecule of phloroglucinol.

XV,aIV,6. *Phloroglucinol dihydrate,* $C_6H_3(OH)_3 \cdot 2H_2O$, is also orthorhombic but with a tetramolecular unit of the dimensions:

$$a_0 = 6.73(1) \text{ A.}; \quad b_0 = 13.58(2) \text{ A.}; \quad c_0 = 8.09(1) \text{ A.}$$

After some uncertainty the space group was finally chosen as V_h^{16} (*Pnma*) with atoms in the positions:

C(1): (4*c*) $\pm(u\ ^1/_4\ v;\ u+^1/_2,^1/_4,^1/_2-v)$
with $u = 0.383$, $v = 0.218$
C(4): (4*c*) with $u = 0.390$, $v = -0.137$
O(1): (4*c*) with $u = 0.380$, $v = 0.383$

The other atoms are in:

$$(8d) \quad \pm(xyz;\ x,^1/_2-y,z;\ x+^1/_2,y,^1/_2-z;\ ^1/_2-x,y+^1/_2,z+^1/_2)$$

with the parameters:

C(2): (8*d*) with $x = 0.385$, $y = 0.160$, $z = 0.128$
C(3): (8*d*) with $x = 0.388$, $y = 0.160$, $z = -0.043$
O(2): (8*d*) with $x = 0.390$, $y = 0.071$, $z = -0.122$
H_2O: (8*d*) with $x = 0.863$, $y = 0.071$, $z = -0.038$

The structure, not defined with great accuracy, is shown in Figure XVAIV,8. In it the mean bond lengths are C–C = 1.41 A. and C–OH = 1.36 A. Hydrogen bonds, with O–H–O = 2.74–2.88 A., tie the molecules

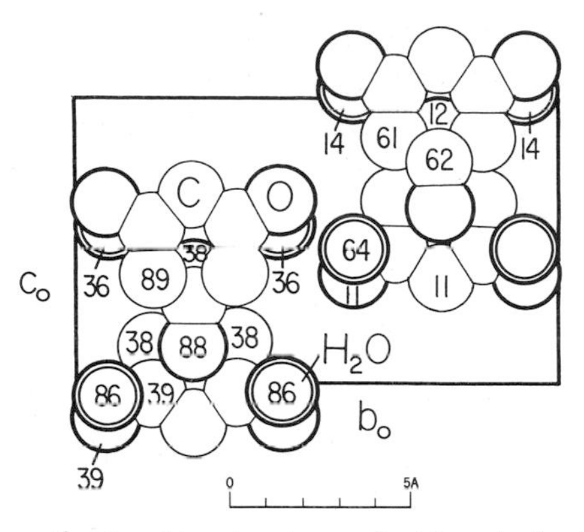

Fig. XVAIV,8. The orthorhombic structure of phloroglucinol dihydrate projected along its a_0 axis.

together. Six of these bonds, involving OH and H_2O, bind the molecules in layers normal to a_0; bonds between water molecules unite the layers.

XV,aIV,7. The red form of *5-methoxy-2-nitrosophenol*, $CH_3OC_6H_3$-(O)NO, is orthorhombic with a tetramolecular unit of the edge lengths:

$$a_0 = 17.11 \text{ A.}; \quad b_0 = 3.80 \text{ A.}; \quad c_0 = 10.67 \text{ A.}$$

Atoms are in

$$(4a) \quad xyz;\ {}^1/_2-x,\bar{y},z+{}^1/_2;\ x+{}^1/_2,{}^1/_2-y,\bar{z};\ \bar{x},y+{}^1/_2,{}^1/_2-z$$

of V^4 ($P2_12_12_1$) with the assigned parameters of Table XVAIV,7.

TABLE XVAIV,7

Parameters of the Atoms in 5-Methoxy-2-nitrosophenol

Atom	x	y	z
C(1)	0.1667	−0.253	0.0400
C(2)	0.1033	−0.451	0.0417
C(3)	0.2050	−0.167	−0.0783
C(4)	0.0567	−0.520	−0.0700
C(5)	0.1583	−0.260	−0.1933
C(6)	0.0900	−0.460	−0.1833
C(7)	0.1017	−0.455	0.2633
N	0.1900	−0.193	−0.2933
O(1)	0.0600	−0.500	0.1450
O(2)	0.1500	−0.273	−0.4000
O(3)	0.2667	0.000	−0.0833

The structure (Fig. XVAIV,9) is composed of molecules having the bond dimensions of Figure XVAIV,10. Held together by hydrogen bonds between quinoid and nitroso oxygens, with O–H–O = 2.63 A., they form chains running along the c_0 axis. Between the chains the atomic separations are 3.41 A. and more.

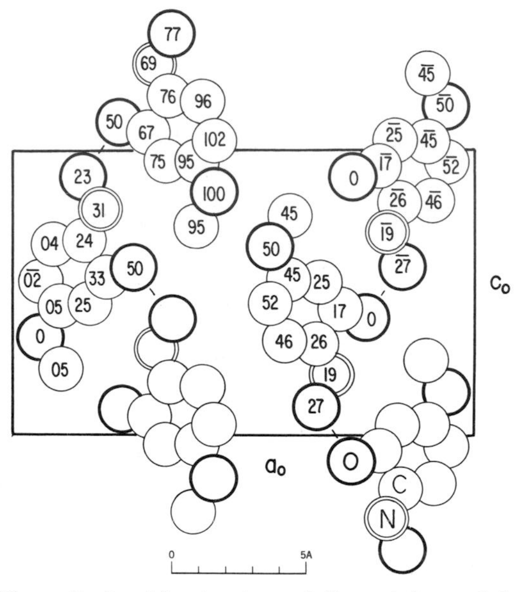

Fig. XVAIV,9. The orthorhombic structure of the red form of 5-methoxy-2-nitrosophenol projected along its b_0 axis.

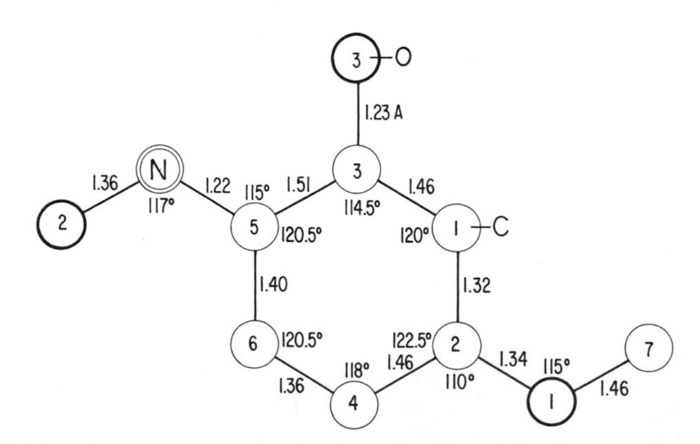

Fig. XVAIV,10. Bond dimensions in the molecule of 5-methoxy-2-nitrosophenol.

XV,aIV,8. The monoclinic crystals of *2-chloro-5-nitrobenzoic acid*, $ClC_6H_3(NO_2)COOH$, possess a tetramolecular unit of the dimensions:

$$a_0 = 5.86(2) \text{ A.}; \quad b_0 = 5.13(2) \text{ A.}; \quad c_0 = 26.65(5) \text{ A.}$$
$$\beta = 97°54'$$

The space group is C_{2h}^5 ($P2_1/c$) with atoms in the positions:

$$(4e) \quad \pm(xyz;\ x,{}^1/_2-y,z+{}^1/_2)$$

The parameters are those of Table XVAIV,8.

TABLE XVAIV,8

Parameters of the Atoms in $ClC_6H_3(NO_2)COOH$

Atom	x	y	z
C(1)	−0.2551	0.1807	0.1006
C(2)	−0.4479	0.0736	0.1162
C(3)	−0.5264	0.1526	0.1598
C(4)	−0.4142	0.3503	0.1889
C(5)	−0.2209	0.4581	0.1734
C(6)	−0.1431	0.3791	0.1300
C(7)	−0.1520	0.0996	0.0550
O(1)	−0.1858	−0.1173	0.0367
O(2)	−0.0226	0.2753	0.0384
O(3)	0.0806	0.7441	0.1022
O(4)	−0.1861	0.7523	0.2391
N	−0.1002	0.6646	0.2034
Cl	−0.6040	−0.1761	0.0822

The structure, shown in Figure XVAIV,11, has molecules with the bond dimensions of Figure XVAIV,12. In a molecule the benzene ring and its attached chlorine and nitrogen atoms are coplanar but the carboxyl C(7) is 0.05 A. outside this plane. The NO_2 group is twisted 7° and the CO_2 group 23° with respect to this plane. Pairs of molecules are tied together by hydrogen bonds between carboxyl groups (O–H–O = 2.613 A.). The nitro groups of neighboring molecules also face one another, with N–O = 2.988 A. and O–O = 3.128 and 3.198 A.

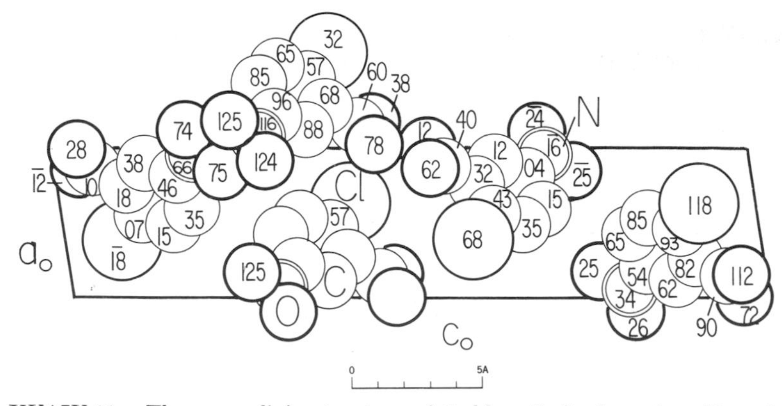

Fig. XVAIV,11. The monoclinic structure of 2-chloro-5-nitrobenzoic acid projected along its b_0 axis.

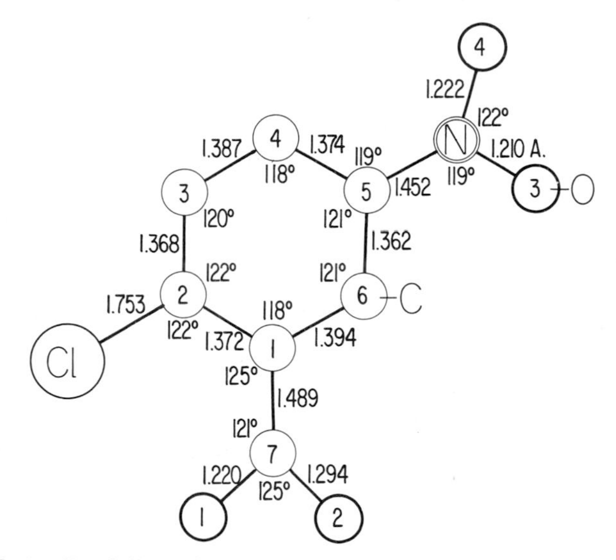

Fig. XVAIV,12. Bond dimensions in the molecule of 2-chloro-5-nitrobenzoic acid.

XV,aIV,9. The tetramolecular unit of the monoclinic crystals of *2,6-dimethyl benzoic acid*, $C_6H_3(CH_3)_2COOH$, has the dimensions:

$$a_0 = 15.24(1) \text{ A.}; \quad b_0 = 4.04(1) \text{ A.}; \quad c_0 = 13.16(1) \text{ A.}$$
$$\beta = 94°8'$$

The space group is C_{2h}^5 ($P2_1/a$) with atoms in the positions:

$$(4e) \qquad \pm(xyz;\ x+{}^1/_2,{}^1/_2-y,z)$$

The parameters are listed in Table XVAIV,9.

TABLE XVAIV,9

Parameters of the Atoms in $C_6H_3(CH_3)_2COOH$

Atom	x	y	z
C(1)	0.4025(4)	−0.0302(16)	0.2289(4)
C(2)	0.4530(4)	−0.1414(16)	0.3156(4)
C(3)	0.4129(4)	−0.1505(19)	0.4079(5)
C(4)	0.3265(5)	−0.0455(21)	0.4161(5)
C(5)	0.2791(4)	0.0699(20)	0.3287(6)
C(6)	0.3155(4)	0.0736(18)	0.2344(5)
C(7)	0.5471(4)	−0.2644(19)	0.3142(5)
C(8)	0.4455(3)	−0.0155(17)	0.1278(4)
C(9)	0.2595(5)	0.2080(23)	0.1422(6)
O(1)	0.5163(3)	0.1251(16)	0.1223(3)
O(2)	0.4054(3)	−0.1647(16)	0.0527(3)

The structure (Fig. XVAIV,13) is composed of molecules which, except for the oxygen atoms, are planar and have the bond dimensions of Figure XVAIV,14. Their carboxyl radicals are turned by 53°31′ about the bond C(1)–C(8) to put oxygens 0.885 A. on either side of the molecular plane.

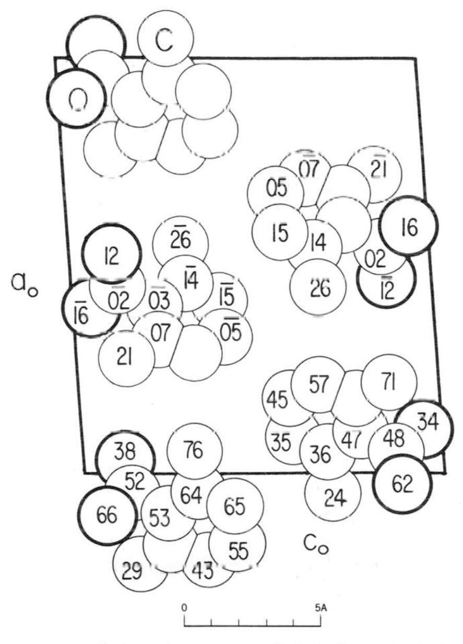

Fig. XVAIV,13. The monoclinic structure of 2,6-dimethyl benzoic acid projected along its b_0 axis.

In the crystal the molecules are joined into pairs by O–H–O bonds of length 2.67 A. The next longer intermolecular distances are O–O = 3.376 A., C–O = 3.412 A., and C–C = 3.670 A.

XV,aIV,10. The monoclinic crystals of *2-amino-3-methyl benzoic acid*, $NH_2CH_3C_6H_3COOH$, have a tetramolecular unit of the dimensions:

$$a_0 = 11.4803(17) \text{ A.}; \quad b_0 = 4.0445(14) \text{ A.}; \quad c_0 = 15.7923(36) \text{ A.}$$
$$\beta = 91°18(2)'$$

Atoms are in the positions

$$(4e) \quad \pm(xyz;\ x,{}^1/_2-y,z+{}^1/_2)$$

of C_{2h}^5 ($P2_1/c$). The parameters are those of Table XVAIV,10.

TABLE XVAIV,10

Parameters of the Atoms in $NH_2CH_3C_6H_3COOH$

Atom	x	y	z
C(1)	0.2859(1)	0.5817(3)	0.1014(1)
C(2)	0.2901(1)	0.5197(3)	0.1896(1)
C(3)	0.1955(1)	0.3511(3)	0.2267(1)
C(4)	0.1029(1)	0.2507(4)	0.1758(1)
C(5)	0.0980(1)	0.3118(4)	0.0889(1)
C(6)	0.1889(1)	0.4749(4)	0.0526(1)
C(7)	0.3799(1)	0.7579(4)	0.0594(1)
C(8)	0.1976(2)	0.2833(4)	0.3203(1)
N	0.3825(1)	0.6125(4)	0.2402(1)
O(1)	0.3658(1)	0.7879(4)	−0.0234(1)
O(2)	0.4665(1)	0.8745(3)	0.0963(1)
H(1)	0.032	0.129	0.203
H(2)	0.032	0.239	0.051
H(3)	0.189	0.515	−0.009
H(4)	0.374	0.582	0.297
H(5)	0.438	0.735	0.219
H(6)	0.198	0.480	0.355
H(7)	0.131	0.149	0.337
H(8)	0.268	0.143	0.337
H(9)	0.430	0.899	−0.044

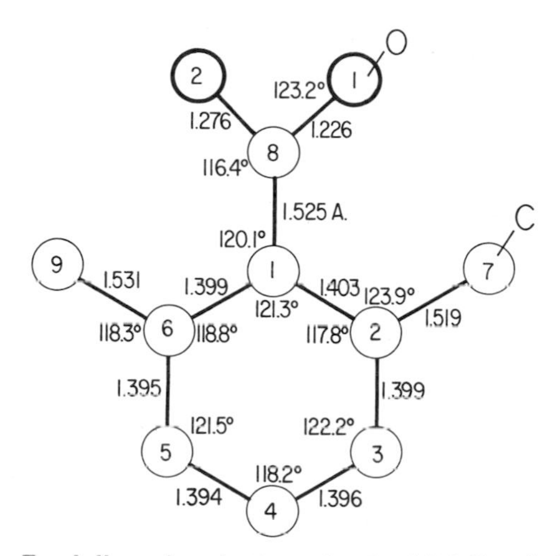

Fig. XVAIV,14. Bond dimensions in the molecule of 2,6-dimethyl benzoic acid.

Molecules in the resulting structure (Fig. XVAIV,15) have the bond dimensions of Figure XVAIV,16. They are very nearly planar, the benzene and methyl carbons being strictly coplanar. Other atoms, except the methyl hydrogens, are within 0.07 A. of this plane. Carboxyl oxygens of adjacent molecules are hydrogen-bonded to one another, with O–H–O = 2.646 A. There is considered to be an internal hydrogen bond N–H–O = 2.706 A.; the other amino hydrogen atom does not form bonds with neighboring molecules.

XV,aIV,11. *Propargyl-2-bromo-3-nitrobenzoate,* $C_6H_3(NO_2)BrCo_2CH_2$-C≡CH, forms monoclinic crystals whose bimolecular unit has the dimensions:

$$a_0 = 4.07 \text{ A.}; \quad b_0 = 17.99 \text{ A.}; \quad c_0 = 7.24 \text{ A.}; \quad \beta = 91°30'$$

Atoms are in the positions

$$(2a) \quad xyz;\ \bar{x},y+{}^1/_2,\bar{z}$$

of the space group C_2^2 ($P2_1$). The determined parameters are those of Table XVAIV,11; calculated hydrogen parameters are given in the original.

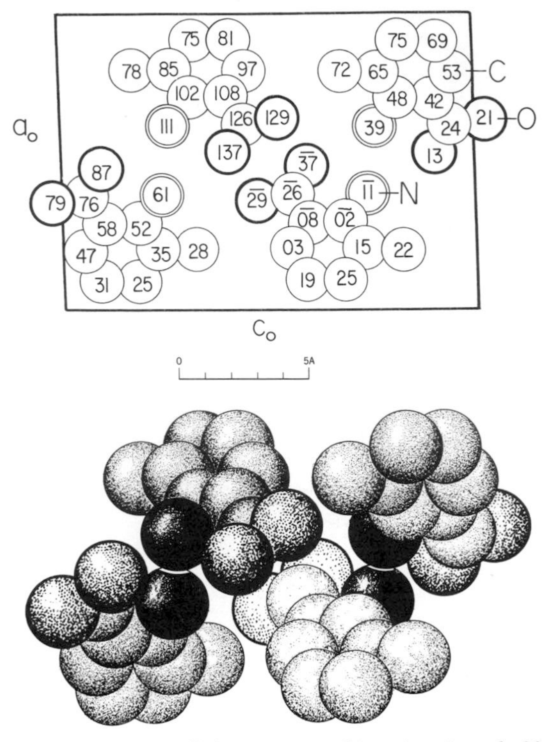

Fig. XVAIV,15a (top). The monoclinic structure of 2-amino-3-methyl benzoic acid projected along its b_0 axis.

Fig. XVAIV,15b (bottom). A packing drawing of the monoclinic structure of 2-amino-3-methyl benzoic acid viewed along its b_0 axis. The nitrogen atoms are black; the oxygens are heavily outlined and coarsely dotted. Atoms of carbon are finely dot shaded.

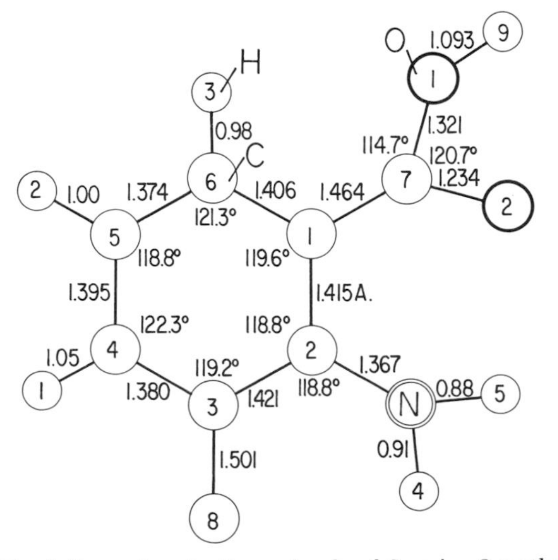

Fig. XVAIV,16. Bond dimensions in the molecule of 2-amino-3-methyl benzoic acid.

TABLE XVAIV,11

Parameters of the Atoms in Propargyl-2-bromo-3-nitrobenzoate

Atom	x	y	z
C(1)	0.4939	0.2413	0.6970
C(2)	0.4177	0.2028	0.5340
C(3)	0.5008	0.1291	0.5307
C(4)	0.6529	0.0924	0.6738
C(5)	0.7300	0.1335	0.8335
C(6)	0.6415	0.2066	0.8445
C(7)	0.4147	0.3238	0.7127
O(8)	0.2545	0.3369	0.8647
O(9)	0.4976	0.3710	0.6103
C(10)	0.1682	0.4142	0.8994
C(11)	0.0000	0.4163	0.0716
C(12)	−0.1371	0.4187	0.2127
N(13)	0.4127	0.0840	0.3649
O(14)	0.2322	0.0326	0.3916
O(15)	0.5189	0.1009	0.2227
Br	0.2085	0.2500	0.3336

The structure (Fig. XVAIV,17) is composed of molecules having the bond dimensions of Figure XVAIV,18. The shortest intermolecular separations are C–O = 3.02 and 3.11 A.

XV,aIV,12. *Noradrenaline hydrochloride*, $C_8H_{11}O_3N \cdot HCl$, is orthorhombic with a tetramolecular unit of the edge lengths:

$$a_0 = 8.580 \text{ A.}; \quad b_0 = 19.120 \text{ A.}; \quad c_0 = 5.775 \text{ A.}$$

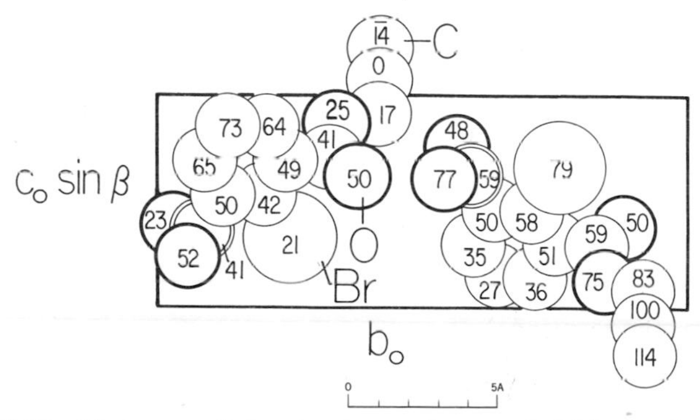

Fig. XVAIV,17. The monoclinic structure of propargyl-2-bromo-3-nitrobenzoate projected along its a_0 axis. The nitrogen atoms are doubly ringed.

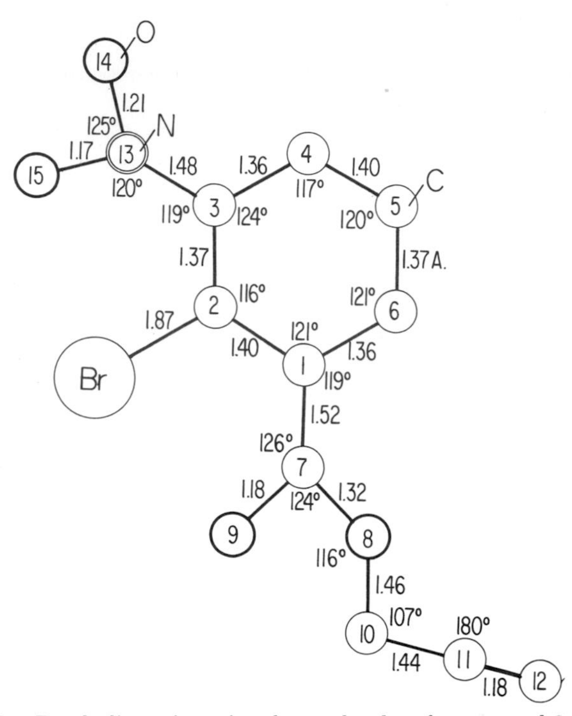

Fig. XVAIV,18. Bond dimensions in the molecule of propargyl-2-bromo-3-nitrobenzoate.

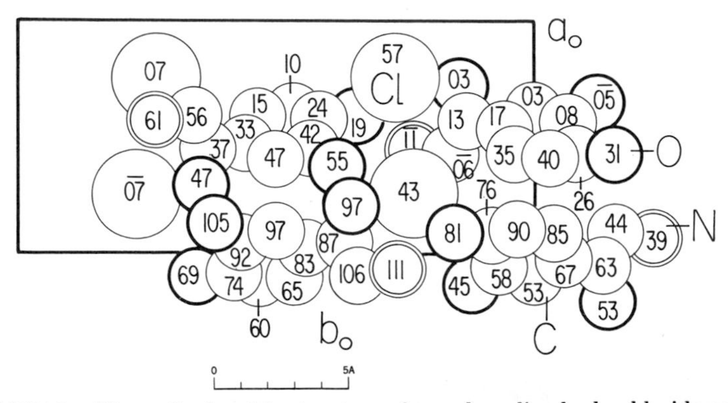

Fig. XVAIV,19. The orthorhombic structure of noradrenaline hydrochloride projected along its c_0 axis.

Atoms are in the positions

$$(4a) \quad xyz;\ {}^1/_2-x,\bar{y},z+{}^1/_2;\ x+{}^1/_2,{}^1/_2-y,\bar{z};\ \bar{x},y+{}^1/_2,{}^1/_2-z$$

of V^4 ($P2_12_12_1$). Their parameters are stated in Table XVAIV,12.

TABLE XVAIV,12
Parameters of the Atoms in Noradrenaline Hydrochloride

Atom	x	y	z
C(1)	0.5614(9)	0.5359(4)	0.1456(14)
C(2)	0.5988(9)	0.4666(4)	0.0987(15)
C(3)	0.5359(8)	0.4143(3)	0.2357(13)
C(4)	0.4361(9)	0.4318(3)	0.4165(13)
C(5)	0.4044(8)	0.5022(3)	0.4653(14)
C(6)	0.4675(8)	0.5545(3)	0.3254(12)
C(7)	0.4371(9)	0.6317(3)	0.3741(13)
C(8)	0.5494(8)	0.6548(3)	0.5630(16)
N	0.5335(7)	0.7314(3)	0.6083(11)
O(1)	0.5649(6)	0.3441(2)	0.1907(10)
O(2)	0.3702(8)	0.3810(3)	0.5521(12)
O(3)	0.2839(5)	0.6431(2)	0.4656(10)
Cl	0.7577(2)	0.2687(1)	0.5685(3)
H(1)	0.605(11)	0.569(5)	0.013(18)
H(2)	0.658(11)	0.446(4)	−0.054(18)
H(3)	0.601(11)	0.332(5)	0.314(18)
H(4)	0.318(13)	0.383(5)	0.714(20)
H(5)	0.336(10)	0.513(5)	0.616(18)
H(6)	0.458(11)	0.663(4)	0.226(17)
H(7)	0.219(11)	0.687(4)	0.562(17)
H(8)	0.496(12)	0.630(5)	0.748(19)
H(9)	0.650(11)	0.637(5)	0.519(19)
H(10)	0.448(11)	0.745(5)	0.706(17)
H(11)	0.607(11)	0.746(5)	0.695(17)
H(12)	0.556(11)	0.756(5)	0.461(18)

The structure (Fig. XVAIV,19) is built up of molecules possessing the bond dimensions of Figure XVAIV,20. In them the catechol part of the

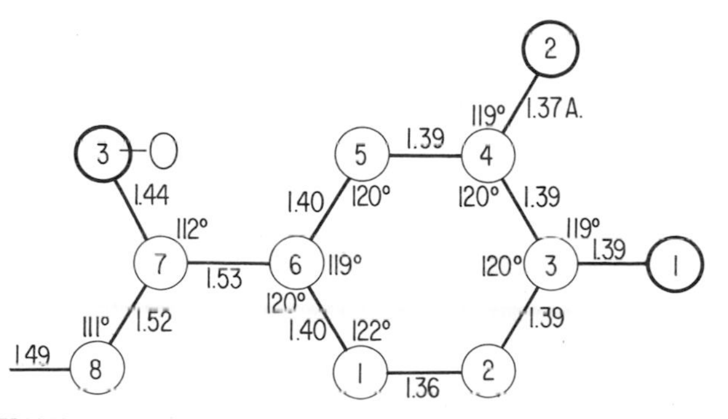

Fig. XVAIV,20. Bond dimensions in the molecule of noradrenaline.

molecule is planar. They are interconnected by hydrogen bonds O–H–O = 2.77 A. and N–H–O = 2.89 A. as well as by attractions to the chlorine atom. The atomic separations involving chlorine range upwards from an O–Cl = 3.10 A.

XV,aIV,13. Crystals of *p-amino salicylic acid*, $C_6H_3NH_2OHCOOH$, are monoclinic with a tetramolecular unit of the dimensions:

$$a_0 = 7.28 \text{ A.}; \quad b_0 = 3.82 \text{ A.}; \quad c_0 = 25.33 \text{ A.}; \quad \beta = 103°$$

All atoms are in general positions of C_{2h}^5 ($P2_1/c$):

$$(4e) \quad \pm(xyz;\ x,{}^1/_2-y,z+{}^1/_2)$$

Their parameters are given in Table XVAIV,13.

TABLE XVAIV,13

Parameters of the Atoms in *p*-Amino Salicylic Acid

Atom	x	y	z
C(1)	0.6343	0.329	0.0932
C(2)	0.6473	0.407	0.1485
C(3)	0.4960	0.543	0.1855
C(4)	0.3204	0.617	0.1742
C(5)	0.3020	0.542	0.1197
C(6)	0.4547	0.404	0.0821
C(7)	0.7892	0.190	0.0554
O(1)	0.8130	0.346	0.1642
O(2)	0.9412	0.125	0.0547
O(3)	0.7547	0.131	0.0060
N	0.1665	0.754	0.2112

The resulting structure is shown in Figure XVAIV,21. Molecules having the bond dimensions of Figure XVAIV,22 are held together by hydrogen bonds involving acidic groups only—there is no hydrogen bond from the amino groups. Atom O(1) of the carboxyl group is considered to be hydrogen-bonded to the phenolic oxygen of the same molecule and to a carboxylic O(2) of an adjacent molecule. These O(2) atoms have only this one hydrogen linkage.

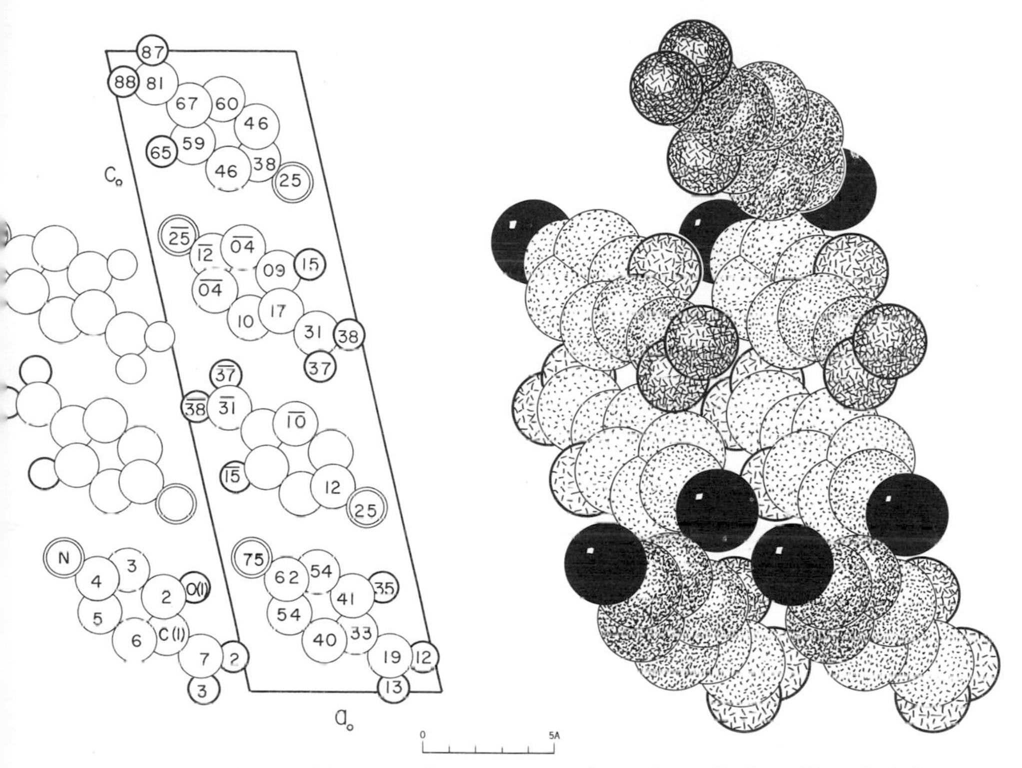

Fig. XVAIV,21a (left). The monoclinic structure of p-amino salicylic acid projected along its b_0 axis. Left hand axes.

Fig. XVAIV,21b (right). A packing drawing of the monoclinic structure of p-amino salicylic acid viewed along its b_0 axis. The nitrogen atoms are black; the oxygens are heavily outlined and line shaded. The carbon atoms are dot shaded.

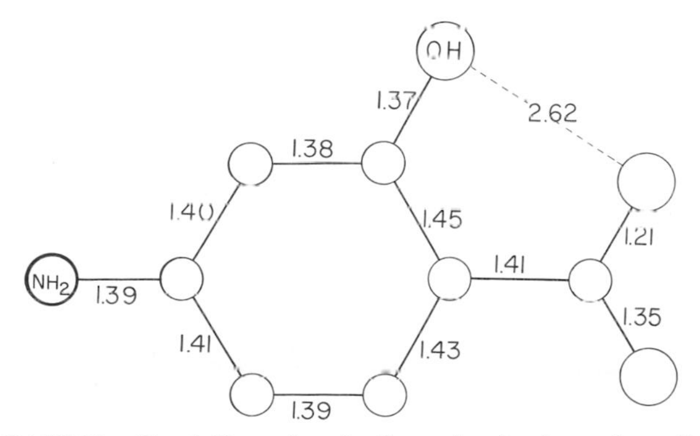

Fig. XVAIV,22. Bond dimensions in the molecule of p-amino salicylic acid.

XV,aIV,14. The monoclinic crystals of *5-chlorosalicylaldoxime,* C_6H_3-Cl(OH)CHNOH, have a tetramolecular unit of the dimensions:

$$a_0 = 14.35 \text{ A.}; \quad b_0 = 3.90 \text{ A.}; \quad c_0 = 13.69 \text{ A.}; \quad \beta = 100°0'$$

The space group is C_{2h}^5 ($P2_1/c$) with atoms in the positions:

$$(4e) \quad \pm(xyz;\ x,{}^1/_2-y,z+{}^1/_2)$$

The assigned parameters are those of Table XVAIV,14.

TABLE XVAIV,14

Parameters of the Atoms in 5-Chlorosalicylaldoxime

Atom	x	y	z
C(1)	0.3146	0.2002	0.3082
C(2)	0.2832	0.0482	0.3875
C(3)	0.1916	−0.0914	0.3813
C(4)	0.1289	−0.0417	0.2893
C(5)	0.1613	0.1026	0.2099
C(6)	0.2515	0.2407	0.2167
C(7)	0.4103	0.3421	0.3113
O(1)	0.3428	−0.0012	0.4736
O(2)	0.5591	0.4344	0.3774
N	0.4713	0.3032	0.3863
Cl	0.0819	0.1535	0.0958

The structure appears in Figure XVAIV,23. Its molecule is planar except for the nitrogen and O(2) atoms which are shifted by such a rotation about

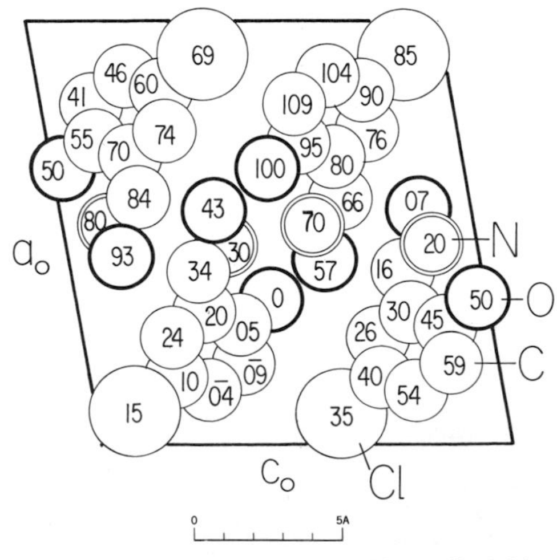

Fig. XVAIV,23. The monoclinic structure of 5-chlorosalicylaldoxime projected along its b_0 axis.

C(1)–C(7) that nitrogen is 0.09 A. and O(2) is 0.13 A. outside of the plane. There is an O(1)–O(2) separation of 2.83 A. between molecules suggestive of a hydrogen bond.

XV,aIV,15. Crystals of *bis(5-chlorosalicylaldoximato)copper(II)*, $[OC_6H_3(Cl)CHNOH]_2Cu$, are triclinic. Their unimolecular cell has the dimensions:

$$a_0 = 13.38 \text{ A.}; \quad b_0 = 3.84 \text{ A.}; \quad c_0 = 7.31 \text{ A.}$$
$$\alpha = 81°30'; \quad \beta = 98°3'; \quad \gamma = 98°3'$$

The copper atom is in the origin $(1a)$ 000 of C_i^1 $(P\bar{1})$; all other atoms are in the general positions $(2i)$ $\pm(xyz)$, their parameters being those of Table XVAIV,15. The values there stated for hydrogen are based on the assumption that C H = 1.04 A.

TABLE XVAIV,15

Parameters of Atoms in $[OC_6H_3(Cl)CHNOH]_2Cu$

Atom	x	y	z
Cl	0.5101	0.8156	0.2634
O(1)	0.0741	0.3041	0.1695
O(2)	0.0079	0.0404	−0.3347
N	0.1090	0.0900	−0.1601
C(1)	0.2352	0.3883	0.0419
C(2)	0.1731	0.4130	0.1867
C(3)	0.2175	0.5665	0.3452
C(4)	0.3195	0.6894	0.3679
C(5)	0.3797	0.6552	0.2318
C(6)	0.3389	0.5144	0.0737
C(7)	0.1982	0.2355	−0.1241
H(1)	0.173	0.589	0.449
H(2)	0.352	0.812	0.484
H(3)	0.386	0.499	−0.029
H(4)	0.249	0.249	−0.222
H(5)	0.024	−0.143	−0.270

The structure, shown in Figure XVAIV,24, has molecules with the bond dimensions of Figure XVAIV,25. The benzene ring and its attached chlorine, O(1) and C(7) atoms are coplanar but nitrogen departs 0.10 A. and O(2) 0.08 A. from this plane. The copper atom is 0.38 A. outside the plane and therefore the planes of the two chelating groups are displaced

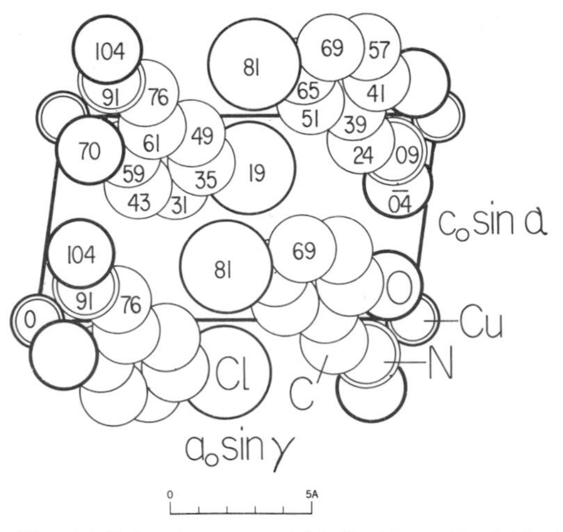

Fig. XVAIV,24. The triclinic structure of bis(5-chlorosalicylaldoximato)copper projected along its b_0 axis.

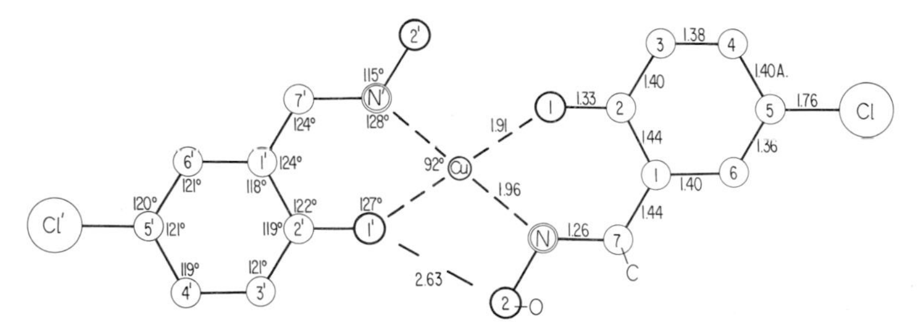

Fig. XVAIV,25. Bond dimensions in the molecules chelated to a copper atom in bis(5-chlorosalicylaldoximato)copper.

0.76 A. with respect to one another. The slightly rectangular coordination of copper is supplied by nitrogen and O(1) atoms. There are also two more distant neighbors, with Cu–O(1) = 3.01 A.; the O(2)–Cu–O(2′) line makes 80° with the Cu–O(1) bond. All other intermolecular distances exceed 3.40 A.

XV,aIV,16. Crystals of *bis(N-isopropyl-3-methylsalicylaldiminato)nickel*, $[OC_6H_3(CH_3)CHNC_3H_7]_2Ni$, are monoclinic with a bimolecular unit of the dimensions:

$$a_0 = 11.209(2) \text{ A.}; \quad b_0 = 9.979(4) \text{ A.}; \quad c_0 = 9.520(4) \text{ A.}$$
$$\beta = 107°31(4)'$$

The nickel atoms are in the positions:

$$(2a) \quad 000;\ 0\ {}^1/_2\ {}^1/_2$$

of C_{2h}^5 ($P2_1/c$). All other atoms are in:

$$(4e) \quad \pm(xyz;\ x,{}^1/_2-y,z+{}^1/_2)$$

with the parameters of Table XVAIV,16.

TABLE XVAIV,16

Parameters of Atoms in Bis(N-isopropyl-3-methylsalicylaldiminato)nickel

Atom	x	y	z
O	0.1685(2)	0.4917(2)	0.5967(2)
N	−0.0349(2)	0.5779(3)	0.6678(2)
C(1)	0.1804(3)	0.6381(3)	0.7982(3)
C(2)	0.2355(3)	0.5577(3)	0.7125(3)
C(3)	0.3672(3)	0.5479(4)	0.7538(3)
C(4)	0.4381(3)	0.6200(4)	0.8737(4)
C(5)	0.3842(4)	0.7014(4)	0.9561(4)
C(6)	0.2577(4)	0.7100(4)	0.9123(4)
C(7)	0.0491(3)	0.6362(3)	0.7728(3)
C(8)	−0.1645(3)	0.5797(4)	0.6832(3)
C(9)	−0.1648(4)	0.5174(7)	0.8305(5)
C(10)	−0.2203(5)	0.7168(6)	0.6557(7)
C(11)	0.4287(3)	0.4589(6)	0.6692(5)
H(4)	0.527(5)	0.606(5)	0.900(5)
H(5)	0.444(5)	0.754(6)	0.042(6)
H(6)	0.210(4)	0.761(5)	0.973(5)
H(7)	0.028(3)	0.691(4)	0.849(4)
H(8)	−0.215(4)	0.518(4)	0.612(5)
H(9,1)	−0.248(7)	0.504(5)	0.820(8)
H(9,2)	−0.109(8)	0.431(9)	0.825(9)
H(9,3)	−0.132(7)	0.591(8)	0.894(8)
H(10,1)	−0.306(8)	0.715(8)	0.658(8)
H(10,2)	−0.177(6)	0.781(7)	0.738(7)
H(10,3)	−0.194(8)	0.758(9)	0.558(9)
H(11,1)	0.412(6)	0.482(5)	0.559(8)
H(11,2)	0.518(6)	0.446(6)	0.699(7)
H(11,3)	0.398(5)	0.372(7)	0.655(6)

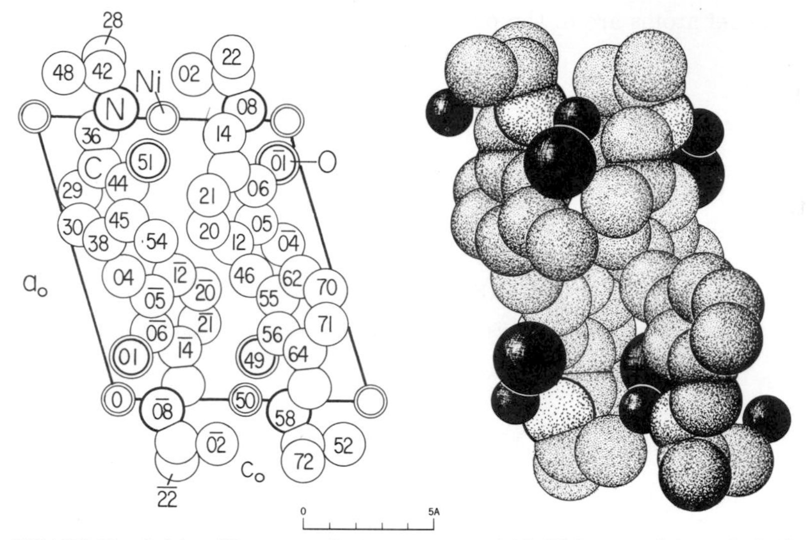

Fig. XVAIV,26a (left). The monoclinic structure of bis(*N*-isopropyl-3-methylsalicylaldiminato)nickel projected along its b_0 axis.

Fig. XVAIV,26b (right). A packing drawing of the monoclinic structure of bis(*N*-isopropyl-3-methylsalicylaldiminato)nickel seen along its b_0 axis. The nickel atoms are the small circles; the oxygen atoms are the large black circles. The nitrogen atoms are heavily outlined and coarsely dotted; the atoms of carbon are finely dotted.

The resulting structure (Fig. XVAIV,26) has molecules with the bond dimensions of Figure XVAIV,27. Coordination about the nickel atom in this case is square. The salicylaldimine part of the molecule is nearly planar, the benzene ring itself being planar to about 0.01 A. The isopropyl radical is inclined about 79° to the rest. The plane defined by the nickel atom and its chelated nitrogen and oxygen atoms and that of the salicylaldimine portion of the molecule make an angle of 16.8° with one another. The closest intermolecular distance is a C–C = 3.382 A.

The palladium compound, *bis(N-isopropyl-3-methylsalicylaldiminato)-palladium*, has two known modifications, one of which is isostructural with the above. Its unit has the dimensions:

$$a_0 = 11.297(1) \text{ A.}; \quad b_0 = 9.755(1) \text{ A.}; \quad c_0 = 9.844(1) \text{ A.}$$
$$\beta = 107°51'$$

The parameters are given in Table XVAIV,17. The bond dimensions of its molecule are those of Figure XVAIV,28. The structure of the second, tetragonal, form is described in paragraph **XV,aIV,17.**

TABLE XVAIV,17

Parameters of Atoms in Monoclinic Bis(*N*-isopropyl-3-methylsalicylaldiminato)palladium

Atom	*x*	*y*	*z*
O	0.1827(4)	0.5002(11)	0.5955(5)
N	−0.0310(7)	0.5797(9)	0.6754(8)
C(1)	0.1876(9)	0.6327(10)	0.7998(10)
C(2)	0.2446(8)	0.5633(9)	0.7133(9)
C(3)	0.3749(9)	0.5512(9)	0.7531(11)
C(4)	0.4432(12)	0.6207(14)	0.8735(13)
C(5)	0.3901(14)	0.6933(16)	0.9563(15)
C(6)	0.2635(13)	0.7025(13)	0.9230(13)
C(7)	0.0559(10)	0.6360(11)	0.7778(11)
C(8)	−0.1600(10)	0.5846(14)	0.6867(10)
C(9)	−0.1662(9)	0.5321(46)	0.8231(21)
C(10)	−0.2128(16)	0.7295(21)	0.6568(23)
C(11)	0.4357(12)	0.4685(15)	0.6641(15)
H(4)	0.519(9)	0.614(9)	0.878(9)
H(5)	0.431(12)	0.721(14)	0.029(13)
H(6)	0.229(9)	0.738(10)	0.983(9)
H(7)	0.034(8)	0.676(9)	0.860(10)
H(8)	0.196(7)	0.497(13)	0.604(9)
H(9,1)	−0.251(11)	0.535(11)	0.832(11)
H(9,2)	−0.132(10)	0.479(16)	0.860(12)
H(9,3)	−0.110(13)	0.596(16)	0.907(15)
H(10,1)	−0.284(12)	0.736(14)	0.670(13)
H(10,2)	−0.159(7)	0.785(7)	0.713(6)
H(10,3)	−0.172(11)	0.745(12)	0.565(12)
H(11,1)	0.414(11)	0.539(13)	0.568(14)
H(11,2)	0.550(19)	0.466(22)	0.702(19)
H(11,3)	0.405(8)	0.363(11)	0.666(9)

XV,aIV,17. The tetragonal form of *bis(N-isopropyl-3-methylsalicylaldiminato)palladium*, $[OC_6H_3(CH_3)CHNC_3H_7]_2Pd$, has a unit containing eight molecules. Its edge lengths are:

$$a_0 = 18.178(1) \text{ A.}, \qquad c_0 = 12.879 \text{ A.}$$

The space group is C_{4h}^6 ($I4_1/a$) with the palladium atoms in:

(8*c*) 000; 0 ½ 0; ¼ ¼ ¼; ¾ ¼ ¼; B.C.

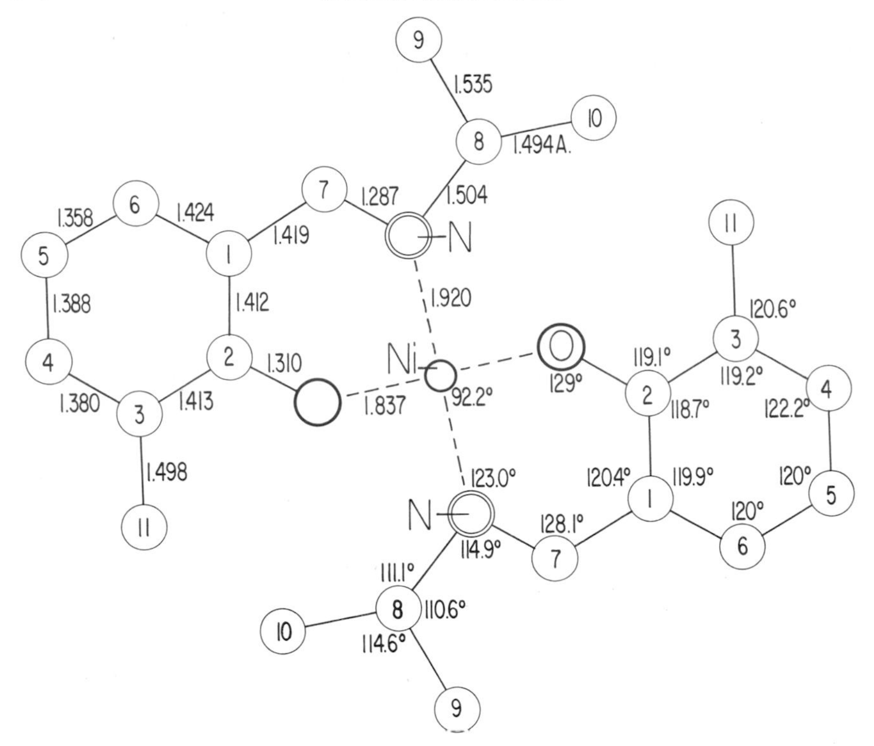

Fig. XVAIV,27. Bond dimensions in the two molecules that chelate with the nickel atom in bis(*N*-isopropyl-3-methylsalicylaldiminato)nickel.

All other atoms are in the general positions:

(16*f*) $\pm(xyz;\ x,y+{}^1/_2,\bar{z};\ y+{}^1/_4,{}^3/_4-x,{}^3/_4-z;\ y+{}^1/_4,{}^1/_4-x,z+{}^1/_4)$; B.C.

Their parameters are listed in Table XVAIV, 18.

TABLE XVAIV,18

Parameters of the Atoms in the Tetragonal Form of Bis(*N*-isopropyl-3-methylsalicylaldiminato)palladium

Atom	x	y	z
O	0.2927(1)	0.2557(2)	0.1081(2)
N	0.1821(2)	0.3362(2)	0.2197(3)
C(1)	0.2096(2)	0.3453(3)	0.0343(3)
C(2)	0.2687(2)	0.2940(2)	0.0291(3)
C(3)	0.3041(2)	0.2838(2)	−0.0673(3)
C(4)	0.2798(3)	0.3205(3)	−0.1530(4)
C(5)	0.2207(3)	0.3696(3)	−0.1494(4)
C(6)	0.1871(3)	0.3810(3)	−0.0567(4)
C(7)	0.1731(2)	0.3631(2)	0.1279(3)
C(8)	0.1380(3)	0.3700(3)	0.3046(4)

(*continued*)

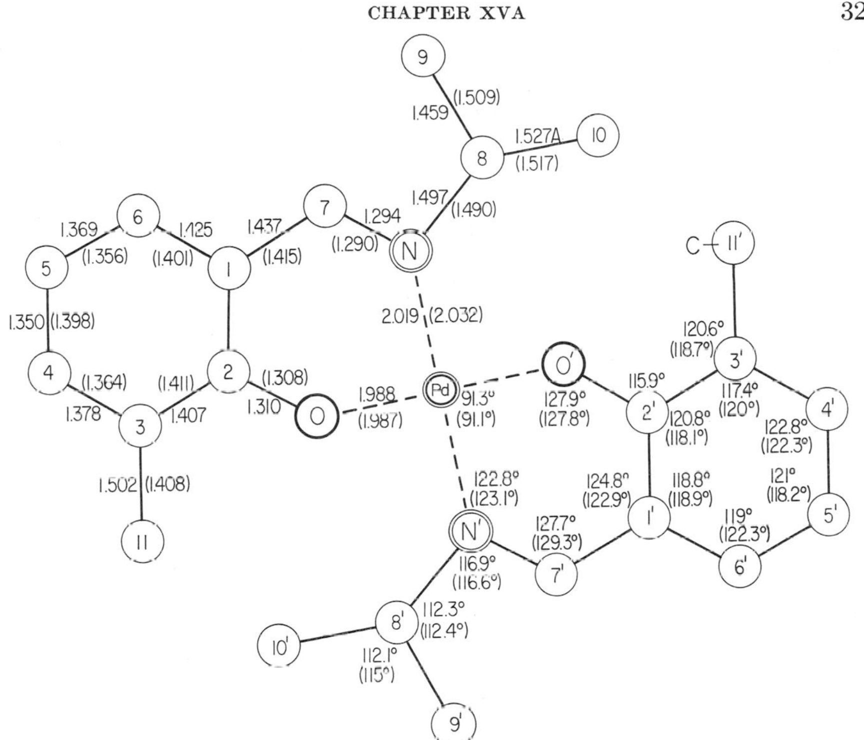

Fig. XVAIV,28. Bond dimensions in the molecule of bis(*N*-isopropyl-3-methylsalicylaldiminato)palladium. The dimensions of these molecules in the tetragonal form (**XV,aIV,17**) are given in parentheses.

TABLE XVAIV,18 (*continued*)

Atom	x	y	z
C(9)	0.1452(5)	0.4527(4)	0.3072(6)
C(10)	0.0606(5)	0.3419(7)	0.2969(9)
C(11)	0.3682(3)	0.2313(4)	−0.0732(5)
H(4)	0.296(2)	0.312(2)	−0.213(3)
H(5)	0.203(3)	0.396(3)	−0.203(4)
H(6)	0.147(3)	0.416(3)	−0.049(4)
H(7)	0.138(2)	0.399(2)	0.122(3)
H(8)	0.157(2)	0.351(2)	0.364(3)
H(9,1)	0.127(4)	0.470(3)	0.372(5)
H(9,2)	0.206(4)	0.470(4)	0.306(6)
H(9,3)	0.125(4)	0.472(4)	0.253(6)
H(10,1)	0.043(3)	0.354(3)	0.335(4)
H(10,2)	0.032(4)	0.351(4)	0.226(6)
H(10,3)	0.051(4)	0.288(5)	0.274(8)
H(11,1)	0.391(3)	0.235(3)	−0.136(4)
H(11,2)	0.415(3)	0.249(3)	−0.019(5)
H(11,3)	0.355(2)	0.184(2)	−0.056(4)

As can be seen from Figure XVAIV,28, the molecules of this form have substantially the same bond dimensions as in the monoclinic modification (**XV,aIV,16**).

XV,aIV,18. Crystals of *bis(N-isopropyl-3-ethylsalicylaldiminato)nickel*, $[OC_6H_3(C_2H_5)CHNC_3H_7]_2Ni$, are orthorhombic with a unit that contains eight molecules and has the edge lengths:

$$a_0 = 26.947(2) \text{ A.}; \quad b_0 = 19.883(1) \text{ A.}; \quad c_0 = 8.820(1) \text{ A.}$$

Atoms are in the positions

$$(8c) \quad \pm(xyz;\ {}^1/_2-x,y+{}^1/_2,z;\ x,{}^1/_2-y,z+{}^1/_2;\ x+{}^1/_2,y,{}^1/_2-z)$$

of V_h^{15} (*Pbca*). Parameters are listed in Table XVAIV,19.

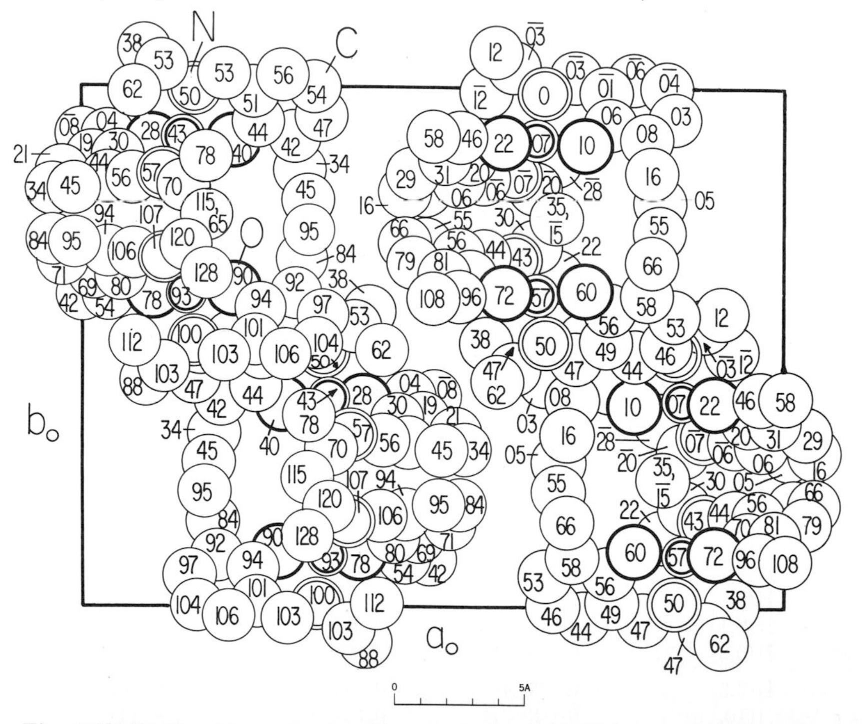

Fig. XVAIV,29. The orthorhombic structure of bis(*N*-isopropyl-3-ethylsalicylaldiminato)nickel projected along its c_0 axis. The nickel atoms are the smaller doubly ringed circles.

TABLE XVAIV,19

Parameters of the Atoms in $(C_{12}H_{16}NO)_2Ni$

Atom	x	y	z
Ni	0.3529(1)	0.3945(1)	0.4272(1)
O(1)	0.4048(1)	0.3850(2)	0.2842(5)
O(2)	0.2848(1)	0.3805(2)	0.3976(5)
N(1)	0.3839(2)	0.3323(3)	0.5748(7)
N(2)	0.3452(2)	0.4867(3)	0.5039(6)
C(1)	0.4628(2)	0.3244(3)	0.4375(8)
C(2)	0.4487(2)	0.3586(3)	0.3040(8)
C(3)	0.4840(2)	0.3637(3)	0.1865(9)
C(4)	0.5311(3)	0.3390(4)	0.2093(10)
C(5)	0.5448(3)	0.3073(4)	0.3416(11)
C(6)	0.5119(3)	0.2994(4)	0.4518(10)
C(7)	0.4294(3)	0.3121(4)	0.5615(10)
C(8)	0.3552(3)	0.3053(4)	0.7057(9)
C(9)	0.3233(4)	0.2483(8)	0.6520(15)
C(10)	0.3253(5)	0.3584(7)	0.7779(15)
C(11)	0.4677(3)	0.3960(6)	0.0394(12)
C(12)	0.5028(6)	0.3957(8)	−0.0836(20)
C(2,1)	0.2543(3)	0.4826(3)	0.5133(7)
C(2,2)	0.2480(2)	0.4189(3)	0.4418(8)
C(2,3)	0.1985(2)	0.3968(4)	0.4192(9)
C(2,4)	0.1595(3)	0.4376(5)	0.4707(11)
C(2,5)	0.1688(3)	0.4974(5)	0.5443(10)
C(2,6)	0.2135(3)	0.5182(4)	0.5651(10)
C(2,7)	0.3030(3)	0.5112(4)	0.5345(9)
C(2,8)	0.3890(3)	0.5313(4)	0.5270(10)
C(2,9)	0.4079(5)	0.5564(8)	0.3779(16)
C(2,10)	0.4275(5)	0.4979(6)	0.6194(18)
C(2,11)	0.1892(3)	0.3312(6)	0.3425(13)
C(2,12)	0.1819(6)	0.2740(8)	0.4481(23)

The structure (Fig. XVAIV,29) is made up of molecules which have the bond dimensions of Figure XVAIV,30. The salicylaldimine groups are planar to within less than 0.1 A. but the two planes about a metal atom are sharply inclined towards one another. The angle between the planes O(1)–Ni–N(1) and O(2)–Ni–N(2) is 85.3° and in consequence the nickel atom, lying outside these planes, has a flattened tetrahedral coordination.

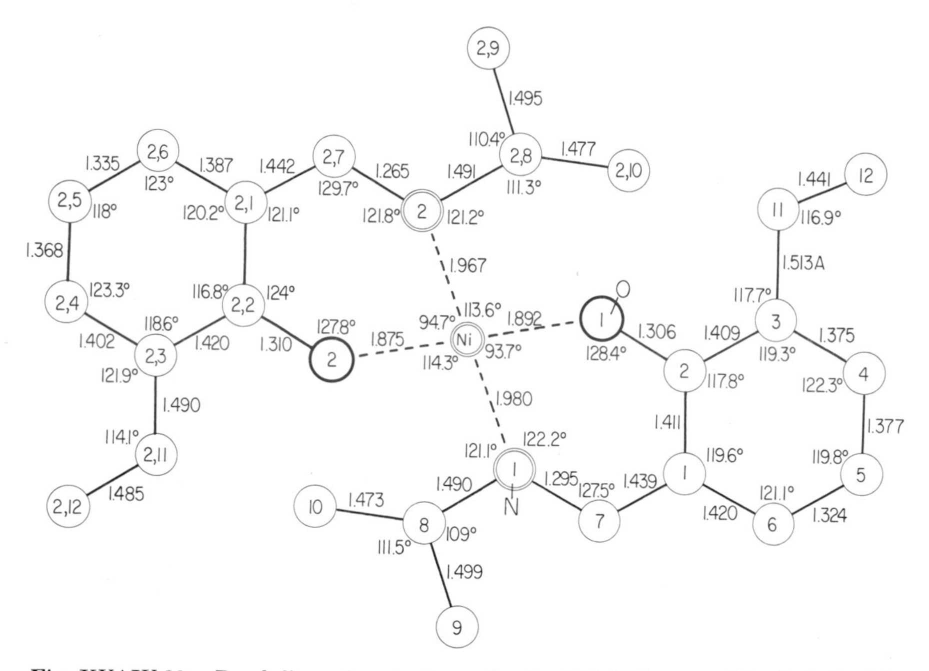

Fig. XVAIV,30. Bond dimensions in the molecule of bis(*N*-isopropyl-3-ethylsalicylaldiminato)nickel.

The various C–H separations, not shown in Figure XVAIV,30, lie between 0.77 and 1.16 A.; their parameters are given in the original article.

XV,aIV,19. The crystals of *bis(N-isopropyl-3-ethylsalicylaldiminato)-palladium*, $[OC_6H_3(C_2H_5)CHNC_3H_7]_2Pd$, have a symmetry and structure different from those of the nickel compound. Their symmetry is monoclinic and the bimolecular unit has the dimensions:

$$a_0 = 10.672(2) \text{ A.}; \quad b_0 = 13.063(2) \text{ A.}; \quad c_0 = 7.998(1) \text{ A.}$$
$$\beta = 98.09(1)^\circ$$

The space group is C_{2h}^5 in the orientation $P2_1/a$. Palladium atoms are in:

$$(2a) \quad 000; \ ^1/_2 \ ^1/_2 \ 0$$

All other atoms including those of hydrogen are in

$$(4e) \quad \pm(xyz; x+{}^1/_2, {}^1/_2-y, z)$$

with the parameters of Table XVAIV,20.

TABLE XVAIV,20

Parameters of Atoms in $(C_{12}H_{16}NO)_2Pd$

Atom	x	y	z
O	−0.1138(2)	0.0072(1)	0.1767(2)
N	0.0950(2)	0.1269(2)	0.0938(3)
C(1)	−0.0111(2)	0.1431(2)	0.3466(3)
C(2)	−0.1012(2)	0.0643(2)	0.3127(3)
C(3)	−0.1853(3)	0.0454(2)	0.4338(3)
C(4)	−0.1755(3)	0.1049(2)	0.5771(4)
C(5)	−0.0874(3)	0.1830(3)	0.6088(4)
C(6)	−0.0064(3)	0.2012(2)	0.4964(4)
C(7)	0.0776(3)	0.1689(2)	0.2340(3)
C(8)	0.1967(3)	0.1706(2)	0.0049(4)
C(9)	0.1919(4)	0.2864(3)	−0.0087(6)
C(10)	0.3227(4)	0.1325(4)	0.0897(7)
C(11)	−0.2830(3)	−0.0393(3)	0.4046(4)
C(12)	−0.4052(4)	−0.0041(4)	0.3014(7)
H(4)	−0.226(3)	0.087(2)	0.657(4)
H(5)	−0.077(3)	0.219(3)	0.710(5)
H(6)	0.055(3)	0.256(3)	0.514(4)
H(7)	0.131(3)	0.223(3)	0.272(4)
H(8)	0.180(3)	0.148(3)	−0.111(5)
H(9,1)	0.247(4)	0.305(3)	−0.074(6)
H(9,2)	0.105(4)	0.306(3)	−0.064(5)
H(9,3)	0.208(4)	0.319(3)	0.099(6)
H(10,1)	0.375(4)	0.154(3)	0.032(5)
H(10,2)	0.342(4)	0.157(3)	0.207(6)
H(10,3)	0.323(4)	0.057(4)	0.103(5)
H(11,1)	0.247(3)	−0.100(3)	0.351(4)
H(11,2)	−0.304(3)	−0.060(2)	0.513(4)
H(12,1)	−0.464(4)	−0.051(4)	0.293(6)
H(12,2)	−0.440(5)	0.050(5)	0.377(7)
H(12,3)	−0.389(6)	0.018(4)	0.174(8)

The structure is shown in Figure XVAIV,31. Its centrosymmetric molecules have the bond dimensions of Figure XVAIV,32. Proposed C–H separations lie between 0.83 and 1.09 A. In contrast to the nickel compound, the benzene ring and salicylaldimine group are coplanar and the palladium atom does not depart by more than about 0.02 A. from this

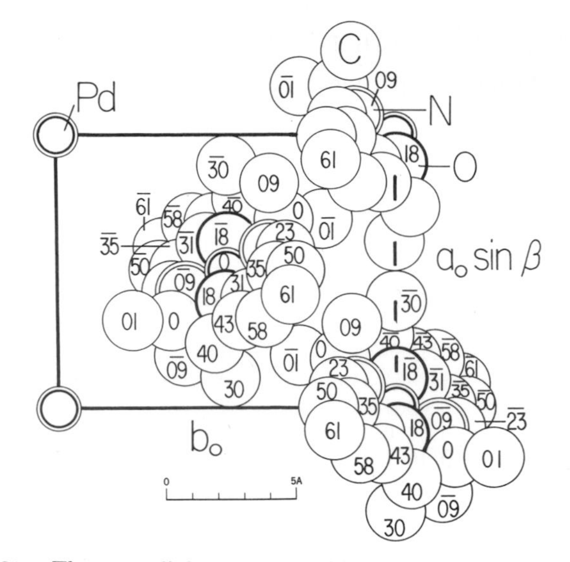

Fig. XVAIV,31. The monoclinic structure of bis(N-isopropyl-3-ethylsalicylaldiminato) palladium projected along its c_0 axis.

plane. Both the isopropyl and the ethyl groups are nearly at right angles to it (83 and 87°). The two chelated molecular groups are parallel, their two planes being displaced 0.42 A. with respect to one another. The closest intermolecular separation is a C(12)–C(10) = 3.619 A. between methyl groups.

XV,aIV,20. Crystals of *N-β-diethylaminoethyl-5-chlorosalicylaldimine nickel*, $Ni[OC_6H_3ClCHNC_2H_4N(C_2H_5)_2]_2$, are monoclinic with a tetramolecular unit of the dimensions:

$$a_0 = 12.68 \text{ A.}; \quad b_0 = 21.77 \text{ A.}; \quad c_0 = 12.05 \text{ A.}; \quad \beta = 122°48'$$

Their atoms are in the positions

$$(4e) \quad \pm(xyz;\ x,{}^1/_2-y,z+{}^1/_2)$$

of the space group C_{2h}^5 ($P2_1/c$). Parameters are listed in Table XVAIV,21.

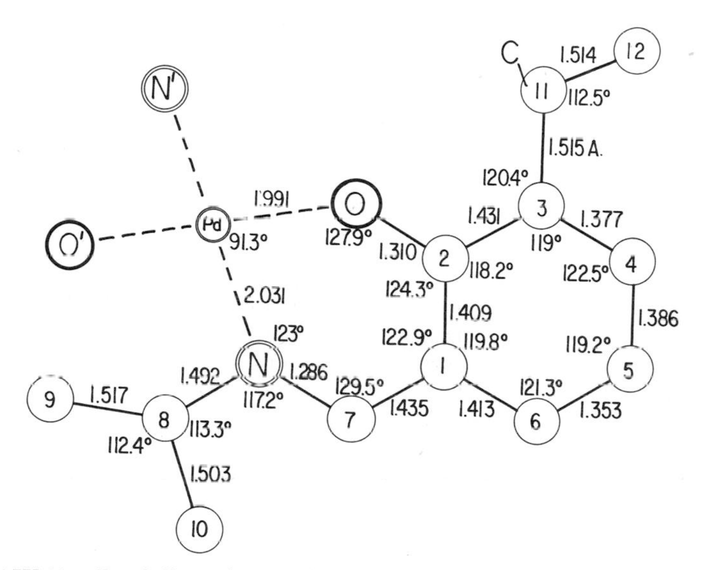

Fig. XVAIV,32. Bond dimensions in the molecule of bis(N-isopropyl-3-ethylsalicylaldiminato)palladium.

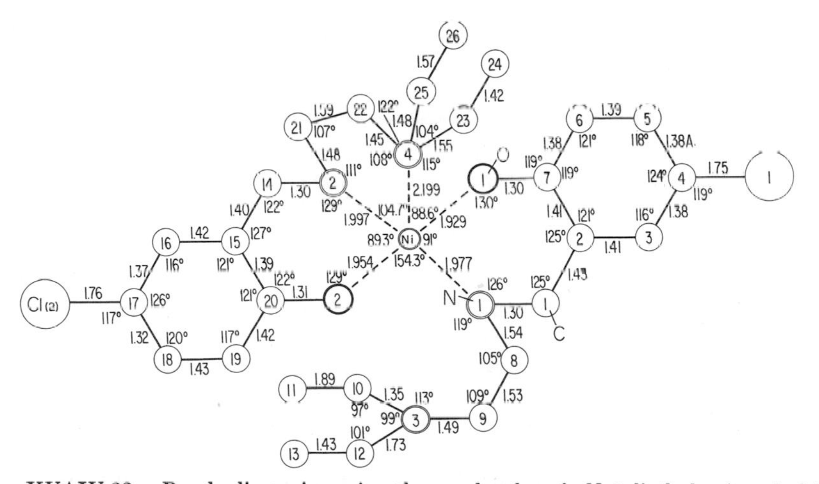

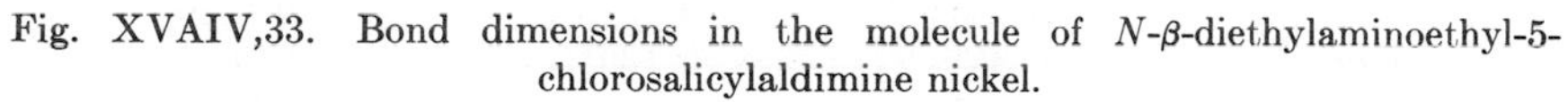

Fig. XVAIV,33. Bond dimensions in the molecule of N-β-diethylaminoethyl-5-chlorosalicylaldimine nickel.

TABLE XVAIV,21

Parameters of the Atoms in *N*-β-Diethylaminoethyl-5-chlorosalicylaldimine Nickel

Atom	*x*	*y*	*z*
Ni	0.0566(3)	0.1785(2)	−0.1288(3)
Cl(1)	−0.0217(7)	0.0451(4)	0.3698(7)
Cl(2)	−0.4099(7)	0.3776(4)	−0.6402(7)
O(1)	0.1256(13)	0.1777(7)	0.0585(13)
O(2)	−0.0634(12)	0.2436(6)	−0.1625(12)
N(1)	−0.0461(15)	0.1051(7)	−0.1544(16)
N(2)	0.0293(15)	0.1943(7)	−0.3058(16)
N(3)	−0.3331(22)	0.0896(12)	−0.4990(23)
N(4)	0.2363(17)	0.1424(9)	−0.0879(18)
C(1)	−0.0623(19)	0.0803(9)	−0.0667(20)
C(2)	0.0013(18)	0.0991(9)	0.0682(18)
C(3)	−0.0366(19)	0.0670(9)	0.1418(20)
C(4)	0.0258(19)	0.0810(9)	0.2744(19)
C(5)	0.1201(20)	0.1242(10)	0.3367(20)
C(6)	0.1585(19)	0.1534(9)	0.2618(20)
C(7)	0.0948(19)	0.1448(10)	0.1271(20)
C(8)	−0.1259(21)	0.0771(10)	−0.2938(21)
C(9)	−0.2519(23)	0.1108(11)	−0.3599(23)
C(10)	−0.3618(31)	0.0294(17)	−0.5079(32)
C(11)	−0.3657(31)	0.0128(16)	−0.6644(33)
C(12)	−0.4807(31)	0.1201(16)	−0.5599(31)
C(13)	−0.4606(30)	0.1829(16)	−0.5779(32)
C(14)	−0.0511(20)	0.2312(9)	−0.3985(19)
C(15)	−0.1367(17)	0.2655(8)	−0.3854(17)
C(16)	−0.2188(19)	0.2993(9)	−0.5014(19)
C(17)	−0.3023(21)	0.3371(10)	−0.4954(22)
C(18)	−0.3084(22)	0.3461(11)	−0.3903(23)
C(19)	−0.2276(23)	0.3128(11)	−0.2716(23)
C(20)	−0.1407(20)	0.2723(10)	−0.2727(20)
C(21)	0.1158(24)	0.1572(12)	−0.3258(25)
C(22)	0.2503(27)	0.1626(13)	−0.1936(29)
C(23)	0.3548(27)	0.1691(14)	0.0370(29)
C(24)	0.3477(25)	0.2332(14)	0.0526(26)
C(25)	0.2374(24)	0.0767(12)	−0.0567(24)
C(26)	0.3562(26)	0.0404(13)	−0.0289(27)

The structure is made up of molecules that have the bond dimensions of Figure XVAIV,33. About the nickel atom the coordination is fivefold and approximately that of a square pyramid with N(1) at the apex. The N(3)–C(12) separation as given in the original is clearly in error.

The isostructural cobalt compound has:

$$a_0 = 13.20 \text{ A.}; \quad b_0 = 21.86 \text{ A.}; \quad c_0 = 12.15 \text{ A.}; \quad \beta = 126°25'$$

XV,aIV,21. The monoclinic crystals of *2,5-dichloroaniline*, $C_6H_3Cl_2NH_2$, have a tetramolecular unit of the dimensions:

$$a_0 = 13.237(7) \text{ A.}; \quad b_0 = 3.892(6) \text{ A.}; \quad c_0 = 18.80(2) \text{ A.}$$
$$\beta = 135°13(11)'$$

All atoms are in the positions

$$(4e) \quad \pm(xyz;\ x,{}^1/_2-y,z+{}^1/_2)$$

of $C_{2h}^5(P2_1/c)$. The parameters are stated in Table XVAIV,22.

TABLE XVAIV,22

Parameters of the Atoms in $C_6H_3Cl_2NH_2$

Atom	x	y	z
Cl(1)	0.5367(2)	0.3879(8)	0.4148(2)
Cl(2)	−0.1084(3)	0.0186(9)	0.0672(2)
C(1)	0.2860(8)	0.346(3)	0.2173(6)
C(2)	0.3566(10)	0.280(3)	0.3182(7)
C(3)	0.2843(11)	0.140(3)	0.3376(7)
C(4)	0.1418(9)	0.059(3)	0.2631(8)
C(5)	0.0725(10)	0.118(3)	0.1639(7)
C(6)	0.1413(10)	0.262(3)	0.1402(7)
N	0.3586(10)	0.475(3)	0.1943(7)

The structure is shown in Figure XVAIV,34. Its molecules, having the bond dimensions of Figure XVAIV,35, are planar to 0.01 A. Between them the shortest atomic separations are: Cl–Cl = 3.369 A., N–N = 3.313 A. and Cl–N = 3.540 A.

XV,aIV,22. The orthorhombic crystals of *2-chloro-4-nitroaniline*, $ClC_6H_3(NH_2)NO_2$, possess a tetramolecular unit of the edge lengths:

$$a_0 = 11.25 \text{ A.}; \quad b_0 = 16.85 \text{ A.}; \quad c_0 = 3.87 \text{ A.}$$

The space group is C_{2v}^9 ($Pna2_1$) with atoms in the positions:

$$(4a) \quad xyz;\ \bar{x},\bar{y},z+{}^1/_2;\ {}^1/_2-x,y+{}^1/_2,z+{}^1/_2;\ x+{}^1/_2,{}^1/_2-y,z$$

Parameters are stated in Table XVAIV,23.

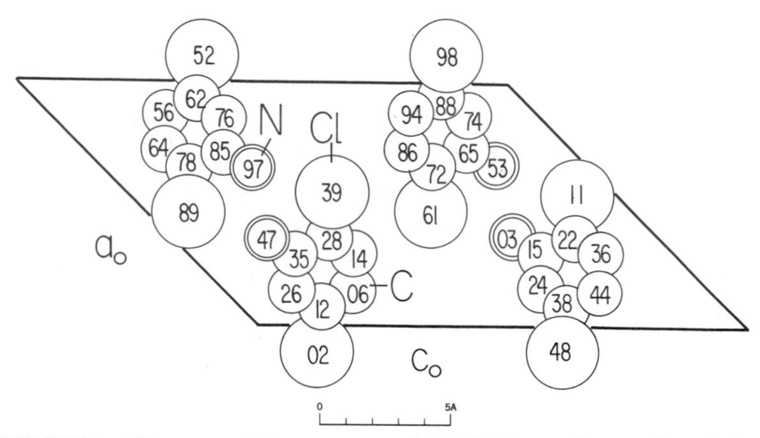

Fig. XVAIV,34. The monoclinic structure of 2,5-dichloroaniline projected along its b_0 axis.

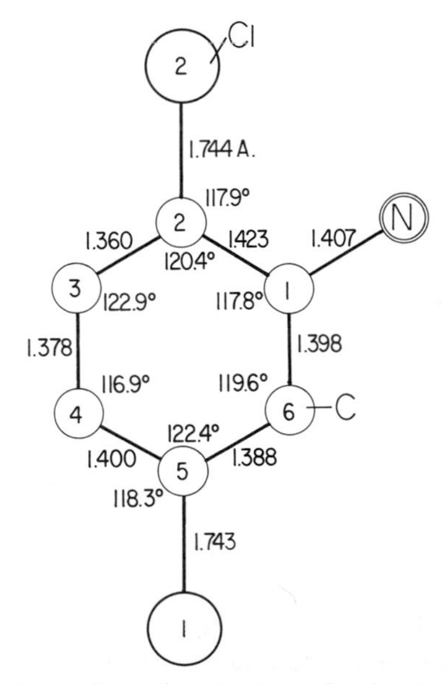

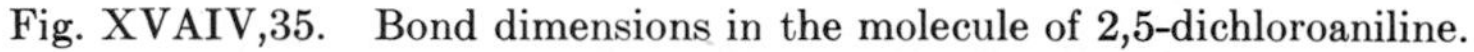
Fig. XVAIV,35. Bond dimensions in the molecule of 2,5-dichloroaniline.

TABLE XVAIV,23

Parameters of the Atoms in $ClC_6H_3(NH_2)NO_2$

Atom	x	y	z
C(1)	0.1849	0.4534	0.1159
C(2)	0.2947	0.4641	0.2762
C(3)	0.3676	0.4014	0.3793
C(4)	0.3295	0.3247	0.2942
C(5)	0.2217	0.3095	0.1339
C(6)	0.1513	0.3740	0.0414
N(1)	0.1101	0.5147	0.0218
N(2)	0.4019	0.2573	0.4090
O(1)	0.4937	0.2714	0.5727
O(2)	0.3707	0.1895	0.3287
Cl	0.3396	0.5614	0.3790

The molecules of this structure (Fig. XVAIV,36) have the bond dimensions of Figure XVAIV,37. Their benzene ring, carbon and N(1) atoms are coplanar to within 0.02 A. The N(2) is 0.08 A. outside this plane and the plane defined by it and its two oxygens makes an angle of 4°20′ with the first. Between molecules the shortest interatomic distance is an N–O = 3.05 A.

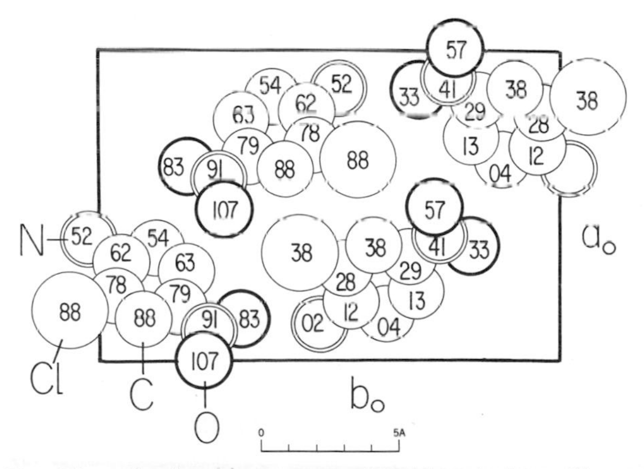

Fig. XVAIV,36. The orthorhombic structure of 2-chloro-4-nitroaniline projected along its c_0 axis.

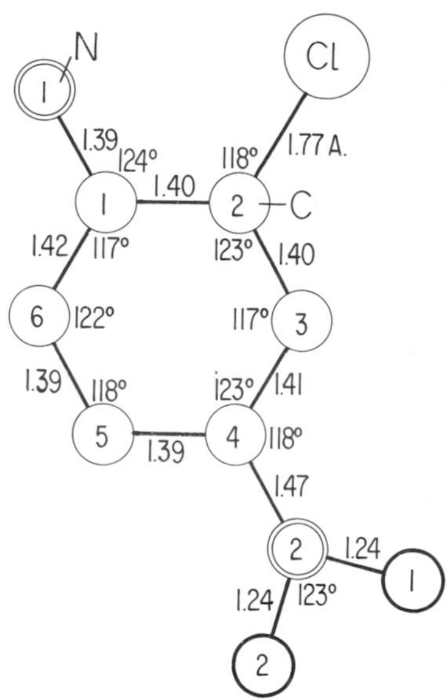

Fig. XVAIV,37. Bond dimensions in the molecule of 2-chloro-4-nitroaniline.

XV,aIV,23. According to a preliminary note, the crystals of *p-bromo-m-nitro-N-methylaniline*, $BrC_6H_3(NO_2)NHCH_3$, are monoclinic with a tetramolecular unit of the dimensions:

$$a_0 = 13.18(5) \text{ A.}; \quad b_0 = 4.04(1) \text{ A.}; \quad c_0 = 15.81(3) \text{ A.}$$
$$\beta = 99°$$

The space group is C_{2h}^5 ($P2_1/c$) with atoms in the positions:

$$(4e) \quad \pm(xyz;\ x,{}^1/_2-y,z+{}^1/_2)$$

Parameters are stated to have the approximate values of Table XVAIV,24.

TABLE XVAIV,24
Parameters of the Atoms in $BrC_6H_3(NO_2)NHCH_3$

Atom	x	y	z
Br	0.452	0.324	0.153
C(1)	0.350	0.125	0.065
C(2)	0.369	0.153	−0.013
C(3)	0.300	0.034	−0.083
C(4)	0.207	−0.110	−0.066
C(5)	0.190	−0.135	0.015
C(6)	0.262	−0.015	0.083
N(1)	0.240	−0.048	0.167
O(1)	0.299	0.021	0.232
O(2)	0.160	−0.176	0.177
N(2)	0.141	−0.236	−0.137
C(7)	0.045	−0.423	−0.121

XV,aIV,24. Years ago a structure was described for a complex formed by a molecule of *p-iodoaniline* and one of *trinitrobenzene*, $[IC_6H_4NH_2]$-$[C_6H_3(NO_2)_3]$. Its crystals are monoclinic with a tetramolecular unit of the dimensions:

$$a_0 = 7.42 \text{ A.}; \quad b_0 = 7.39 \text{ A.}; \quad c_0 = 28.3 \text{ A.}; \quad \beta = 103°25'$$

All atoms are in the general positions

$$(4e) \quad \pm(xyz;\ x,{}^1/_2-y,z+{}^1/_2)$$

of the space group C_{2h}^5 ($P2_1/c$). The assigned parameters are those of Table XVAIV,25. This proposed arrangement gives acceptable atomic separations, though there is a short intermolecular O–C distance of 3.25 A.

TABLE XVAIV,25

Parameters of the Atoms in the *p*-Iodoaniline-*S*-Trinitrobenzene Complex

Atom	x	y	z
	Trinitrobenzene molecule		
C(1)	0.498	0.637	0.161
C(2)	0.420	0.605	0.112
C(3)	0.439	0.735	0.078
C(4)	0.535	0.895	0.093
C(5)	0.613	0.927	0.142
C(6)	0.594	0.797	0.176
N(1)	0.479	0.507	0.195
N(2)	0.361	0.703	0.028
N(3)	0.709	0.087	0.157
O(1)	0.548	0.535	0.238
O(1*a*)	0.395	0.366	0.182
O(2)	0.277	0.562	0.015
O(2*a*)	0.378	0.817	0.998
O(3)	0.778	0.115	0.200
O(3*a*)	—	—	—
	Iodoaniline molecule		
C(7)	0.003	0.642	0.162
C(8)	0.090	0.802	0.182
C(9)	0.121	0.949	0.153
C(10)	0.065	0.935	0.106
C(11)	0.976	0.775	0.087
C(12)	0.946	0.627	0.115
I	0.110	0.153	0.064
NH_2	0.974	0.491	0.191

XV,aIV,25. The complex *tricarbonyl chromium anisole-1,3,5-trinitrobenzene*, $C_{16}H_{11}CrN_3O_{10}$, has monoclinic symmetry. Its tetramolecular unit possesses the dimensions:

$$a_0 = 10.10 \text{ A.}; \quad b_0 = 13.42 \text{ A.}; \quad c_0 = 13.87 \text{ A.}; \quad \beta = 101°45'$$

The space group is C_{2h}^5 ($P2_1/c$) with atoms in the positions:

$$(4e) \quad \pm(xyz;\ x,{}^1/_2-y,z+{}^1/_2)$$

The parameters are listed in Table XVAIV,26.

TABLE XVAIV,26

Parameters of the Atoms in Tricarbonyl Chromium Anisole-1,3,5-Trinitrobenzene

Atom	x	y	z
Cr	0.4480	0.2538	0.9156
C(1)	−0.0439	0.2221	0.9579
C(2)	−0.0644	0.3229	0.9402
C(3)	−0.1173	0.3500	0.8447
C(4)	−0.1539	0.2812	0.7662
C(5)	−0.1382	0.1817	0.7896
C(6)	−0.0787	0.1492	0.8846
N(7)	0.0082	0.1890	0.0576
O(8)	0.0187	0.0987	0.0712
O(9)	0.0483	0.2498	0.1196
N(10)	−0.1386	0.4546	0.8209
O(11)	−0.0997	0.5151	0.8851
O(12)	−0.1913	0.4779	0.7397
N(13)	−0.1681	0.1078	0.7123
O(14)	−0.2096	0.1383	0.6290
O(15)	−0.1560	0.0224	0.7336
C(16)	0.3289	0.1982	0.0252
C(17)	0.3249	0.3055	0.0260
C(18)	0.2793	0.3561	0.9346
C(19)	0.2408	0.3042	0.8466
C(20)	0.2418	0.2006	0.8479
C(21)	0.2918	0.1474	0.9377
O(22)	0.3625	0.1594	0.1153
C(23)	0.3863	0.0516	0.1224
C(24)	0.5429	0.3671	0.9058
O(25)	0.6113	0.4371	0.9018
C(26)	0.5100	0.2045	0.8141
O(27)	0.5517	0.1665	0.7493
C(28)	0.5903	0.1975	0.9924
O(29)	0.6904	0.1553	0.0402

The structure that results consists of sheets of the planar anisole and trinitrobenzene components extending roughly normal to the a_0 axis. Between these components of a sheet and binding the sheets to one another are pairs of $Cr(CO)_3$ trigonal pyramids having their apical chromium atoms pointing in opposite directions. The chromium atoms are centered over the benzene rings of the anisole components they face, and are almost centered with respect to their adjacent trinitrobenzene rings. They are closest to carbon atoms of the anisole ring with an average Cr–C = 2.23 A.; the separation carbonyl carbon–chromium averages 1.79 A.

XV,aIV,26. Crystals of the α modification of *2-chloro-p-benzoquinone-4-oxime acetate,* $OC_6H_3(Cl)NOC(O)CH_3$, are monoclinic. At room and low temperatures they have a tetramolecular unit of the dimensions:

$$a_0 = 7.50 \text{ A.}; \quad b_0 = 19.20 \text{ A.}; \quad c_0 = 6.09 \text{ A.}$$
$$\beta = 96°36' \quad (20°\text{C.})$$
$$a_0 = 7.28 \text{ A.}; \quad b_0 = 19.04 \text{ A.}; \quad c_0 = 6.06 \text{ A.}$$
$$\beta = 96°36' \quad (-140°\text{C.})$$

Atoms are in the positions

$$(4e) \quad \pm(xyz;\ x+{}^1/_2,{}^1/_2-y,z)$$

of the space group C_{2h}^5 ($P2_1/a$). At 140°C. the parameters were determined to be those of Table XVAIV,27.

TABLE XVAIV,27

Parameters of the Atoms in α-2-Chloro-p-benzoquinone-4-oxime Acetate

Atom	x	y	z
Cl	−0.016	0.168	0.226
O(1)	0.175	0.204	0.659
O(2)	−0.170	0.425	0.020
O(3)	−0.272	0.537	0.047
N	−0.085	0.438	0.236
C(1)	0.119	0.256	0.562
C(2)	0.019	0.253	0.338
C(3)	−0.045	0.306	0.212
C(4)	−0.022	0.379	0.317
C(5)	0.072	0.384	0.547
C(6)	0.152	0.328	0.655
C(7)	−0.267	0.483	−0.057
C(8)	−0.356	0.469	−0.292

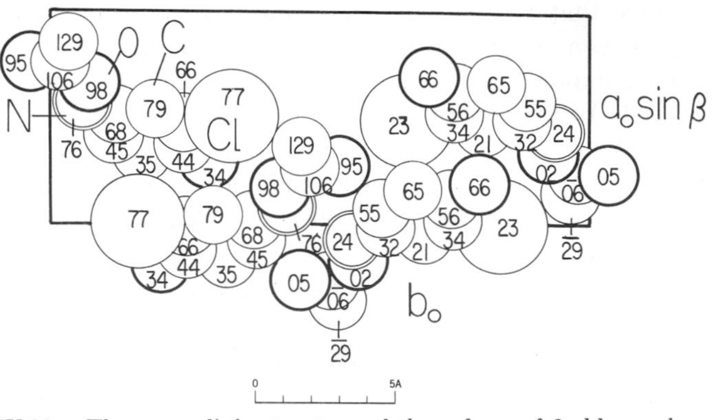

Fig. XVAIV,38. The monoclinic structure of the α form of 2-chloro-*p*-benzoquinone-4-oxime acetate projected along its c_0 axis.

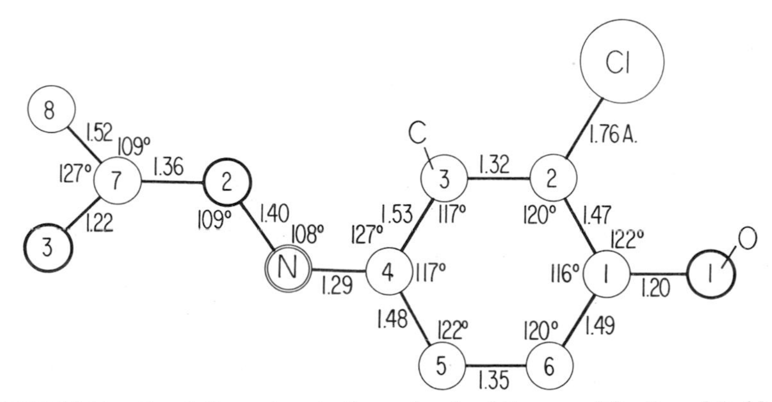

Fig. XVAIV,39. Bond dimensions in the molecule of the α modification of 2-chloro-*p*-benzoquinone-4-oxime acetate.

The structure, shown in Figure XVAIV,38, is composed of molecules having the bond dimensions of Figure XVAIV,39. Their benzene rings are not strictly planar, the plane through one half being tilted by 6° to the plane through the other. The chlorine atom is *syn* with respect to the oxime acetate part of the molecule. The molecules lie in layers, interatomic distances between them ranging upwards from 3.26 A.

The corresponding bromine compound is isostructural with a unit which at room temperature has the dimensions:

$$a_0 = 7.72 \text{ A.}; \quad b_0 = 19.37 \text{ A.}; \quad c_0 = 6.06 \text{ A.}; \quad \beta = 99°24'$$

XV,aIV,27. The β form of *2-chloro-p-benzoquinone-4-oxime acetate*, $OC_6H_3(Cl)NOC(O)CH_3$, is also monoclinic with a unit not greatly different from that of the α modification. Its tetramolecular cell has the dimensions:

$$a_0 = 7.18(5) \text{ A.};\quad b_0 = 19.95(5) \text{ A.};\quad c_0 = 6.80(5) \text{ A.}$$
$$\beta - 113° \quad (20°\text{C.})$$
$$a_0 = 7.14 \text{ A.};\quad b_0 = 19.88 \text{ A.};\quad c_0 = 6.80 \text{ A.}$$
$$\beta = 113° \quad (-140°\text{C.})$$

The space group is, as with the α form, C_{2h}^5 but has the different axial orientation $P2_1/n$. Atoms are in the positions:

$$(4e)\quad \pm(xyz;\ x+{}^1/_2,{}^1/_2-y,z+{}^1/_2)$$

The parameters as determined at $-140°$C. are those of Table XVAIV,28.

TABLE XVAIV,28

Parameters of the Atoms in β-2-Chloro-p-benzoquinone-4-oxime Acetate

Atom	x	y	z
Cl	0.148	0.066	0.100
O(1)	0.136	0.135	0.475
O(2)	0.158	0.374	−0.089
O(3)	0.140	0.390	−0.421
N	0.164	0.306	−0.136
C(1)	0.128	0.175	0.339
C(2)	0.138	0.153	0.138
C(3)	0.156	0.194	−0.008
C(4)	0.135	0.269	0.017
C(5)	0.133	0.296	0.214
C(6)	0.113	0.251	0.358
C(7)	0.157	0.416	−0.246
C(8)	0.190	0.490	−0.172

The molecule of this structure (Fig. XVAIV,40) which in contrast to the α form has its chlorine *anti* to the oxime acetate part, possesses the bond dimensions of Figure XVAIV,41. In this arrangement, as in the α form, the benzene ring is not strictly planar and the acetyl group O(2)C(7)-C(8)O(3) is rotated about O(2)–C(7) out of the approximate plane of the rest.

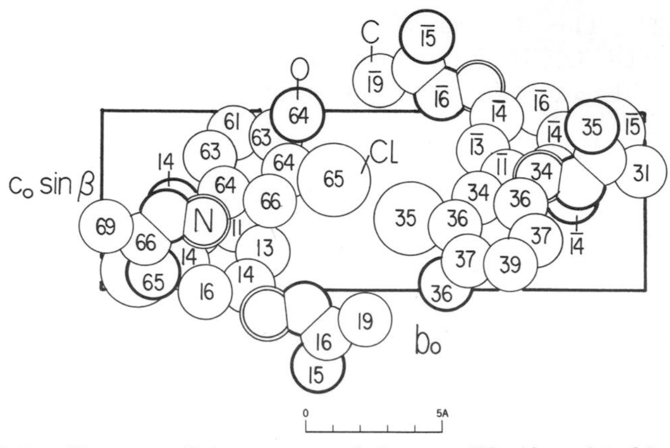

Fig. XVAIV,40. The monoclinic structure of the β modification of 2-chloro-*p*-benzoquinone-4-oxime acetate projected along its a_0 axis.

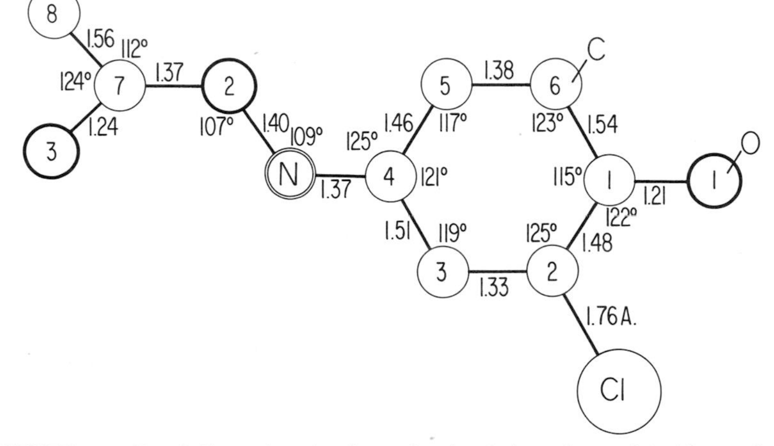

Fig. XVAIV,41. Bond dimensions in the molecule of the β form of 2-chloro-*p*-benzoquinone-4-oxime acetate.

XV,aIV,28. Crystals of *β-5-n-propoxy-o-quinone-2-oxime*, (C_3H_7O)-$C_6H_3(O)NOH$, are monoclinic. They have a tetramolecular unit of the dimensions:

$$a_0 = 10.66(1) \text{ A.}; \quad b_0 = 12.67(1) \text{ A.}; \quad c_0 = 6.79(2) \text{ A.}$$
$$\beta = 95°36(20)' \quad (20°\text{C.})$$
$$a_0 = 10.64(1) \text{ A.}; \quad b_0 = 12.57(2) \text{ A.}; \quad c_0 = 6.63(1) \text{ A.}$$
$$\beta = 96°31(30)' \quad (-120°\text{C.})$$

The space group is C_{2h}^5 ($P2_1/c$) with all atoms in the positions:

$$(4e) \quad \pm(xyz;\ x,{}^1/_2-y,z+{}^1/_2)$$

Parameters are stated in Table XVAIV,29.

TABLE XVAIV,29

Parameters of the Atoms in β-5-*n*-Propoxy-*o*-quinone-2-oxime

Atom	x	y	z
C(1)	0.2845	0.0241	0.3010
C(2)	0.3864	0.9640	0.2672
C(3)	0.5026	0.0140	0.2431
C(4)	0.5101	0.1315	0.2575
C(5)	0.3988	0.1909	0.2952
C(6)	0.2900	0.1398	0.3170
C(7)	0.1486	0.8723	0.3023
C(8)	0.0139	0.8497	0.3431
C(9)	0.9167	0.9034	0.1928
N	0.6133	0.1873	0.2383
O(1)	0.1686	0.9861	0.3233
O(2)	0.7185	0.1318	0.2061
O(3)	0.6012	0.9622	0.2109
H(1)	0.929	0.872	0.062
H(2)	0.815	0.887	0.200
H(3)	0.928	0.976	0.153
H(4)	0.997	0.890	0.456
H(5)	0.005	0.230	0.357
H(6)	0.200	0.842	0.423
H(7)	0.155	0.848	0.154
H(8)	0.392	0.888	0.267
H(9)	0.405	0.270	0.316
H(10)	0.198	0.163	0.333
H(11)	0.694	0.065	0.267

In this structure (Fig. XVAIV,42) the molecules have the bond dimensions of Figure XVAIV,43. They are planar except for the terminal methyl C(9) which is 1.20 A. and the C(7) which is 0.09 A. outside the plane of the rest. Between molecules there is a short O(3)–C(3) = 3.12 A. and C–C = 3.25 and 3.36 A.

The similar halogen ethoxy compounds are isostructural. Their units, which are pseudo-orthorhombic, have the dimensions:

β-$(BrC_2H_4O)C_6H_3(O)NOH$:

$$a_0 = 10.86 \text{ A.}; \quad b_0 = 12.73 \text{ A.}; \quad c_0 = 6.76 \text{ A.}; \quad \beta = 90° \quad (20°\text{C.})$$

β-$(ClC_2H_4O)C_6H_3(O)NOH$:

$$a_0 = 10.70 \text{ A.}; \quad b_0 = 12.28 \text{ A.}; \quad c_0 = 6.59 \text{ A.}; \quad \beta = 90° \quad (-120°\text{C.})$$

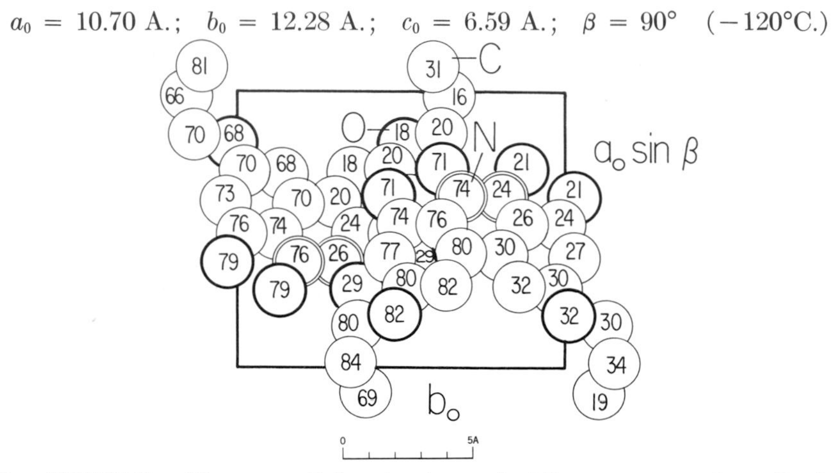

Fig. XVAIV,42. The monoclinic structure of β-5-n-propoxy-o-quinone-2-oxime projected along its c_0 axis.

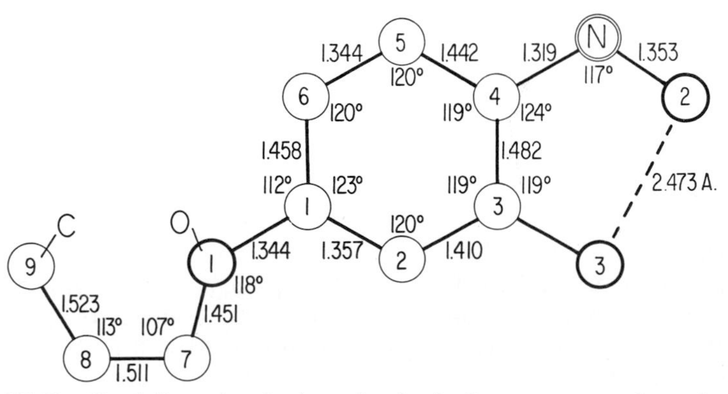

Fig. XVAIV,43. Bond dimensions in the molecule of β-5-n-propoxy-o-quinone-2-oxime.

V. Compounds containing a Tetra-substituted Benzene Ring

XV,aV,1. Crystals of the β (room temperature) form of *1,2,4,5-tetrabromobenzene*, $C_6H_2Br_4$, are monoclinic with a bimolecular unit of the dimensions:

$$a_0 = 10.323(1) \text{ A.}; \quad b_0 = 10.705(1) \text{ A.}; \quad c_0 = 4.018(4) \text{ A.}$$
$$\beta = 102°22(6)'$$

Atoms are in the positions

$$(4e) \quad \pm(xyz; x+{}^1/_2,{}^1/_2-y,z)$$

of the space group C_{2h}^5 ($P2_1/a$). The determined parameters are as follows (values for the chloro compound, see below, in parentheses):

Atom	x	y	z
Br(1)	0.0891 (0.0915)	0.2930 (0.2805)	−0.0272
Br(2)	0.2951 (0.2985)	0.0784 (0.0690)	0.4246
C(1)	0.1229 (0.132)	0.0343 (0.029)	0.1680
C(2)	0.0376 (0.038)	0.1243 (0.125)	−0.0113
C(3)	0.900 (0.083)	−0.0897 (−0.092)	0.1950

The structure, shown in Figure XVAV,1, has molecules that are planar (except that C(1) is 0.03 A. outside the best plane through the other atoms) and the bond dimensions of Figure XVAV,2. They are stacked in the c_0 direction. In the stacks the shortest intermolecular distances are

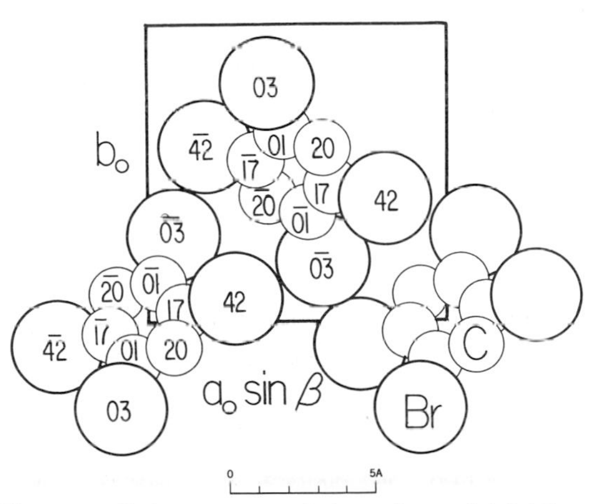

Fig. XVAV,1. The monoclinic structure of the β form of 1,2,4,5-tetrabromobenzene projected along its c_0 axis.

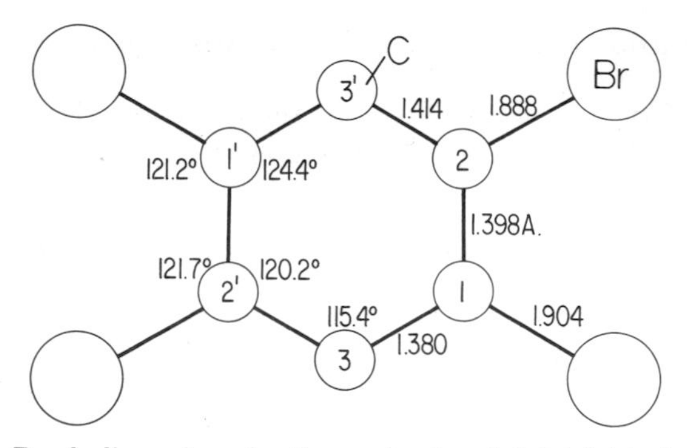

Fig. XVAV,2. Bond dimensions in the molecule of 1,2,4,5-tetrabromobenzene in crystals of its β form.

Br–Br = 4.02 A., Br–C = 3.81 A., and C–C = 3.72 A. Packing of the stacks is rather loose, and in this direction the shortest separation is Br–Br = 4.10 A.

An earlier partial study suggested a structure for the chlorine compound *1,2,4,5-tetrachlorobenzene* like that just described. Exchanging the a_0 and c_0 axes to agree with those used for the bromine compound:

$$a_0 = 9.725(5)\text{ A.};\quad b_0 = 10.602(5)\text{ A.};\quad c_0 = 3.850(5)\text{ A.}$$
$$\beta = 103.28(10)^\circ$$

Such of the parameters as were determined are given in parentheses above. At reduced temperatures (ca. 150°K) there is an apparently related triclinic modification. For it the unit has the dimensions:

$$a_0 = 9.60\text{ A.};\quad b_0 = 10.59\text{ A.};\quad c_0 = 3.76\text{ A.}$$
$$\alpha = 95^\circ;\quad \beta = 102.5^\circ;\quad \gamma = 92.5^\circ$$

XV,aV,2. The γ modification of *1,2,4,5-tetrabromobenzene*, γ-$C_6H_2Br_4$, stable above 46.5°C., is monoclinic. Its tetramolecular unit has the dimensions:

$$a_0 = 10.00(1)\text{ A.};\quad b_0 = 11.18(1)\text{ A.};\quad c_0 = 4.07(2)\text{ A.}$$
$$\beta = 103^\circ 48(20)'$$

Atoms are in the positions

$$(4e)\quad \pm(xyz;\ x+{}^1/_2,{}^1/_2-y,z)$$

of C_{2h}^5 ($P2_1/a$). The determined parameters are as follows:

Atom	x	y	z
Br(1)	0.0949(3)	0.2796(2)	0.0651(8)
Br(2)	0.3026(2)	0.0651(2)	0.4800(8)
C(1)	0.1240(16)	0.0285(13)	0.2017(58)
C(2)	0.0397(17)	0.1218(14)	0.0277(58)
C(3)	0.0802(20)	−0.0881(16)	0.1870(69)

This structure is so close to that of the β form (**XV,aV,1**) that its molecular distribution can be seen by reference to Figure XVAV,1. Bond dimensions are those of Figure XVAV,3. Between molecules the shortest Br–Br = 4.07 A., Br–C = 3.81 A., and C–C = ca. 3.65 A. As in the other modification the molecules are in stacks parallel to [001]; they make 38°25′ to the (001) plane.

XV,aV,3. A partial structure has been described for *2,4,6-trichlorobromobenzene*, $C_6H_2Cl_3Br$. Its tetragonal, tetramolecular unit has the dimensions:

$$a_0 = 14.28 \text{ A.}; \quad c_0 = 3.99 \text{ A.} \quad \text{(as corrected, see SR 23, 660)}$$

The space group, chosen as V_d^3 ($P\bar{4}2_1m$), places atoms in the positions:

$$(4e) \quad u,u+{}^1/_2,v;\ \bar{u},{}^1/_2-u,v;\ u+{}^1/_2,\bar{u},\bar{v};\ {}^1/_2-u,u,\bar{v}$$

$$(8f) \quad xyz;\ \bar{x}\bar{y}z;\ {}^1/_2-x,y+{}^1/_2,\bar{z};\ x+{}^1/_2,{}^1/_2-y,\bar{z};$$
$$\bar{y}x\bar{z};\ y\bar{x}\bar{z};\ y+{}^1/_2,x+{}^1/_2,z;\ {}^1/_2-y,{}^1/_2-x,z$$

Assigned parameters are listed in Table XVAV,1, those for z being chosen from geometrical considerations only.

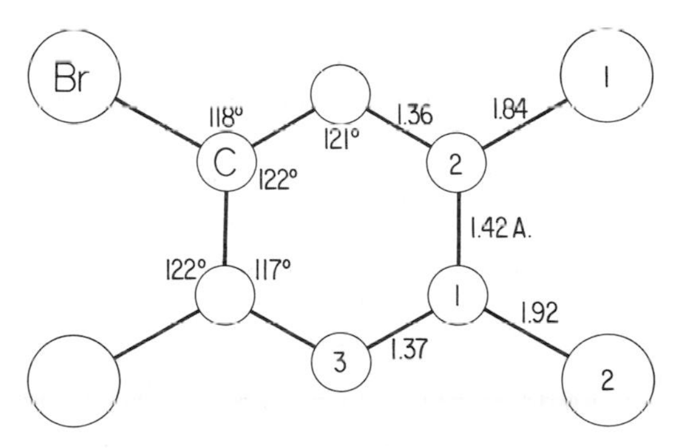

Fig. XVAV,3. Bond dimensions in the molecule of 1,2,4,5-tetrabromobenzene in crystals of its high temperature, γ, modification.

TABLE XVAV,1

Parameters of the Atoms in $C_6H_2Cl_3Br$

Atom	Position	x	y	z
C(1)	(4*e*)	0.2229	0.2771	−0.038
C(2)	(8*f*)	0.1954	0.1907	0.193
C(3)	(8*f*)	0.1299	0.2566	0.308
C(4)	(4*e*)	0.3442	0.1558	−0.436
Br(1)	(4*e*)	0.1391	0.3609	0.145
Cl(2)	(8*f*)	0.1478	0.0834	0.253
Cl(4)	(4*e*)	0.4171	0.0829	0.353

XV,aV,4. Crystals of *1,2,4,5-tetramethylbenzene* (durene), $C_6H_2(CH_3)_4$, are monoclinic with a bimolecular unit of the dimensions:

$$a_0 = 11.57 \text{ A.}; \quad b_0 = 5.77 \text{ A.}; \quad c_0 = 7.03 \text{ A.}; \quad \beta = 113°18'$$

Atoms are in the general positions of C_{2h}^5 ($P2_1/a$):

$$(4e) \quad \pm(xyz;\ x+{}^1/_2,{}^1/_2-y,z)$$

Their parameters, established long ago, are as follows:

Atom	x	y	z
C(1)	0.188	0.314	0.267
C(2)	0.093	0.157	0.127
C(3)	0.037	−0.005	0.212
C(4)	−0.055	−0.162	0.090
C(5)	−0.108	−0.325	0.194

The structure, illustrated in Figure XVAV,4, has molecules which are planar within the limit of accuracy of the determination.

XV,aV,5. The orthorhombic crystals of *nitromesitylene*, $C_6H_2(CH_3)_3$-(NO_2), have a tetramolecular unit of the edge lengths:

$$a_0 = 15.14(4) \text{ A.}; \quad b_0 = 8.41(2) \text{ A.}; \quad c_0 = 7.26(2) \text{ A.}$$

Atoms are in the positions

$$(4a) \quad xyz;\ \bar{x},\bar{y},z+{}^1/_2;\ {}^1/_2-x,y+{}^1/_2,z+{}^1/_2;\ x+{}^1/_2,{}^1/_2-y,z$$

of the space group C_{2v}^9 ($Pna2_1$). The chosen parameters are those of Table XVAV,2.

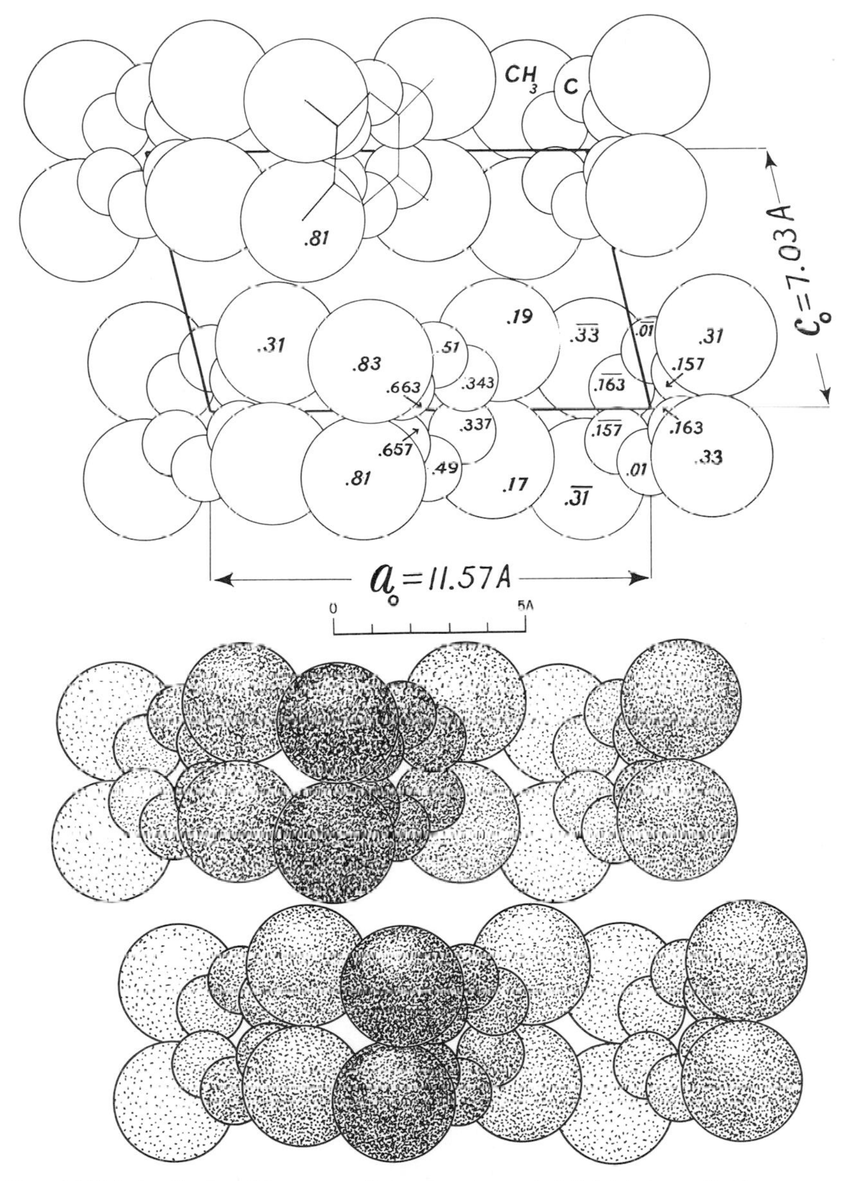

Fig. XVAV,4a (top). The monoclinic structure of 1,2,4,5-tetramethylbenzene projected along its b_0 axis. Left hand axes.

Fig. XVAV,4b (bottom). A packing drawing of the monoclinic structure of 1,2,4,5-tetramethyl-benzene viewed along its b_0 axis. The methyl groups are larger than the benzene carbons.

TABLE XVAV,2

Parameters of the Atoms in Nitromesitylene

Atom	x	y	z
C(1)	−0.1222	0.2377	0.0000
C(2)	−0.0407	0.1734	0.0000
C(3)	−0.0145	0.0159	0.0000
C(4)	−0.0884	−0.0814	0.0000
C(5)	−0.1756	−0.0386	0.0000
C(6)	−0.1926	0.1221	0.0000
C(7)	−0.1493	0.4109	0.0000
C(8)	0.0797	−0.0431	0.0000
C(9)	−0.2500	−0.1568	0.0000
N	0.0321	0.2903	0.0000
O(1)	0.0429	0.3790	0.1308
O(2)	0.0810	0.3084	−0.1308

The molecule thus defined, with the bond dimensions of Figure XVAV,5, is planar except for the oxygen atoms. Lying 0.95 A. on either side of the molecular plane, they result in an NO_2 group that is twisted through 66.4° with respect to it.

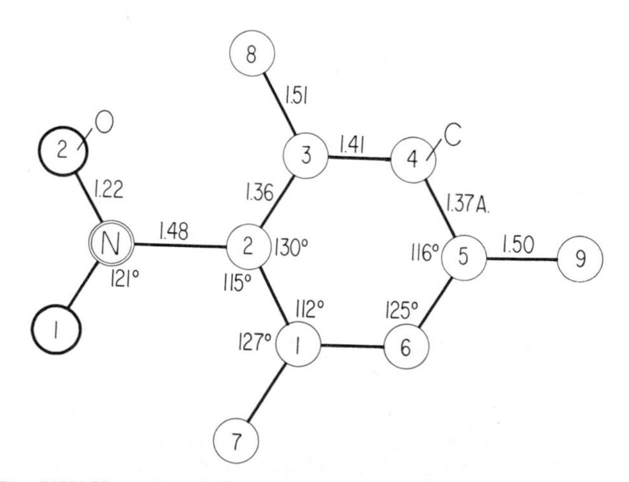

Fig. XVAV,5. Bond dimensions in the molecule of nitromesitylene.

XV,aV,6. The monoclinic crystals of *N-methyl-N-2,4,6-tetranitroaniline* (tetryl), $C_6H_2(NO_2)_3[N(CH_3)(NO_2)]$, have a tetramolecular unit of the dimensions:

$$a_0 = 14.1290(19) \text{ A.}; \quad b_0 = 7.3745(13) \text{ A.}; \quad c_0 = 10.6140(20) \text{ A.}$$
$$\beta = 95.071(17)°$$

Atoms are in the positions

$$(4e) \quad \pm(xyz;\ x,{}^1/_2-y,z+{}^1/_2)$$

of the space group C_{2h}^5 ($P2_1/c$). The parameters are those of Table XVAV,3.

TABLE XVAV,3

Parameters of the Atoms in Tetryl

Atom	x	y	z
C(1)	0.7368(2)	0.0415(4)	0.4240(3)
C(2)	0.8168(2)	0.1372(4)	0.4741(3)
C(3)	0.8802(2)	0.0655(4)	0.5672(3)
C(4)	0.8605(2)	−0.1000(4)	0.6162(3)
C(5)	0.7828(2)	−0.2025(4)	0.5728(3)
C(6)	0.7244(2)	−0.1303(4)	0.4727(3)
C(7)	0.5723(3)	0.1683(6)	0.3694(4)
N(1)	0.8355(2)	0.3260(3)	0.4344(3)
N(2)	0.9235(2)	−0.1735(3)	0.7239(2)
N(3)	0.6404(2)	−0.2502(4)	0.4160(3)
N(4)	0.6661(2)	0.1130(3)	0.3344(2)
N(5)	0.6929(2)	0.1638(4)	0.2209(3)
O(1)	0.7694(2)	0.4164(3)	0.3885(2)
O(2)	0.9160(2)	0.3790(3)	0.4516(3)
O(3)	0.0044(2)	−0.1161(3)	0.7364(2)
O(4)	0.8898(2)	−0.2814(3)	0.7934(2)
O(5)	0.6040(2)	−0.3369(3)	0.4906(2)
O(6)	0.6343(2)	−0.2559(3)	0.3015(2)
O(7)	0.6348(2)	0.2505(3)	0.1519(2)
O(8)	0.7732(2)	0.1204(3)	0.1953(2)
H(1)	0.930(2)	0.135(3)	0.599(2)
H(2)	0.772(2)	−0.339(4)	0.609(3)
H(3)	0.578(3)	0.296(5)	0.399(4)
H(4)	0.532(2)	0.140(5)	0.314(3)
H(5)	0.560(2)	0.117(4)	0.461(3)

A portion of the structure is shown in Figure XVAV,6. The bond dimensions of its molecule are those of Figure XVAV,7. Its benzene nucleus is planar but the attached nitrogen atoms are about 0.10 A. on either side of this plane and the NO_2 groups of which they form part are rotated through various angles with respect to it.

XV,aV,7. Crystals of *4-bromo-2,6-di(tertiarybutyl)phenol*, $BrC_6H_2(OH)(C_4H_9)_2$, have been described as orthorhombic with a tetramolecular unit of the edge lengths:

$$a_0 = 15.65(5) \text{ A.}; \quad b_0 = 10.30(5) \text{ A.}; \quad c_0 = 8.90(5) \text{ A.}$$

They have been given the space group V^4 ($P2_12_12_1$) with atoms in the positions:

$$(4a) \quad xyz;\ {}^1/_2-x,\bar{y},z+{}^1/_2;\ x+{}^1/_2,{}^1/_2-y,\bar{z};\ \bar{x},y+{}^1/_2,{}^1/_2-z$$

According to a preliminary note, the parameters are those of Table XVAV,4.

TABLE XVAV,4

Parameters of the Atoms in 4-Bromo-2,6-di(tertiarybutyl)phenol

Atom	x	y	z
O	0.687	0.084	0.913
C(1)	0.682	0.117	0.770
C(2)	0.700	0.258	0.538
C(3)	0.730	0.231	0.711
C(4)	0.647	0.186	0.491
C(5)	0.597	0.083	0.693
C(6)	0.595	0.101	0.547
C(7)	0.816	0.291	0.772
C(8)	0.775	0.368	0.634
C(9)	0.861	0.369	0.773
C(10)	0.864	0.183	0.953
C(11)	0.563	0.959	0.776
C(12)	0.516	0.009	0.807
C(13)	0.474	0.934	0.588
C(14)	0.606	0.867	0.822
Br	0.623	0.234	0.280

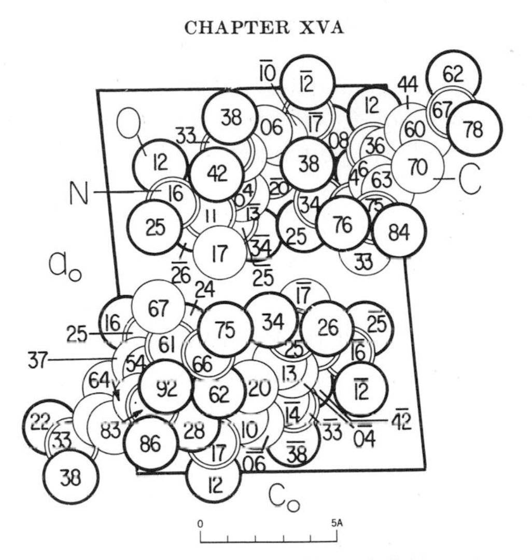

Fig. XVAV,6. The monoclinic structure of *N*-methyl-*N*-2,4,6-tetranitroaniline projected along its b_0 axis.

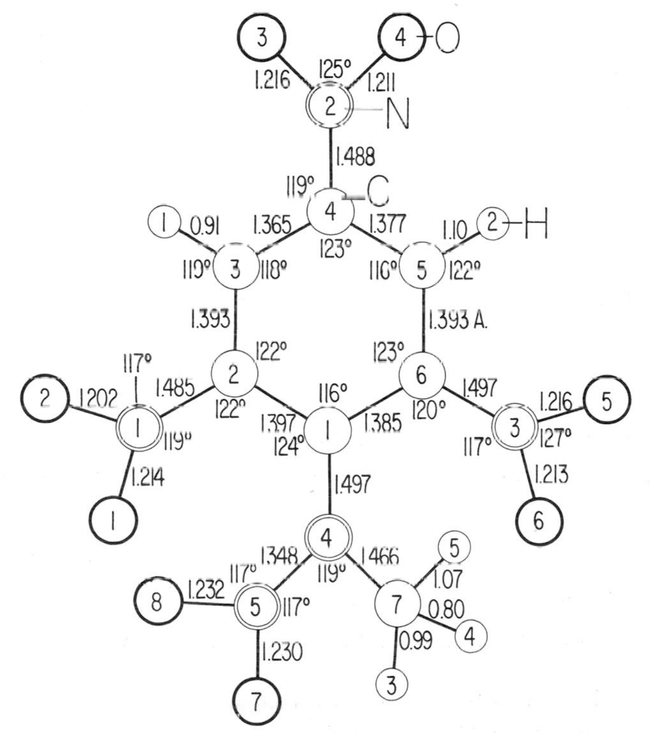

Fig. XVAV,7. Bond dimensions in the molecule of *N*-methyl-*N*-2,4,6-tetranitroaniline.

The corresponding compound having a methyl group in place of the bromine atom, *4-methyl-2,6-di(tertiarybutyl)phenol*, $C_6H_2(OH)(CH_3)(C_4H_9)_2$, is isostructural. Its unit has the edge lengths:

$$a_0 = 15.56(5) \text{ A.}; \quad b_0 = 10.30(5) \text{ A.}; \quad c_0 = 8.80 \text{ A.}$$

Preliminary values for the parameters are those of Table XVAV,5.

TABLE XVAV,5

Parameters of the Atoms in 4-Methyl-2,6-di(tertiarybutyl)phenol

Atom	x	y	z
O	0.698	0.085	0.924
C(1)	0.677	0.124	0.774
C(2)	0.707	0.248	0.541
C(3)	0.729	0.225	0.689
C(4)	0.641	0.200	0.474
C(5)	0.608	0.079	0.700
C(6)	0.585	0.099	0.542
C(7)	0.810	0.292	0.772
C(8)	0.768	0.358	0.933
C(9)	0.858	0.388	0.676
C(10)	0.883	0.183	0.796
C(11)	0.546	0.964	0.779
C(12)	0.514	0.021	0.947
C(13)	0.471	0.931	0.661
C(14)	0.595	0.842	0.811
C(15)	0.621	0.234	0.323

XV,aV,8. Crystals of *3,5-dibromo-p-aminobenzoic acid*, $Br_2C_6H_2(NH_2)COOH$, are orthorhombic. There are eight molecules in a large unit having the edge lengths:

$$a_0 = 22.46(2) \text{ A.}; \quad b_0 = 19.47(2) \text{ A.}; \quad c_0 = 3.94(1) \text{ A.} \quad \text{at } 25°\text{C.}$$
$$a_0 = 22.38(2) \text{ A.}; \quad b_0 = 19.46(2) \text{ A.}; \quad c_0 = 3.88(1) \text{ A.} \quad \text{at } -150°\text{C.}$$

Atoms have been placed in the following positions of V_h^7 (*Pman*):

$(4f) \quad \pm(u\,0\,{}^1/_2;\ u+{}^1/_2,{}^1/_2,{}^1/_2)$

$(4h) \quad \pm(0uv;\ {}^1/_2,u+{}^1/_2,\bar{v})$

$(8i) \quad \pm(xyz;\ x\bar{y}\bar{z};\ x+{}^1/_2,y+{}^1/_2,\bar{z};\ x+{}^1/_2,{}^1/_2-y,z)$

Two crystallographically different types of molecule are present; one of these has an axis, the other a plane of symmetry. The parameters of their atoms, as determined at the two temperatures, are listed in Table XVAV,6.

TABLE XVAV,6

Parameters of the Atoms in $Br_2C_6H_2(NH_2)COOH$ [a]

Atom	Position	x	y	z
Br	(8*i*)	0.1275 (0.1280)	0.2820 (0.2828)	0.172 (0.172)
N	(4*h*)	0	0.3352 (0.3380)	0.038 (0.023)
C(1)	(4*h*)	0	0.2774 (0.2766)	0.209 (0.213)
C(2)	(8*i*)	0.0527 (0.0537)	0.2427 (0.2430)	0.287 (0.294)
C(3)	(8*i*)	0.0550 (0.0548)	0.1776 (0.1805)	0.481 (0.481)
C(4)	(4*h*)	0	0.1510 (0.1490)	0.574 (0.570)
C(5)	(4*h*)	0	0.0854 (0.0850)	0.760 (0.753)
O	(8*i*)	0.0500 (0.0495)	0.0595 (0.0573)	0.843 (0.837)
Br′	(8*i*)	0.2164 (0.2152)	0.1329 (0.1328)	0.812 (0.816)
N′	(4*f*)	0.1615 (0.1609)	0	$^1/_2$
C(1′)	(4*f*)	0.2254 (0.2222)	0	$^1/_2$
C(2′)	(8*i*)	0.2556 (0.2546)	0.0536 (0.0560)	0.635 (0.649)
C(3′)	(8*i*)	0.3198 (0.3173)	0.0569 (0.0557)	0.635 (0.649)
C(4′)	(4*f*)	0.3486 (0.3499)	0	$^1/_2$
C(5′)	(4*f*)	0.4132 (0.4133)	0	$^1/_2$
O′	(8*i*)	0.4425 (0.4428)	0.0495 (0.0506)	0.628 (0.638)

[a] Parameters in parentheses apply to the structure at −150°C.

The bond dimensions of these molecules at room temperature are shown in Figure XVAV,8. In the structure, illustrated in Figure XVAV,9, the molecules are tied together in pairs by hydrogen bonds between carboxyl oxygen atoms, with O–H . . . O = 2.560 or 2.563 A. (at −150°C.). Between molecules there are also N–O separations as short as 2.840 A. Bromine atoms are nearest to other bromines, at a distance of 3.775 A.

XV,aV,9. The monoclinic crystals of *3,5-dichloro-4-nitroaniline*, $C_6H_2Cl_2(NO_2)(NH_2)$, have a tetramolecular unit of the dimensions:

$$a_0 = 7.51 \text{ A.}; \quad b_0 = 14.8 \text{ A.}; \quad c_0 = 7.29 \text{ A.}; \quad \beta = 94°$$

Atoms are in the positions

$(4e)\quad \pm(0\,u\,{}^1/_4;\ {}^1/_2,u+{}^1/_2,{}^1/_4)$

$(8f)\quad \pm(xyz;\ x,\bar{y},z+{}^1/_2;\ x+{}^1/_2,y+{}^1/_2,z;\ x+{}^1/_2,{}^1/_2-y,z+{}^1/_2)$

of the space group C_{2h}^6 ($C2/c$). Parameters in the approximate structure that has been described are listed in Table XVAV,7.

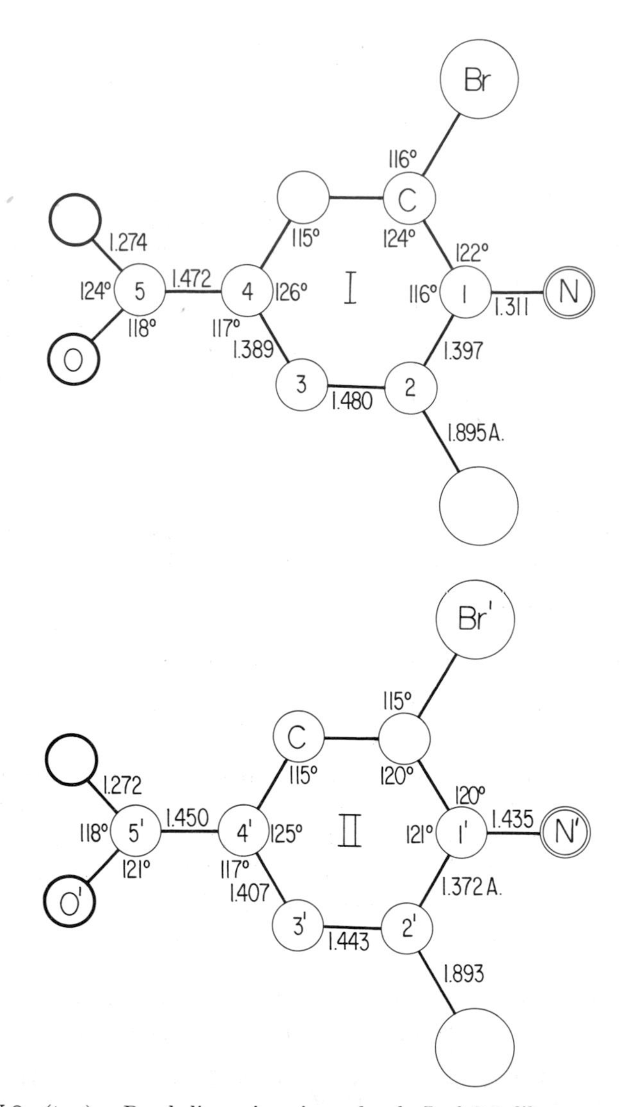

Fig. XVAV,8*a* (top). Bond dimensions in molecule I of 3,5-dibromo-*p*-aminobenzoic acid as found in its crystals.

Fig. XVAV,8b (bottom). Bond dimensions in molecule II of 3,5-dibromo-*p*-aminobenzoic acid.

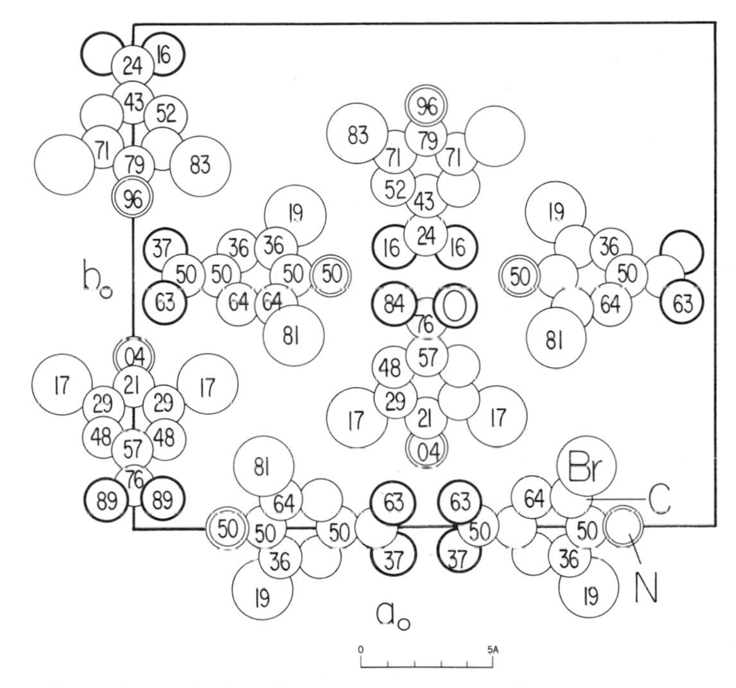

Fig. XVAV,9. The orthorhombic structure of 3,5-dibromo-*p*-aminobenzoic acid projected along its c_0 axis.

TABLE XVAV,7

Parameters of the Atoms in $C_6H_2Cl_2(NO_2)(NH_2)$

Atom	Position	x	y	z
C(1)	(4*c*)	0	−0.140	1/4
C(2)	(8*f*)	—	0.312	0.185
C(3)	(8*f*)	0.353	0.219	—
C(4)	(4*e*)	0	−0.327	1/4
Cl	(8*f*)	0.165	0.166	0.095
N(1)	(4*e*)	0	−0.046	1/4
N(2)	(4*e*)	0	−0.424	1/4
O	(8*f*)	0.390	0.042	0.143

XV,aV,10. Picryl iodide, *2,4,6-trinitro-iodobenzene*, 2,4,6-$C_6H_2(NO_2)_3I$, crystallizes in the tetragonal system. Its tetramolecular unit has the edges:

$$a_0 = 7.03 \text{ A.}, \qquad c_0 = 19.8 \text{ A.}$$

The space group is D_4^4 ($P4_12_1$), or the enantiomorphic D_4^8, with some atoms in special, others in the general positions:

$$(4a) \quad u\,u\,0;\ \bar{u}\,\bar{u}\,{}^1/_2;\ {}^1/_2-u,u+{}^1/_2,{}^1/_4;\ u+{}^1/_2,{}^1/_2-u,{}^3/_4$$

$$(8b) \quad xy\bar{z};\ \bar{x},\bar{y},z+{}^1/_2;\ {}^1/_2-y,x+{}^1/_2,z+{}^1/_4;\ y+{}^1/_2,{}^1/_2-x,z+{}^3/_4;$$
$$yxz;\ \bar{y},\bar{x},{}^1/_2-z;\ x+{}^1/_2,{}^1/_2-y,{}^3/_4-z;\ {}^1/_2-x,y+{}^1/_2,{}^1/_4-z$$

The chosen parameters are those of Table XVAV,8. The original description of structure is accompanied by few data and, especially in view of the dominant effects of the heavy iodine atoms, it must be taken as approximate only.

TABLE XVAV,8

Parameters of the Atoms in Picryl Iodide

Atom	Position	x	y	z
I	(4*a*)	0.045	0.045	0
C(1)	(4*a*)	0.255	0.255	0
C(2)	(8*b*)	0.400	0.250	0.048
C(3)	(8*b*)	0.538	0.388	0.048
N(1)	(8*b*)	0.405	0.099	0.098
N(2)	(4*a*)	0.668	0.668	0
O(1)	(8*b*)	0.790	0.666	0.040
O(2)	(8*b*)	0.303	0.115	0.146
O(3)	(8*b*)	0.511	0.971	0.806

XV,aV,11. The iodine-substituted amino acid *diiodo*-L-*tyrosine dihydrate*, p-$OHC_6H_2I_2CH_2CH(COOH)NH_2 \cdot 2H_2O$, forms monoclinic crystals. Their bimolecular unit has the dimensions:

$$a_0 = 11.162(14) \text{ A.};\quad b_0 = 11.086(23) \text{ A.};\quad c_0 = 5.668(14) \text{ A.}$$
$$\beta = 96°33(15)'$$

Atoms are in the positions

$$(2a) \quad xyz;\ \bar{x},y+{}^1/_2,\bar{z}$$

of the space group C_2^2 ($P2_1$). The determined parameters are listed in Table XVAV,9.

TABLE XVAV,9

Parameters of the Atoms in Di-iodo-L-tyrosine Dihydrate

Atom	x	y	z
I(1)	0.7735(1)	0.0000 —	0.0425(1)
I(2)	0.4839(1)	0.3337(3)	0.6041(2)
C(3)	0.6132(14)	0.1644(17)	0.2929(25)
C(4)	0.6206(17)	0.0822(18)	0.1146(32)
C(5)	0.5107(21)	0.0450(21)	−0.0294(40)
C(6)	0.3927(16)	0.0913(17)	0.0069(29)
C(7)	0.3916(14)	0.1754(17)	0.2041(26)
C(8)	0.4924(16)	0.2061(17)	0.3317(30)
C(9)	0.2747(15)	0.0595(16)	−0.1557(29)
C(10)	0.2235(14)	−0.0508(15)	−0.0560(26)
C(11)	0.1767(13)	−0.0278(14)	0.1917(25)
N(12)	0.1063(15)	−0.0749(18)	−0.2335(31)
O(13)	0.7072(16)	0.1917(17)	0.4222(30)
O(14)	0.1958(10)	−0.1070(10)	0.3383(18)
O(15)	0.1041(13)	0.0562(13)	0.2112(24)
O(16)	0.0262(12)	0.2680(13)	−0.0076(24)
O(17)	0.9763(14)	0.1337(16)	0.5612(25)

The structure, shown in Figure XVAV,10, has molecules with the bond dimensions of Figure XVAV,11. They are planar except for I(1) and C(9) which are about 0.1 A. on either side of the plane. In the crystal the molecules are held together by a complex system of hydrogen bonds of lengths ranging upwards from 2.75 A. These bonds are between oxygen and other oxygens (both carboxyl and water) and nitrogen.

XV,aV,12. Crystals of *2,3-dichloro-1,4-benzoquinone*, $C_6H_2O_2Cl_2$, are tetragonal with a tetramolecular unit of the edge lengths:

$$a_0 = 5.650(1) \text{ A.}, \qquad c_0 = 21.220(12) \text{ A.}$$

Atoms are in the positions

$$(8b) \quad xyz;\ \bar{x},\bar{y},z+1/2;\ 1/2-y,x+1/2,z+1/4;\ y+1/2,1/2-x,z+3/4;$$
$$yx\bar{z};\ \bar{y},\bar{x},1/2-z;\ 1/2-x,y+1/2,1/4-z;\ x+1/2,1/2-y,3/4-z$$

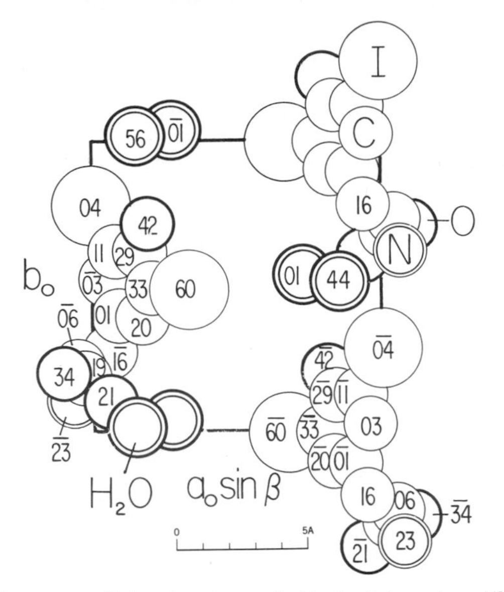

Fig. XVAV,10. The monoclinic structure of diiodo-L-tyrosine dihydrate projected along its c_0 axis.

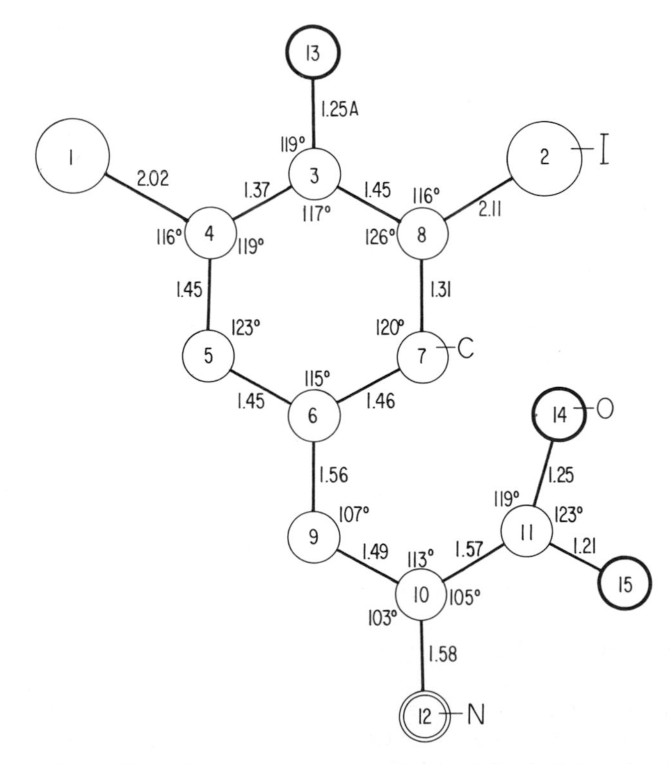

Fig. XVAV,11. Bond dimensions in the molecule of diiodo-L-tyrosine dihydrate.

of the space group D_4^4 ($P4_12_12$). The following parameters have been found for them:

Atom	x	y	z
Cl	0.1612	−0.0154	0.0673
O	−0.1084	−0.4196	0.1112
C(6)	−0.3863	−0.4559	0.0280
C(2)	−0.1774	−0.3524	0.0595
C(3)	−0.0730	−0.1485	0.0280

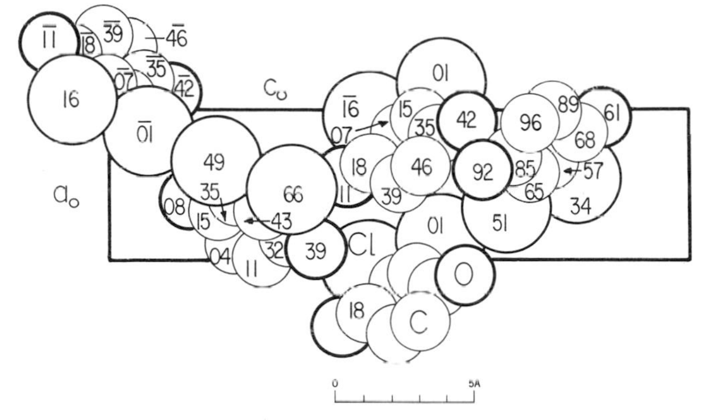

Fig. XVAV,12. The tetragonal structure of 2,3-dichloro-1,4-benzoquinone projected along an a_0 axis.

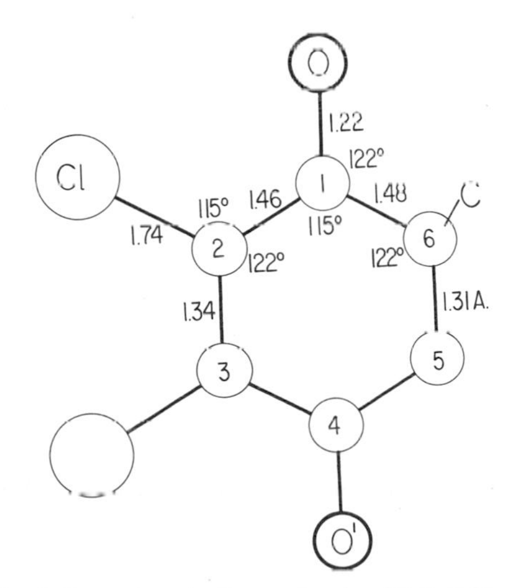

Fig. XVAV,13. Bond dimensions in the molecule of 2,3-dichloro-1,4-benzoquinone.

The structure (Fig. XVAV,12) has molecules with the bond dimensions of Figure XVAV,13. The shortest intermolecular distances in the crystal are Cl–O = 2.99 A. and C–O = 3.20 A.

XV,aV,13. The monoclinic crystals of *2,3-dimethyl-1,4-benzoquinone*, $C_6H_2(CH_3)_2O_2$, possess a tetramolecular unit of the dimensions:

$$a_0 = 9.01(1) \text{ A.}; \quad b_0 = 12.38(1) \text{ A.}; \quad c_0 = 7.36(2) \text{ A.}$$
$$\beta = 118.0(2)^\circ$$

The space group is C_{2h}^5 ($P2_1/c$) with atoms in the positions:

$$(4e) \quad \pm(xyz;\ x,{}^1/_2-y,z+{}^1/_2)$$

The parameters are those of Table XVAV,10.

TABLE XVAV,10

Parameters of the Atoms in 2,3-Dimethyl-1,4-benzoquinone

Atom	x	y	z
O(1)	−0.1628	−0.1769	0.1924
C(2)	−0.0557	−0.1088	0.2333
C(3)	−0.0973	0.0064	0.1981
C(4)	−0.2828	0.0321	0.0940
C(5)	0.0229	0.0800	0.2507
C(6)	−0.0039	0.1995	0.2272
C(7)	0.2001	0.0440	0.3496
O(8)	0.3124	0.1106	0.4058
C(9)	0.2394	−0.0696	0.3750
C(10)	0.1213	−0.1419	0.3223
H(11)	−0.328	−0.039	0.076
H(12)	−0.297	0.071	0.153
H(13)	−0.302	0.100	0.045
H(14)	0.099	0.237	0.259
H(15)	−0.014	0.229	0.321
H(16)	−0.101	0.208	0.110
H(17)	0.367	−0.091	0.437
H(18)	0.135	−0.216	0.330

The structure is made up of molecules which have the bond dimensions of Figure XVAV,14. The shortest intermolecular contacts not involving hydrogen atoms are C–O = 3.39, 3.47, and 3.49 A.

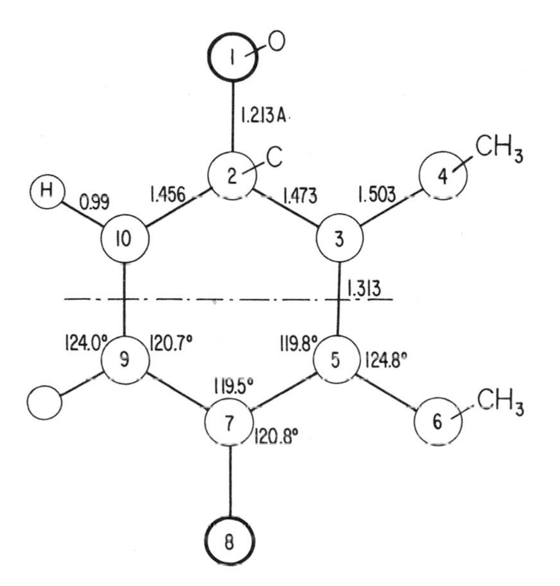

Fig. XVAV,14. Bond dimensions in the molecule of 2,3-dimethyl-1,4-benzoquinone.

XV,aV,14. The monoclinic crystals of *2,5-dichloro-1,4-benzoquinone*, $C_6H_2O_2Cl_2$, have a bimolecular unit of the dimensions:

$$a_0 = 10.103(13) \text{ A.}; \quad b_0 = 5.469(2) \text{ A.}; \quad c_0 = 6.018(5) \text{ A.}$$
$$\beta = 92°14(3)'$$

All atoms are in the positions

$$(4e) \quad \pm(xyz;\ x+{}^1/_2,{}^1/_2-y,z)$$

of C_{2h}^5 ($P2_1/a$). Their parameters are as follows:

Atom	x	y	z
Cl	0.1315	0.2385	−0.4079
O	0.2198	−0.1900	−0.1544
C(1)	0.1155	−0.1051	−0.0822
C(2)	0.0531	0.1173	−0.1876
C(3)	−0.0560	0.2175	−0.1075

The structure is shown in Figure XVAV,15. Its planar, centrosymmetric molecules have the bond dimensions of Figure XVAV,16. Between them the shortest distances are Cl–O = C–O = 3.12 A.

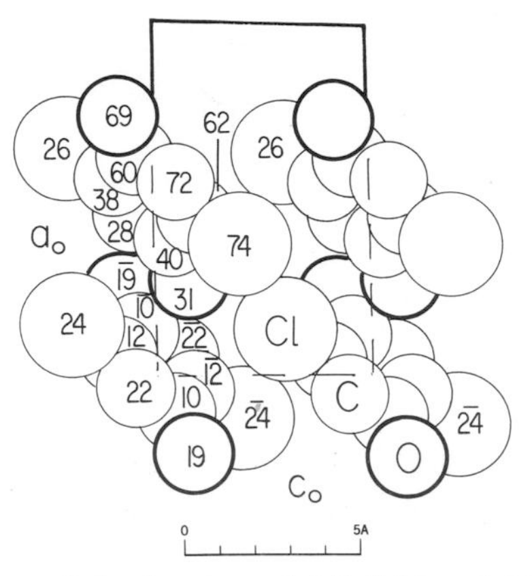

Fig. XVAV,15. The monoclinic structure of 2,5-dichloro-1,4-benzoquinone projected along its b_0 axis.

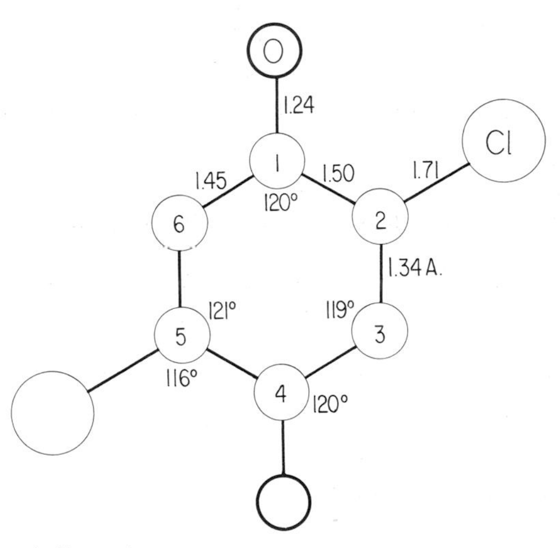

Fig. XVAV,16. Bond dimensions in the molecule of 2,5-dichloro-1,4-benzinoquone.

The corresponding *bromine derivative*, $C_6H_2O_2Br_2$, is isostructural. For it:

$$a_0 = 10.214(11) \text{ A.}; \quad b_0 = 5.654(3) \text{ A.}; \quad c_0 = 6.173(4) \text{ A.}$$
$$\beta = 91°58(5)'$$

There is also an isostructural mixed *2,5-chlorobromo-1,4-benzoquinone*, $C_2H_2O_2ClBr$, which has the cell:

$$a_0 = 10.573(21) \text{ A.}; \quad b_0 = 5.499(2) \text{ A.}; \quad c_0 = 6.099(4) \text{ A.}$$
$$\beta = 93°13(7)'$$

As in many other chlorine–bromine derivatives there appears to be disorder in the way chlorine and bromine together occupy one set of (4*e*).

The parameters for these compounds follow, those for the chlorobromo derivative being in parentheses:

Atom	x	y	z
Br[Cl,Br]	0.1365 (0.1354)	0.2500 (0.2431)	−0.4139 (−0.4133)
O	0.221 (0.2183)	0.173 (0.1830)	−0.145 (−0.1441)
C(1)	0.115 (0.1142)	−0.096 (−0.0975)	−0.082 (−0.0801)
C(2)	0.057 (0.0565)	0.112 (0.1176)	−0.177 (−0.1833)
C(3)	−0.059 (−0.0544)	0.204 (0.2112)	−0.107 (−0.1046)

In $C_6H_2O_2Br_2$, C–Br = 1.87 A.; in $C_6H_2O_2ClBr$, C–(Cl,Br) = 1.81 A.

XV,aV,15. Triclinic crystals of *2,5-dimethyl-p-benzoquinone*, $C_6H_2O_2(CH_3)_2$, have a bimolecular unit of the dimensions:

$$a_0 = 4.0132(2) \text{ A.}; \quad b_0 = 9.3663(2) \text{ A.}; \quad c_0 = 9.7383(4) \text{ A.}$$
$$\alpha = 93.501(2)°; \quad \beta = 101.356(5)°; \quad \gamma = 98.568(4)°$$

Atoms are in the positions (2*i*) $\pm(xyz)$ of C_i^1 ($P\bar{1}$). Parameters for all atoms, including hydrogen, are listed in Table XVAV,11.

TABLE XVAV,11

Parameters of the Atoms in $C_6H_2O_2(CH_3)_2$

Atom	x	y	z
O(1)	0.1964	0.1067	−0.2214
C(2)	0.0453	0.3131	−0.0294
C(3)	0.1057	0.0586	−0.1191
C(4)	0.0176	0.1553	−0.0092
C(5)	−0.0819	0.0974	0.1014
H(6)	−0.146	0.160	0.181
H(7)	−0.034	0.369	0.051
H(8)	0.299	0.359	−0.029
H(9)	−0.108	0.329	−0.125
O(1′)	0.3097	0.6169	0.7194
C(2′)	0.4691	0.8133	0.5230
C(3′)	0.4006	0.5646	0.6188
C(4′)	0.4889	0.6549	0.5063
C(5′)	0.5824	0.5919	0.3961
H(6′)	0.647	0.650	0.315
H(7′)	0.539	0.865	0.439
H(8′)	0.634	0.864	0.615
H(9′)	0.219	0.829	0.527

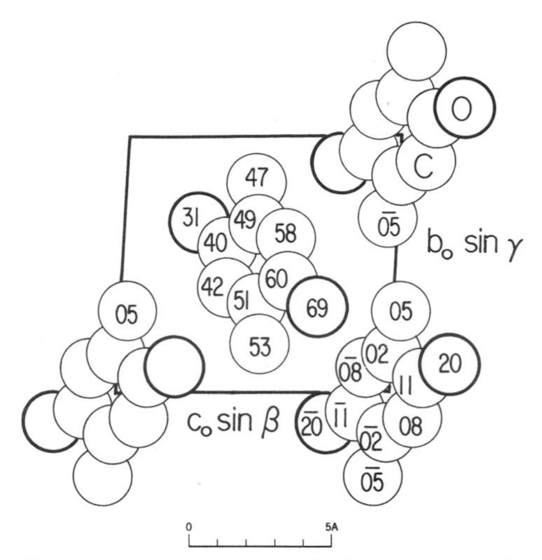

Fig. XVAV,17. The triclinic structure of 2,5-dimethyl-*p*-benzoquinone projected along its a_0 axis.

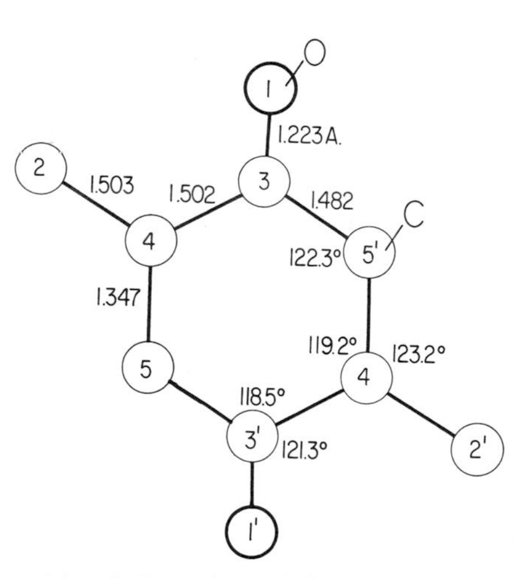

Fig. XVAV,18. Averaged bond dimensions of the molecules of 2,5-dimethyl-*p*-benzoquinone in its crystals.

The resulting structure, shown in Figure XVAV,17, contains two crystallographically different molecules each having a center of symmetry. Their bond dimensions are given in Figure XVAV,18. This study has been the basis for an especially detailed evaluation of electron distribution within the molecules.

XV,aV,16. The monoclinic crystals of *2,6-dimethyl-1,4-benzoquinone*, $C_6H_2(CH_3)_2O_2$, have a tetramolecular unit of the dimensions:

$$a_0 = 17.098(7) \text{ A.};\quad b_0 = 3.976(10) \text{ A.};\quad c_0 = 10.662(5) \text{ A.}$$
$$\beta = 93.51(1)°$$

All atoms are in the positions

$$(4e)\quad \pm(xyz;\ x+{}^1/_2,{}^1/_2-y,z)$$

of C_{2h}^5 ($P2_1/a$). Parameters, including those for hydrogen, are listed in Table XVAV,12.

TABLE XVAV,12

Parameters of the Atoms in 2,6-Dimethyl-1,4-benzoquinone

Atom	x	y	z
O(1)	0.3752	−0.2009	0.0852
C(2)	0.3751	−0.0513	0.1855
C(3)	0.4500	0.0548	0.2515
C(4)	0.5239	−0.0176	0.1881
C(5)	0.4482	0.2055	0.3632
C(6)	0.3754	0.2653	0.4239
O(7)	0.3758	0.4049	0.5268
C(8)	0.3022	0.1607	0.3572
C(9)	0.3003	0.0118	0.2448
C(10)	0.2261	−0.1029	0.1744
H(11)	0.525	0.141	0.122
H(12)	0.567	0.068	0.246
H(13)	0.522	−0.279	0.163
H(14)	0.496	0.309	0.407
H(15)	0.219	0.106	0.107
H(16)	0.185	−0.053	0.223
H(17)	0.236	−0.304	0.130
H(18)	0.253	0.211	0.397

The molecules in this structure (Fig. XVAV,19) have the bond dimensions of Figure XVAV,20. Except for hydrogen the molecule is planar to within 0.15 A. The separations C–H lie between 0.88 and 1.10 A.

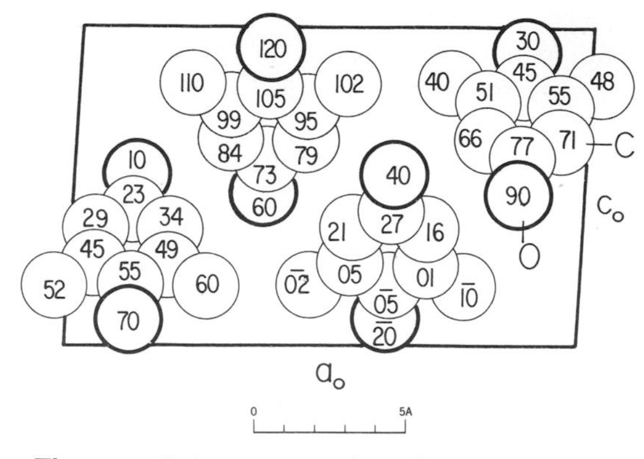

Fig. XVAV,19. The monoclinic structure of 2,6-dimethyl-1,4-benzoquinone projected along its b_0 axis.

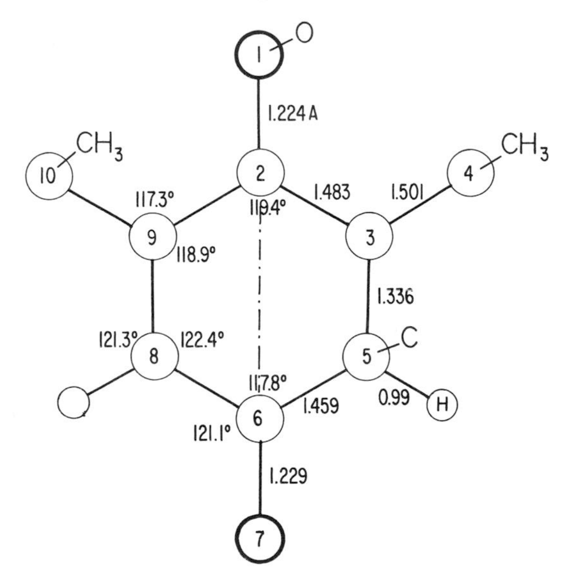

Fig. XVAV,20. Bond dimensions in the molecule of 2,6-dimethyl-1,4-benzoquinone.

XV,aV,17. Crystals of the *α-2-chloro-5-methyl-p-benzoquinone-4-oxime*, $C_6H_2(O)Cl(CH_3)NOH$, are monoclinic with a tetramolecular unit of the dimensions:

$$a_0 = 3.85 \text{ A.}; \quad b_0 = 13.30 \text{ A.}; \quad c_0 = 14.15 \text{ A.}; \quad \beta = 92°12'$$

Atoms are in the positions

$$(4e) \quad \pm(xyz;\ x,{}^1/_2-y,z+{}^1/_2)$$

of C_{2h}^5 ($P2_1/c$). Their parameters are stated in Table XVAV,13.

TABLE XVAV,13

Parameters of the Atoms in $C_6H_2(O)Cl(CH_3)NOH$

Atom	x	y	z
C(1)	0.622	0.1824	0.2062
C(2)	0.764	0.1224	0.1284
C(3)	0.878	0.1630	0.0520
C(4)	0.990	0.2694	0.0508
C(5)	0.850	0.3301	0.1291
C(6)	0.703	0.2890	0.2040
C(7)	0.918	0.4432	0.1220
N	0.121	0.3120	−0.0194
O(1)	0.496	0.1429	0.2739
O(2)	0.200	0.2516	−0.0893
Cl	0.646	−0.0063	0.1301

The structure (Fig. XVAV,21) is composed of molecules having the bond dimensions of Figure XVAV,22. They are tied together into strings along the c_0 direction by O–H–O bonds of length 2.69 A. Between these strings the shortest atomic separations are O–Cl = 3.40 A., C–O = 3.42 A., and C–N = 3.50 A.

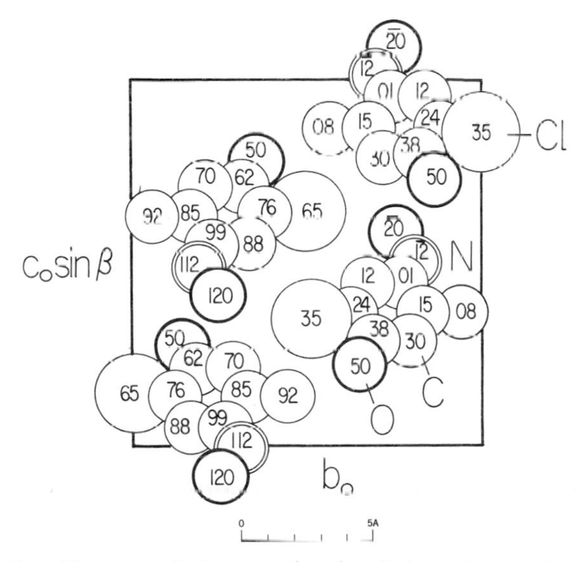

Fig. XVAV,21. The monoclinic crystals of α-2-chloro-5-methyl-p-benzoquinone-4-oxime projected along its a_0 axis.

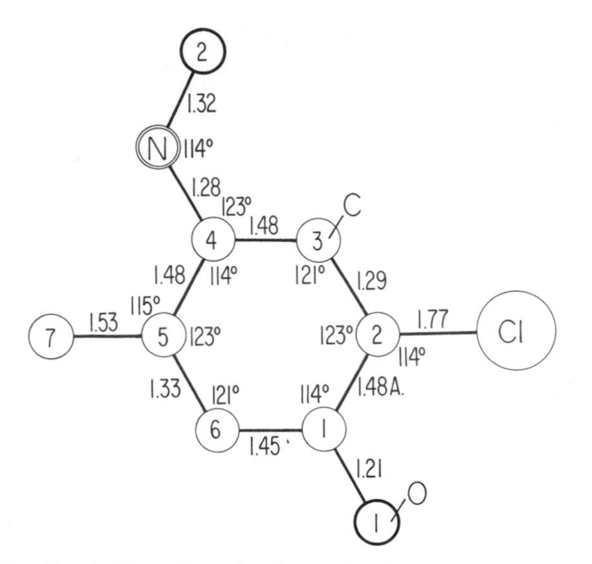

Fig. XVAV,22. Bond dimensions in the molecule of α-2-chloro-5-methyl-p-benzoquinone-4-oxime.

The corresponding bromine derivative is isostructural with:

$$a_0 = 3.91 \text{ A.}; \quad b_0 = 13.6 \text{ A.}; \quad c_0 = 14.2 \text{ A.}; \quad \beta = 90°$$

Parameters have not been established.

VI. Compounds containing a Penta-substituted Benzene Ring

XV,aVI,1. According to a preliminary description, the α form of *1-bromo-2,3,5,6-tetramethylbenzene*, $C_6HBr(CH_3)_4$, is orthorhombic with a tetramolecular unit of the edge lengths:

$$a_0 = 14.62(5) \text{ A.}; \quad b_0 = 5.43(2) \text{ A.}; \quad c_0 = 12.05(4) \text{ A.}$$

Atoms are in the positions

$$(4a) \qquad xyz;\ {}^1/_2-x,\bar{y},z+{}^1/_2;\ x+{}^1/_2,{}^1/_2-y,\bar{z};\ \bar{x},y+{}^1/_2,{}^1/_2-z$$

of the space group V^4 $(P2_12_12_1)$. The parameters have been given as those of Table XVAVI,1.

TABLE XVAVI,1

Parameters of the Atoms in $BrC_6H(CH_3)_4$

Atom	x	y	z
Br	0.2325	0.0698	0.0618
C(1)	0.334	0.054	0.150
C(2)	0.409	−0.125	0.137
C(3)	0.487	−0.130	0.207
C(4)	0.490	0.035	0.290
C(5)	0.420	0.188	0.310
C(6)	0.342	0.216	0.246
C(7)	0.390	−0.275	0.033
C(8)	0.565	−0.305	0.186
C(9)	0.425	0.381	0.407
C(10)	0.262	0.395	0.267

The higher temperature (β) form is monoclinic. Its tetramolecular unit has the dimensions:

$$a_0 = 16.07(5) \text{ A.}; \quad b_0 = 5.80(2) \text{ A.}; \quad c_0 = 11.30(1) \text{ A.}$$
$$\beta = 110°45(30)'$$

The space group is C_{2h}^5 ($P2_1/a$) and parameters have been stated for x and z but not for y.

XV,aVI,2. According to a preliminary note, crystals of *1,3-dichloro-2,4,6-trinitrobenzene*, $C_6HCl_2(NO_2)_3$, are orthorhombic. Their tetramolecular unit has the edge lengths:

$$a_0 = 5.93 \text{ A.}; \quad b_0 = 14.98 \text{ A.}; \quad c_0 = 10.91 \text{ A.}$$

The space group is V_h^{14} ($Pbcn$) with atoms in the positions:

$$(4c) \quad \pm(0\,u\,{}^1/_4;\ {}^1/_2,u+{}^1/_2,{}^1/_4)$$
$$(8d) \quad \pm(xyz;\ x+{}^1/_2,y+{}^1/_2,{}^1/_2-z;\ {}^1/_2-x,y+{}^1/_2,z;\ x,\bar{y},z+{}^1/_2)$$

Parameters are stated in Table XVAVI,2.

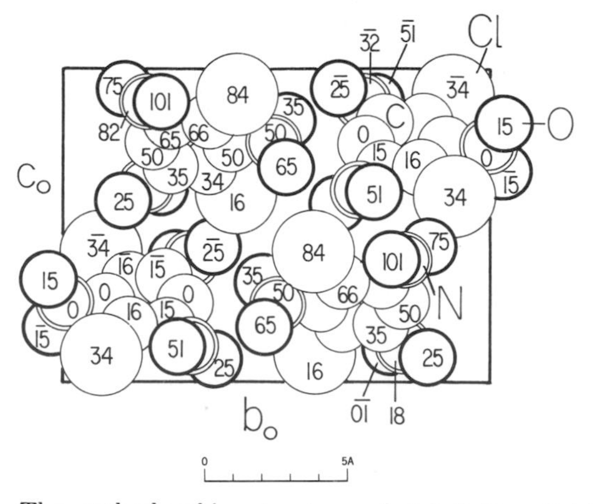

Fig. XVAVI,1. The orthorhombic structure of 1,3-dichloro-2,4,6-trinitrobenzene projected along its a_0 axis.

TABLE XVAVI,2

Positions and Parameters of the Atoms in $C_6HCl_2(NO_2)_3$

Atom	Position	x	y	z
Cl(3)	(8*d*)	0.3374(5)	0.0904(1)	0.0906(2)
C(2)	(4*c*)	0	0.1070(8)	$^1/_4$
C(3)	(8*d*)	0.1601(16)	0.1520(5)	0.1776(7)
C(4)	(8*d*)	0.1546(15)	0.2435(5)	0.1798(7)
C(5)	(4*c*)	0	0.2894(8)	$^1/_4$
N(2)	(4*c*)	0	0.0104(7)	$^1/_4$
N(4)	(8*d*)	0.3189(15)	0.2970(5)	0.1111(6)
O(2)	(8*d*)	0.1480(13)	−0.0265(4)	0.3064(6)
O(4)	(8*d*)	0.5119(12)	0.2686(4)	0.1028(6)
O(4′)	(8*d*)	0.2520(11)	0.3665(4)	0.0662(6)
H	(4*c*)	0	0.360(10)	$^1/_4$

The structure is shown in Figure XVAVI,1. Its molecules, with a twofold axis of symmetry through the C(2) and C(5) atoms, have the bond dimensions of Figure XVAVI,2. The NO_2 group about N(2) is turned through 75° and that about N(4) through 37° with respect to the plane of the ring.

XV,aVI,3. The triclinic crystals of *N,3-dimethyl-4-bromo-2,6-dinitroaniline*, $C_6H(CH_3)(NO_2)_2(NHCH_3)Br$, have a bimolecular unit of the dimensions.:

$a_0 = 6.14(2)$ A.; $b_0 = 11.22(4)$ A.; $c_0 = 8.28(2)$ A.
$\alpha = 77.3(3)$ °; $\beta = 98.3(4)$°; $\gamma = 109.7(3)$°

The space group, chosen as C_i^1 ($P\bar{1}$), puts atoms in the positions: ($2i$) $\pm(xyz)$. Parameters are listed in Table XVAVI,3.

TABLE XVAVI,3

Parameters of the Atoms in $C_6H(CH_3)(NO_2)_2(NHCH_3)Br$

Atom	x	y	z
Br	0.23829(14)	0.37062(5)	0.17131(8)
C(1)	0.4863(12)	0.8111(4)	0.1180(7)
C(2)	0.2783(14)	0.7277(4)	0.1997(6)
C(3)	0.0675(13)	0.6715(4)	0.1253(7)
C(4)	0.0514(14)	0.7022(4)	0.9506(7)
C(5)	0.2291(14)	0.7793(4)	0.8618(6)
C(6)	0.4459(12)	0.8331(4)	0.9429(6)
C(7)	0.2487(19)	0.1294(7)	0.6322(8)
C(8)	0.1311(16)	0.4183(6)	0.7750(9)
O(1)	0.2093(12)	0.7327(6)	0.4695(7)
O(2)	0.4092(13)	0.6124(4)	0.4327(6)
O(3)	0.1831(11)	0.0442(5)	0.1042(5)
O(4)	0.4269(11)	0.0625(5)	0.3108(4)
N(1)	0.3130(12)	0.1359(4)	0.8032(6)
N(2)	0.3001(12)	0.6878(4)	0.3824(6)
N(3)	0.3767(11)	0.0863(4)	0.1633(6)
H(N,1)	0.811	0.914	0.116
H(5,1)	0.209	0.800	−0.271
H(7,1)	0.753	0.776	0.440
H(7,2)	0.637	0.919	0.417
H(7,3)	0.929	0.927	0.373
H(8,1)	−0.199	0.639	0.303
H(8,2)	−0.076	0.501	0.304
H(8,3)	−0.280	0.551	0.139

The structure, shown in Figure XVAVI,3, has molecules with the bond dimensions of Figure XVAVI,4. The benzene ring and directly attached atoms are coplanar to within 0.1 A. and the NO_2 group attached to C(6) is twisted by only 6° from this plane. As a result the O(3) atom is 2.61 A.

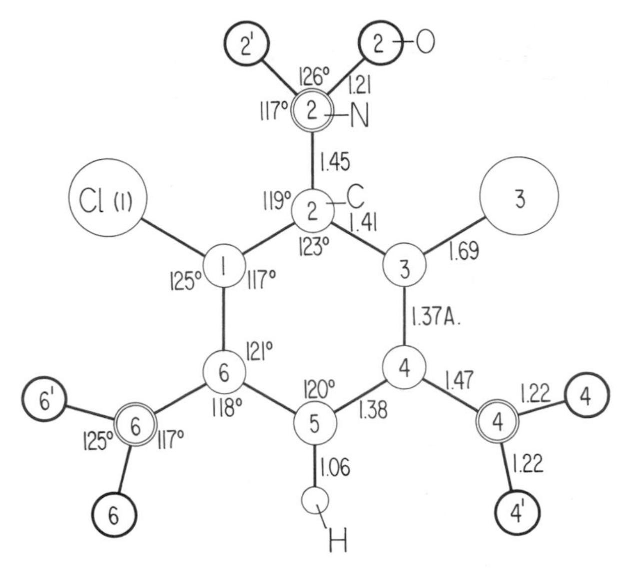

Fig. XVAVI,2. Bond dimensions in the molecule of 1,3-dichloro-2,4,6-trinitrobenzene.

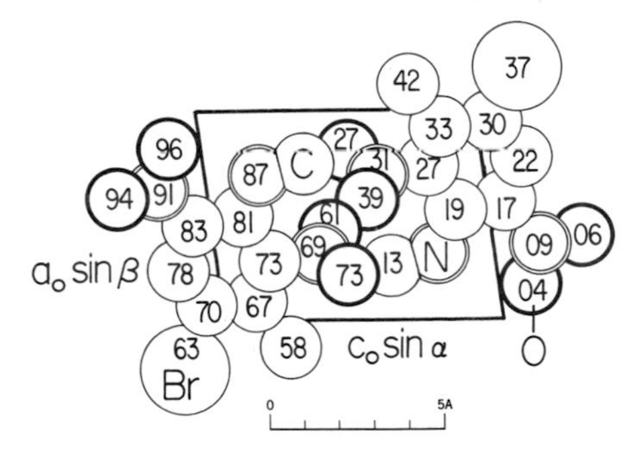

Fig. XVAVI,3. The triclinic structure of *N*,3-dimethyl-4-bromo-2,6-dinitroaniline projected along its b_0 axis.

from the amino N(1) to provide what has been considered an intramolecular hydrogen bond. The twist of the NO_2 group attached to C(2) is, on the contrary, 73°. Between molecules there are atomic separations as short as O–O = 2.68 A.

XV,aVI,4. The monoclinic crystals of *1,3-diamino-2,4,6-trinitrobenzene*, $C_6H(NH_2)_2(NO_2)_3$, possess a bimolecular unit of the dimensions:

$$a_0 = 7.30(1) \text{ A.}; \quad b_0 = 5.20(1) \text{ A.}; \quad c_0 = 11.63(2) \text{ A.}$$
$$\beta = 95.9(3)°$$

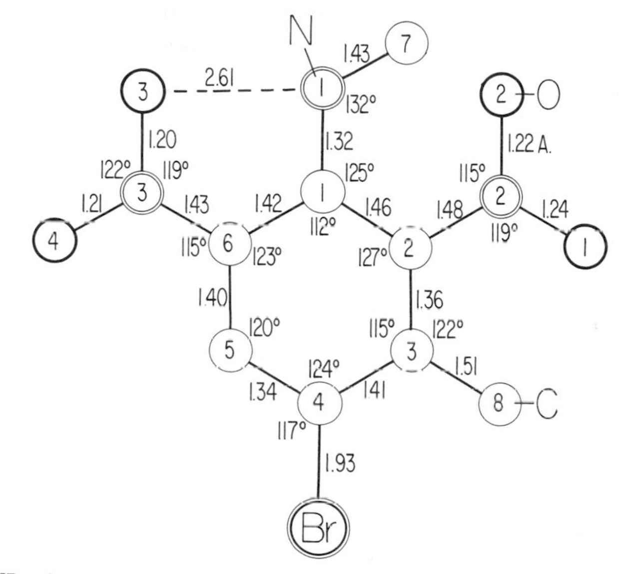

Fig. XVAVI,4. Bond dimensions in the molecule of N,3-dimethyl-4-bromo-2,6-dinitroaniline.

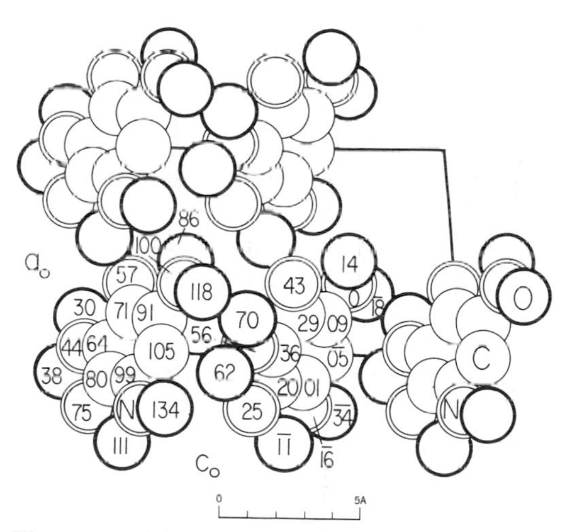

Fig. XVAVI,5. The monoclinic structure of 1,3-diamino-2,4,6-trinitrobenzene projected along its b_0 axis.

The space group is the low symmetry C_s^2 (Pc) with atoms in the positions:

$$(2a) \quad xyz;\ x,\bar{y},z+1/2$$

Parameters are given in Table XVAVI,4.

TABLE XVAVI,4

Parameters of the Atoms in $C_6H(NH_2)_2(NO_2)_3$

Atom	x	y	z
C(1)	0.1760	0.7060(27)	0.0015
C(2)	0.0168(21)	0.6447(28)	−0.0814(14)
C(3)	−0.1520(25)	0.8007(29)	−0.0860(15)
C(4)	−0.1533(21)	0.9932(27)	−0.0029(14)
C(5)	−0.0046(24)	0.0516(27)	0.0756(14)
C(6)	0.1529(23)	0.9062(25)	0.0773(14)
N(1)	0.3279(21)	0.5667(24)	0.0030(14)
N(2)	0.0276(22)	0.4379(24)	−0.1595(15)
N(3)	−0.2922(22)	0.7495(26)	−0.1632(14)
N(4)	−0.3080(22)	0.1599(24)	0.0041(14)
N(5)	0.3080(22)	0.9981(25)	0.1640(14)
O(1)	0.1680(21)	0.2956(22)	−0.1566(14)
O(2)	−0.1050(21)	0.3830(25)	−0.2309(14)
O(3)	−0.4556(20)	0.1120(22)	−0.0581(14)
O(4)	−0.3013(20)	0.3385(20)	0.0728(13)
O(5)	0.2824(21)	0.1835(22)	0.2182(13)
O(6)	0.4483(21)	0.8637(24)	0.1724(14)

The structure (Fig. XVAVI,5) has molecules that are essentially planar, the maximum departure of any atom from the plane of the benzene ring

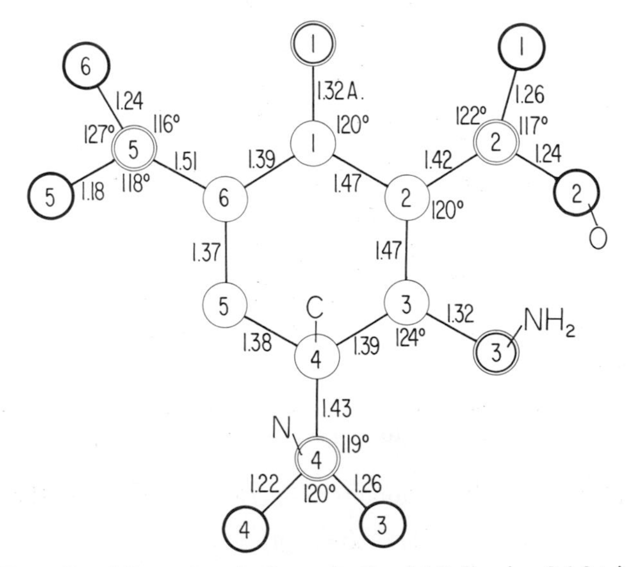

Fig. XVAVI,6. Bond dimensions in the molecule of 1,3-diamino-2,4,6-trinitrobenzene.

being less than 0.2 A. Bond dimensions in these molecules are to be seen in Figure XVAVI,6. There appear to be intramolecular hydrogen bonds between the amino nitrogen and the oxygen atoms of adjacent nitro groups; their separations are shown in the figure. Between molecules there also appear to be longer hydrogen bonds involving the N(1) amino groups and oxygens, with N–H–O = 2.97 or 2.99 A.

XV,aVI,5. There are four molecules in the monoclinic unit of *2,3,4,6-tetranitroaniline*, $C_6HNH_2(NO_2)_4$. For it the dimensions are:

$$a_0 = 7.27(1) \text{ A.}; \quad b_0 = 11.06(2) \text{ A.}; \quad c_0 = 12.27(2) \text{ A.}$$
$$\beta = 98°48(18)'$$

Atoms are in the positions

$$(4e) \quad \pm(xyz;\ x,{}^1/_2-y,z+{}^1/_2)$$

of C_{2h}^5 ($P2_1/c$) with the parameters listed in Table XVAVI,5.

TABLE XVAVI,5

Parameters of the Atoms in 2,3,4,6-Tetranitroaniline

Atom	x	y	z
C(1)	0.7532(5)	0.3357(3)	0.2208(3)
C(2)	0.5893(5)	0.4094(4)	0.2004(3)
C(3)	0.4453(5)	0.3834(4)	0.1192(3)
C(4)	0.4541(5)	0.2841(4)	0.0515(3)
C(5)	0.6127(6)	0.2143(4)	0.0623(3)
C(6)	0.7587(5)	0.2401(3)	0.1431(3)
N(1)	0.8867(5)	0.3577(4)	0.3031(3)
N(2)	0.5837(5)	0.5209(3)	0.2646(3)
N(3)	0.2815(5)	0.4655(3)	0.1039(3)
N(4)	0.2953(5)	0.2434(4)	−0.0254(3)
N(6)	0.9187(5)	0.1591(3)	0.1489(3)
O(2,1)	0.6364(4)	0.5170(3)	0.3636(2)
O(2,2)	0.5262(5)	0.6118(3)	0.2146(3)
O(3,1)	0.1903(4)	0.4682(4)	0.1786(3)
O(3,2)	0.2551(5)	0.5215(3)	0.0185(3)
O(4,1)	0.1419(4)	0.2826(4)	−0.0150(3)
O(4,2)	0.3225(5)	0.1718(4)	−0.0977(3)
O(6,1)	0.9133(5)	0.0794(3)	0.0795(3)
O(6,2)	0.0504(4)	0.1752(3)	0.2211(3)
H(5)	0.6125(72)	0.1431(51)	0.0098(41)
H(1,1)	0.9919(77)	0.3028(56)	0.3157(43)
H(1,2)	0.8721(74)	0.4194(59)	0.3555(45)

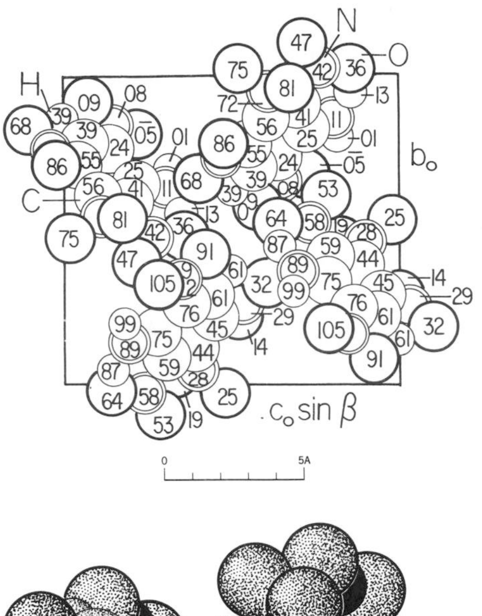

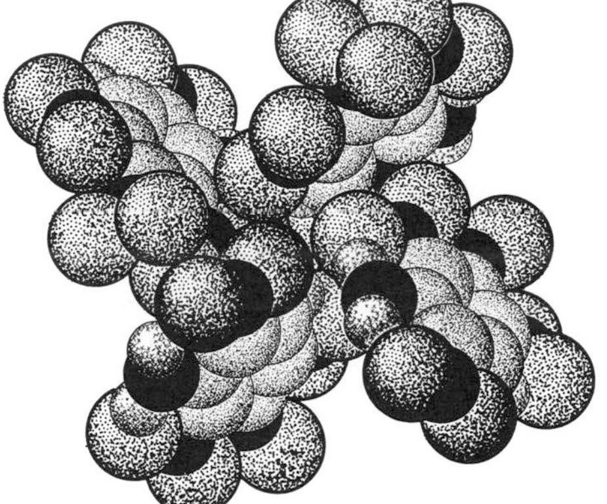

Fig. XVAVI,7a (top). The monoclinic structure of 2,3,4,6-tetranitroaniline projected along its a_0 axis.

Fig. XVAVI,7b (bottom). A packing drawing of the monoclinic structure of 2,3,4,6-tetranitroaniline viewed along its a_0 axis. The nitrogen atoms are black; the oxygens are heavily outlined and coarsely dotted. The atoms of carbon are equally large but lightly dot shaded; the small cross shaded atoms are hydrogen.

The structure, as shown in Figure XVAVI,7, is built up of molecules having the bond dimensions of Figure XVAVI,8. The benzene ring is planar and its attached nitrogen atoms are from 0.08 to 0.22 A. outside this plane. The nitro groups are rotated by angles of 45° [for N(2)], 64° [for N(3)], 19° [for N(4)] and 3° [for N(6)] about their N–C bonds, the direction of rotation being clockwise except for the N(6) group.

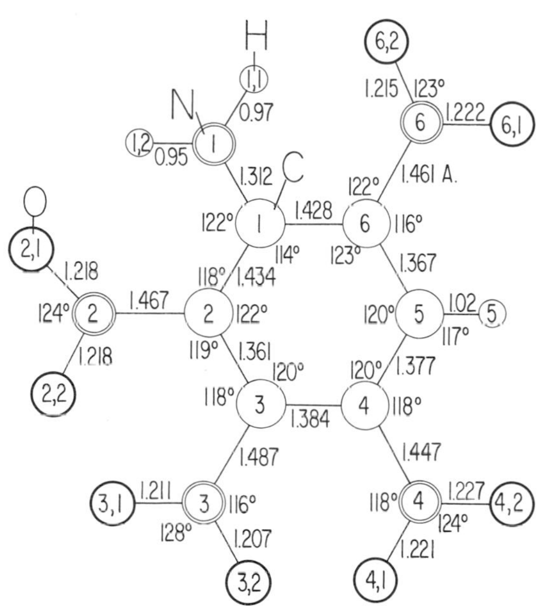

Fig. XVAVI,8. Bond dimensions in the molecule of 2,3,4,6-tetranitroaniline.

VII. Compounds containing a Hexa-substituted Benzene Ring

XV,aVII,1. The monoclinic crystals of *hexachlorobenzene*, C_6Cl_6, have a bimolecular unit of the dimensions:

$$a_0 = 8.08(2) \text{ A.}; \quad b_0 = 3.87(1) \text{ A.}; \quad c_0 = 16.65(3) \text{ A.}$$
$$\beta = 117.0(2)°$$

All atoms are in the positions

$$(4e) \quad \pm(xyz;\ x,{}^1/_2-y,z+{}^1/_2)$$

of C_{2h}^5 ($P2_1/c$). According to the latest and most complete study, the parameters are those of Table XVAVII,1.

TABLE XVAVII,1
Parameters of the Atoms in C_6Cl_6

Atom	x	y	z
C(1)	0.183	0.111	0.030
C(2)	0.126	−0.015	0.092
C(3)	−0.058	−0.125	0.062
Cl(1)	0.408	0.245	0.066
Cl(2)	0.279	−0.033	0.204
Cl(3)	−0.128	−0.278	0.138

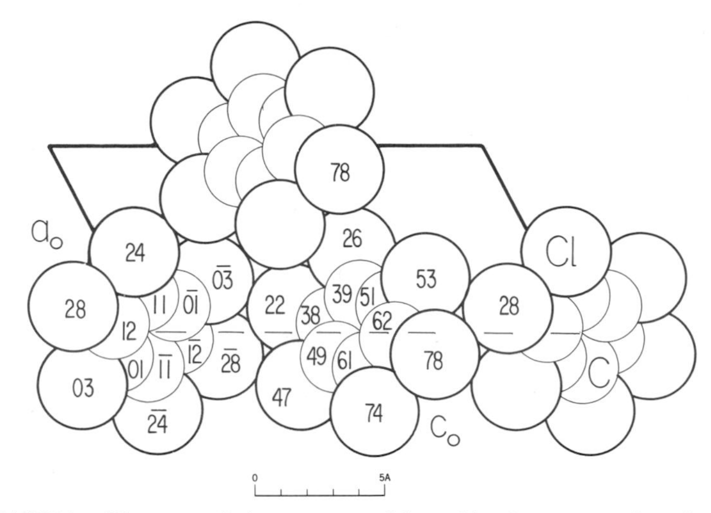

Fig. XVAVII,1. The monoclinic structure of hexachlorobenzene projected along its b_0 axis.

The structure, shown in Figure XVAVII,1, has molecules that are planar to within 0.02 A.; in them C–C = 1.40 A. and C–Cl = 1.71 A. Between them the shortest atomic separation is Cl–Cl = 3.51 A.

Four other hexa-substituted benzenes are known to be isostructural though in each case there is statistical disorder in the way the substituents are oriented within the crystal. Two of these compounds are *pentachlorobromobenzene*, C_6Cl_5Br, with

$$a_0 = 8.18 \text{ A.}; \quad b_0 = 3.87 \text{ A.}; \quad c_0 = 16.61 \text{ A.}; \quad \beta = 116°40'$$

and *pentabromochlorobenzene*, C_6Br_5Cl, with

$$a_0 = 8.34 \text{ A.}; \quad b_0 = 4.00 \text{ A.}; \quad c_0 = 17.14 \text{ A.}; \quad \beta = 116°45'$$

For both these crystals the x-ray data do not discriminate between the chlorine and bromine atoms (the molecules have an apparent center of symmetry). As a result sometimes one and sometimes the other of the two halogens are to be thought of as having the X parameters listed in Table XVAVII,2.

TABLE XVAVII,2

Parameters of the Atoms in C_6Cl_5Br and C_6Br_5Cl (in parentheses)

Atom	x	y	z
C(1)	0.181 (0.176)	0.108 (0.118)	0.030 (0.029)
C(2)	0.124 (0.121)	0.063 (−0.054)	0.090 (0.088)
C(3)	−0.058 (−0.055)	−0.139 (−0.134)	0.061 (0.060)
X(1)	0.407 (0.408)	0.243 (0.273)	0.068 (0.067)
X(2)	0.279 (0.282)	0.142 (0.126)	0.202 (0.204)
X(3)	−0.128 (−0.127)	−0.311 (−0.309)	0.137 (0.138)

Another compound with this structure is *tetrabromo-1,3-dimethylbenzene*, $C_6Br_4(CH_3)_2$. Its unit has the dimensions:

$$a_0 = 8.25(5) \text{ A.}; \quad b_0 = 3.96(2) \text{ A.}; \quad c_0 = 17.24(7) \text{ A.}$$
$$\beta = 117°8(8)'$$

The atomic parameters have been determined to be those of Table XVAVII,3, with sometimes a bromine atom and sometimes a methyl radical in an X position.

TABLE XVAVII,3

Parameters of the Atoms in $C_6Br_4(CH_3)_2$

Atom	x	y	z
C(1)	0.183	0.086	0.068
C(2)	0.124	−0.078	0.188
C(3)	−0.070	−0.038	0.118
X(1)	0.414	0.232	0.144
X(2)	0.278	−0.057	0.414
X(3)	−0.135	−0.275	0.272

Still another compound with this arrangement is *1,2-dichloro-tetramethylbenzene*, $C_6Cl_2(CH_3)_4$. For it the unit has the dimensions:

$$a_0 = 8.16(2) \text{ A.}; \quad b_0 = 3.98(1) \text{ A.}; \quad c_0 = 16.95(3) \text{ A.}$$
$$\beta = 116.4(2)°$$

The parameters attributed to it are listed in Table XVAVII,4, the X positions again being occupied sometimes by chlorine and sometimes by a methyl group.

TABLE XVAVII,4

Parameters of the Atoms in $C_6Cl_2(CH_3)_4$

Atom	x	y	z
C(1)	0.192	0.103	0.072
C(2)	0.128	0.000	0.183
C(3)	−0.067	−0.103	0.114
X(1)	0.408	0.258	0.156
X(2)	0.256	0.000	0.397
X(3)	−0.145	−0.258	0.250

XV,aVII,2. The association product *hexabromobenzene-1,2,4,5-tetrabromobenzene*, $C_6Br_6 \cdot C_6H_2Br_4$, forms monoclinic crystals with a bimolecular unit of the dimensions:

$$a_0 = 17.80(2) \text{ A.}; \quad b_0 = 4.01(4) \text{ A.}; \quad c_0 = 14.42(2) \text{ A.}$$
$$\beta = 111°13(7)'$$

Atoms are in the positions

$$(4e) \quad \pm(xyz;\ x+{}^1/_2,{}^1/_2-y,z)$$

of the space group C_{2h}^5 ($P2_1/a$). The determined parameters are stated in Table XVAVII,5.

TABLE XVAVII,5

Parameters of the Atoms in $C_6Br_6 \cdot C_6H_2Br_4$

Atom	x	y	z
Br(1)	0.0942(2)	0.1630(15)	−0.1530(3)
Br(2)	0.1908(2)	−0.1922(14)	0.0628(3)
Br(3)	0.0968(3)	−0.3591(12)	0.2144(3)
C(1)	0.0380(15)	0.059(9)	−0.0649(19)
C(2)	0.0846(18)	−0.114(10)	0.0276(22)
C(3)	0.0395(18)	−0.142(11)	0.0918(22)
Br(4)	0.1560(2)	0.2220(12)	0.4533(3)
Br(5)	0.1782(2)	0.5524(15)	0.6767(3)
C(4)	0.0686(19)	0.389(11)	0.4831(22)
C(5)	0.0787(18)	0.529(10)	0.5751(21)
C(6)	0.0118(18)	0.631(11)	0.5967(21)

The structure that results is shown in Figure XVAVII,2. It consists of alternate stacks (along b_0) of the two kinds of molecule, each of which is planar and has the expected dimensions.

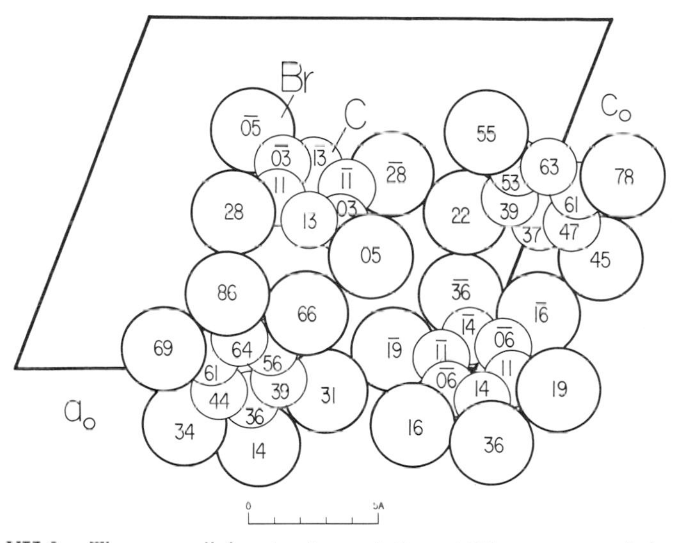

Fig. XVAVII,2. The monoclinic structure of the addition compound hexabromobenzene-1,2,4,5-tetrabromobenzene projected along its b_0 axis.

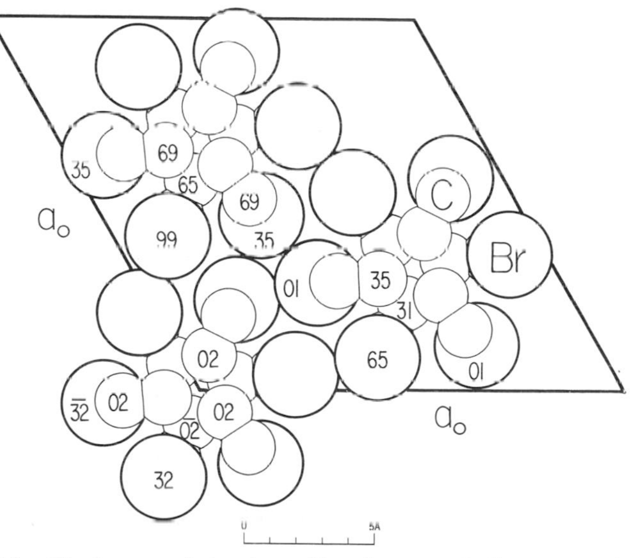

Fig. XVAVII,3. The hexagonal structure of hexabromomethylbenzene projected along its c_0 axis.

XV,aVII,3. *Hexabromomethylbenzene,* $C_6(CH_2Br)_6$, has rhombohedral symmetry. For it the unimolecular rhombohedron has the dimensions:

$$a_0 = 9.62 \text{ A.}, \qquad \alpha = 116°36'$$

The corresponding trimolecular cell referred to hexagonal axes has the edges:

$$a_0' = 16.41(2) \text{ A.}, \qquad c_0' = 5.38(2) \text{ A.}$$

The space group is C_{3i}^2 ($R\bar{3}$) with atoms, in terms of this hexagonal unit, in the positions:

$$(18f) \quad \pm(xyz;\ \bar{y},x-y,z;\ y-x,\bar{x},z); \quad \text{rh}$$

According to a recent redetermination, the parameters are as follows:

Atom	x	y	z
C(1)	0.0945	0.0150	−0.0165
C(2)	0.1950	0.0375	−0.0190
Br	0.2440	0.0402	0.3175

The structure is shown in Figure XVAVII,3. The molecule is approximately planar except for the bromine atoms which lie far from this plane. In the ring, C–C = 1.39 A. and between these carbon atoms and CH_2 the distance is 1.53 A.; CH_2–Br = 1.98 A. Between molecules the shortest Br–Br = 3.72 A. These new atomic positions differ materially from those found many years ago in the original investigation.

XV,aVII,4. *Hexamethylbenzene,* $C_6(CH_3)_6$, is triclinic. Its unimolecular cell has the dimensions:

$$a_0 = 8.92 \text{ A.};\quad b_0 = 8.86 \text{ A.};\quad c_0 = 5.30 \text{ A.}$$
$$\alpha = 44°27';\quad \beta = 116°43';\quad \gamma = 119°34'$$

Atoms, in general positions $(2i)$ $\pm(xyz)$ of C_i^1 ($P\bar{1}$) have the parameters of Table XVAVII,6. This results in planar molecules which are almost exactly in the a_0b_0 plane. In view of the close approach of γ to 120° the molecules thus form a nearly perfect hexagonal net. Succeeding layers along c_0 are displaced so that molecular centers overlie points between molecules. In a molecule the ring C–C = 1.39 A., and the C(ring)–CH_3 = 1.53 A. Between molecules the C–C distances range upwards from 3.70 A.

TABLE XVAVII,6

Parameters of the Atoms in Hexamethylbenzene

Atom	x	y	z
C(1)	0.372	0.234	−0.016
C(2)	0.144	0.379	−0.007
C(3)	−0.227	0.146	0.008
C(4)	0.177	0.111	−0.007
C(5)	0.069	0.180	−0.003
C(6)	−0.108	0.069	0.004

Hexaethylbenzene, $C_6(C_2H_5)_6$, has a unit of similar size with

$$a_0 = 9.90 \text{ A.}; \quad b_0 = 9.84 \text{ A.}; \quad c_0 = 6.10 \text{ A.}$$
$$\alpha = 58°5'; \quad \beta = 103°54'; \quad \gamma = 123°43'$$

Its structure must be closely related to that of the methyl derivative.

XV,aVII,5. Crystals of *hexamethylbenzene–chromium tricarbonyl*, $(CH_3)_6C_6 \cdot Cr(CO)_3$, are orthorhombic with a large unit that contains eight molecules and has the edge lengths:

$$a_0 = 13.67(3) \text{ A.}; \quad b_0 = 13.53(3) \text{ A.}; \quad c_0 = 15.27(3) \text{ A.}$$

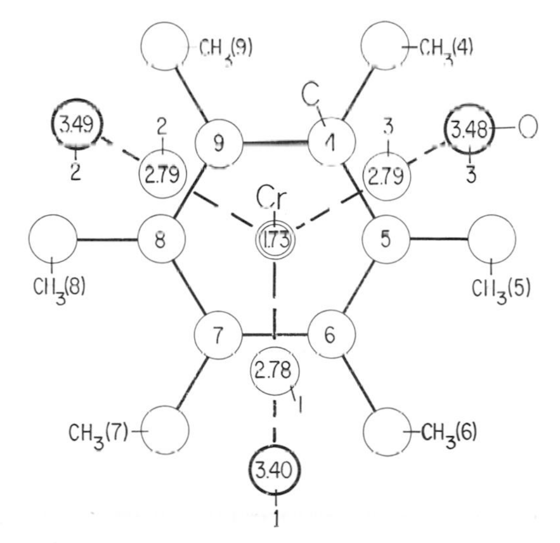

Fig. XVAVII,4. A drawing to show the principal bond dimensions in the molecule of hexamethylbenzene-chromium tricarbonyl.

Atoms are in the positions

$$(8c) \quad \pm(xyz;\ {}^1/_2-x,y+{}^1/_2,z;\ x,{}^1/_2-y,z+{}^1/_2;\ x+{}^1/_2,y,{}^1/_2-z)$$

of V_h^{15} (*Pbca*). The parameters are listed in Table XVAVII,7.

TABLE XVAVII,7

Parameters of the Atoms in $(CH_3)_6C_6 \cdot Cr(CO)_3$

Atom	x	y	z
Cr	0.1927(1)	0.2431(1)	0.0770(1)
C(1)	0.1994(10)	0.3381(8)	−0.0074(7)
O(1)	0.2070(8)	0.3953(7)	−0.0615(5)
C(2)	0.2888(7)	0.3062(10)	0.1391(7)
O(2)	0.3477(6)	0.3474(7)	0.1790(6)
C(3)	0.1042(8)	0.3163(9)	0.1340(7)
O(3)	0.0421(6)	0.3645(7)	0.1696(6)
C(4)	0.1358(7)	0.1057(7)	0.1366(6)
C(5)	0.0816(8)	0.1223(7)	0.0564(7)
C(6)	0.1384(8)	0.1366(8)	−0.0238(6)
C(7)	0.2390(10)	0.1300(8)	−0.0204(7)
C(8)	0.2908(6)	0.1120(7)	0.0586(7)
C(9)	0.2358(8)	0.0989(7)	0.1370(6)
$CH_3(4)$	0.0745(8)	0.0908(9)	0.2220(8)
$CH_3(5)$	−0.0275(8)	0.1286(9)	0.0567(9)
$CH_3(6)$	0.0796(10)	0.1540(9)	−0.1062(7)
$CH_3(7)$	0.3014(9)	0.1389(9)	−0.1039(7)
$CH_3(8)$	0.3987(9)	0.0991(9)	0.0599(7)
$CH_3(9)$	0.2918(8)	0.0795(10)	0.2218(8)

The molecules that constitute this structure have their chromium atoms centered over benzene rings. The atoms of the ring and its attached methyl groups are essentially coplanar. The general shape of the molecule can be seen from Figure XVAVII,4 where numbers within the various atomic circles give the distances of atoms above the plane of the ring. The distances of the methyl carbon atoms from their attached ring carbon atoms range between 1.486 and 1.564 A.

XV,aVII,6. The complex *chloranil–hexamethylbenzene*, $C_6Cl_4O_2 \cdot C_6(CH_3)_6$, forms monoclinic crystals which have two molecules in a unit of the dimensions:

$$a_0 = 7.30 \text{ A.}; \quad b_0 = 8.64 \text{ A.}; \quad c_0 = 15.26 \text{ A.}; \quad \beta = 106°$$

The space group is C_{2h}^5 ($P2_1/c$) with atoms in the positions:

$$(4e) \quad \pm(xyz;\ x,{}^1/_2-y,z+{}^1/_2)$$

The original assignment of parameters gave molecules that depart far from planarity but a more recent reworking of the original data leads to molecules that are planar within the rather large limits of error of the determination. The parameters from this recalculation are stated in Table XVAVII,8.

TABLE XVAVII,8

Parameters of the Atoms in $C_6Cl_4O_2 \cdot C_6(CH_3)_6$

Atom	x	y	z
C(1)	0.071(5)	−0.140(6)	0.028(2)
C(2)	0.056(7)	−0.024(5)	0.093(3)
C(3)	−0.011(6)	0.124(5)	0.063(3)
Cl(1)	−0.024(2)	0.260(2)	0.142(1)
Cl(2)	0.141(2)	−0.335(2)	0.071(1)
O	0.095(4)	−0.058(4)	0.181(2)
C(4)	0.576(6)	−0.132(6)	0.057(3)
C(5)	0.542(8)	0.007(5)	0.098(3)
C(6)	0.479(6)	0.130(6)	0.037(3)
CH_3(1)	0.446(6)	0.302(7)	0.086(3)
CH_3(2)	0.654(6)	−0.265(8)	0.116(4)
CH_3(3)	0.590(7)	0.034(8)	0.202(4)

The resulting molecular arrangement is indicated in Figure XVAVII,5.

XV,aVII,7. *Tetrachloro-p-benzoquinone* (chloranil), $C_6Cl_4O_2$, forms monoclinic crystals. Their bimolecular unit has the dimensions:

$$a_0 = 8.708(8) \text{ A.}; \quad b_0 = 5.755(5) \text{ A.}; \quad c_0 = 8.603(8) \text{ A.}$$
$$\beta = 105°51(8)'$$

Atoms are in the positions

$$(4e) \quad \pm(xyz;\ x+{}^1/_2,{}^1/_2-y,z)$$

of C_{2h}^5 ($P2_1/a$). The determined parameters are listed in Table XVAVII,9.

TABLE XVAVII,9

Parameters of the Atoms in Chloranil

Atom	x	y	z
Cl(1)	0.0244	0.2282	−0.3218
Cl(2)	0.1904	0.2177	0.3295
O	0.1640	0.3843	0.0071
C(1)	0.0058	0.0981	−0.1501
C(2)	0.0824	0.0923	0.1528
C(3)	0.0912	0.2080	0.0020

The structure, illustrated in Figure XVAVII,6, has molecules with the averaged bond dimensions of Figure XVAVII,7. Their rings are planar but the chlorine and oxygen are 0.06 A. on either side of the plane. The shortest intermolecular distances are C–O = 2.85, 3.06, and 3.14 A.

An independent, and slightly earlier, study was made of both chloranil and the isostructural *bromanil*, $C_6Br_4O_2$. Allowing for a different choice of axial angle the results for the chlorine compound are similar to the foregoing. For bromanil, using the conventional axes, the unit has the dimensions:

$$a_0 = 8.62(1) \text{ A.}; \quad b_0 = 6.17(1) \text{ A.}; \quad c_0 = 9.03(1) \text{ A.}$$
$$\beta = 105.8(4)°$$

The parameters are listed in Table XVAVII,10.

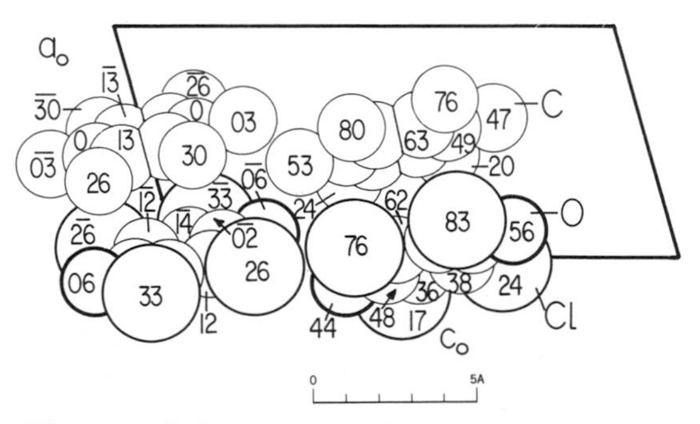

Fig. XVAVII,5. The monoclinic structure of the complex chloranil-hexamethylbenzene projected along its b_0 axis.

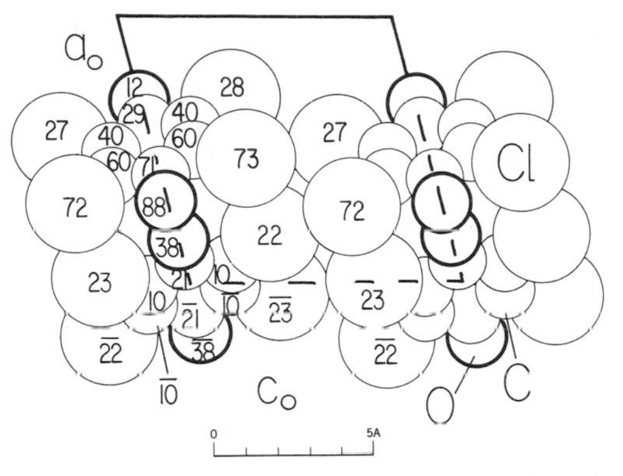

Fig. XVAVII,6. The monoclinic structure of chloranil projected along its b_0 axis.

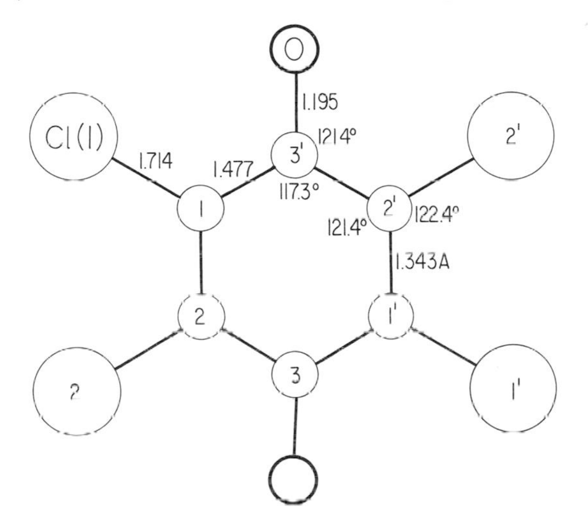

Fig. XVAVII,7. Averaged bond dimensions in the molecule of chloranil.

TABLE XVAVII,10

Parameters of the Atoms in Bromanil

Atom	x	y	z
Br(1)	0.0315	0.2203	−0.3188
Br(2)	0.1970	0.2178	0.3298
O	0.1801	0.3081	0.0087
C(1)	0.0027	0.0634	−0.1429
C(2)	0.0840	0.0620	0.1105
C(3)	0.0983	0.1638	0.0047

XV,aVII,8. The triclinic crystals of *2,3,5,6-tetramethyl-1,4-benzoquinone* (duroquinone), $C_6(CH_3)_4O_2$, possess a bimolecular unit of the dimensions:

$$a_0 = 6.9246(6) \text{ A.}; \quad b_0 = 8.8514(3) \text{ A.}; \quad c_0 = 9.5363 \text{ A.}$$
$$\alpha = 119.742(7)°; \quad \beta = 116.587(5)°; \quad \gamma = 72.033(4)°$$

A less oblique unit, used in the determination of structure and for the following description, is body-centered and has the dimensions:

$$a_0' = 6.925 \text{ A.}; \quad b_0' = 9.249 \text{ A.}; \quad c_0' = 14.285 \text{ A.}$$
$$\alpha' = 90.84°; \quad \beta' = 89.73°; \quad \gamma = 99.57°$$

The space group is C_i^1 and in terms of the second unit the parameters for the two different molecules in the cell are those of Table XVAVII,11. The experimental bond lengths in these molecules are as follows:

Bond	Mol. A	Mol. B
O(1)–C(2)	1.217 A.	1.217 A.
C(2)–C(3)	1.471	1.482
C(2)–C(4′)	1.480	1.479
C(3)–C(4)	1.327	1.322
C(3)–C(5)	1.499	1.499
C(4)–C(6)	1.495	1.503

TABLE XVAVII,11

Parameters of the Atoms in Duroquinone[a]

Atom	x′	y′	z′
O(1)	0.0425 (0.5496)	0.1968 (0.2209)	0.1353 (−0.1154)
C(2)	0.0233 (0.5265)	0.1051 (0.1180)	0.0726 (−0.0620)
C(3)	0.0343 (0.4934)	0.1520 (−0.0348)	−0.0256 (−0.0985)
C(4)	0.0115 (0.4687)	0.0530 (−0.1460)	−0.0945 (−0.0402)
C(5)	0.0707 (0.4904)	0.3135 (−0.0544)	−0.0435 (−0.2029)
C(6)	0.0196 (0.4343)	0.0920 (−0.3034)	−0.1958(−0.0728)
H(7)	0.088 (0.515)	0.368 (0.039)	0.014 (−0.233)
H(8)	−0.038 (0.365)	0.342 (−0.106)	−0.076 (−0.223)
H(9)	0.186 (0.589)	0.340 (−0.110)	−0.081 (−0.223)
H(10)	−0.021 (0.419)	0.006 (−0.368)	−0.234 (−0.020)
H(11)	0.151 (0.542)	0.135 (−0.324)	−0.214 (−0.110)
H(12)	−0.065 (0.317)	0.162 (−0.324)	−0.208(−0.110)

[a] Values for Molecule B in parentheses.

XV,aVII,9. The monoclinic crystals of *hexanitrobenzene*, $C_6(NO_2)_6$, have a tetramolecular unit of the dimensions:

$$a_0 = 13.22(3)\ \text{A.};\quad b_0 = 9.13(4)\ \text{A.};\quad c_0 = 9.68(3)\ \text{A.}$$
$$\beta = 95°30(30)'$$

The space group is C_{2h}^6 in the orientation $I2/c$. All atoms are thus in the positions:

$$(8f)\quad \pm(xyz;\ x,\bar{y},z+{}^1/_2);\quad \text{B.C.}$$

Their parameters are stated in Table XVAVII,12.

TABLE XVAVII,12

Parameters of the Atoms in $C_6(NO_2)_6$

Atom	x	y	z
C(1)	0.044	0.117	0.291
C(2)	0.087	0.251	0.338
C(3)	0.043	0.385	0.298
N(1)	0.093	−0.018	0.344
N(2)	0.183	0.251	0.431
N(3)	0.090	0.516	0.345
O(1)	0.108	−0.114	0.254
O(2)	0.107	−0.034	0.485
O(3)	0.254	0.178	0.393
O(4)	0.182	0.320	0.539
O(5)	0.179	0.537	0.324
O(6)	0.037	0.614	0.396

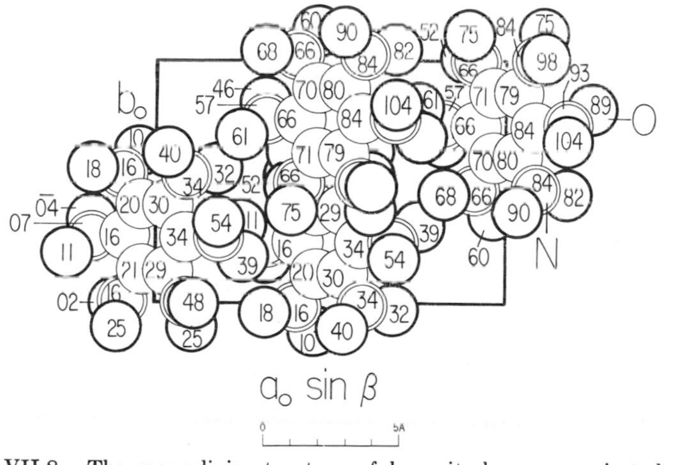

Fig. XVAVII,8. The monoclinic structure of hexanitrobenzene projected along its c_0 axis.

The structure is shown in Figure XVAVII,8. In its molecules, which have the bond dimensions of Figure XVAVII,9, the benzene ring is planar and the attached nitrogen atoms are coplanar with it to within about 0.04 A. The nitro groups are, however, turned through about 53° with respect to this plane. Between molecules the shortest atomic separations are O–O = 3.02 A., C–O = 3.24 A., and N–O = 3.36 A.

XV,aVII,10. Crystals of *1,3,5-triamino-2,4,6-trinitrobenzene*, $C_6(NH_2)_3$-$(NO_2)_3$, are triclinic with a bimolecular cell of the dimensions:

$$a_0 = 9.010(3) \text{ A.}; \quad b_0 = 9.028(3) \text{ A.}; \quad c_0 = 6.812(3) \text{ A.}$$
$$\alpha = 108°35(1)'; \quad \beta = 91°42(2)'; \quad \gamma = 119°58(1)'$$

Their atoms in the positions $(2i)$ $\pm(xyz)$ of C_i^1 $(P\bar{1})$ have the parameters of Table XVAVII,13.

TABLE XVAVII,13

Parameters of the Atoms in $C_6(NH_2)_3(NO_2)_3$

Atom	x	y	z
C(1)	0.5334(6)	0.1657(7)	0.2569(10)
C(2)	0.3733(6)	0.0027(7)	0.2485(9)
C(3)	0.2155(7)	0.0072(7)	0.2486(10)
C(4)	0.2147(6)	0.1666(7)	0.2511(9)
C(5)	0.3764(7)	0.3219(7)	0.2444(11)
C(6)	0.5380(6)	0.3269(7)	0.2522(9)
N(1)	0.6926(5)	0.1686(6)	0.2702(8)
N(2)	0.3689(7)	−0.1468(7)	0.2388(10)
N(3)	0.0565(5)	−0.1494(6)	0.2467(9)
N(4)	0.0712(7)	0.1725(7)	0.2580(10)
N(5)	0.3741(6)	0.4761(6)	0.2344(8)
N(6)	0.6849(6)	0.4709(6)	0.2516(9)
O(1)	0.8334(5)	0.3070(6)	0.2824(9)
O(2)	0.6930(5)	0.0298(6)	0.2645(8)
O(3)	0.0510(6)	−0.2923(6)	0.2373(10)
O(4)	−0.0812(5)	−0.1498(6)	0.2525(9)
O(5)	0.2381(5)	0.4789(6)	0.2304(9)
O(6)	0.5128(5)	0.6173(5)	0.2385(8)
H(1)	0.467	−0.143(7)	0.235(9)
H(2)	0.272	−0.238(8)	0.254(10)
H(3)	−0.021(8)	0.085(9)	0.252(10)
H(4)	0.071(10)	0.288(11)	0.256(12)
H(5)	0.680(8)	0.566(9)	0.247(11)
H(6)	0.796(12)	0.459(12)	0.231(14)

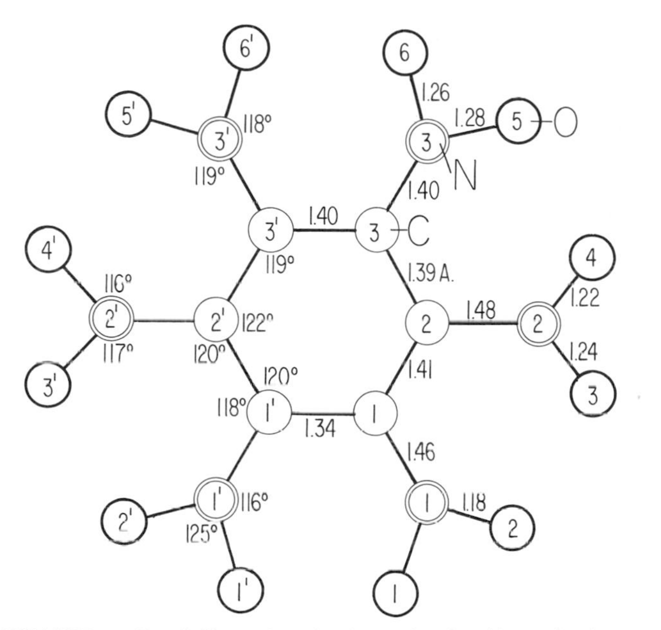

Fig. XVAVII,9. Bond dimensions in the molecule of hexanitrobenzene.

The resulting structure (Fig. XVAVII,10) is composed of molecules possessing the bond lengths of Figure XVAVII,11. The molecule as a whole is essentially planar though the planes of the NO_2 groups make angles of between 1.5 and 4° with that of the benzene ring. The short atomic separations of 2.48–2.51 A. between the amino nitrogen and the nitro oxygens are considered to represent intramolecular hydrogen bonds. The molecules themselves are held together by similar but longer N–H–O bonds of 2.93–3.00 A. In this structure the C–C bonds in the benzene ring are unusually

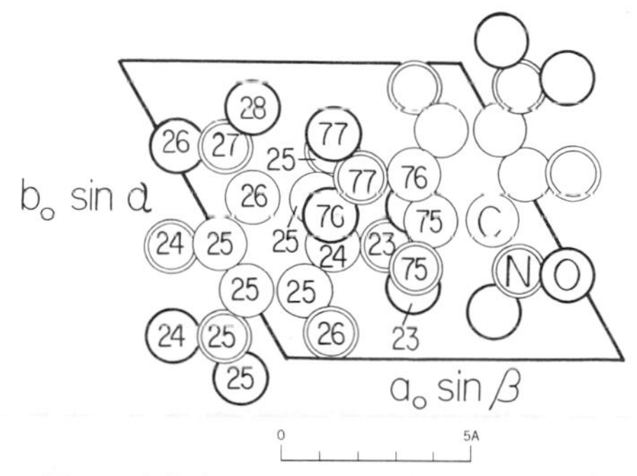

Fig. XVAVII,10. The triclinic structure of 1,3,5-triamino-2,4,6-trinitrobenzene projected along its c_0 axis.

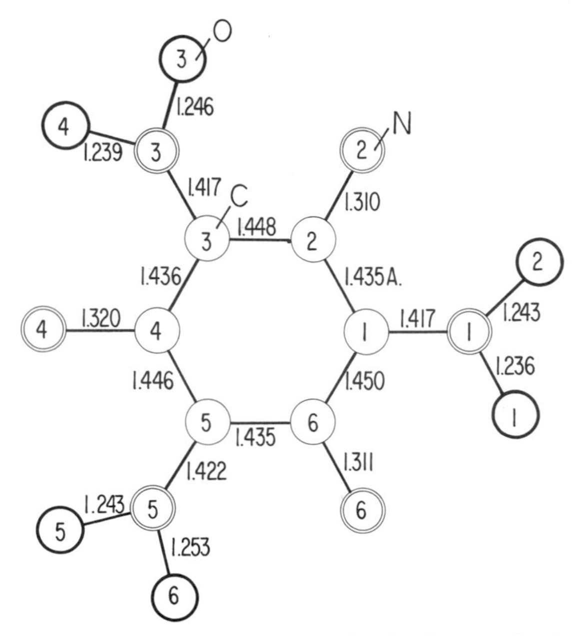

Fig. XVAVII,11. Bond dimensions in the molecule of 1,3,5-triamino-2,4,6-trinitrobenzene.

long (1.44 A. on the average) while the C–N bonds are much shorter for C–NH_2 (1.32 A.) than for C–NO_2 (1.42 A.). The N–H distances range between 0.76 and 1.06 A.

XV,aVII,11. *Benzene hexacarboxylic* (mellitic) *acid*, $C_6(COOH)_6$, forms orthorhombic crystals. Their units, containing eight molecules, have the edges:

$$a_0 = 8.14 \text{ A.}; \quad b_0 = 16.50 \text{ A.}; \quad c_0 = 19.05 \text{ A.}$$

The space group is V_h^{10} (*Pccn*) with atoms in the positions:

$(4c) \quad \pm(^1/_4\ ^1/_4\ u;\ ^1/_4,^1/_4,u+^1/_2)$
$(4d) \quad \pm(^1/_4\ ^3/_4\ u;\ ^1/_4,^3/_4,u+^1/_2)$
$(8e) \quad \pm(xyz;\ x+^1/_2,y+^1/_2,\bar{z};\ ^1/_2-x,y,z+^1/_2;\ x,^1/_2-y,z+^1/_2)$

The two crystallographically different molecules in the structure have the parameters of Table XVAVII,14. Values corresponding to assumed positions for the hydrogen atoms are given in the original article.

TABLE XVAVII,14

Positions and Parameters of the Atoms in $C_6(COOH)_6$

Atom	Position	x	y	z
		Molecule A		
C(1)	(4c)	1/4	1/4	0.3709
C(2)	(8e)	0.2567	0.1776	0.3350
C(3)	(8e)	0.2567	0.1762	0.2625
C(4)	(4c)	1/4	1/4	0.2245
C(7)	(4c)	1/4	1/4	0.4522
C(8)	(8e)	0.2640	0.0980	0.3753
C(9)	(8e)	0.2640	0.0970	0.2217
C(10)	(4c)	1/4	1/4	0.1453
O(1)	(8e)	0.1274	0.2207	0.4807
O(2)	(8e)	0.2268	0.0360	0.3455
O(3)	(8e)	0.1164	0.0724	0.1995
O(8)	(8e)	0.3384	0.1018	0.4357
O(9)	(8e)	0.4012	0.0631	0.2146
O(10)	(8e)	0.3139	0.1884	0.1162
		Molecule B		
C(1′)	(4d)	1/4	3/4	0.4747
C(2′)	(8e)	0.7500	0.1773	0.4880
C(3′)	(8e)	0.7500	0.1770	0.4159
C(4′)	(4d)	1/4	3/4	0.6198
C(7′)	(4d)	1/4	3/4	0.3957
C(8′)	(8e)	0.7500	0.0973	0.5263
C(9′)	(8e)	0.7500	0.0983	0.3732
C(10′)	(4d)	1/4	3/4	0.6992
O(1′)	(8e)	0.6373	0.2117	0.6319
O(2′)	(8e)	0.6582	0.0433	0.5060
O(3′)	(8e)	0.6336	0.0822	0.3384
O(8′)	(8e)	0.8488	0.0932	0.5788
O(9′)	(8e)	0.8830	0.0565	0.3790
O(10′)	(8e)	0.8576	0.2077	0.2717

The atomic arrangement, shown in Figure XVAVII,12, contains molecules having the bond dimensions of Figure XVAVII,13. Carboxyl groups of different molecules face one another throughout the structure to furnish hydrogen bonds of lengths O H . . . O = 2.64–2.66 A.

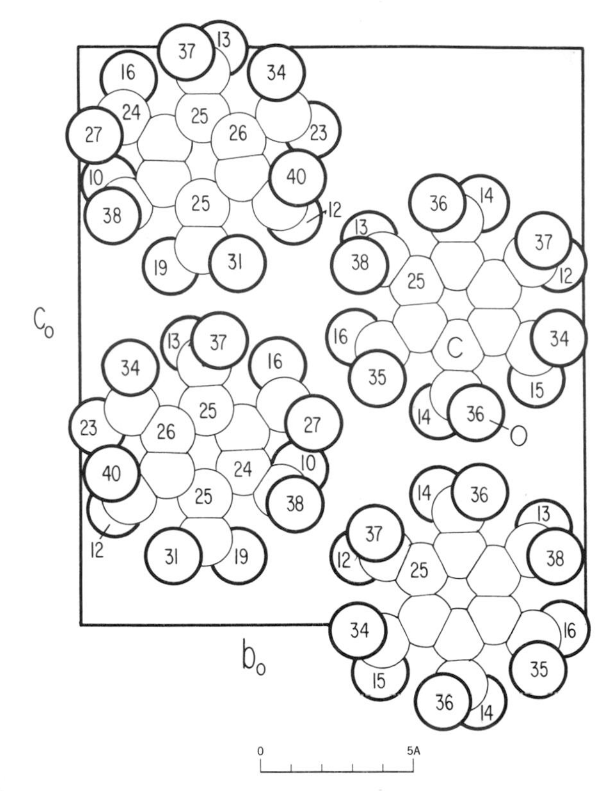

Fig. XVAVII,12. A portion of the orthorhombic structure of mellitic acid projected along its a_0 axis. Only the molecules in the lower half of the cell are shown.

XV,aVII,12. The monoclinic unit of *pentachlorophenol*, Cl_5C_6OH, contains eight molecules and has the dimensions:

$$a_0 = 29.11(3) \text{ A.}; \quad b_0 = 4.930(5) \text{ A.}; \quad c_0 = 12.09(2) \text{ A.}$$
$$\beta = 93°38(4)'$$

Atoms are in the positions

$$(8f) \quad \pm(xyz;\ x,\bar{y},z+{}^1/_2;\ x+{}^1/_2,y+{}^1/_2,z;\ x+{}^1/_2,{}^1/_2-y,z+{}^1/_2)$$

of C_{2h}^6 ($C2/c$). The determined parameters are those of Table XVAVII,15.

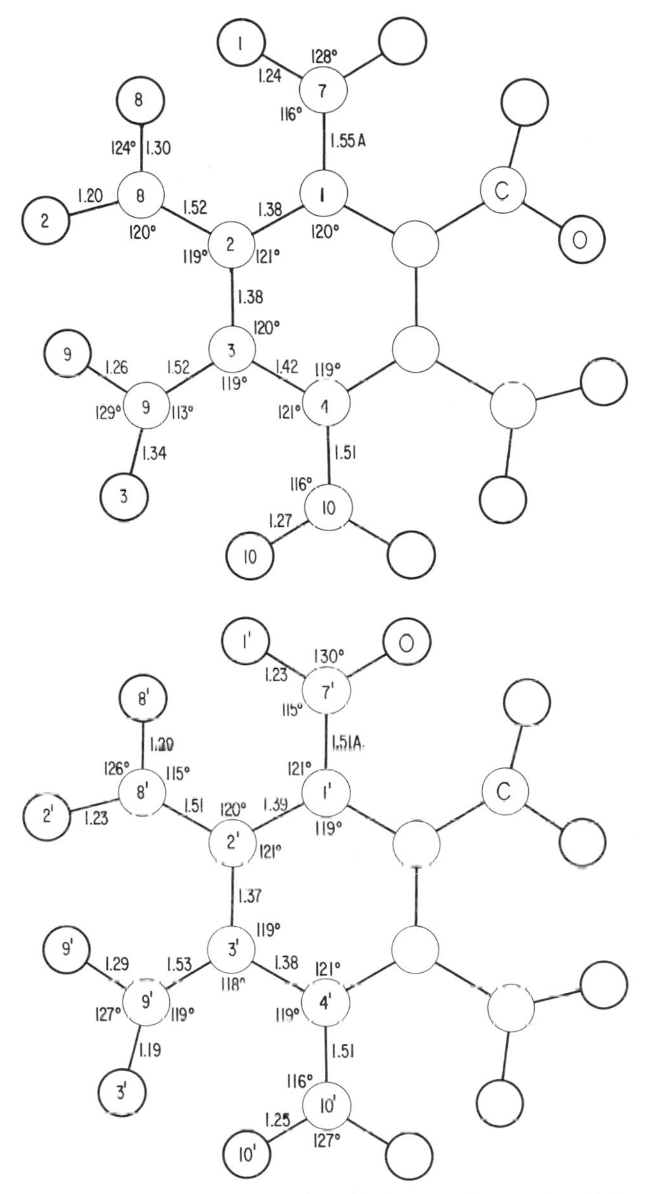

Fig. XVAVII,13a (top). Bond dimensions in the A molecule of mellitic acid. This molecule has a twofold axis of symmetry through the atoms C(7), C(1), C(4) and C(10).
Fig. XVAVII,13b (bottom). Bond dimensions in the B molecule of mellitic acid. Like the A molecule, this also has a twofold axis.

TABLE XVAVII,15

Parameters of the Atoms in Cl_5C_6OH

Atom	x	y	z
Cl(1)	0.20067	−0.1068	0.0497
Cl(2)	0.09574	−0.1544	−0.0132
Cl(3)	0.02768	0.2183	0.0996
Cl(4)	0.06470	0.6168	0.2803
Cl(5)	0.16953	0.6583	0.3419
C(1)	0.1793	0.277	0.1905
C(2)	0.1624	0.096	0.1116
C(3)	0.1163	0.074	0.0838
C(4)	0.0862	0.234	0.1362
C(5)	0.1009	0.416	0.2158
C(6)	0.1489	0.437	0.2448
O	0.2243	0.288	0.2168

The structure, illustrated in Figure XVAVII,14, is built of molecules with the bond dimensions of Figure XVAVII,15. Carbon atoms of the benzene nucleus and Cl(5) are exactly coplanar, while the oxygen and the other chlorine atoms are as much as 0.06 A. outside this plane. Between molecules there is an O–O = 2.97 A. and Cl–O = 3.28 A.; the shortest Cl–Cl = 3.43 A.

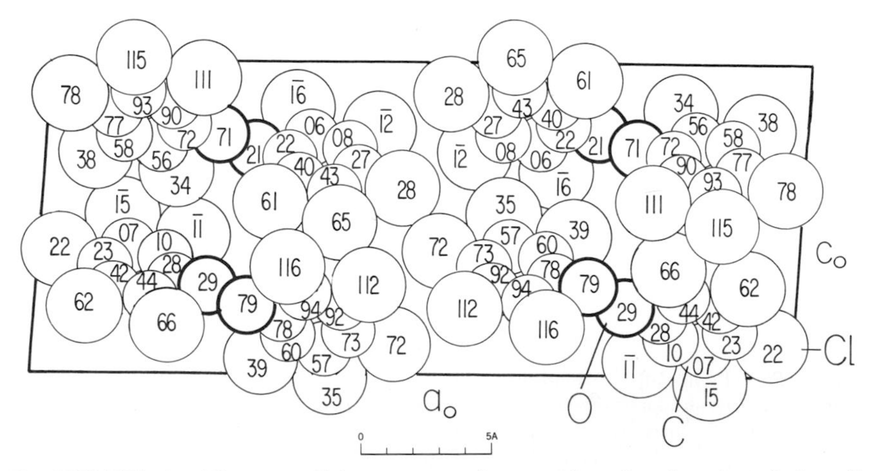

Fig. XVAVII,14. The monoclinic structure of pentachlorophenol projected along its b_0 axis.

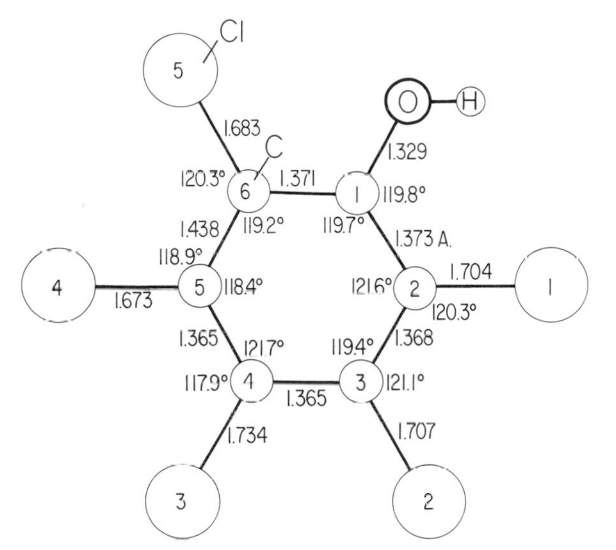

Fig. XVAVII,15. Bond dimensions in the molecule of pentachlorophenol.

XV,aVII,13. *Tetrachlorohydroquinone*, $C_6Cl_4(OH)_2$, has been studied by both x-ray and neutron diffraction. Its monoclinic crystals possess a bimolecular unit of the dimensions:

$$a_0 = 8.214(17)\ \text{A.}; \quad b_0 = 4.849(10)\ \text{A.}; \quad c_0 = 12.141(25)\ \text{A.}$$
$$\beta = 123°49(10)'$$

The space group is C_{2h}^5 ($P2_1/c$) with atoms in the positions:

$$(4e) \quad \pm(xyz;\ x,{}^1/_2-y,z+{}^1/_2)$$

The older parameters obtained by x-ray diffraction are given in square brackets in Table XVAVII,16 along with those from the neutron investigation.

TABLE XVAVII,16

Parameters of the Atoms in $C_6Cl_4(OH)_2$

Atom	x	y	z
Cl(1)	0.1523(4) [0.1520]	−0.4140(12) [−0.4172]	0.2259(3) [0.2264]
Cl(2)	−0.2992(4) [−0.3009]	−0.3519(13) [−0.3516]	0.0180(3) [0.0176]
C(1)	0.0656(5) [0.0624]	−0.1882(14) [−0.187]	0.1004(3) [0.0975]
C(2)	−0.1334(5) [−0.1329]	−0.1562(13) [−0.156]	0.0080(3) [0.0096]
C(3)	0.2014(5) [0.2001]	−0.0308(17) [−0.030]	0.0924(4) [0.0915]
O	0.3945(11) [0.3944]	−0.0426(19) [−0.0539]	0.1807(5) [0.1803]
H	0.4299(20) —	−0.2081(45) —	0.2317(12) —

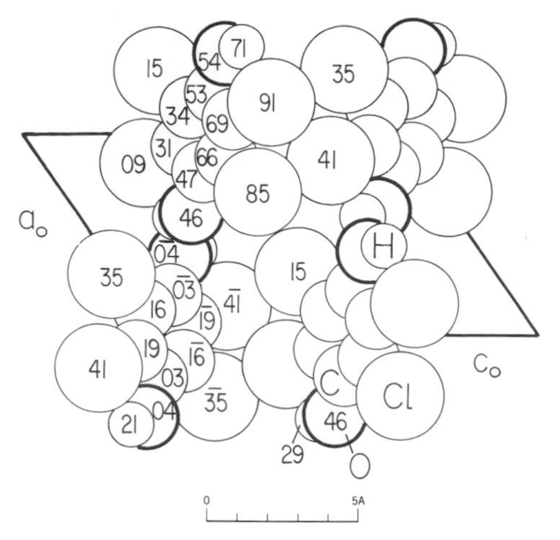

Fig. XVAVII,16. The monoclinic structure of tetrachlorohydroquinone projected along its b_0 axis.

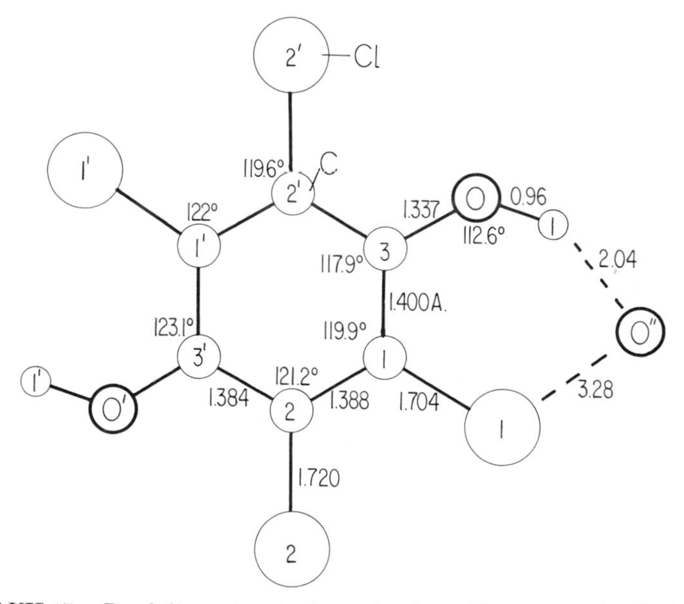

Fig. XVAVII,17. Bond dimensions in the molecule and between molecules in crystals of tetrachlorohydroquinone.

The structure (Fig. XVAVII,16) contains planar molecules having the bond dimensions of Figure XVAVII,17. Between molecules there is the O–H–Cl distance shown in the figure as well as short Cl–Cl separations of 3.37 and 3.43 A.

XV,aVII,14. The monoclinic crystals of *chloranilic acid*, $C_6Cl_2O_2(OH)_2$, possess a bimolecular unit of the dimensions:

$$a_0 = 10.025(2) \text{ A.}; \quad b_0 = 5.544(1) \text{ A.}; \quad c_0 = 7.566(2) \text{ A.}$$
$$\beta = 122.9°$$

The space group is C_{2h}^5 ($P2_1/a$) with atoms in the positions:

$$(4e) \quad \pm(xyz;\ x+{}^1/_2,{}^1/_2-y,z)$$

Their parameters are as follows:

Atom	x	y	z
Cl	0.4414(1)	0.8583(2)	0.2747(2)
C(1)	0.4714(5)	0.6670(8)	0.1209(6)
C(2)	0.5834(4)	0.4718(8)	0.2238(5)
C(3)	0.6100(5)	0.3051(8)	0.0895(6)
O(1)	0.6612(4)	0.4395(6)	0.4141(4)
O(2)	0.7155(4)	0.1312(6)	0.1883(5)
H	0.74	0.10	0.32

The centrosymmetric molecules in the resulting structure (Fig. XVAVII,18), have the bond dimensions of Figure XVAVII,19. They are planar except for the oxygen atoms which lie up to 0.04 A. from the ring plane. In the crystal they are stacked one above another along c_0 with a shortest distance in this direction of 3.26 A. In the a_0c_0 plane they are tied together by hydrogen bonds of length 2.769 A. between O(1) and O(2) atoms. Other intermolecular contacts in this plane are O(1)–O(1) = 3.167 A., O(2)–Cl = 3.363 A., O(1)–Cl = 3.458 A. and Cl–Cl = 3.335 A.

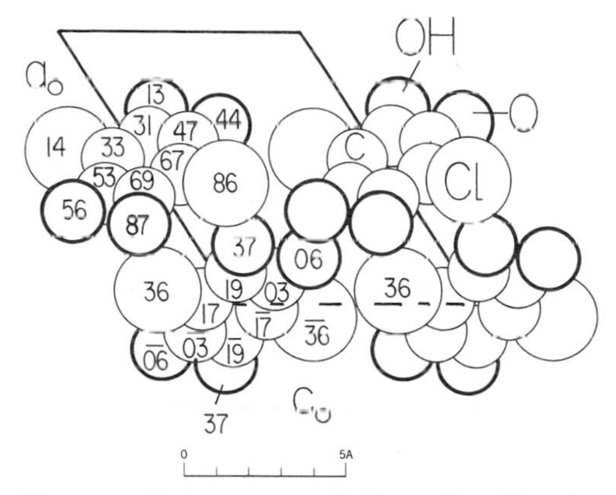

Fig. XVAVII,18. The monoclinic structure of chloranilic acid projected along its b_0 axis.

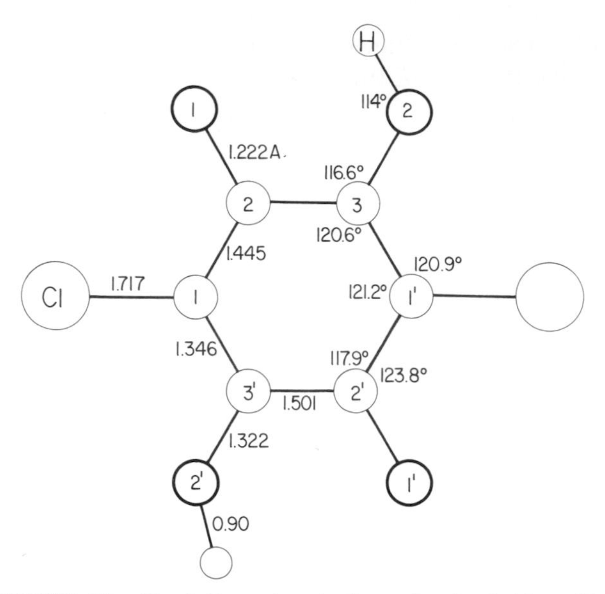

Fig. XVAVII, 19. Bond dimensions in the molecule of chloranilic acid.

XV,aVII,15. Monoclinic *chloranilic acid dihydrate*, $C_6Cl_2O_2(OH)_2 \cdot 2H_2O$, has a bimolecular unit of the dimensions:

$$a_0 = 8.617(1) \text{ A.}; \quad b_0 = 10.386(1) \text{ A.}; \quad c_0 = 5.203(1) \text{ A.}$$
$$\beta = 104.8°$$

Its atoms, placed in the positions

$$(4e) \quad \pm(xyz;\ x,{}^1/_2-y,z+{}^1/_2)$$

of C_{2h}^5 ($P2_1/c$), have the parameters listed in Table XVAVII,17.

TABLE XVAVII,17

Parameters of the Atoms in $C_6Cl_2O_2(OH)_2 \cdot 2H_2O$

Atom	x	y	z
Cl	0.2188(2)	0.1420(1)	0.4972(2)
C(1)	0.0952(6)	0.0673(5)	0.2262(10)
C(2)	0.8290(6)	0.0133(5)	0.9331(10)
C(3)	0.9350(6)	0.0832(5)	0.1679(10)
O(1)	0.6832(4)	0.0278(4)	0.8916(8)
O(2)	0.8651(5)	0.1578(4)	0.3100(9)
O(w)	0.5529(5)	0.1843(4)	0.2605(9)

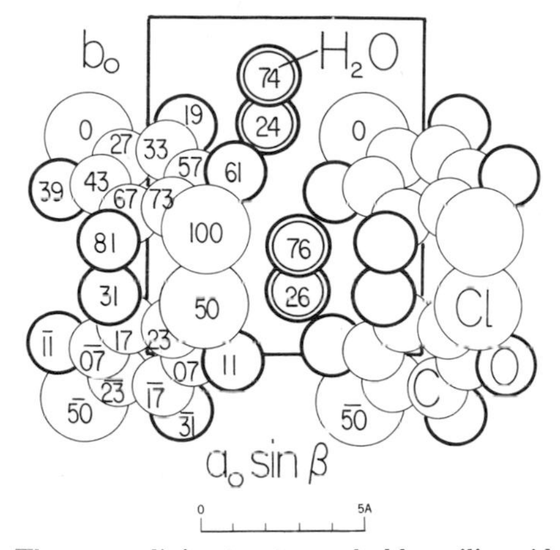

Fig. XVAVII, 20. The monoclinic structure of chloranilic acid dihydrate projected along its c_0 axis.

The structure is shown in Figure XVAVII,20. Except that the chlorine atoms are about 0.04 A. on either side of the plane of the rest, the molecules have the same dimensions as in the anhydrous acid (**XV,aVII,14**). As in this other structure, the molecules are in stacks with a separation of 3.23 A. Normal to the stacking the shortest interatomic distances (other than those involving water) are O(2)–O(2) = 3.23 A. and Cl–Cl = 3.435 A. The water molecules, in chains along c_0, are tied by hydrogen bonds to one another [O(w)–O(w) = 2.938 A.] and to the quinone oxygens [O(2)–O(w) = 2.651 A. and O(1)–O(w) = 2.966 A.].

XV,aVII,16. Crystals of *ammonium chloranilate monohydrate*, $(NH_4)_2$-$C_6Cl_2O_4 \cdot H_2O$, are monoclinic. The dimensions of their tetramolecular unit are:

$$a_0 = 16.988(3)\text{ A.};\quad b_0 = 4.780(1)\text{ A.};\quad c_0 = 14.101(3)\text{ A.}$$
$$\beta = 118.01(2)^\circ$$

Water oxygens are in the positions

$$\text{O(w)}: (4e)\quad \pm(0\,u\,{}^1/_4;\ {}^1/_2,u+{}^1/_2,{}^1/_4)\qquad \text{with } u = 0.0484(12)$$

of C_{2h}^6 $(C2/c)$. All other atoms are in

$$(8f)\quad \pm(xyz;\ x,\bar{y},z+{}^1/_2;\ x+{}^1/_2,y+{}^1/_2,z;\ x+{}^1/_2,{}^1/_2-y,z+{}^1/_2)$$

with the parameters of Table XVAVII,18.

TABLE XVAVII,18

Parameters of the Atoms in Ammonium Chloranilate Monohydrate

Atom	x	y	z
Cl	0.2253(1)	0.2970(3)	0.3235(1)
C(1)	0.2396(3)	0.5462(11)	0.4203(4)
C(2)	0.1704(3)	0.5791(12)	0.4476(4)
C(3)	0.1812(3)	0.2016(12)	0.0324(4)
O(1)	0.0997(3)	0.4435(10)	0.4083(4)
O(2)	0.1174(3)	0.1727(9)	0.0528(3)
N	0.4386(3)	0.2122(13)	0.3726(5)

The structure (Fig. XVAVII,21) contains anions that are planar except for oxygen atoms 0.01 and 0.035 A. outside this plane; the bond dimensions are those of Figure XVAVII,22. The NH_4^+ ions have four oxygen neighbors at distances between 2.792 and 2.956 A. and two chlorines 3.379 and 3.480 A. away. The water oxygens have, besides the two nitrogen atoms at 2.890 A., two O(1) atoms at a distance of 2.805 A.

XV,aVII,17. *Nitranilic acid hexahydrate* (hydronium nitranilate), $C_6O_2(NO_2)_2(OH)_2 \cdot 6H_2O$, crystallizes with monoclinic symmetry. Its bimolecu-

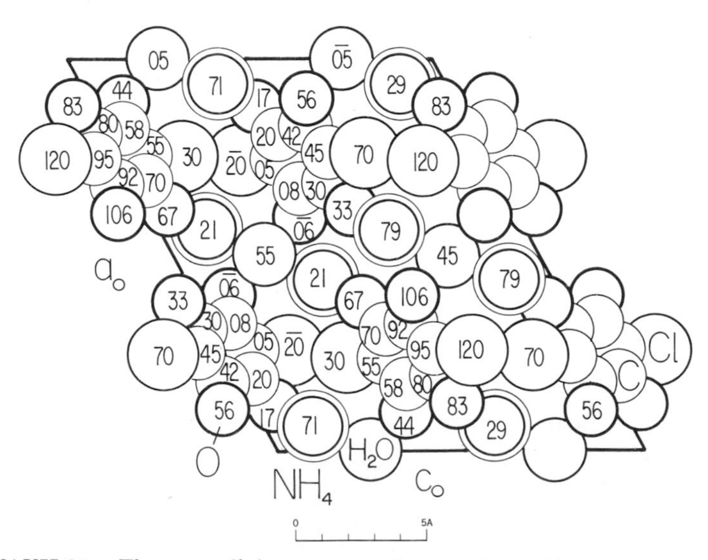

Fig. XVAVII,21. The monoclinic structure of ammonium chloranilate monohydrate projected along its b_0 axis.

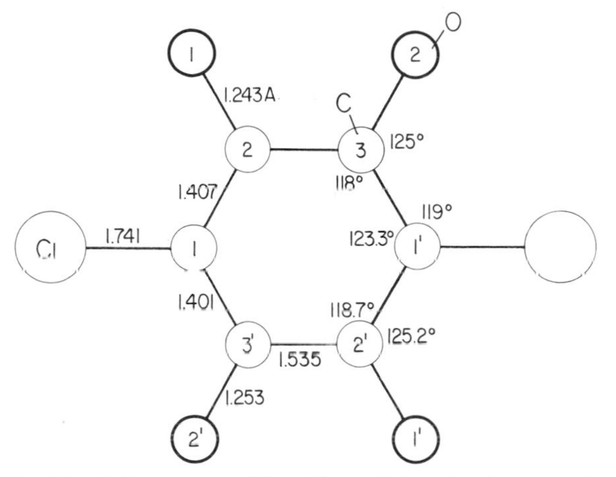

Fig. XVAVII,22. Bond dimensions of the chloranilate anion in ammonium chloranilate monohydrate.

lar unit has the dimensions:

$$a_0 = 3.657(2) \text{ A.}; \quad b_0 = 19.399(6) \text{ A.}; \quad c_0 = 9.184(3) \text{ A.}$$
$$\beta = 94.28(3)°$$

The space group is C_{2h}^5 ($P2_1/c$) with atoms in the positions:

$$(4e) \quad \pm(xyz;\ x,{}^1/_2-y,z+{}^1/_2)$$

The determined parameters are listed in Table XVAVII,19.

TABLE XVAVII,19

Parameters of the Atoms in $C_6O_2(NO_2)_2(OH)_2 \cdot 6H_2O$

Atom	x	y	z
C(1)	0.1528(18)	0.0627(3)	0.0699(5)
C(2)	0.0154(18)	0.0110(3)	0.1590(5)
C(3)	0.1403(18)	0.0563(2)	0.9148(5)
O(1)	0.4926(14)	0.1613(2)	0.0690(4)
O(2)	0.1952(15)	0.1426(2)	0.2555(4)
O(3)	0.0101(13)	0.0120(2)	0.2932(4)
O(4)	0.2411(14)	0.0999(2)	0.8304(4)
N	0.2877(15)	0.1253(2)	0.1348(4)
O(w,1)	0.2223(15)	0.0887(2)	0.5283(3)
O(w,2)	0.9843(15)	0.2054(2)	0.5542(4)
O(w,3)	0.6484(15)	0.2304(2)	0.8081(4)

In this structure (Fig. XVAVII,23) the water molecules form endless chains running in the c_0 direction. The O(w,1) thus are bound to nitranilate oxygens and O(w,2) atoms, while O(w,3) atoms are connected to nitranilate and O(w,2). Except for a short O(w,1)–H–O(w,2) = 2.443 A., the hydrogen bonds range in lengths from 2.687–3.062 A. Bond dimensions of the nitranilate group are the same as those of the anion in ammonium nitranilate (**XV,aVII,18**) and this is considered reason for viewing the present compound as a hydronium salt. In the crystal the nitranilate groups are stacked one above another along the a_0 axis, with a perpendicular distance between them of 3.306 A.

XV,aVII,18. *Ammonium nitranilate,* $(NH_4)_2C_6O_4(NO_2)_2$, forms triclinic crystals whose unimolecular cell has the dimensions:

$$a_0 = 4.712(1) \text{ A.}; \quad b_0 = 7.000(2) \text{ A.}; \quad c_0 = 7.847(2) \text{ A.}$$
$$\alpha = 111.14(2)°; \quad \beta = 93.45(2)°; \quad \gamma = 102.24(2)°$$

Atoms, in the general positions of C_i^1 $(P\bar{1})$: $(2i) \pm (xyz)$, have the parameters of Table XVAVII,20.

TABLE XVAVII,20

Parameters of the Atoms in $(NH_4)_2C_6O_4(NO_2)_2$

Atom	x	y	z
C(1)	0.7173(14)	0.3367(8)	0.4419(6)
C(2)	0.8627(16)	0.3929(9)	0.3050(6)
C(3)	0.8400(16)	0.4372(9)	0.6341(6)
O(1)	0.2979(13)	0.1527(8)	0.5029(6)
O(2)	0.3443(13)	0.0917(8)	0.2218(6)
O(3)	0.7748(12)	0.3203(8)	0.1393(5)
O(4)	0.7309(12)	0.4031(7)	0.7603(5)
N(1)	0.4429(12)	0.1878(7)	0.3886(6)
N(2)	0.2298(13)	0.2205(8)	0.8956(6)

The centrosymmetric anions in this structure have the bond dimensions of Figure XVAVII,24. The benzene ring is planar with attached atoms lying slightly outside. The nitrogen atom is 0.06 A. and one of its oxygen atoms 0.2 A. from the plane; the NO_2 groups thus are twisted from the plane by 6.1°. The other oxygens are about 0.03 A. on one side or the other

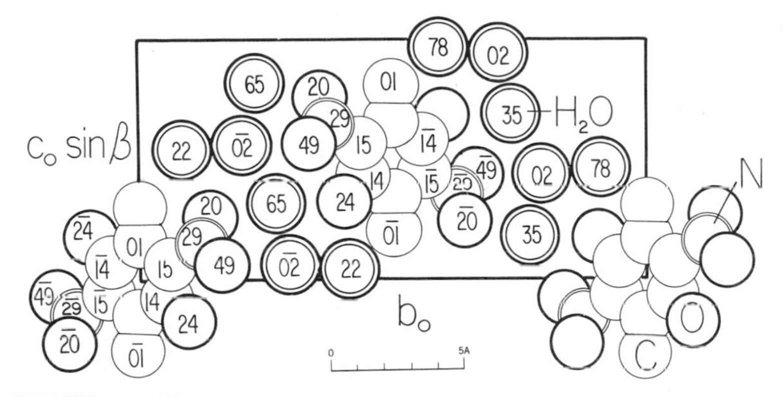

Fig. XVAVII,23. The monoclinic structure of nitranilic acid hexahydrate projected along its a_0 axis.

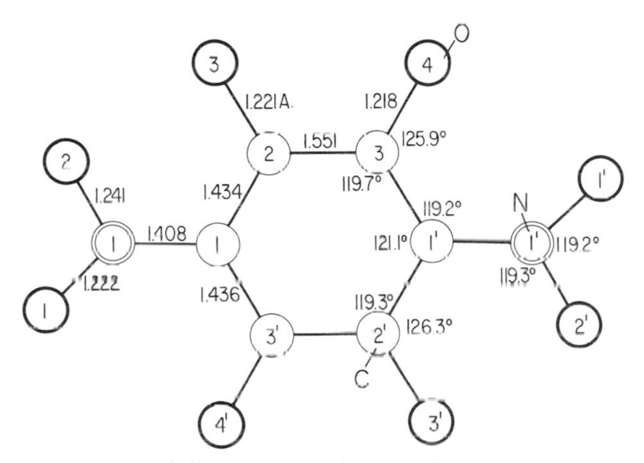

Fig. XVAVII,24. The bond dimensions in the nitranilate anion in its ammonium salt.

of the plane. In the structure the anions are stacked one above another along a_0 at a distance of 3.222 A.; between anions lying in the same b_0c_0 plane, there are O–O separations as short as 3.134 A. Nine oxygens are coordinated with each NH_4^+ cation at distances between 2.887 and 3.207 A.; some of these may involve hydrogen bonds.

XV,aVII,19. *Tetrahydroxy-p-benzoquinone dihydrate*, $C_6O_2(OH)_4 \cdot 2H_2O$, forms monoclinic crystals which have a bimolecular cell of the dimensions:

$$a_0 = 5.226(2) \text{ A.}; \quad b_0 = 5.118(2) \text{ A.}; \quad c_0 = 15.502(8) \text{ A.}$$
$$\beta = 103°53(1)'$$

The space group is C_{2h}^5 ($P2_1/c$) with all atoms in the positions:

$$(4e) \quad \pm(xyz;\ x,{}^1/_2-y,z+{}^1/_2)$$

The parameters are those of Table XVAVII,21.

TABLE XVAVII,21
Parameters of the Atoms in $C_6O_2(OH)_4 \cdot 2H_2O$

Atom[a]	x	y	z
C(1)	0.6587	0.2205	0.0310
C(2)	0.4690	0.1243	0.0809
C(3)	0.3204	−0.0858	0.0512
O(1)	0.7978	0.4118	0.0572
O(2)	0.4669	0.2661	0.1536
O(3)	0.1428	−0.1762	0.0939
O(4)	0.8523	0.6547	0.2418
H(2)	0.310	0.225	0.170
H(3)	0.070	0.680	0.075
H(4)	−0.070	−0.190	0.240
H(4′)	0.750	0.540	0.185

[a] For carbon, $\sigma(x) = 0.0006$, $\sigma(y) = 0.0007$, $\sigma(z) = 0.0002$. For oxygen, $\sigma(x) = 0.0005$, $\sigma(y) = 0.0005$ [for O(4) it is 0.0006], $\sigma(z) = 0.0002$.

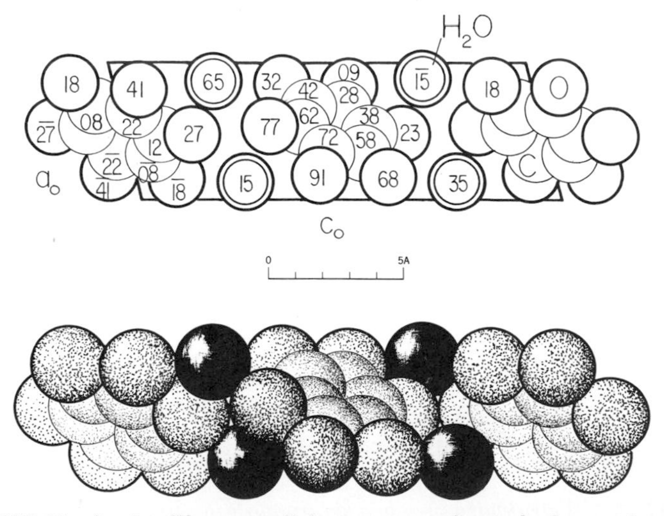

Fig. XVAVII,25a (top). The monoclinic structure of tetrahydroxy-*p*-benzoquinone dihydrate projected along its b_0 axis.

Fig. XVAVII,25b (bottom). A packing drawing of the monoclinic structure of tetrahydroxy-*p*-benzoquinone dihydrate viewed along its b_0 axis. The water molecules are black; the other oxygen atoms are heavily outlined and coarsely dotted. The carbon atoms, of the same size, are lightly outlined and dot shaded.

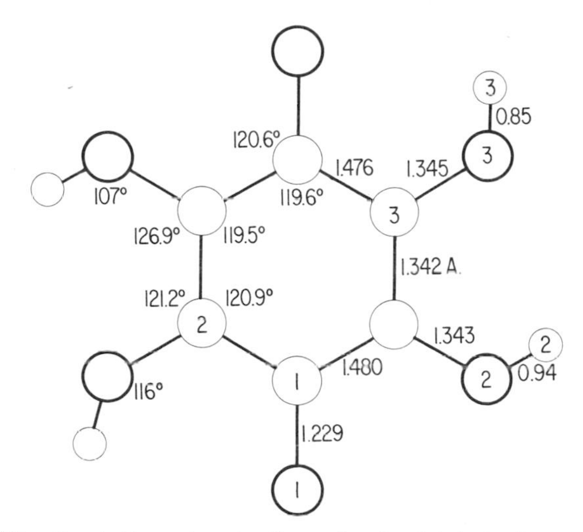

Fig. XVAVII,26. Bond dimensions in the molecule of tetrahydroxy-*p*-benzoquinone.

In this structure (Fig. XVAVII,25) the molecule has the bond dimensions of Figure XVAVII,26. It is planar to 0.01 A. though H(2) is 0.26 A. and H(3) 0.10 A. from this plane. The water molecules, in strings running along the b_0 direction, are tied together by hydrogen bonds of length 2.968 A.; these strings are in a plane that bisects the c_0 edge of the cell. The organic molecules, in sheets parallel to the a_0b_0 plane, are tied together by the hydrogen bonds O(1) . . . H–O(3) = 2.744 A. These molecules and water are further united by hydrogen bonds involving the O(2) atoms; their lengths are 2.654 and 2.926 A. Close van der Waals approaches are C–C = 3.331 A., C–O = 3.088 A. and O–O = 3.055 A.

XV,aVII,20. The monoclinic crystals of *tetrachloro-1,4-bis(triethylstannyloxy)benzene*, $C_6Cl_4[OSn(C_2H_5)_3]_2$, possess a bimolecular unit of the dimensions:

$$a_0 = 8.72(2) \text{ A.}; \quad b_0 = 11.59(3) \text{ A.}; \quad c_0 = 12.96(3) \text{ A.}$$
$$\beta = 97.9(2)^\circ$$

Atoms are in the positions

$$(4e) \quad \pm(xyz;\ x+{}^1/_2,{}^1/_2-y,z+{}^1/_2)$$

of C_{2h}^5 in the axial orientation $P2_1/n$. Parameters are listed in Table XVAVII,22.

TABLE XVAVII,22

Parameters of the Atoms in $C_6Cl_4[OSn(C_2H_5)_3]_2$

Atom	x	y	z
Sn	0.118	0.767	0.2675
Cl(1)	0.345	0.988	0.082
Cl(2)	0.757	0.924	0.146
O	0.100	0.933	0.207
C(1)	0.157	0.000	0.028
C(2)	0.060	0.962	0.110
C(3)	0.890	0.962	0.067
C(4)	0.032	0.728	0.411
C(5)	0.942	0.833	0.445
C(6)	0.368	0.755	0.283
C(7)	0.442	0.857	0.350
C(8)	0.978	0.643	0.165
C(9)	0.803	0.617	0.167

The centrosymmetric molecules of this structure (Fig. XVAVII,27) have the bond dimensions of Figure XVAVII,28, the unstated C–C separation in the ethyl radicals being in each case 1.55 A. Distribution about the tin atoms is tetrahedral.

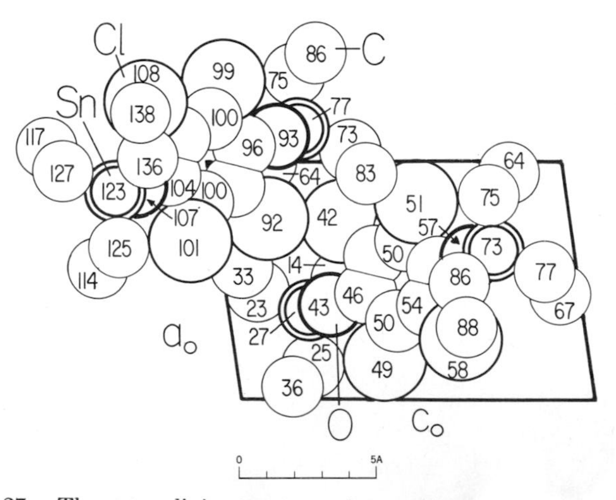

Fig. XVAVII,27. The monoclinic structure of tetrachloro-1,4-bis(triethylstannyloxy) benzene projected along its b_0 axis.

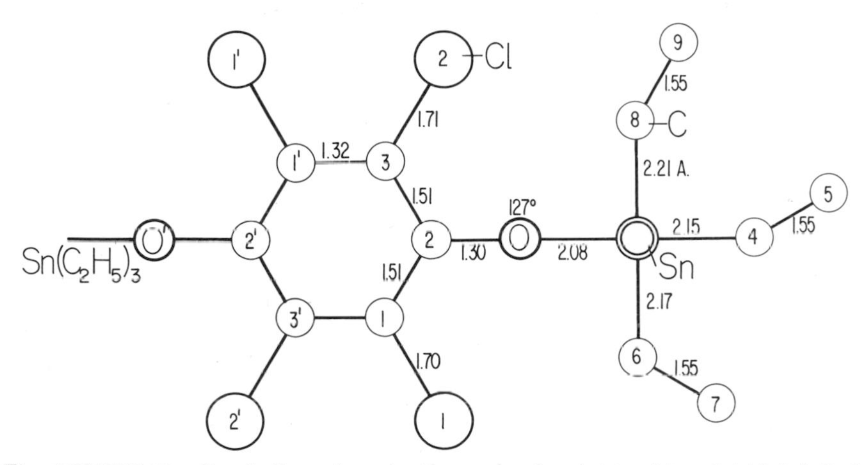

Fig. XVAVII,28. Bond dimensions in the molecule of tetrachloro-1,4-bis(triethylstannyloxy)benzene.

BIBLIOGRAPHY TABLE, CHAPTER XVA

Compound	Paragraph	Literature
Acetanilide	**aII,30**	1941: J&K; 1948: B&C; 1954: B&C; 1966: B
Acetylacetone–mono(*o*-hydroxyanil)-copper(II)	**aIII,104**	1965: B&H
Acetylsalicylic acid	**aIII,87**	1935: W,W,K&T; 1937: N&W; 1964: W
Aluminum bromide–benzene	**aI,5**	1961; E,T&W
Aluminum chloride–benzoyl chloride	**aII,12**	1966: R&B
p-Aminobenzamide	**aIII,28**	1965: A&D
m-Aminobenzenearsonic acid	**aIII,116**	1962: S
p-Aminobenzenearsonic acid	**aIII,117**	1961: S
p-Aminobenzoic acid	**aIII,48**	1939: P,K&T; 1966: A,SC&D; 1967: L&M
2-Amino-3-methyl benzoic acid	**aIV,10**	1963: B&M
1,3-di-Amino-2,4,6-trinitrobenzene	**aVI,4**	1967: H
1,3,5-tri-Amino-2,4,6-trinitrobenzene	**aVII,10**	1965: C&L
p-Aminophenol	**aIII,59**	1927: C; 1951: B
o-Aminophenol hydrochloride	**aIII,60**	1965: C&R
p-Amino salicylic acid	**aIV,13**	1951: C,G&L; 1954: B,G&L

(*continued*)

BIBLIOGRAPHY TABLE, CHAPTER XVA *(continued)*

Compound	Paragraph	Literature
Ammonium acid *o*-carboxybenzenesulfonate	**aIII,115**	1967: O
Ammonium acid disalicylate monohydrate	**aIII,78**	1954: D&S
Ammonium acid phthalate	**aIII,115**	1957: O&P
Ammonium chloranilate monohydrate	**aVII,16**	1967: A
Ammonium hydrogen di-*p*-chlorobenzoate	**aIII,41**	1963: M&S
Ammonium hydrogen dicinnamate	**aII,20**	1963: B,M&S
Ammonium nitranilate	**aVII,18**	1964: J&A; 1967: A
Aniline hydrobromide	**aII,27**	1948: N,W&T; 1961: N,W&T
Aniline hydrochloride	**aII,26**	1949: B
Aniline *p*-thiocyanate	**aIII,27**	1965: I&Z
Benzamide	**aII,28**	1959: P&W
Benzamide–hydrogen triiodide	**aII,29**	1964: R,K&R
Benzene	**aI,1**	1923: B; 1924: E; 1925: B; 1928; C; 1929: B&N; 1930: B&N: 1932: C; 1954: C&S; K; 1955: C,C&S; 1958: C; C,C&S; 1964: B,C&W
Benzene–bromine (1:1)	**aI,2**	1958: H&S
Benzene–chlorine (1:1)	**aI,2**	1959: H&S
Benzenediazonium chloride	**aII,32**	1963: R
Benzenediazonium tribromide	**aII,33**	1962: A&R
Benzene dichloroiodide	**aII,3**	1953: A&vS
Benzene hexacarboxylic acid	**aVII,11**	1960: B&G; D; 1961: D
Benzene seleninic acid	**aII,49**	1954: B&MC
Benzoic acid	**aII,7**	1922: B; 1955: S,R&G
Benzoquinone	**aIII,38**	1932: C; 1934: R; 1935: R; 1960: T
p-Benzoquinone-*p*-bromophenol	**aIII,56**	1967: S&W
p-Benzoquinone-2*p*-bromophenol	**aIII,57**	1967: S&W
p-Benzoquinone-*p*-chlorophenol	**aIII,56**	1967: S&W
p-Benzoquinone-2*p*-chlorophenol	**aIII,57**	1967: S&W
(+)-2-Benzylglutamic acid hydrobromide dihydrate	**aII,36**	1967: A,S&K
3,5-di-Bromo-*p*-aminobenzoic acid	**aV,8**	1962: P; 1965: P
p-di-Bromobenzene	**aIII,1**	1932: H; 1942: B&C; B,C&L; 1948: C&B
1,3,5-tri-Bromobenzene	**aIV,1**	1960: M&P

(continued)

BIBLIOGRAPHY TABLE, CHAPTER XVA *(continued)*

Compound	Paragraph	Literature
1,2,4,5-tetra-Bromobenzene	aV,1,2	1958: G&H; 1960: G&H; 1964: G&H
hexa-Bromobenzene-1,2,4,5-tetrabromobenzene	aVII,2	1964: G&H
o-Bromobenzoic acid	aIII,40	1962: F&S
2,5-di-Bromo-1,4-benzoquinone	aV,14	1964: W,R&H; 1966: R,H&W
tetra-Bromo-*p*-benzoquinone	aVII,7	1939: C; 1940: C; 1944: N&N; 1961: U
2-Bromo-*p*-benzoquinone-4-oxime acetate	aIV,26	1961: F,MG&R
penta-Bromochlorobenzene	aVII,1	1966: K,B&S
tetra-Bromo-1,3-dimethyl benzene	aVII,1	1961: S&S
β-5-*n*-Bromoethoxy-*o*-quinone-2-oxime	aIV,28	1964: R
hexa-Bromomethylbenzene	aVII,3	1935: B,T&W; 1965: M
α-2-Bromo-5-methyl-*p*-benzoquinone-4-oxime	aV,17	1960: R&F
m-Bromonitrobenzene	aIII,13	1963: C&T
p-Bromo-*m*-nitro-*N*-methylaniline	aIV,23	1967: C,D,K,R&R
p-Bromophenyl boric acid	aIII,7	1958: Z&G
4-Bromo-2,6-di(tertiary-butyl) phenol	aV,7	1964: M&R
1-Bromo-2,3,5,6-tetramethylbenzene	aVI,1	1965: C,B,M&M
tetra-*n*-Butylammonium benzoate hydrate	aII,11	1962: B,J&MM
bis(*N*-*t*-Butylsalicylaldiminato)-copper(II)	aIII,102	1965: C,H&W; 1966: C,H&W
bis-*N*-Butylsalicylaldiminato palladium(II)	aIII,101	1964: F,P&S
Carbon tetrabromide *p*-xylene	aIII,34	1962: S&T
tri-Carbonyl chromium anisole-1,3,5-trinitrobenzene	aIV,25	1966: C,MP&S
tri-Carbonyl chromium-*o* toluidine	aIII,26	1967: C,MP&S
Catechol	aIII,66	1926: C; S; 1964: T; 1965: C&L; 1966: B
Cesium acid phthalate	aIII,52	1957: O&P
Chloranil-hexamethylbenzene	aVII,6	1955: H&W; 1962: J&M; W&H
Chloranilic acid	aVII,14	1967: A
Chloranilic acid dihydrate	aVII,15	1967: A
p-Chloroacetanilide	aIII,30	1966: S

(continued)

BIBLIOGRAPHY TABLE, CHAPTER XVA *(continued)*

Compound	Paragraph	Literature
α-Chloroacetophenone	**aII,13**	1966: B&M
p-Chloroaniline	**aIII,22**	1966: P; S; T,W&Z
2,5-di-Chloroaniline	**aIV,21**	1963: S,S&J
anti-p-Chlorobenzaldoxime	**aIII,9**	1957: J; 1964: F,L&J
syn-p-Chlorobenzaldoxime	**aIII,8**	1950: J; 1957: J; 1964: F,L&J
p-di-Chlorobenzene	**aIII,1,2**	1932: H; 1952: C,B&B; 1957: H&C; 1959: F,G&B
1,3,5-tri-Chlorobenzene	**aIV,1**	1960: M&P
1,2,4,5-tetra-Chlorobenzene	**aV,1**	1958: D,P,C&J; 1965: H
hexa-Chlorobenzene	**aVII,1**	1931: L; 1958: T&W; 1961: S&S
p-Chlorobenzene seleninic acid	**aIII,114**	1956: B&MC
o-Chlorobenzoic acid	**aIII,40**	1961: F&S
2,3-di-Chloro-1,4-benzoquinone	**aV,12**	1965: R,H&W; 1966: R,H&W
2,5-di-Chloro-1,4-benzoquinone	**aV,14**	1966: R,H&W
tetra-Chloro-*p*-benzoquinone	**aVII,7**	1939: C; 1940: C; 1961: U; 1962: C,J&S
2-Chloro-*p*-benzoquinone-4-oxime acetate	**aVI,26,27**	1961: F,MG&R
p-Chlorobromobenzene	**aIII,1**	1932: H; 1947: K
2,4,6-tri-Chlorobromobenzene	**aV,3**	1959: S&SS
penta-Chlorobromobenzene	**aVII,1**	1966: K,B&S
2,5-Chlorobromo-1,4-benzoquinone	**aV,14**	1966: R,H&W
trans-di-Chloro(*cis*-2-butene)*S*-α-phenethylamine platinum(II)	**aII,35**	1965: G&P
α,5(2′-Chloroethoxy)*o*-quinone-2-oxime	**aIII,11**	1966: vO&R
β-5-*n*-Chloroethoxy-*o*-quinone-2-oxime	**aIV,28**	1964: R
tetra-Chloro-1,4-bis(triethyl-stannyloxy)benzene	**aVII,20**	1961: W
tetra-Chlorohydroquinone	**aVII,13**	1962: S; 1967: S&C
p-Chloroiodoxybenzene	**aIII,6**	1948: A
α-2-Chloro-5-methyl-*p*-benzo-quinone-4-oxime	**aV,17**	1960: R&F
2-Chloro-4-nitroaniline	**aIV,22**	1965: MP&S
3,5-di-Chloro-4-nitroaniline	**aV,9**	1955: Z&G
m-Chloronitrobenzene	**aIII,13**	1965: G
p-Chloronitrobenzene	**aIII,12**	1962: M&T
2-Chloro-5-nitrobenzoic acid	**aIV,8**	1962: F&S
penta-Chlorophenol	**aVII,12**	1962: S

(continued)

BIBLIOGRAPHY TABLE, CHAPTER XVA *(continued)*

Compound	Paragraph	Literature
di-Chloro-di(*o*-phenylene bis di-methylarsine)platinum(II)	**aIII,119**	1964: S
p-Chlorophenylmethyl sulfone	**aIII,105**	1961: R&T; 1964: R&T
bis(5-Chlorosalicylaldoximato)-copper(II)	**aIV,15**	1964: O,L&B
5-Chlorosalicylaldoxime	**aIV,14**	1961: S,P&T
1,2-di-Chloro-tetramethylbenzene	**aVII,1**	1958: T&W
1,3-di-Chloro-2,4,6-trinitrobenzene	**aVI,2**	1967: H&D
Chromium dibenzene	**aI,3**	1960: J; 1963: C,D&W; J; 1964: I
Chromium tricarbonyl benzene	**aI,4**	1959: C&A; 1961: A; 1965: B&D
Cobalt dimercury hexathio-cyanate–benzene	**aI,9**	1964: G&D
Cobalt(II) bis(*o*-phenylene bis dimethylarsine) diperchlorate	**aIII,118**	1967: E&R
Cobaltous chloride-di-*p*-toluidine	**aIII,24**	1957: M
Copper benzoate trihydrate	**aII,9**	1963: K,O&W
Copper salicylate tetrahydrate	**aIII,79**	1960: H&M
p-Cresol	**aIII,58**	1966: B
Cuprous aluminum chloride–benzene	**aI,7**	1963: T&A; 1966: T&A
7,7,8,8-tetra-Cyanoquinodimethane	**aIII,31**	1965: L,S&T
7,7,8,8 tetra Cyanoquinodi-methane–*N*,*N*,*N'*,*N'*-tetra-methyl *p* phenylene-diamine (1:1)	**aIII,32**	1965: H
Ephedrine hydrochloride	**aII,40**	1933: G&N; 1954: P
N-*β*-di-Ethylaminoethyl-5-chloro-salicylaldimine cobalt	**aIV,20**	1965: S,O&DV
N-*β*-di-Ethylaminoethyl-5-chloro-salicylaldimine nickel	**aIV,20**	1965: S,O&DV; 1966: O,DV&S
hexa-Ethylbenzene	**aVII,4**	1935: P&G
di-Ethyl(salicylaldehydato) thallium(III)	**aIII,86**	1967: M&T
bis(*N*-Ethylsalicylaldiminato)-copper(II)	**aIII,95**	1966: P,B&G; 1967: B,C,H&W; P,B&G
bis(*N*-Ethylsalicylaldiminato)-nickel(II)	**aIII,97**	1967: S,K&V
bis(*N*-Ethylsalicylaldiminato)-palladium(II)	**aIII,96**	1964: F,P&S
di-Ethyl terephthalate	**aIII,55**	1949: B

(continued)

BIBLIOGRAPHY TABLE, CHAPTER XVA *(continued)*

Compound	Paragraph	Literature
Ferric chloride-*o*-methoxy-phenyldiazonium chloride	**aIII,37**	1965: P,B&PK
Fluorobenzene	**aII,1**	1961: H
o-Fluorobenzoic acid	**aIII,39**	1961: K; 1966: F&I; K&D
Glycyl phenylalanylglycine hemihydrate	**aII,43**	1961: M&G
Glycyl-L-tyrosine hydrobromide monohydrate	**aIII,75**	1953: S&W
Glycyl-L-tyrosine hydrochloride monohydrate	**aIII,75**	1952: T; 1953: S&W
γ-Hydroquinone	**aIII,69**	1921: B; B&J; 1926: C; 1927: C; 1949: K; 1966: MM
Hydroquinone–acetone	**aIII,72**	1959: L&W
m-Hydroxybenzamide	**aIII,62**	1966: K,S&K
tetra-Hydroxy-*p*-benzoquinone dihydrate	**aVII,19**	1965: K
bis(*N*-2-Hydroxyethylsalicylaldiminato)copper(II)	**aIII,98**	1966: B,H,K&W
bis(α-Hydroxy-α-phenylbutyramidine)copper(II) dihydrate	**aII,37**	1967: I&M
p-Iodoaniline-*S*-trinitrobenzene complex	**aIV,24**	1943: P,H&C
p-di-Iodobenzene	**aIII,3**	1933: H,M,M&J; 1959: L&S; 1960: L&S
p-Iodobenzonitrile	**aIII,4**	1965: S&B
bis(*p*-Iodo-*N*,*N*-dimethylaniline) hydrochloride triiodide	**aIII,21**	1959: A&H; 1962: A
di-Iodo-di(*o*-phenylene bis dimethylarsine) nickel	**aIII,120**	1964: S
di-Iodo-di(*o*-phenylene bis dimethylarsine) palladium	**aIII,120**	1962: S
di-Iodo-di(*o*-phenylene bis dimethylarsine) platinum	**aIII,120**	1962: S
p-Iodonitrosobenzene	**aIII,5**	1956: W
di-Iodo-L-tyrosine dihydrate	**aV,11**	1967: H&S
bis(*N*-Isopropyl-3-ethylsalicylaldiminato)nickel	**aIV,18**	1967: B&L
bis(*N*-Isopropyl-3-ethylsalicylaldiminato)palladium	**aIV,19**	1967: B&L
4-Isopropylidineaminophenol	**aIII,61**	1953: H&P
bis(*N*-Isopropyl-3-methylsalicylaldiminato)nickel	**aIV,16**	1966: B&L

(continued)

BIBLIOGRAPHY TABLE, CHAPTER XVA *(continued)*

Compound	Paragraph	Literature
bis(*N*-Isopropyl-3-methylsalicylaldiminato)palladium	**aIV,16,17**	1967: J&L
bis(*N*-Isopropylsalicylaldiminato)-copper(II)	**aIII,99**	1966: O&S
bis(*N*-Isopropylsalicylaldiminato) nickel	**aIII,99**	1964: F,O,L&S
Magnesium benzenesulfonate hexahydrate	**aII,44**	1947: B&N; 1948: B&N
Magnesium *p*-toluenesulfonate hexahydrate	**aIII,113**	1946: H; 1957: H
Metanilic acid	**aIII,107**	1965: H&M
1,4-di-Methoxy benzene	**aIII,36**	1950: G,P&R
5-Methoxy-2-nitrosophenol	**aIV,7**	1959: B,C&M
(+)-*m*-Methoxyphenoxypropionic acid·(−)-*m*-bromophenoxy-propionic acid quasi racemate	**aIII,35**	1966: K&K
N-Methyl acetanilide	**aII,31**	1941: J&K; 1967: P
bis(*N*-γ-di-Methylaminopropyl-salicylaldiminato)nickel	**aIII,100**	1967: DV&O
mono-Methylammonium acid phthalate	**aIII,115**	1957: O&P
m-Methylbenzamide	**aIII,29**	1963: O,N,T,S&K
1,2,4,5-tetra-Methylbenzene	**aV,4**	1933: R
hexa-Methylbenzene	**aVII,4**	1929: L; 1939: B&R
hexa-Methylbenzene–chromium tricarbonyl	**aVII,5**	1965: B&D
Methyl benzoate–chromium tricarbonyl	**aII,10**	1967: C,MP&S
2,6-di-Methyl benzoic acid	**aIV,9**	1967: A,MC&GB
2,3-di-Methyl-1,4-benzoquinone	**aV,13**	1967: R
2,5-di-Methyl-*p*-benzoquinone	**aV,15**	1964: R&S; 1967: H&R
2,6-di-Methyl-1,4-benzoquinone	**aV,16**	1967: R&S
2,3,5,6-tetra-Methyl-1,4-benzo-quinone	**aVII,8**	1967: R,S&U
bis(tri-Methylbenzyl ammonium)-copper(II) tetrachloride	**aII,34**	1967: B,D&V
Methyl-*m*-bromocinnamate	**aIII,49**	1965: L&S
Methyl-*p*-bromocinnamate	**aIII,50**	1965: L&S
N,3-di-Methyl-4-bromo-2,6-dinitroaniline	**aVI,3**	1967: A,I&L
2-Methyl-3-bromophenol	**aIV,2**	1966: M

(continued)

BIBLIOGRAPHY TABLE, CHAPTER XVA *(continued)*

Compound	Paragraph	Literature
N-Methyl-*p*-chlorobenzaldoxime	**aIII,10**	1964: F,L&J
4-Methyl-2,6-(tertiary-butyl)phenol	**aV,7**	1964: M&R
p-Methyliodoxybenzene	**aIII,6**	1948: A
N-Methyl-2-methylsulfonyl-2-phenylsulfonyl vinylidineamine	**aII,47**	1957: B&W
N,*N*-di-Methyl-*p*-nitroaniline	**aIII,20**	1965: M&T
2,3-di-Methylphenol	**aIV,3**	1966: B,GP&V; M; 1967: B, GP&V
2,5-di-Methylphenol	**aIV,4**	1965: GP
Methylphenyl sulfone	**aII,45**	1965: V
di-Methylphenylsulfonium perchlorate	**aII,48**	1964: LC&T
bis(*N*-Methylsalicylaldiminato)-cobalt	**aIII,94**	1966: O,DV&S
α-bis(*N*-Methylsalicylaldiminato)-copper(II)	**aIII,93**	1960: M&vS; 1961: L,S,M,S&F
bis(*N*-Methylsalicylaldiminato)-manganese	**aIII,94**	1966: O,DV&S
bis(*N*-Methylsalicylaldiminato)-nickel	**aIII,92,93**	1958: F&S; 1959: F,P&S; 1967: F&L
bis(*N*-Methylsalicylaldiminato)zinc	**aIII,94**	1966: O,DV&S
N-Methyl-*N*-2,4,6-tetranitroaniline	**aV,6**	1967: C
S-Methylthiouronium-*p*-chlorobenzoate	**aIII,42**	1963: K&W
Nickel cyanide ammonia–benzene	**aI,10**	1952:R&P
Nitranilic acid hexahydrate	**aVII,17**	1967: A
p-Nitroaniline	**aIII,19**	1947: A&R; 1948: A&R; 1956: A&R; D&T; 1961: T,G&D
2,3,4,6-tetra-Nitroaniline	**aVI,5**	1966: D,S&H
o-Nitrobenzaldehyde	**aIII,17**	1964: C; C&S
Nitrobenzene	**aII,2**	1959: T
m-di-Nitrobenzene	**aIII,14**	1930: H&S; 1931: H&H; 1940: B&G; 1946: A; 1947: G&L; 1961: T; 1966: T&W
p-di-Nitrobenzene	**aIII,15**	1930: H&S; 1935: B; J,K&H; 1947: A&R; L; 1950: A; 1961: T
hexa-Nitrobenzene	**aVII,9**	1966: A,S&D
p-Nitrobenzene seleninic acid	**aIII,114**	1956: B&MC
o-Nitrobenzoic acid	**aIII,43**	1940: T,K&P; 1967: K,F&S; S,T&P

(continued)

BIBLIOGRAPHY TABLE, CHAPTER XVA *(continued)*

Compound	Paragraph	Literature
p-Nitrobenzoic acid	**aIII,45**	1940: T,K&P; 1965: S&P; 1966: S&P
2,4,6-tri-Nitroiodobenzene	**aV,10**	1933: H&R; 1940: H&P
Nitromesitylene	**aV,5**	1959: T
o-Nitroperoxybenzoic acid	**aIII,47**	1965: S,B&C
p-Nitrophenol	**aIII,63,64**	1939: P,S&B; 1954: T; 1958: C,S&G; 1965: C&S
p-Nitrophenyl trinitride	**aIII,16**	1964: M&M; 1965: M,M&S
bis(Nitrosophenyl hydroxylaminate)copper(II)	**aII,38**	1963: S&S
tris(Nitrosophenyl hydroxylaminate)iron(III)	**aII,39**	1965: vdH,M,D&MG
Noradrenaline hydrochloride	**aIV,12**	1967: C&B
Orthanilic acid	**aIII,106**	1967: H&M
Phenol	**aII,4**	1951: K&K; 1954: W&vS; 1960: S,W&vS; 1963: S; 1967: GP
Phenol clathrates	**aII,6**	1958: vS,H&S
Phenol hemihydrate	**aII,5**	1960: M&vS
Phenylalanine hydrochloride	**aII,42**	1963: G&V; 1964: G; V&G
Phenylarsenic bis(diethyldithiocarbamate)	**aII,52**	1965: B; 1967: B
Phenylarsonic acid	**aII,51**	1959: S; 1960: S
bis(1 Phenyl-1,3-butanedionato)-copper	**aII,15**	1966: H,P&B
trans-bis(1-Phenyl-1,3-butanedionato)palladium(II)	**aII,16**	1967: H,P&B
bis(1-Phenyl-1,3-butanedionato)-vanadyl	**aII,18**	1965: H,B&P
mono(*o*-Phenylene bis dimethylarsine)iron tricarbonyl	**aIII,121**	1967: B&B
β-Phenylethylamine hydrobromide	**aII,41**	1961: T
β-Phenylethylamine hydrochloride	**aII,41**	1961: T
Phenylethynyl(*n*-amylamine)gold	**aII,24**	1967: C&S
Phenylethynyl(isopropylamine)gold	**aII,24**	1967: C&S
Phenylethynyl(*n*-nonylamine)gold	**aII,24**	1967: C&S
trans-bis(Phenylethynyl)bis(triethylphosphine)nickel(II)	**aII,25**	1967: S,C&A
Phenylethynyl(trimethylphosphine)copper(I)	**aII,23**	1966: C&S
Phenylethynyl(trimethylphosphine)silver	**aII,22**	1966: C&S

(continued)

BIBLIOGRAPHY TABLE, CHAPTER XVA (*continued*)

Compound	Paragraph	Literature
bis(3-Phenyl-2,4-pentanedionato)-copper	**aII,17**	1965: C,S&B
Phenylpropiolic acid	**aII,21**	1955: R
Phenyl bis(thiourea) tellurium(II) chloride	**aII,50**	1966: F&M
Phloroglucinol	**aIV,5**	1965: MM
Phloroglucinol dihydrate	**aIV,6**	1938: B&A; 1943: B&R; C; 1957: W&P
o-Phthalic acid	**aIII,51**	1921: B&J; 1925: B&B; 1954: C; vS; V&MG; 1957: N&J
Potassium acid dibenzoate	**aII,8**	1954: S,S&S
Potassium acid disalicylate monohydrate	**aIII,78**	1954: D&S
Potassium acid phthalate	**aIII,52**	1957: O&P; 1965: O
Potassium hydrogen bis-acetylsalicylate	**aIII,88**	1967: M&S
Potassium hydrogen bis(*p*-hydroxybenzoate) monohydrate	**aIII,76**	1951: S&S
Potassium hydrogen bis(phenylacetate)	**aII,14**	1948: S; 1949: S; 1957: B&C; 1960: B&C
Potassium hydrogen di-*p*-chlorobenzoate	**aIII,41**	1963: M&S
Potassium hydrogen dicinnamate	**aII,20**	1963: B,M&S
Potassium hydrogen di-*p*-nitrobenzoate	**aIII,46**	1961: S&S
Potassium α-hydroxybenzyl sulfonate	**aII,46**	1967: K,A,S&K
Potassium-*o*-nitrophenol hemihydrate	**aIII,65**	1960: T&D; 1961: R
Potassium *p*-nitrophenyl dicyanomethide	**aIII,18**	1967: S&B
Propargyl-2-bromo-3-nitrobenzoate	**aIV,11**	1966: C,MP&S
β-5-*n*-Propoxy-*o*-quinone-2-oxime	**aIV,28**	1964: R
Quinhydrone	**aIII,70,71**	1937: A; 1958: M,O&N; 1965: S
Quinol clathrates	**aIII,73**	1947: P&P; 1948: P&P; 1950: P
Resorcinol	**aIII,67,68**	1921: B&J; 1922: B; 1928: S; 1934: R; 1935: R; 1936: R; 1938: R&U; 1955: B&C; 1956: B&C
Rubidium acid disalicylate monohydrate	**aIII,78**	1954: D&S

(*continued*)

BIBLIOGRAPHY TABLE, CHAPTER XVA *(continued)*

Compound	Paragraph	Literature
Rubidium acid phthalate	**aIII,52**	1957: O&P
Rubidium hydrogen bis-acetyl-salicylate	**aIII,88**	1967: G&W; M&S
Rubidium hydrogen bis(*p*-hydroxy-benzoate) monohydrate	**aIII,76**	1951: S&S
Rubidium hydrogen bis(phenyl-acetate)	**aII,14**	1949: S
Rubidium hydrogen di-*p*-chloro-benzoate	**aIII,41**	1963: M&S
Rubidium hydrogen di-*o*-nitro-benzoate	**aIII,44**	1961: S&S
bis(Salicylaldehydato)copper(II)	**aIII,83,84**	1963: B,G&MC; 1964: MK,W&H; 1965: H,MK&W
bis(Salicylaldehydato)nickel dihydrate	**aIII,85**	1961: S,L&B
bis(Salicylaldiminato)copper(II)	**aIII,89**	1966: B,H&W
bis(Salicylaldiminato)nickel	**aIII,89**	1959: S&L
bis(Salicylaldoximato)copper(II)	**aIII,90**	1964: J&L
bis(Salicylaldoximato)nickel	**aIII,91**	1956: M,G&L; 1958: S; 1967: S,L&J
Salicylamide	**aIII,81**	1953: MC&H; 1964: S,T&K
Salicylic acid	**aIII,77**	1922: B; 1951: C; 1958: C; 1965: S&J
Salicylic aldehyde	**aIII,82**	1953: BM
N-Salicylideneglycinatoaquo-copper(II) hemihydrate	**aIII,103**	1967: U,A,S&K
Silver perchlorate–benzene	**aI,6**	1950: R&G; 1958: S&R; 1964: S
Silver tetrachloroaluminate–benzene	**aI,8**	1966: T&A
Sodium acid phthalate	**aIII,115**	1957: O&P
Styrene–palladous chloride	**aII,19**	1955: H&B
Sulfanilamide	**aIII,109,110,111**	1941: W; 1942: W; 1964: A&D; 1965: A&D; OC&M; 1967: OC&M
Sulfanilamide monohydrate	**aIII,112**	1965: A&D
p-Sulfanilic acid monohydrate	**aIII,108**	1962: R&M
Terephthalic acid	**aIII,53,54**	1967: B&B
p-bis(Tertiary-butyl)benzene	**aIII,33**	1951: M
Thallium acid phthalate	**aIII,52**	1957: O&P
p-Toluene seleninic acid	**aIII,114**	1956: B&MC
p-Toluidine	**aIII,23**	1935: W; 1963: B

(continued)

BIBLIOGRAPHY TABLE, CHAPTER XVA *(continued)*

Compound	Paragraph	Literature
bis(*p*-Toluidinium)hexachloro-rhenate(IV)	**aIII,25**	1967: A&M
L-Tyrosine hydrobromide	**aIII,74**	1958: S; 1959: S
L-Tyrosine hydrochloride	**aIII,74**	1959: S
Zinc benzenesulfonate hexahydrate	**aII,44**	1947: B&N; 1948: B&N
Zinc salicylate dihydrate	**aIII,80**	1958: K,A&S
Zinc-*p*-toluenesulfonate hexa-hydrate	**aIII,113**	1957: H

BIBLIOGRAPHY, CHAPTER XVA

1921

Becker, K., and Jancke, W., "X-Ray Spectroscopic Investigations with Organic Compounds I and II," *Z. Physik. Chem.*, **99**, 242.

Bragg, W. H., "The Structure of Organic Crystals," *Proc. Phys. Soc. London*, **34**, 33.

1922

Bragg, W. H., "The Significance of Crystal Structure," *Trans. Chem. Soc. (London)*, **121**, 2766.

1923

Broomé, B., "X-Ray Observations upon Solid Benzene," *Physik. Z.*, **24**, 124.

1924

Eastman, E. D., "X-Ray Diffraction Patterns from Crystalline and Liquid Benzene," *J. Am. Chem. Soc.*, **46**, 917.

1925

Bragg, W. H., and Bragg, W. L., *X-Rays and Crystal Structure*, 5th ed., G. Bell & Son, London.

Broomé, B. H., "Laue Photographs of Crystalline Benzene," *Z. Krist.*, **62**, 325.

1926

Caspari, W. A., "The Crystal Structure of Catechol," *J. Chem. Soc.*, **1926**, 573.

Caspari, W. A., "The Crystal Structure of Quinol I," *J. Chem. Soc.*, **1926**, 2944.

Sarkar, A. N., "X-Ray Examination of the Crystal Structure of Certain Compounds," *Phil. Mag.*, **2**, 1153.

1927

Caspari, W. A., "The Crystallography of Some Simple Benzene Derivatives," *Phil. Mag.*, **4**, 1276.

Caspari, W. A., "The Crystal Structure of Quinol II," *J. Chem. Soc.*, **1927**, 1093.

1928

Cox, E. G., "The Crystalline Structure of Benzene," *Nature*, **122**, 401.

Sarkar, A. N., "X-Ray Examination of the Crystal Structure of Resorcinol," *Proc. 15th Indian Sci. Cong.*, **1928**, 92.

1929

Bruni, G., and Natta, G., "The Crystalline Form of Thiophene and its Solid Solutions with Benzene," *Rec. Trav. Chim.*, **48**, 860; *Atti Accad. Nazl. Lincei, Rend., Cl. Sci. Fis. Mat. Nat.*, **11**, 934 (1930).

Lonsdale, K., "The Structure of the Benzene Ring in $C_6(CH_3)_6$," *Proc. Roy. Soc. (London)*, **123A**, 494.

Lonsdale, K., "X-Ray Evidence on the Structure of the Benzene Nucleus," *Trans. Faraday Soc.*, **25**, 352.

1930

Bruni, G., and Natta, G., "The Crystal Structure of Benzene and its Relation to that of Thiophene II," *Atti Accad. Nazl. Lincei, Rend., Cl. Sci. Fis. Mat. Nat.*, **11**, 1058.

Hertel, E., and Schneider, K., "Secondary Valence and Crystal Structure," *Z. Physik. Chem.*, **7B**, 188.

1931

Hendricks, S. B., and Hilbert, G. E., "The Molecular Association, the Apparent Symmetry of the Benzene Ring, and the Structure of the Nitro Group in Crystalline *m*-Dinitrobenzene. The Valences of Nitrogen in Some Organic Compounds," *J. Am. Chem. Soc.*, **53**, 4280.

Lonsdale, K., "An X-Ray Analysis of the Structure of Hexachlorobenzene, Using the Fourier Method," *Proc. Roy. Soc. (London)*, **133A**, 536.

1932

Caspari, W. A., "Crystallography of the Simpler Quinones," *Proc. Roy. Soc. (London)*, **136A**, 82.

Cox, E. G., "Crystalline Structure of Benzene," *Proc. Roy. Soc. (London)*, **135A**, 491.

Halla, F., "Structure Determination by Means of Weissenberg Photographs," *Z. Krist.*, **82**, 316.

Hendricks, S. B., "*p*-Bromochlorobenzene and its Congeners: Various Equivalent Points in Molecular Lattices," *Z. Krist.*, **84**, 85.

1933

Gossner, B., and Neff, H., "Crystals of Hydrochlorides, Hydrobromides and Hydroiodides of Ephedrine and Pseudoephedrine," *Z. Krist.*, **85**, 370; **86**, 32.

Hendricks, S. B., Maxwell, L. R., Mosley, V. L., and Jefferson, M. E., "X-Ray and Electron Diffraction of Iodine and the Diiodobenzenes," *J. Chem. Phys.*, **1**, 549.

Hertel, E., and Römer, G. H., "The Fine Structure of Trinitrobenzene Derivatives," *Z. Physik. Chem.*, **22B**, 267.

Robertson, J. M., "The Crystal Structure of Durene," *Proc. Roy. Soc. (London)*, **141A**, 594; **142A**, 659.

1934

Robertson, J. M., "Orientation of Molecules in *p*-Benzoquinone Crystals by X-Ray Analysis," *Nature*, **134**, 138.

Robertson, J. M., "The Space Group of Resorcinol, $C_6H_6O_2$," *Z. Krist.*, **89**, 518.

1935

Banerjee, K., "Determination of the Atomic Positions in Paradinitrobenzene by Fourier Analysis Method," *Phil. Mag.* [7], **18,** 1004.
Beintema, J., Terpstra, P., and Weerden, W. J. van, "The Crystallography of Hexabromohexamethylbenzene," *Rec. Trav. Chim.*, **54,** 962.
James, R. W., King, G., and Horrocks, H., "The Crystal Structure of Paradinitrobenzene," *Proc. Roy. Soc. (London)*, **153A,** 225.
Pal, H. K., and Guha, A. C., "Crystal Structure of Hexaethylbenzene, $C_6(C_2H_5)_6$," *Z. Krist.*, **92A,** 392.
Robertson, J. M., "A Molecular Map of Resorcinol," *Nature*, **136,** 755.
Robertson, J. M., "The Structure of Benzoquinone. A Quantitative X-Ray Investigation," *Proc. Roy. Soc. (London)*, **150A,** 106.
Watanabé, S., Watanabé, A., Kozu, S., and Takané, K., "Cell Dimensions and Space Group of Acetylsalicylic Acid," *Proc. Imp. Acad. (Tokyo)*, **11,** 381.
Wyart, J., "The Crystal Structure of *p*-Toluidine," *Compt. Rend.*, **200,** 1862.

1936

Robertson, J. M., "The Structure of Resorcinol. A Quantitative X-Ray Investigation," *Proc. Roy. Soc. (London)*, **157A,** 79.

1937

Anderson, J. S., "Structure of Organic Molecular Compounds," *Nature*, **140,** 583.
Nitta, I., and Watanabé, T., "The Unit Cell and Space Group of Acetylsalicylic Acid," *Sci. Papers Inst. Phys. Chem. Res. (Tokyo)*, **31,** 125.

1938

Banerjee, K., and Ahmad, R., "Structure of Aromatic Compounds IV. Space Group and Atomic Arrangements of Phloroglucine Dihydrate," *Indian J. Phys.*, **12,** 249.
Robertson, J. M., and Ubbelohde, A. R., "A New Form of Resorcinol I. Structure Determination," *Proc. Roy. Soc. (London)*, **167A,** 122.

1939

Brockway, L. O., and Robertson, J. M., "The Crystal Structure of Hexamethylbenzene and the Length of the Methyl Group Bond to Aromatic Carbon Atoms," *J. Chem. Soc.*, **1939,** 1324.
Chorgade, S. L., "The Crystal Structure of Chloranil, $C_6Cl_4O_2$," *Z. Krist.*, **101A,** 418.
Prasad, M., Kapadia, M. R., and Thakar, V. C., "Unit Cell of the Crystals of *p*-Aminobenzoic Acid," *J. Univ. Bombay*, **8,** 123.
Prasad, M., Shankar, J., and Baljekar, P. N., "Space Group Determination of the Crystals of *p*-Nitrophenol (Metastable), Phenacetin and Tribenzylamine," *J. Indian Chem. Soc.*, **16,** 357.

1940

Banerjee, K., and Ganguly, M., "Determination of the Structure of *m*-Dinitrobenzene by Patterson-Fourier Summation," *Indian J. Phys.*, **14,** 231.

Chorgade, S. L., "The Crystal Structure of Bromanil, $C_6Br_4O_2$," *Z. Krist.*, **102A,** 112.

Huse, G., and Powell, H. M., "The Crystal Structure of Picryl Iodide," *J. Chem. Soc.*, **1940,** 1398.

Thakar, V. C., Kapadia, M. R., and Prasad, M., "Space Groups of Crystals, of *o*-, *m*-, and *p*-Nitrobenzoic Acids," *J. Indian Chem. Soc.*, **17,** 555.

1941

Joshi, R. H., and Kapadia, M. R., "X-Ray Analysis of Some Organic Compounds," *J. Univ. Bombay*, **10,** 35.

Watanabé, A., "Polymorphism of Sulfanilamide," *Naturwissenschaften*, **29,** 116.

1942

Bezzi, S., and Croatto, U., "Determination of the Structure of *p*-Dibromobenzene," *Gazz. Chim. Ital.*, **72,** 318; *Atti Reale Ist. Veneto Sci., Pt. II, Cl. Sci. Mat. Nat.* (*Venice*), **101,** 219, 237.

Watanabé, A., "Polymorphism in Sulfanilamides II. Crystallographic and X-Ray Studies of Polymorphic Forms," *J. Pharm. Soc. Japan*, **62,** 503.

1943

Bose, C. R., and Ranjitkumar, S., "Axial Lengths of Phloroglucinol Dihydrate Crystals," *Indian J. Phys.*, **17,** 163.

Chorgade, S. L., "Crystal Structures and Space Groups of Some Aromatic Crystals I. Phloroglucinol Dihydrate," *Proc. Natl. Acad. Sci. India*, **13A,** 261.

Powell, H. M., Huse, G., and Cooke, P. W., "Structure of Molecular Compounds I. Crystal Structure of *p*-Iodoaniline-*s*-trinitrobenzene," *J. Chem. Soc.*, **1943,** 153.

1944

Neuhaus, A., and Noll, W., "Partial Isomorphous Systems VIII. Oriented Growth of Organic Substances on Typical Metals," *Naturwissenschaften*, **32,** 76.

1946

Archer, E. M., "The Crystal Structure of *m*-Dinitrobenzene," *Proc. Roy. Soc.* (*London*), **188A,** 51.

Hargreaves, A., "Crystal Structure of Zinc *p*-Toluene-sulphonate," *Nature*, **158,** 620.

1947

Abrahams, S. C., and Robertson, J. M., "Crystal Structures of *p*-Dinitrobenzene and *p*-Nitroaniline," *Nature*, **160,** 569.

Broomhead, J. M., and Nicol, A. D., "Crystal Structures of Zinc and Magnesium Benzenesulfonates," *Nature*, **160,** 795.

Gregory, N. W., and Lassettre, E. N., "Crystal Structure of *m*-Dinitrobenzene," *J. Am. Chem. Soc.*, **69**, 102.

Klug, A., "Crystal Structure of *p*-Bromochlorobenzene," *Nature*, **160**, 570.

Llewellyn, F. J., "The Crystal Structure of *p*-Dinitrobenzene," *J. Chem. Soc.*, **1947**, 884.

Palin, D. E., and Powell, H. M., "Structure of Molecular Compounds III. Crystal Structure of Addition Complexes of Hydroquinone with Certain Volatile Compounds," *J. Chem. Soc.*, **1947**, 208.

1948

Abrahams, S. C., and Robertson, J. M., "The Crystal Structure of *p*-Nitroaniline," *Acta Cryst.*, **1**, 252.

Archer, E. M., "The Crystal Structure of *p*-Chloroiodoxybenzene," *Acta Cryst.*, **1**, 64.

Broomhead, J. M., and Nicol, A. D., "Crystal Structures of Zinc and Magnesium Benzenesulfonates," *Acta Cryst.*, **1**, 88.

Brown, C. J., and Corbridge, D. E. C., "Crystal Structure of Acetanilide: Use of Polarized Infrared Radiation," *Nature*, **162**, 72.

Croatto, U., and Bezzi, S., "A System Capable of Improving the Results of Structural Investigations of Crystals by the Fourier Method of Analysis VII. Its Application to the Determination of the Structure of *p*-Dibromobenzene," *Gazz. Chim. Ital.*, **79**, 240.

Nitta, I., Watanabé, T., and Taguchi, I., "Crystal Structure of Aniline Hydrobromide," *X-Sen*, **5**, 31.

Palin, D. E., and Powell, H. M., "Structure of Molecular Compounds V. Clathrate Compound of Quinol and Methanol," *J. Chem. Soc.*, **1948**, 571.

Palin, D. E., and Powell, H. M., "Structure of Molecular Compounds VI. The β-Type Clathrate Compounds of Quinol," *J. Chem. Soc.*, **1948**, 815.

Speakman, J. C., "Crystal Structures of the Acid Salts of Some Monocarboxylic Acids," *Nature*, **162**, 695.

1949

Bailey, M., "The Crystal Structure of Diethyl Terephthalate," *Acta Cryst.*, **2**, 120.

Brown, C. J., "The Crystal Structure of Aniline Hydrochloride," *Acta Cryst.*, **2**, 228.

Kitaigorodskii, A. I., "Crystallochemistry of Aromatic Compounds VII. Pseudosymmetry of the Internal Structure of Crystals and Its Application to the Structure Analysis of Some Organic Compounds (Acenaphthene, α-Naphthol, γ-Hydroquinone)," *Izv. Akad. Nauk SSSR, Otdel. Khim. Nauk*, **1949**, 263.

Speakman, J. C., "The Crystal Structure of the Acid Salts of Some Monobasic Acids I. Potassium Hydrogen Bisphenylacetate," *J. Chem. Soc.*, **1949**, 3357.

1950

Abrahams, S. C., "The Crystal Structure of *p*-Dinitrobenzene," *Acta Cryst.*, **3**, 194.

Goodwin, T. H., Przybylska, M., and Robertson, J. M., "The Crystal and Molecular Structure of 1,4-Dimethoxybenzene," *Acta Cryst.*, **3**, 279.

Jerslev, B., "Crystal Structure of *syn-p*-Chlorobenzaldoxime," *Nature*, **166**, 741.

Powell, H. M., "The Structure of Molecular Compounds IX. A Compound of Xenon and Quinol," *J. Chem. Soc.*, **1950**, 468.

Rundle, R. E., and Goring, J. H., "Structure of the Silver Perchlorate–Benzene Complex," *J. Am. Chem. Soc.*, **72**, 5337.

1951

Brown, C. J., "Crystal Structure of *p*-Aminophenol," *Acta Cryst.*, **4,** 100.

Coccia, S., Giacomello, G., and Liquori, A. M., "Structure of 4-Amino-2-hydroxybenzoic Acid," *Ric. Sci.*, **21,** 205.

Cochran, W., "Crystal Structure of Salicylic Acid," *Acta Cryst.*, **4,** 376.

Kitaigorodskii, A. I., and Kozhin, V. M., "Crystal Structure of Phenol," *Zh. Fiz. Khim.*, **25,** 1261.

Magdoff, B. S., "Crystal Structure of *p*-Di-tertiary-butylbenzene," *Acta Cryst.*, **4,** 176.

Skinner, J. M., and Speakman, J. C., "The Crystal Structure of Acid Salts of Some Monobasic Acids II. Potassium Hydrogen Di-*p*-hydroxybenzoate Hydrate," *J. Chem. Soc.*, **1951,** 185.

1952

Croatto, U., Bezzi, S., and Bua, E., "The Crystal Structure of *p*-Dichlorobenzene," *Acta Cryst.*, **5,** 825.

Raynor, J. H., and Powell, H. M., "Structure of Molecular Compounds X. Crystal Structure of the Compound of Benzene with an Ammonia–Nickel Cyanide Complex, *J. Chem. Soc.*, **1952,** 319.

Tranter, T. C., "Unit Cell Dimensions and Space Groups of Synthetic Peptides," *Acta Cryst.*, **5,** 843.

1953

Archer, E. M., and Schalkwyck, T. G. D. van, "The Crystal Structure of Benzene Iododichloride," *Acta Cryst.*, **6,** 88.

Bourré-Maladière, P., "The Crystal Structure of Salicyl Aldehyde," *Compt. Rend.*, **237,** 825.

Cochran, W., "The Crystal and Molecular Structure of Salicylic Acid," *Acta Cryst.*, **6,** 260.

Holmes, D. R., and Powell, H. M., "The Crystal Structure of 4-Isopropylidene Aminophenol," *Acta Cryst.*, **6,** 256.

McCrone, W. C., and Hinch, R. J., Jr., "Crystallographic Data. Salicylamide (*o*-Hydroxybenzamide)," *Anal. Chem.*, **25,** 1277.

Smits, D. W., and Wiebenga, E. H., "Crystal Structure of Glycyl-L-tyrosine Hydrochloride," *Acta Cryst.*, **6,** 531.

1954

Bertinotti, F., Giacomello, G., and Liquori, A. M., "Crystal and Molecular Structure of *p*-Amino Salicylic Acid," *Acta Cryst.*, **7,** 808.

Brown, C. J., and Corbridge, D. E. C., "The Crystal Structure of Acetanilide," *Acta Cryst.*, **7,** 709.

Bryden, J. H., and McCullough, J. D., "The Crystal Structure of Benzene Seleninic Acid," *Acta Cryst.*, **7,** 833.

Chakraburtty, D. M., "Cell Dimensions of Phthalic Acid Crystal," *Sci. Culture* (*India*), **19,** 505.

Cox, E. G., and Smith, J. A. S., "Crystal Structure of Benzene at −3° C.," *Nature*, **173,** 75.

Downie, T. C., and Speakman, J. C., "The Crystal Structures of the Acid Salts of Some Monobasic Acids IV. Ammonium Hydrogen Disalicylate Hydrate," *J. Chem. Soc.*, **1954,** 787.

Kozhin, V. M., "Crystal Structure of Benzene," *Zh. Fiz. Khim.*, **28,** 566.

Phillips, D. C., "The Crystal and Molecular Structures of Ephedrine Hydrochloride," *Acta Cryst.*, **7,** 159.

Schalkwyk, T. G. D. van, "The Crystal Structure of Phthalic Acid," *Acta Cryst.*, **7,** 775.

Skinner, J. M., Stewart, G. M. D., and Speakman, J. C., "The Crystal Structure of the Acid Salts of Some Monobasic Acids III. Potassium Hydrogen Dibenzoate," *J. Chem. Soc.*, **1954,** 180.

Toussaint, J., "X-Ray Crystallographic Study of *p*-Nitrophenol (Metastable)," *Bull. Soc. Roy. Sci. Liège*, **23,** 24.

Veenendaal, A. L., and MacGillavry, C. H., "Phthalic Acid at Low Temperature," *Acta Cryst.*, **7,** 775.

Wehrhahn, O. J., and Stackelberg, M. von, "Crystal Structure of Phenol and Occlusion Compounds of Phenol," *Naturwissenschaften*, **41,** 358.

1955

Bacon, G. E., and Curry, N. A., "The Hydrogen Atoms in α-Resorcinol," *Nature*, **175,** 894.

Cox, E. G., Cruickshank, D. W. J., and Smith, J. A. S., "Crystal Structure of Benzene: A New Type of Systematic Error in Precision X-Ray Crystal Analysis," *Nature*, **175,** 766.

Harding, T. T., and Wallwork, S. C., "The Structures of Molecular Compounds Exhibiting Polarization Bonding II. The Crystal Structure of the Chloranil–Hexamethylbenzene Complex," *Acta Cryst.*, **8,** 787.

Holden, J. R., and Baenziger, N. C., "The Crystal Structure of Styrene-palladium Chloride," *J. Am. Chem. Soc.*, **77,** 4987.

Rollett, J. S., "The Crystal Structure of Phenylpropiolic Acid," *Acta Cryst.*, **8,** 487.

Sim, G. A., Robertson, J. M., and Goodwin, T. H., "The Crystal and Molecular Structure of Benzoic Acid," *Acta Cryst.*, **8,** 157.

Zhdanov, G. S., and Gol'der, G. A., "X-Ray Investigation of the Structure of 3,5-Dichloro-4-nitroaniline," *Zh. Fiz. Khim.*, **29,** 1248.

1956

Abrahams, S. C., and Robertson, J. M., "The Structure of *p*-Nitroaniline," *Acta Cryst.*, **9,** 966.

Bacon, G. E., and Curry, N. A., "A Study of α-Resorcinol by Neutron Diffraction," *Proc. Roy. Soc.* (*London*), **235A,** 552.

Bryden, J. H., and McCullough, J. D., "The Crystal Structure of *p*-Chlorobenzeneseleninic Acid," *Acta Cryst.*, **9,** 528.

Donohue, J., and Trueblood, K. N., "The Crystal Structure of *p*-Nitroaniline," *Acta Cryst.*, **9,** 960.

Merritt, L. L., Jr., Guare, C., and Lessor, A. E., Jr., "The Crystal Structure of Nickel Salicylaldoxime," *Acta Cryst.*, **9,** 253.

Webster, M. S., "An X-Ray Examination of the Crystal Structure of *p*-Iodonitrosobenzene," *J. Chem. Soc.*, **1956,** 2841.

1957

Bacon, G. E., and Curry, N. A., "A Neutron Diffraction Study of Potassium Hydrogen Bis-phenylacetate," *Acta Cryst.*, **10,** 524.

Bullough, R. K., and Wheatley, P. J., "The Stereochemistry of Molecules Containing the C=C=N Group II. The Crystal Structure of *N*-Methyl-2-methylsulfonyl-2-phenylsulfonylvinylidineamine," *Acta Cryst.*, **10,** 233.

Hargreaves, A., "The Crystal Structure of Zinc *p*-Toluenesulfonate Hexahydrate," *Acta Cryst.*, **10,** 191.

Housty, J., and Clastre, J., "Crystal Structure of the Triclinic Form of *p*-Dichlorobenzene," *Acta Cryst.*, **10,** 695.

Jerslev, B., "Crystal Structure of Oximes," *Nature*, **180,** 1410.

Malinovskii, T. I., "X-Ray Investigation of Bis(*p*-toluidine) Cobalt Dichloride," *Kristallografiya*, **2,** 734.

Nowacki, W., and Jaggi, H., "The Crystal Structure of Phthalic Acid," *Z. Krist.*, **109,** 272.

Okaya, Y., and Pepinsky, R., "The Crystal Structure of Ammonium Acid Phthalate," *Acta Cryst.*, **10,** 324.

Wallwork, S. C., and Powell, H. M., "The Crystal Structure of Phloroglucinol Dihydrate," *Acta Cryst.*, **10,** 48.

1958

Coppens, P., Schmidt, G. M. J., and Gillis, J., "Crystal Structure of the Stable (α) Modification of *p*-Nitrophenol," *Koninkl. Ned. Akad. Wetenschap., Proc.*, **61B,** 196.

Cox, E. G., "Crystal Structure of Benzene," *Rev. Mod. Phys.*, **30,** 159.

Cox, E. G., Cruickshank, D. W. J., and Smith, J. A. S., "The Crystal Structure of Benzene at $-3°$," *Proc. Roy. Soc. (London)*, **247A,** 1.

Dean, C., Pollak, M., Craven, B. M., and Jeffrey, G. A., "A Nuclear Quadrupole Resonance and X-Ray Study of the Crystal Structure of 1,2,4,5-Tetrachlorobenzene," *Acta Cryst.*, **11,** 710.

Frasson, E., and Sacconi, L., "X-Ray Structural Investigation of the Diamagnetic Form of Bis(*N*-methylsalicylaldimine)nickel(II)," *J. Inorg. Nuclear Chem.*, **8,** 443.

Gafner, G., and Herbstein, F. H., "Planarity of 1,2,4,5-Tetrabromobenzene Molecule," *Mol. Phys.*, **1,** 412.

Hassel, O., and Strømme, K. O., "Structure of the Crystalline Compound Benzene–Bromine (1:1)," *Acta Chem. Scand.*, **12,** 1146.

Klug, H. P., Alexander, L. E., and Sumner, G. G., "The Crystal Structure of Zinc Salicylate Dihydrate," *Acta Cryst.*, **11,** 41.

Matsuda, H., Osaki, K., and Nitta, I., "Crystal Structure of Quinhydrone," *Bull. Chem. Soc. Japan*, **31,** 611.

Shugam, E. A., "The Crystal Structure of Nickel Salicylaldoximate," *Trudy Vses. Nauch.-Issled. Inst. Khim. Reaktivov*, **1958,** 53.

Smith, H. G., and Rundle, R. E., "The Silver Perchlorate–Benzene Complex, $C_6H_6 \cdot AgClO_4$, Crystal Structure and Charge-transfer Energy," *J. Am. Chem. Soc.*, **80,** 5075.

Srinavasan, R., "The Crystal Structure of L-Tyrosine Hydrobromide," *Current Sci. (India)*, **27,** 46.

Stackelberg, M. v., Hoverath, A., and Scheringer, C., "The Structure of Clathrate Compounds of Phenol," *Z. Elektrochem.*, **62,** 123.

Tulinsky, A., and White, J. G., "Rigid-body Torsional Vibrations in Three Typical Members of a Class of Benzene Derivatives," *Acta Cryst.*, **11,** 7.

Zvonkova, Z. V., and Glushkova, V. P., "Crystal Structure of *p*-Bromophenylboric Acid," *Kristallografiya*, **3,** 559.

1959

Allmann, R., and Hellner, E., "Triiodide Ions in an Iodine Addition Compound of *p*-Iodo-*N*,*N*-dimethylaniline," *Naturwissenschaften*, **46,** 557.

Bartindale, G. W. R., Crowder, M. M., and Morley, K. A., "The Direct Determination, by Optical Transform Methods, of the Structure of the Red Form of 5-Methoxy-2-nitrosophenol II. Results and Chemical Discussion," *Acta Cryst.*, **12,** 111.

Corradini, P., and Allegra, G., "Structural Study of [$Cr(C_6H_6)(CO)_3$]," *Atti Accad. Nazl. Lincei, Rend., Classe Sci. Fis. Mat. Nat.*, **26,** 511; *J. Am. Chem. Soc.*, **81,** 2271.

Frasson, E., Garbuglio, C., and Bezzi, S., "Structure of the Monoclinic Form of *p*-Dichlorobenzene at Low Temperature," *Acta Cryst.*, **12,** 126.

Frasson, E. Panattoni, C., and Sacconi, L., "Studies in Coordination Chemistry V. Structure of the Diamagnetic bis-(*N*-Methylsalicylaldimine)–Nickel Complex," *J. Phys. Chem.*, **63,** 1908.

Hassel, O., and Strømme, K. O., "Crystal Structure of the Addition Compound Benzene–Chlorine (1:1)," *Acta Chem. Scand.*, **13,** 1781.

Lee, J. D., and Wallwork, S. C., "The Crystal Structure of the Molecular Compound Formed by Quinol and Acetone," *Acta Cryst.*, **12,** 210.

Liang, T.-C., and Struchkov, Y. T., "Crystal Structure of *p*-Diiodobenzene," *Izv. Akad. Nauk SSSR, Otdel. Khim. Nauk*, **1959,** 2095.

Penfold, P. R., and White, J. C. B., "The Crystal and Molecular Structure of Benzamide," *Acta Cryst.*, **12,** 130.

Shimada, A., "The Crystal Structure of Phenylarsonic Acid," *Bull. Chem. Soc. Japan*, **32,** 309.

Srinavasan, R., "The Crystal Structures of L-Tyrosine Hydrohalides I. L-Tyrosine Hydrobromide," *Proc. Indian Acad. Sci.*, **49A,** 340.

Srinavasan, R., "Crystal Structures of L-Tyrosine Hydrohalides II. L-Tyrosine Hydrochloride," *Proc. Indian Acad. Sci.*, **50A,** 19.

Stewart, J. M., and Lingafelter, E. C., "The Crystal Structure of *bis*(Salicylaldiminato)-nickel(II) and -copper(II)," *Acta Cryst.*, **12,** 842.

Struchkov, Y. T., and Solenova-Siderova, S. L., "Crystal Structure of 2,4,6-Trichlorobromobenzene," *Vestn. Moskov. Univ., Ser. Mat. Mekh. Astron. Fiz. Khim.*, **14,** 157.

Trotter, J., "The Crystal Structures of Some Mesitylene and Durene Derivatives II. Nitromesitylene," *Acta Cryst.*, **12,** 605.

Trotter, J., "The Crystal Structure of Nitrobenzene at −30°," *Acta Cryst.*, **12,** 884.

1960

Bacon, G. E., and Curry, N. A., "A Neutron Diffraction Study of Potassium Hydrogen Bis-Phenylacetate," *Acta Cryst.*, **13,** 717.

Bezjak, A., and Grdenič, D., "Crystal Structure of Mellitic Acid," *Nature*, **185,** 756.

Darlow, S. F., "Crystal Structure of Mellitic Acid," *Nature*, **186,** 542.

Gafner, G., and Herbstein, F. H., "The Crystal and Molecular Structures of Overcrowded Halogenated Compounds II. *β*-1,2,4,5-Tetrabromobenzene," *Acta Cryst.*, **13,** 706.

Hanic, F., and Michalov, J., "The Crystal Structure of Copper Salicylate Tetrahydrate, $Cu(OHC_6H_4COO)_2 \cdot 4H_2O$," *Acta Cryst.*, **13**, 299.

Liang, T.-C., and Struchkov, Y. T., "Crystal Structure of a Solid Solution of 91.3% *p*-Diiodobenzene and 8.7% *p*-Dibromobenzene," *Izv. Akad. Nauk SSSR, Otdel. Khim. Nauk*, **1960**, 1010.

Meuthen, B., and Stackelberg, M. v., "Crystal Structure of the α Modification of the Copper(II) Salicylaldehyde-methylimine Complex," *Z. Anorg. Allgem. Chem.*, **305**, 279.

Meuthen, B., and Stackelberg, M. v., "The Crystal Structure of Phenol Hydrate," *Z. Elektrochem.*, **64**, 387.

Milledge, H. J., and Pant, L. M., "Structures of 1,3,5-Trichlorobenzene at 20° and −183° and of 1,3,5-Tribromobenzene at 20°," *Acta Cryst.*, **13**, 285.

Romers, C., and Fischmann, E., "Isomerism of Benzoquinones and Monoximes (Nitrosophenols) VIII. The Crystal Structure of α-2-Chloro- and α-2-Bromo-5-methyl-*p*-benzoquinone-4-oxime," *Acta Cryst.*, **13**, 809.

Scheringer, C., Wehrhahn, O. J., and Stackelberg, M. v., "The Crystal Structure of Phenol," *Z. Elektrochem.*, **64**, 381.

Shimada, A., "The Crystal Structure of Phenylarsonic Acid," *Bull. Chem. Soc. Japan*, **33**, 301.

Toussaint, J., and Dejace, J., "Crystal Structure of Potassium *o*-Nitrophenolate Hemihydrate," *Bull. Soc. Roy. Sci. Liège*, **29**, 336.

Trotter, J., "A Three-dimensional Analysis of the Crystal Structure of *p*-Benzoquinone," *Acta Cryst.*, **13**, 86.

1961

Allegra, G., "Crystal Structure of Benzene Chromium Tricarbonyl," *Atti Accad. Nazl. Lincei, Rend., Classe Sci. Fis. Mat. Nat.*, **31**, 241.

Darlow, S. F., "The Crystal Structure of Mellitic Acid (Benzene Hexacarboxylic Acid)," *Acta Cryst.*, **14**, 159.

Eley, D. D., Taylor, J. H., and Wallwork, S. C., "The Crystal Structure of the Complex $Al_2Br_6 \cdot C_6H_6$," *J. Chem. Soc.*, **1961**, 3867.

Ferguson, G., and Sim, G. A., "X-Ray Studies of Molecular Overcrowding II. The Crystal and Molecular Structure of *o*-Chlorobenzoic Acid," *Acta Cryst.*, **14**, 1262.

Fischmann, E., MacGillavry, C. H., and Romers, C., "Isomerism of Benzoquinone-Monoximes (Nitrosophenols) X. The Crystal Structures of α-2-Bromo- and α-2-Chloro-*p*-benzoquinone-4-oxime Acetate," *Acta Cryst.*, **14**, 753.

Fischmann, E., MacGillavry, C. H., and Romers, C., "Isomerism of Benzoquinone–Monoximes (Nitrosophenols) XI. The Crystal Structure of β-2-Chloro-*p*-benzoquinone-4-oxime Acetate," *Acta Cryst.*, **14**, 759.

Henshaw, D. E., "Structural Constants of the Monohalobenzenes," *Acta Cryst.*, **14**, 1080.

Krausse, J., "Crystal Structure of *o*-Fluorobenzoic Acid," *Z. Chem.*, **1**, 92.

Lingafelter, E. C., Simmons, G. L., Morosin, B., Scheringer, C., and Freiburg, C., "The Crystal Structure of the α Form of Bis(*N*-methylsalicylaldiminato)copper," *Acta Cryst.*, **14**, 1222.

Marsh, R. E., and Glusker, J. P., "The Crystal Structure of Glycyl-phenylalanylglycine," *Acta Cryst.*, **14**, 110.

Nitta, I., Watanabé, T., and Taguchi, I., "The Crystal Structure of Orthorhombic Aniline Hydrobromide, $C_6H_5NH_3Br$," *Bull. Chem. Soc. Japan*, **34**, 1405.

Rérat, C., and Tsoucaris, G., "The Structure of *p*-Chlorophenyl Methyl Sulfone," *Compt. Rend.*, **252,** 1806.

Richards, J. P. G., "The Crystal Structure of Potassium Salt of *o*-Nitrophenol Hemihydrate," *Z. Krist.*, **116,** 468.

Shimada, A., "The Crystal Structure of *p*-Aminobenzenearsonic Acid." *Bull. Chem. Soc. Japan*, **34,** 639.

Simonsen, S. H., Pfluger, C. E., and Thompson, C. M., "The Crystal Structures of Some Metallo-organic Chelate Compounds I. The Ligand, 5-Chlorosalicylaldoxime," *Acta Cryst.*, **14,** 269.

Srivastava, H. N., and Speakman, J. C., "Crystal Structure of the Acid Salts of Some Monobasic Acids V. RbH Di-*o*-nitrobenzoate and KH Di-*p*-nitrobenzoate," *J. Chem. Soc.*, **1961,** 1151.

Stewart, J. M., Lingafelter, E. C., and Breazeale, J. D., "The Crystal Structure of Diaquobis(salicylaldehydato)nickel," *Acta Cryst.*, **14,** 888.

Strel'tsova, I. N., and Struchkov, Y. T., "Steric Hindrance and Molecular Conformation IV. Crystal Structure of Tetrabromo-*m*-xylene and Tetrabromo-*o*-xylene," *Izv. Akad. Nauk SSSR, Otd. Khim. Nauk*, **1961,** 250.

Strel'tsova, I. N., and Struchkov, Y. T., "Hindered Rotation and Molecular Conformation V. Crystal Structure of Hexachlorobenzene," *Zh. Strukt. Khim.*, **2,** 312.

Trotter, J., "A Three-dimensional Analysis of the Crystal Structure of *m*-Dinitrobenzene," *Acta Cryst.*, **14,** 244.

Trotter, J., "Crystal and Molecular Structure of *m*-Dinitrobenzene and *p*-Dinitrobenzene," *Can. J. Chem.*, **39,** 1638.

Trueblood, K. N., Goldish, E., and Donohue, J., "A Three-dimensional Refinement of the Crystal Structure of 4-Nitroaniline," *Acta Cryst.*, **14,** 1009.

Tsoucaris, G., "The Structure of Some Quaternary Ammonium Compounds I. Structure of β-Phenylethylamine Hydrochloride," *Acta Cryst.*, **14,** 909.

Ueda, I., "The Crystal Structure of Chloranil and Bromanil," *J. Phys. Soc. Japan*, **16,** 1185.

Wheatley, P. J., "Crystal and Molecular Structure of Tetrachloro-1,4-bistriethylstannyloxybenzene," *J. Chem. Soc.*, **1961,** 5027.

1962

Allmann, R., "X-Ray Structural Determination of the Iodine Addition Compound of a Hydrohalic Acid Salt of *p*-Iodo-*N*,*N*-dimethylaniline," *Z. Krist.*, **117,** 184.

Andresen, O., and Rømming, C., "The Crystal Structure of Benzenediazonium Tribromide," *Acta Chem. Scand.*, **16,** 1882.

Bonamico, M., Jeffrey, G. A., and McMullan, R. K., "Polyhedral Clathrate Hydrates III. Structure of the Tetrabutylammonium Benzoate Hydrate," *J. Chem. Phys.*, **37,** 2219.

Chu, S. S. C., Jeffrey, G. A., and Sakurai, T., "The Crystal Structure of Tetrachloro-*p*-benzoquinone (Chloranil)," *Acta Cryst.*, **15,** 661.

Ferguson, G., and Sim, G. A., "X-Ray Studies of Molecular Overcrowding III. The Crystal and Molecular Structure of *o*-Bromobenzoic Acid," *Acta Cryst.*, **15,** 346.

Ferguson, G., and Sim, G. A., "X-Ray Studies of Molecular Overcrowding IV. The Crystal and Molecular Structure of 2-Chloro-5-nitrobenzoic Acid," *J. Chem. Soc.*, **1962,** 1767.

Jones, N. D., and Marsh, R. E., "On the Crystal Structure of Chloranil–Hexamethylbenzene Complex," *Acta Cryst.*, **15,** 809.

Mak, T. C. W., and Trotter, J., "The Crystal Structure of *p*-Chloronitrobenzene," *Acta Cryst.*, **15**, 1078.

Pant, A. K., "The Crystal Structure of 3,5-Dibromo-*p*-aminobenzoic Acid," *Indian J. Phys.*, **36**, 650.

Rae, A. I. M., and Maslen, E. N., "The Crystal Structure of Sulphanilic Acid Monohydrate," *Acta Cryst.*, **15**, 1285.

Sakurai, T., "A Nuclear Quadrupole Resonance and X-Ray Study of the Crystal Structure of Tetrachlorohydroquinone," *Acta Cryst.*, **15**, 443.

Sakurai, T., "A Nuclear Quadrupole Resonance and X-Ray Study of the Crystal Structure of Pentachlorophenol," *Acta Cryst.*, **15**, 1164.

Shimada, A., "The Crystal Structure of *m*-Aminobenzenearsonic Acid," *Bull. Chem. Soc. Japan*, **35**, 1600.

Stephenson, N. C., "The Crystal Structure of Diiododi-(*o*-phenylenebisdimethylarsine) platinum(II), $Pt(C_6H_4[As(CH_3)_2]_2)_2I_2$," *J. Inorg. Nucl. Chem.*, **24**, 791.

Stephenson, N. C., "The Structure of the Diiododi-(*o*-phenylenebisdimethylarsine) palladium(II) Molecule," *J. Inorg. Nucl. Chem.*, **24**, 797.

Strieter, F. J., and Templeton, D. H., "Crystal Structure of the Carbon Tetrabromide-*p*-xylene Complex," *J. Chem. Phys.*, **37**, 161.

Wallwork, S. C., and Harding, T. T., "The Crystal Structure of the Chloranil–Hexamethylbenzene Complex," *Acta Cryst.*, **15**, 810.

1963

Bertinotti, A., "Crystal Structure Determination of *p*-Toluidine," *Compt. Rend.*, **257**, 4174.

Bevan, J. A., Graddon, D. P., and McConnell, J. F., "Crystal Structures of Bis-(Salicylaldehydato)copper(II) and Bis-(8-Hydroxyquinolinato)copper(II)," *Nature*, **199**, 373.

Brown, G. M., and Marsh, R. E., "The Crystal and Molecular Structure of 2-Amino-3-methylbenzoic Acid," *Acta Cryst.*, **16**, 191.

Bryan, R. F., Mills, H. H., and Speakman, J. C., "The Crystal Structures of the Acid Salts of Some Monobasic Acids VII. Ammonium Hydrogen Dicinnamate," *J. Chem. Soc.*, **1963**, 4350.

Charlton, T. L., and Trotter, J., "The Structure of *m*-Bromonitrobenzene," *Acta Cryst.*, **16**, 313.

Cotton, F. A., Dollase, W. A., and Wood, J. S., "The Crystal Structure and Molecular Structure of Dibenzenechromium," *J. Am. Chem. Soc.*, **85**, 1543.

Gurskaya, G. V., and Vainshtein, B. K., "Crystal Structure of Phenalanine Hydrochloride," *Kristallografiya*, **8**, 368.

Jellinek, F., "The Crystal Structure of Dibenzene Chromium at Room Temperature," *J. Organometal. Chem.*, **1**, 43.

Kennard, O., and Walker, J., "Salts of Amidines and Related Compounds II. An X-Ray Crystallographic Study of *S*-Methylthiouronium *p*-Chlorobenzoate: the Constancy of the Amidinium Carboxylate Grouping and its Importance for Protein Conformations," *J. Chem. Soc.*, **1963**, 5513.

Koizumi, H., Osaki, K., and Watanabé, T., "Crystal Structure of Cupric Benzoate Trihydrate $Cu(C_6H_5COO)_2 \cdot 3H_2O$," *J. Phys. Soc. Japan*, **18**, 117.

Mills, H. H., and Speakman, J. C., "The Crystal Structures of the Acid Salts of Some Monobasic Acids VIII. Potassium (or Ammonium, or Rubidium) Hydrogen Di-*p*-chlorobenzoate," *J. Chem. Soc.*, **1963**, 4355.

Orii, S., Nakamura, T., Takaki, Y., Sasada, Y., and Kakudo, M., "Crystal Structure of *m*-Methylbenzamide," *Bull. Chem. Soc. Japan*, **36**, 788.
Rømming, C., "The Structure of Benzene Diazonium Chloride," *Acta Chem. Scand.*, **17**, 1444.
Sakurai, T., Sundaralingam, M., and Jeffrey, G. A., " A Nuclear Quadrupole Resonance and X-Ray Study of the Crystal Structure of 2,5-Dichloroaniline," *Acta Cryst.*, **16**, 354.
Scheringer, C., "The Crystal Structure of Phenol," *Z. Krist.*, **119**, 273.
Shkol'nikova, L. M., and Shugam, E. A., "X-Ray Analysis of Cupferron Copper(II) (Nitrosophenylhydroxylamine)," *Zh. Strukt. Khim.*, **4**, 380.
Turner, R. W., and Amma, E. L., "Crystal and Molecular Structure of Metal Ion–Aromatic Complexes I. The Cuprous Ion–Benzene Complex, $C_6H_6 \cdot CuAlCl_4$," *J. Am. Chem. Soc.*, **85**, 4046.

1964

Alléaume, M., and Decap, J., "Structure of β-Sulfanilamide," *Compt. Rend.*, **258**, 2111.
Alléaume, M., and Decap, J., "Structure of γ-Sulfanilamide," *Compt. Rend.*, **259**, 3265.
Bacon, G. E., Curry, N. A., and Wilson, S. A., "A Crystallographic Study of Solid Benzene by Neutron Diffraction," *Proc. Roy. Soc. (London)*, **279A**, 98.
Coppens, P., "A Neutron Diffraction Study of 2-Nitrobenzaldehyde and the C—H···O Interaction," *Acta Cryst.*, **17**, 573.
Coppens, P., and Schmidt, G. M. J., "X-Ray Diffraction Analysis of *o*-Nitrobenzaldehyde and Some Substituted *o*-Nitrobenzaldehydes," *Acta Cryst.*, **17**, 222.
Folting, K., Lipscomb, W. N., and Jerslev, B., "The Structure of *N*-Methyl-*p*-chlorobenzaldoxime and Refinement of the Structures of *syn*- and *anti-p*-Chlorobenzaldoxime," *Acta Cryst.*, **17**, 1263.
Fox, M. R., Orioli, P. L., Lingafelter, E. C., and Sacconi, L., "The Crystal Structure of Bis-(*N*-isopropylsalicylaldiminato)nickel(II)," *Acta Cryst.*, **17**, 1159.
Frasson, E., Panattoni, C., and Sacconi, L., "X-Ray Studies of the Bis-*N*-alkylsalicylaldiminates of Bivalent Metals II. Structure of Bis-*N*-ethylsalicylaldiminepalladium," *Acta Cryst.*, **17**, 85.
Frasson, E., Panattoni, C., and Sacconi, L., "X-Ray Studies of the Bis-*N*-alkylsalicylaldiminates of Bivalent Metals III. Structure of Bis-*N*-butylsalicylaldiminepalladium," *Acta Cryst.*, **17**, 477.
Gafner, G., and Herbstein, F. H., "The Crystal and Molecular Structures of Overcrowded Halogenated Compounds V. γ-1,2:4,5-Tetrabromobenzene," *Acta Cryst.*, **17**, 982.
Gafner, G., and Herbstein, F. H., "Molecular Compounds and Complexes I. The Crystal Structure of the Equimolar Molecular Complex of Hexabromobenzene and 1,2,4,5-Tetrabromobenzene," *J. Chem. Soc.*, **1964**, 5290.
Grønbaek, R., and Dunitz, J. D., "Structure of the Hexathiocyanatocobalt(II)dimercury(II)–C_6H_6 Complexes," *Helv. Chim. Acta*, **47**, 1889.
Gurskaya, G. V., "Crystal Structure of the L-Phenylalanine Hydrochloride. Refinement and Discussion of Its Structure," *Kristallografiya*, **9**, 839.
Ibers, J. A., "Structure of Dibenzene Chromium," *J. Chem. Phys.*, **40**, 3129.
Jarski, M. A., and Lingafelter, E. C., "The Crystal Structure of Bis(salicylaldoximato) copper(II)," *Acta Cryst.*, **17**, 1109.
Lopez-Castro, A., and Truter, M. R., "Crystal Structure of Dimethylphenylsulphonium Perchlorate," *Acta Cryst.*, **17**, 465.

McKinnon, A. J., Waters, T. N., and Hall, D., "The Color Isomerism and Structure of Copper Coordination Compounds VII. The Crystal Structure of Bis(salicylaldehydato)copper(II)," *J. Chem. Soc.*, **1964**, 3290.

Maze, M., and Rérat, C., "The Structure of 4-Bromo-2,6-di-tert-butylphenol and 4-Methyl-2,6-di-tert-butylphenol," *Compt. Rend.*, **259**, 4612.

Mugnoli, A., and Mariani, C., "Molecular Structure of *p*-Nitrophenylazide," *Gazz. Chim. Ital.*, **94**, 665.

Orioli, P. L., Lingafelter, E. C., and Brown, B. W., "The Crystal Structure of Bis(5-chlorosalicylaldoximato)copper(II)," *Acta Cryst.*, **17**, 1113.

Rabinovich, D., and Schmidt, G. M. J., "Topochemistry V. The Crystal Structure of 2,5-Dimethyl-1,4-benzoquinone," *J. Chem. Soc.*, **1964**, 2030.

Reddy, J. M., Knox, K., and Robin, M. B., "Crystal Structure of $HI_32C_6H_5CONH_2$:A Model of the Starch–Iodine Complex," *J. Chem. Phys.*, **40**, 1082.

Rérat, C., and Tsoucaris, G., "Structure of *p*-Chlorophenyl Methyl Sulfone," *Bull. Soc. Franç. Minéral. Crist.*, **87**, 100.

Romers, C., "Isomerism of Benzoquinone Monoximes (Nitrosophenols) XII. The Crystal Structure of β-5-*n*-Propoxy-*o*-quinone-2-oxime," *Acta Cryst.*, **17**, 1287.

Sasada, Y., Takano, T., and Kakudo, M., "Crystal Structure of Salicylamide," *Bull. Chem. Soc. Japan*, **37**, 940.

Smith, H. G., "Molecular Motion in the Silver Perchlorate–Benzene Complex," *J. Chem. Phys.*, **40**, 2412.

Stephenson, N. C., "Crystal and Molecular Structure of Diiododi-(*o*-phenylenebisdimethylarsine)nickel(II)," *Acta Cryst.*, **17**, 592.

Stephenson, N. C., "The Crystal and Molecular Structure of Dichlorodi-(*o*-phenylenebisdimethylarsine)platinum(II)," *Acta Cryst.*, **17**, 1517.

Talapatra, S. K., "Crystal Structure of Pyrocatechol," *Indian J. Phys.*, **38**, 379.

Vainshtein, B. K., and Gurskaya, G. V., "X-Ray Study of the Structure of L-Phenylalanine Hydrochloride," *Dokl. Akad. Nauk SSSR*, **156**, 312.

Weiss, R., Rees, B., and Haser, R., "Radiocrystallographic Study of the 2,5-Dibromo-*p*-benzoquinone," *Compt. Rend.*, **259**, 1734.

Wheatley, P. J., "The Crystal and Molecular Structure of Aspirin," *J. Chem. Soc., Suppl.*, **1964**, 6036.

1965

Alléaume, M., and Decap, J., "Three-dimensional Refinement of β-Sulfanilamide," *Acta Cryst.*, **18**, 731.

Alléaume, M., and Decap, J., "Three-dimensional Refinement of γ-Sulfanilamide," *Acta Cryst.*, **19**, 934.

Alléaume, M., and Decap, J., "Structure of *p*-Aminobenzamide," *Compt. Rend.*, **260**, 5790.

Alléaume, M., and Decap, J., "Structure of Sulfanilamide Monohydrate," *Compt. Rend.*, **261**, 4111.

Bailey, M. F., and Dahl, L. F., "The Structure of Hexamethylbenzenechromium Tricarbonyl with Comments on the Dibenzene Chromium Structure," *Inorg. Chem.*, **4**, 1298.

Bailey, M. F., and Dahl, L. F., "Three-dimensional Crystal Structure of Benzenechromium Tricarbonyl with Further Comments on the Dibenzene Chromium Structure," *Inorg. Chem.*, **4**, 1314.

Bally, R., "Structure of Phenylarsine Bis(diethyldithiocarbamate)," *Compt. Rend.*, **261**, 3617.

Barclay, G. A., and Hoskins, B. F., "The Crystal Structure of Acetylacetone-mono(*o*-hydroxyanil(copper(II)," *J. Chem. Soc.*, **1965**, 1979.

Cady, H. H., and Larson, A. C., "The Crystal Structure of 1,3,5-Triamino-2,4,6-trinitrobenzene," *Acta Cryst.*, **18**, 485.

Carmichael, J. W., Steinrauf, L. K., and Belford, R. L., "Bis(3-phenyl-2,4-pentanedionato)copper I. Molecular and Crystal Structure," *J. Chem. Phys.*, **43**, 3959.

Cesur, A. F., and Richards, J. P. G., "Hydrogen Bonding in *o*-Aminophenol Hydrochloride," *Z. Krist.*, **122**, 283.

Charbonneau, G., Baudour, J., Messager, J.-C., and Meinnel, J., "Polymorphism of 1-Bromo-2,3,5,6-tetramethylbenzene," *Bull. Soc. Franç. Minéral Crist.*, **88**, 147.

Cheeseman, T. P., Hall, D., and Waters, T. N., "Stereochemistry of Copper in Bis(*N*-tert-butylsalicylaldiminato)copper(II)," *Nature*, **205**, 494.

Clastre, J., and Lamarque, A., "Crystal Structure of Pyrocatechol," *Compt. Rend.*, **260**, 2518.

Coppens, P., and Schmidt, G. M. J., "The Crystal Structure of the α-Modification of *p*-Nitrophenol near 90°K," *Acta Cryst.*, **18**, 62.

Coppens, P., and Schmidt, G. M. J., "The Crystal Structure of the Metastable (β) Modification of *p*-Nitrophenol," *Acta Cryst.*, **18**, 654.

Ganis, P., and Pedone, C., "A Preliminary Crystal Structure Analysis of the *trans*-Dichloro(*cis*-2-butene)[(*S*)-α-phenylethylamine]platinum(II)," *Ric. Sci., Rend.*, **8A**, 1462.

Gillier-Pandraud, H., "Crystal Structure of 2,5-Dimethylphenol," *Bull. Soc. Chim. France*, **1965**, 3267.

Gopalakrishna, E. M., "X-Ray Structure Analysis of *m*-Chloronitrobenzene," *Z. Krist.*, **121**, 378.

Hall, D., McKinnon, A. J., and Waters, T. N., "The Color Isomerism and Structure of Copper Coordination Compounds VIII. The Crystal Structure of a Second Crystalline Form of Bis(salicylaldehydato)copper(II)," *J. Chem. Soc.*, **1965**, 425.

Hall, S. R., and Maslen, E. N., "The Crystal Structure of Metanilic Acid," *Acta Cryst.*, **18**, 301.

Hanson, A. W., "The Crystal Structure of the 1:1 Complex of 7,7,8,8-Tetracyanoquinodimethane, and *N,N,N',N'*-Tetramethyl-*p*-phenylenediamine," *Acta Cryst.*, **19**, 610.

Helm, D. van der, Merritt, L. L., Jr., Degeilh, R., and MacGillavry, C. H., "The Crystal Structure of Iron Cupferron $Fe(O_2N_2C_6H_5)_3$," *Acta Cryst.*, **18**, 355.

Herbstein, F. H., "Twinned Crystals II. α-1,2:4,5- Tetrachlorobenzene," *Acta Cryst.*, **18**, 997.

Hon,P.-K., Belford, R. L., and Pfluger, C. E., "Bis(1-phenyl-1,3-butanedionato) vanadyl I. Molecular and Crystal Structure of the *cis* Form," *J. Chem. Phys.*, **43**, 1323.

Isakov, I. V., and Zvonkova, Z. V., "Crystal Structure of *p*-Thiocyanatoaniline," *Kristallografiya*, **10**, 194.

Klug, H. P., "The Crystal Structure of Tetrahydroxy-*p*-benzoquinone," *Acta Cryst.*, **19**, 983.

Leiserowitz, L., and Schmidt, G. M. J., "Topochemistry XI. The Crystal Structure of Methyl *m*- and *p*-Bromocinnamates," *Acta Cryst.*, **18**, 1058.

Long, R. E., Sparks, R. A., and Trueblood, K. N., "The Crystal and Molecular Structure of 7,7,8,8-Tetracyanoquinodimethane," *Acta Cryst.*, **18**, 932.

Maartmann-Moe, K., "The Crystal and Molecular Structure of Phloroglucinol," *Acta Cryst.*, **19,** 155.

Mak, T. C. W., and Trotter, J., "The Crystal and Molecular Structure of *N,N*-Dimethyl-*p*-nitroaniline," *Acta Cryst.*, **18,** 68.

Marsau, M. P., "New Determination of the Crystal Structure of Hexa(bromomethyl) benzene," *Acta Cryst.*, **18,** 851.

McPhail, A. T., and Sim, G. A., "X-Ray Studies of Molecular Overcrowding V. The Crystal and Molecular Structure of 2-Chloro-4-nitroaniline," *J. Chem. Soc.*, **1965,** 227.

Mugnoli, A., Mariani, C., and Simonetta, M., "Crystal, Molecular and Electronic Structure of *p*-Nitrophenyl Azide," *Acta Cryst.*, **19,** 367.

O'Connor, B. H., and Maslen, E. N., "The Crystal Structure of α-Sulphanilamide," *Acta Cryst.*, **18,** 363.

Okaya, Y., "The Crystal Structure of Potassium Acid Phthalate, $KC_6H_4COOH \cdot COO$," *Acta Cryst.*, **19,** 879.

Pant, A. K., "The Crystal Structure of 3,5-Dibromo-*p*-aminobenzoic Acid at Room Temperature (Approx. 25°C.) and at −150°C.," *Acta Cryst.*, **19,** 440.

Polynova, T. N., Bokii, N. G., and Porai-Koshits, M. A., "Crystal Structure of Diazonium Double Salts I. Structure of the Double Salt of Ferric Chloride and *o*-Methoxyphenyl Diazonium Chloride," *Zh. Strukt. Khim.*, **6,** 878.

Rees, B., Haser, R., and Weiss, R., "X-Ray Study of 2,3-Dichloro-*p*-benzoquinone," *Compt. Rend.*, **261,** 450.

Sacconi, L., Orioli, P. L., and DiVaira, M., "The Structure of Five-coordinated High-spin Complexes of Nickel(II) and Cobalt(II) with *N*-β-Diethylaminoethyl-5-chlorosalicylaldimine," *J. Am. Chem. Soc.*, **87,** 2059.

Sakore, T. D., and Pant, L. M., "Preliminary Structure Analysis of *p*-Nitrobenzoic Acid," *Indian J. Pure Appl. Phys.*, **3,** 143.

Sakurai, T., "The Crystal Structure of the Triclinic Modification of Quinhydrone," *Acta Cryst.*, **19,** 320.

Sax, M., Beurskens, P., and Chu, S., "The Crystal Structure of *o*-Nitroperoxybenzoic Acid," *Acta Cryst.*, **18,** 252.

Schlemper, E. O., and Britton, D., "The Crystal Structure of *p*-Iodobenzonitrile," *Acta Cryst.*, **18,** 419.

Sundaralingam, M., and Jensen, L. H., "Refinement of the Structure of Salicylic Acid," *Acta Cryst.*, **18,** 1053.

Vorontsova, L. G., "Crystal Structure of Methyl Phenyl Sulfone," *Kristallografiya*, **10,** 187.

1966

Akopyan, Z. A., Struchkov, Y. T., and Dashevskii, V. G., "Crystal and Molecular Structure of Hexanitrobenzene," *Zh. Strukt. Khim.*, **7,** 408.

Alléaume, M., Salas-Ciminago, G., and Decap, J., "Structure of *p*-Aminobenzoic Acid (PAB)," *Compt. Rend.*, **262C,** 416.

Baker, E. N., Hall, D., and Waters, T. N., "The Colour Isomerism and Structure of Copper Coordination Compounds X. The Crystal Structure of Bis-salicylaldiminatocopper(II)," *J. Chem. Soc.*, **1966A,** 680.

Barrans, Y., and Maisseu, J.-J., "Crystalline Structure of α-Chloroacetophenone," *Compt. Rend.*, **262C,** 91.

Bois, C., "Crystalline Structure of *p*-Cresol," *Bull. Soc. Chim. France*, **1966**, 4016.

Boyko, E. R., Hall, D., Kinloch, M. E., and Waters, T. N., "The Crystal Structure of Bis(*N*-2-hydroxyethylsalicylaldiminato)copper(II)," *Acta Cryst.*, **21**, 614.

Braun, R. L., and Lingafelter, E. C., "The Crystal Structure of Bis-(*N*-isopropyl-3-methylsalicylaldiminato)nickel," *Acta Cryst.*, **21**, 546.

Brown, C. J., "Further Refinement of the Crystal Structure of Acetanilide," *Acta Cryst.*, **21**, 442.

Brown, C. J., "The Crystal Structure of Catechol," *Acta Cryst.*, **21**, 170.

Brusset, H., Gillier Pandraud, H., and Viossat, C., "Crystal Structure of 2,3-Dimethylphenol," *Compt. Rend.*, **263C**, 53.

Calabrese, J. C., McPhail, A. T., and Sim, G. A., "Hydrogen Bonds I. The Ethynyl–H···O Interaction in Crystals of Propargyl 2-Bromo-3-nitrobenzoate," *J. Chem. Soc.*, **1966B**, 1235.

Carter, O. L., McPhail, A. T., and Sim, G. A., Metal–Carbonyl and Metal–Nitrosyl Complexes II. Crystal and Molecular Structure of the Tricarbonylchromium-anisole-1,3,5-trinitrobenzene Complex," *J. Chem. Soc.*, **1966A**, 822.

Cheeseman, T. P., Hall, D., and Waters, T. N., "The Colour Isomerism and Structure of Copper Coordination Compounds XI. The Crystal Structure of Bis-(*N*-*t*-butyl-salicylaldiminato)copper(II)," *J. Chem. Soc.*, **1966A**, 685.

Corfield, P. W. R., and Shearer, H. M. M., "The Crystal Structure of Phenylethynyl (trimethylphosphine)silver(I)," *Acta Cryst.*, **20**, 502.

Corfield, P. W. R., and Shearer, H. M. M., "The Crystal Structure of Phenylethynyl (trimethylphosphine)copper(I)," *Acta Cryst.*, **21**, 957.

Dickinson, C., Stewart, J. M., and Holden, J. R., "A Direct Determination of the Crystal Structure of 2,3,4,6-Tetranitroaniline," *Acta Cryst.*, **21**, 663.

Ferguson, G., and Islam, K. M. S., "Crystal and Molecular Structure of *o*-Fluorobenzoic Acid," *Acta Cryst.*, **21**, 1000.

Foss, O., and Marøy, K., "The Crystal Structure of Phenylbis(thiourea)tellurium(II) Chloride," *Acta Chem. Scand.*, **20**, 123.

Hon, P.-K., Pfluger, C. E., and Belford, R. L., "The Molecular and Crystal Structure of Bis(1-phenyl-1,3-butanedionato)copper," *Inorg. Chem.*, **5**, 516.

Karle, I. L., and Karle, J., "The Crystal Structure of the Quasiracemate from (+)-*m*-Methoxyphenoxypropionic Acid and ( )-*m*-Bromophenoxypropionic Acid," *J. Am. Chem. Soc.*, **88**, 24.

Katsube, Y., Sasada, Y., and Kakudo, M., "The Crystal Structure of *m*-Hydroxybenz-amide," *Bull. Chem. Soc. Japan*, **39**, 2576.

Khotsyanova, T. L., Babushkina, T. A., and Semin, G. K., "Study of Statistically Dis-ordered $C_6Cl_{6-n}Br_n$ Crystals by X-Ray Analysis and Nuclear Quadrupole Resonance II. Crystal Structure of Pentachlorobromobenzene and Pentabromochlorobenzene and the Nature of Their Statistical Disorder," *Zh. Strukt. Khim.*, **7**, 634.

Krausse, J., and Dunken, H., "Crystal and Molecular Structure of *o*-Fluorobenzoic Acid," *Acta Cryst.*, **20**, 67.

Maartmann-Moe, K., "The Crystal Structure of γ-Hydroquinone," *Acta Cryst.*, **21**, 979.

Maze, M., "Crystal Structures of 2,3-Dimethylphenol and 2-Methyl-3-bromophenol," *Compt. Rend.*, **262C**, 830.

Oijen, J. W. L. van, and Romers, C., "Isomerism of Benzoquinone Monoximes (Nitro-sophenols) XIV. The Crystal Structure of α-5-(2′-Chloroethoxy)-*o*-quinone 2-Oxime at 180°," *Acta Cryst.*, **20**, 169.

Orioli, P. L., DiVaira, M., and Sacconi, L., "Five-coordinated Structure of Bis(*N*-methylsalicylaldiminato)zinc(II). Isomorphism with the Cobalt(II) and Manganese(II) Analogs," *Inorg. Chem.*, **5**, 400.

Orioli, P. L., and Sacconi, L., "Crystal and Molecular Structure of Bis(*N*-isopropylsalicylaldiminato)copper(II)," *J. Am. Chem. Soc.*, **88**, 277.

Palm, J. H., "The Crystal Structure of *p*-Chloroaniline," *Acta Cryst.*, **21**, 473.

Panattoni, C., Bombieri, G., and Graziani, R., "Tetrahedral Coordination of Copper in Bis(*N*-ethylsalicylaldiminato)copper(II)," *Ric. Sci.*, **36**, 191.

Rasmussen, S. E., and Broch, N. C., "Stereochemistry of a Friedel-Crafts Intermediate. The Crystal Structure of the Aluminum Chloride–Benzoyl Chloride Complex," *Acta Chem. Scand.*, **20**, 1351.

Rees, B., Haser, R., and Weiss, R., "Dihalogenated Derivatives of 1,4-Benzoquinone I. Crystalline and Molecular Structures of 2,5-Dibromo-, 2,5-Chlorobromo-, and 2,5-Dichloro-1,4-benzoquinones," *Bull. Soc. Chim. France*, **1966**, 2658.

Rees, B., Haser, R., and Weiss, R., "Dihalogenated Derivatives of 1,4-Benzoquinone II. Crystalline and Molecular Structures of 2,3-Dichloro-1,4-benzoquinone," *Bull. Soc. Chim. France*, **1966**, 2666.

Sakore, T. D., and Pant, L. M., "The Structure of *p*-Nitrobenzoic Acid," *Acta Cryst.*, **21**, 715.

Sarma, V. R., "Crystal Structure of *p*-Chloroaniline," *Indian J. Pure Appl. Phys.*, **4**, 226.

Subramanian, E., "The Crystal Structure of *p*-Chloroacetanilide," *Z. Krist.*, **123**, 222.

Trotter, J., Whitlow, S. H., and Zobel, T., "The Crystal and Molecular Structure of *p*-Chloroaniline," *J. Chem. Soc.*, **1966A**, 353.

Trotter, J., and Williston, C. S., "Bond Length and Thermal Vibrations in *m*-Dinitrobenzene," *Acta Cryst.*, **21**, 285.

Turner, R. W., and Amma, E. L., "Metal Ion–Aromatic Complexes III. The Crystal and Molecular Structure of $C_6H_6 \cdot CuAlCl_4$," *J. Am. Chem. Soc.*, **88**, 1877.

Turner, R. W., and Amma, E. L., Metal Ion–Aromatic Complexes IV. Five-coordinate Silver(I) in $C_6H_6 \cdot AgAlCl_4$," *J. Am. Chem. Soc.*, **88**, 3243.

1967

Abrahamsson, S., Innes, M., and Lamm, B., "The Crystal Structure of *N*,3-Dimethyl-4-bromo-2,6-dinitroaniline," *Acta Chem. Scand.*, **21**, 224.

Adman, E., and Margulis, T. N., "Crystal Structure of Bis(*p*-toluidinium) Hexachlororhenate(IV)," *Inorg. Chem.*, **6**, 210.

Anca, R., Martínez-Carrera, S., and García-Blanco, S., "The Crystal Structure of 2,6-Dimethylbenzoic Acid," *Acta Cryst.*, **23**, 1010.

Andersen, E. K., "The Crystal and Molecular Structure of Hydroxyquinones and Salts of Hydroxyquinones I. Chloranilic Acid," *Acta Cryst.*, **22**, 188.

Andersen, E. K., "The Crystal and Molecular Structure of Hydroxyquinones and Salts of Hydroxyquinones II. Chloranilic Acid Dihydrate," *Acta Cryst.*, **22**, 191.

Andersen, E. K., "The Crystal and Molecular Structure of Hydroxyquinones and Salts of Hydroxyquinones III. Ammonium Chloranilate Monohydrate," *Acta Cryst.*, **22**, 196.

Andersen, E. K., "The Crystal and Molecular Structure of Hydroxyquinones and Salts of Hydroxyquinones IV. Ammonium Nitranilate," *Acta Cryst.*, **22**, 201.

Andersen, E. K., "The Crystal and Molecular Structure of Hydroxyquinones and Salts of Hydroxyquinones V. Hydronium Nitranilate, Nitranilic Acid Hexahydrate," *Acta Cryst.*, **22,** 204.

Ashida, T., Sasada, Y., and Kakudo, M., "The Crystal Structure and the Absolute Configuration of (+)-2-Benzylglutamic Acid Hydrobromide Dihydrate," *Bull. Chem. Soc. Japan*, **40,** 476.

Bailey, M., and Brown, C. J., "The Crystal Structure of Terephthalic Acid," *Acta Cryst.*, **22,** 387.

Baker, E. N., Clark, G. R., Hall, D., and Waters, T. N., "The Colour Isomerism and Structure of Some Copper Coordination Compounds XIII. Crystal Structure of Bis(*N*-ethylsalicylaldiminato)copper(II)," *J. Chem. Soc.*, **1967A,** 251.

Bally, R., "Crystal Structure of Phenylarsenic Bisdiethyldithiocarbamate," *Acta Cryst.*, **23,** 295.

Bonamico, M., Dessy, G., and Vaciago, A., "The Crystal Structure of Bis-trimethylbenzylammoniumtetrachlorocuprate(II)," *Theor. Chim. Acta*, **7,** 367.

Braun, R. L., and Lingafelter, E. C., "The Crystal Structure of Bis(*N*-isopropyl-3-ethylsalicylaldiminato)nickel," *Acta Cryst.*, **22,** 780.

Braun, R. L., and Lingafelter, E. C., "The Crystal Structure of Bis(*N*-isopropyl-3-ethylsalicylaldiminato)palladium," *Acta Cryst.*, **22,** 787.

Brown, D. S., and Bushnell, G. W., "The Crystal Structure of Mono-*o*-phenylenebisdimethylarsine Tricarbonyl Iron(0)," *Acta Cryst.*, **22,** 296.

Brusset, H., Gillier-Pandraud, H., and Viossat, C., "Crystal Structure of 2,3-Dimethylphenol," *Bull. Soc. Chim. France*, **1967,** 530.

Cady, H. H., "The Crystal Structure of *N*-Methyl-*N*,2,4,6- tetranitroaniline (Tetryl)," *Acta Cryst.*, **23,** 601.

Carlström, D., and Bergin, R., "The Structure of the Catecholamines I. The Crystal Structure of Noradrenaline Hydrochloride,' *Acta Cryst.*, **23,** 313.

Carter, O. L., McPhail, A. T., and Sim, G. A., Metal–Carbonyl and Metal–Nitrosyl Complexes IV. The Crystal and Molecular Structure of Tricarbonylchromium-*o*-toluidine. V. The Crystal and Molecular Structure of the Tricarbonylchromium Derivative of Methyl Benzoate," *J. Chem. Soc.*, **1967A,** 228, 1619.

Chiaroni, A., Dauphin, G., Kergomard, A., Rérat, B., and Rérat, C., "Structure of *p*-Bromo-*m*-nitro-*N*-methylaniline," *Compt. Rend.*, **264C,** 433.

Corfield, P. W. R., and Shearer, H. M. M., "The Crystal Structure of Phenylethynyl (isopropylamine)gold(I)," *Acta Cryst.*, **23,** 156.

DiVaira, M., and Orioli, P. L., "Molecular and Crystal Structure of the Octahedral Nickel(II) Complex with *N*-γ-Dimethylaminopropylsalicylaldimine," *Inorg. Chem.*, **6,** 490.

Einstein, F. W. B., and Rodley, G. A., "The Crystal and Molecular Structure of Diperchlorate Bis-(*o*-phenylenebisdimethylarsine) Cobalt(II)," *J. Inorg. Nucl. Chem.*, **29,** 347.

Fox, M. R., and Lingafelter, E. C., "The Crystal Structure of the Orthorhombic Form of Bis(*N*-methylsalicylaldiminato)nickel," *Acta Cryst.*, **22,** 943.

Gillier-Pandraud, H., "Crystal Structure of Phenol," *Bull. Soc. Chim. France*, **1967,** 1988.

Grimvall, S., and Wengelin, R. F., "The Crystal Structures of the Acid Salts of Some Monobasic Acids XI. Rubidium Hydrogen Diaspirinate [Bis(acetylsalicylate)]," *J. Chem. Soc.*, **1967A,** 968.

Hall, S. R., and Maslen, E. N., "The Crystal Structure of Orthanilic Acid," *Acta Cryst.*, **22,** 216.

Hamilton, J. A., and Steinrauf, L. K., "Crystallographic Studies of Iodine-containing Amino Acids I. Di-iodo-L-tyrosine Dihydrate," *Acta Cryst.*, **23,** 817.

Hirshfeld, F. L., and Rabinovich, D., "Structure and Electron Density of 2,5-Dimethyl-*p*-benzoquinone," *Acta Cryst.*, **23,** 989.

Holden, J. R., "The Structure of 1,3-Diamino-2,4,6-trinitrobenzene, Form I," *Acta Cryst.*, **22,** 545.

Holden, J. R., and Dickinson, C., "The Crystal Structure of 1,3-Dichloro-2,4,6-trinitrobenzene," *J. Phys. Chem.*, **71,** 1129.

Hon, P.-K., Pfluger, C. E., and Belford, R. L., "Bis(1-phenyl-1,3-butanedionato) palladium(II). Crystal and Molecular Structure of the *trans* Form," *Inorg. Chem.*, **6,** 730.

Iball, J., and Morgan, C. H., "The Crystal and Molecular Structure of Bis-α-hydroxy-α-phenylbutyramidinecopper(II)," *J. Chem. Soc.*, **1967A,** 52.

Jain, P. C., and Lingafelter, E. C., "The Crystal Structures of the Tetragonal and Monoclinic Forms of Bis(*N*-isopropyl-3-methylsalicylaldiminato)palladium," *Acta Cryst.*, **23,** 127.

Kurahashi, M., Fukuyo, M., and Shimada, A., "The Crystal and Molecular Structure of *o*-Nitrobenzoic Acid," *Bull. Chem. Soc. Japan*, **40,** 1296.

Kuroda, T., Ashida, T., Sasada, Y., and Kakudo, M., "The Crystal Structure of Benzaldehyde–Potassium Bisulfite Addition Product (Potassium α-Hydroxybenzylsulfonate)," *Bull. Chem. Soc. Japan*, **40,** 1377.

Lai, T. F., and Marsh, R. E., "The Crystal Structure of *p*-Aminobenzoic Acid," *Acta Cryst.*, **22,** 885.

Manojlovič, L., and Speakman, J. C., "The Crystal Structures of the Acid Salts of Some Monobasic Acids XII. Potassium Hydrogen Diaspirinate [Bisacetylsalicylate]," *J. Chem. Soc.*, **1967A,** 971.

Milburn, G. H. W., and Truter, M. R., "The Crystal Structure of Diethyl(salicylaldehydato)thallium(III)," *J. Chem. Soc.*, **1967A,** 648.

O'Connell, A. M., and Maslen, E. N., "X-Ray and Neutron Diffraction Studies of β-Sulfanilamide," *Acta Cryst.*, **22,** 134.

Okaya, Y., "The Crystal Structure of Ammonium *o*-Carboxybenzenesulfonate, $NH_4(C_6H_4COOH \cdot SO_3)$," *Acta Cryst.*, **22,** 104.

Panattoni, C., Bombieri, G., and Graziani, R., "The Crystal Structure of Bis-(*N*-ethylsalicylaldiminato)copper(II)," *Acta Cryst.*, **23,** 537.

Pedersen, B. F., "The Crystal and Molecular Structure of *N*-Methylacetanilide," *Acta Chem. Scand.*, **21,** 1415.

Rabinovich, D., "Topochemistry XIV. The Crystal Structure of 2,3-Dimethyl-1,4-benzoquinone," *J. Chem. Soc.*, **1967B,** 140.

Rabinovich, D., and Schmidt, G. M. J., "Topochemistry XII. The Crystal Structure of 2,6-Dimethyl-1,4-benzoquinone," *J. Chem. Soc.*, **1967B,** 127.

Rabinovich, D., Schmidt, G. M. J., and Ubell, E., "Topochemistry XIII. The Crystal Structure of 2,3,5,6-Tetramethyl-1,4-benzoquinone (Duroquinone)," *J. Chem. Soc.*, **1967B,** 131.

Sakore, T. D., Tavale, S. S., and Pant, L. M., "The Structure of *o*-Nitrobenzoic Acid," *Acta Cryst.*, **22,** 720.

Sass, R. L., and Bugg, C., "The Crystal Structure of Potassium *p*-Nitrophenyldicyanomethide," *Acta Cryst.*, **23,** 282.

Shipley, G. G., and Wallwork, S. C., "Molecular Complexes Exhibiting Polarization Bonding VII. The Crystal Structure of the 2:1 Complexes Formed by *p*-Chlorophenol and *p*-Bromophenol with *p*-Benzoquinone," *Acta Cryst.*, **22**, 585.
Shipley, G. G., and Wallwork, S. C., "Molecular Complexes Exhibiting Polarization Bonding VIII. The Crystal Structures of the 1:1 Complexes Formed by *p*-Chlorophenol and *p*-Bromophenol with *p*-Benzoquinone," *Acta Cryst.*, **22**, 593.
Shkol'nikova, L. M., Kuyazeva, A. N., and Voblikova, V. A., "Crystallochemical Data on Chelates of *N*-Substituted Salicylalimines V. X-Ray Study of Nickel Salicylai-*N*-ethyliminate," *Zh. Strukt. Khim.*, **8**, 94.
Sikka, S. K., and Chidambaram, R., "Neutron Diffraction Study of the Hydrogen Bond System in Tetrachlorohydroquinone," *Acta Cryst.*, **23**, 107.
Spofford, W. A., III, Carfagna, P. D., and Amma, E. L., "The Crystal Structure of *trans*-Bis(phenylethynyl)bis(triethylphosphine)nickel(II)," *Inorg. Chem.*, **6**, 1553.
Srivastava, R. C., Lingafelter, E. C., and Jain, P. C., "Bond Lengths in Bis-salicylaldoximato-nickel," *Acta Cryst.*, **22**, 922.
Ueki, T., Ashida, T., Sasada, Y., and Kakudo, M., "Crystal and Molecular Structure of *N*-Salicylideneglycinatoaquocopper(II) Hemihydrate," *Acta Cryst.*, **22**, 870.

NAME INDEX

A

B

M

N

FORMULA INDEX*

*In this bulk formula index carbon atoms come first and the formulas are arranged in order of the increasing number of these atoms. In each formula carbon is followed by hydrogen, oxygen, nitrogen, etc., each of which is ordered according to increasing number. Names are added only to distinguish between compounds having the same total formula.